Fire Service Manual

Volume 1
Fire Service Technology, Equipment and Media

Communications and Mobilising

HM Fire Service Inspectorate Publications Section
London: TSO

Published by TSO (The Stationery Office) and available from:

Online
www.tso.co.uk/bookshop

Mail, Telephone, Fax & E-mail
TSO
PO Box 29, Norwich, NR3 1GN
Telephone orders/General enquiries: 0870 600 5522
Fax orders: 0870 600 5533
E-mail: customer.services@tso.co.uk
Textphone 0870 240 3701

TSO Shops
123 Kingsway, London, WC2B 6PQ
020 7242 6393 Fax 020 7242 6394
68-69 Bull Street, Birmingham B4 6AD
0121 236 9696 Fax 0121 236 9699
9-21 Princess Street, Manchester M60 8AS
0161 834 7201 Fax 0161 833 0634
16 Arthur Street, Belfast BT1 4GD
028 9023 8451 Fax 028 9023 5401
18-19 High Street, Cardiff CF10 1PT
029 2039 5548 Fax 029 2038 4347
71 Lothian Road, Edinburgh EH3 9AZ
0870 606 5566 Fax 0870 606 5588

TSO Accredited Agents
(see Yellow Pages)

and through good booksellers

Published with the permission of the Office of the Deputy Prime Minister
on behalf of the Controller of Her Majesty's Stationery Office

ISBN 0-11-341185-5
ISBN 978-0-11-341185-5

Cover photograph (upper): Surrey Fire and Rescue Service
Cover photograph (lower): London Fire Brigade
Part-title page photograph: Hertfordshire Fire and Rescue Service

Printed in Great Britain on material containing 75% post-consumer waste and 25% ECF pulp.

Printed in the United Kingdom for TSO

Communications and Mobilising

Foreword

This revision of the Communications and Mobilising manual has been produced to meet publication requirements for reprinting at a time of intense change in the service and, in particular, in the areas of communications and technology.

At this time (2003), it has been possible to update several sections. Some others, which have become outdated and/or where development work is still progressing (e.g., Smart and Swipe Cards and Sub-surface communications), have been removed from this impression.

Where a substantial change has been made to the content, this has been indicated with a vertical red line in the adjacent margin.

Communications and Mobilising

Preface

The first edition Part 5 of the Manual of Firemanship dealing with the subject of communications was issued in 1954. It concluded 'Fire Service communications are intimately related to an intricate field of electrical engineering, which includes telecommunications both by landline and wireless, which is in turn only a small part of the territory covered by electrical science'.

The passage of time and advances in technology have changed every concept of fire service communications from those identified by the writers within the first edition of Part 5. The basic requirements of fire service communications have, however, remained unchanged and are identified in the Fire Services Act 1947. This is still as relevant today as it was when first mandated to fire authorities in 1947, to 'secure efficient arrangements for dealing with calls for the assistance of the fire brigade in case of fire and for summoning members'.

To ensure that the efficiency of fire service communications is maintained to the highest level requires the introduction of modern technology systems, coupled with frequent reviews of individual brigades' practises and procedures. Changes in equipment and procedures become inevitable, because either equipment becomes obsolete, or technical maintenance support is exhausted or overly expensive. New equipment often has advantages over what it replaces, in that it generally incorporates more functionality and flexibility, thus affording greater opportunities for changes in procedures and practises.

It is an impossible task to bring the reader fully up-to-date with the technology that is both available and continually evolving, or indeed to indicate that which may be available in the future. This book is written in non-technical terms and aimed primarily at covering the operational and functional communications requirements of the professional firefighter. This, by necessity, encompasses all the communicating elements from that of *'the originating caller to the incident's conclusion'* via the brigade Control, station call-out and Incident Command structures. The text, diagrams and symbols used, whilst not necessarily conforming to those in other technical publications, have been modified as appropriate to assist the reader. Those who require further technical detail must refer to other publications and technical sources which specialise in the area concerned.

It is anticipated that this book will be invaluable to brigade Communications Officers and all personnel who are or become intimately involved in the planning, procurement, implementation and operation of mobilising systems, communications systems, radio and fixed and mobile communications. As in Book 10 of the Manual of Firemanship, a great deal of emphasis has been placed upon planning principles, and the importance of clearly identifying both the operational requirements and the constraints associated with procurement processes. New technology solutions can be both implemented and beneficial if, as a result of a due planning process, they address and meet the needs and criteria of 'the user'. New technology should not, however, be seen as the driving force and the reason to change for changes sake. This is especially so in areas where an overall simpler solution could be adopted instead.

It is hoped that the information and advice contained within this book will help to ensure that the communications and associated systems used by the fires services of the United Kingdom will at least maintain and ideally improve their present standards of efficiency and reliability.

The Fire Service Inspectorate is greatly indebted to all those who have contributed and assisted (by providing material and information) in the preparation of the edition.

This book replaces the earlier Manual of Firemanship Book 10 Fire Brigade Communications and Mobilising.

HM Fire Service Inspectorate
July 2003

Communications and Mobilising

Contents

Chapter 10 Radio Alerting System 99

Chapter 11 Mobile Data 103

Chapter 12 Breathing Apparatus Telemetry 109

Chapter 13 Potential hazards of using radio equipment 113

Glossary of terms and abbreviations 119

Communications
and Mobilising

Chapter 1 – Regulatory issues

1.1   H. M. Government

Central Government Responsibility for the Fire Service

While fire authorities have statutory responsibility for the provision of fire cover and exercise day-to-day control over activities of their fire brigades, the Deputy Prime Minister – assisted by junior ministerial colleagues – is answerable to Parliament for fire policy in England and Wales. Within the Office of the Deputy Prime Minister (ODPM), the Fire, Health and Safety Directorate advises Ministers on fire matters including the operational efficiency of the fire service and the enforcement of fire safety legislation. The Directorate also assists fire authorities by establishing standards and providing technical guidance.

Fire services in England and Wales are inspected by HM Fire Service Inspectorate (which also inspects the Northern Ireland brigade). The Inspectorate also provides the technical resource for compilation of codes of practice and guides and advises on legislation relating to fire authorities. A separate Inspectorate performs the same functions in Scotland. HM Chief Inspectors provide reports to the relevant Secretary of State.

Structure of the service

In England fire services are provided by: county councils in the shires; joint fire and civil defence authorities (FCDAs) set up under the Local Government Act 1985 in the former metropolitan counties; the London Fire and Emergency Planning Authority (LFEPA), established as a functional body of the Greater London Authority, and combined fire authorities (CFAs) for county council areas affected by local government re-organisation in the period from 1 April 1996 to

1 April 1998. In England there are currently 16 shire fire authorities (including the Council of the Isles of Scilly), 24 CFAs, 6 FCDAs and LFEPA. In Wales there are 3 CFAs. In Northern Ireland fire services are provided by the Fire Authority for Northern Ireland, a non-departmental public body funded by government.

The 6 FCDAs and LFEPA provide both fire and civil defence services, but the bulk of their resources is devoted to fire. The FCDAs and LFEPA are local authorities in their own right, but members are not elected to them directly; they are mandated by the constituent district or borough councils in numbers prescribed in the 1985 Act. CFAS comprise representatives from their constituent county or unitary authorities.

Revenue finance

The fire service in England is financed almost entirely from the council tax and from Formula Grant. Formula Grant comprises National Non-Domestic Rates (NNDR), which are collected locally and passed to central government, and Revenue Support Grant (RSG). In the case of CFAs the grant is provided to the constituent authorities, which are billed by the combined authority.

Following the review of local government finance, Formula Grant is being distributed from 2003–04 using Formula Spending Shares (FSS) – which are what the Government has brought in to replace Standard Spending Assessments (SSAs).

The Government allocates FSS to each service, including fire, in each authority. They are not what the Government thinks needs to be spent on particular services; their purpose is simply to allocate the fixed pot of grant according to the relative needs and circumstances faced by local authorities. Also, the Government does not allocate or ring-fence the grant to particular services – it is simply given out as a lump sum to each authority as a whole.

From April 2003 the new fire FSS specifically takes account of the widening role of the service as a fire and rescue service and increases the share of funding allocated by an indicator for community fire safety. It provides a basic amount per head of population with top-ups for fire risk index (which includes indicators of deprivation and is calibrated against fire and non-fire calls), fire safety, pensions, area of category A risk, coastline and area cost (a measure of wage cost pressures).

In addition, a 'floor' mechanism ensures that every authority gets a reasonable level of grant increase, given the overall distribution and the total level of funding available. Grant 'ceilings' (limits to the size of increases) are used to help pay for the floor. However, since grant is not allocated to services but to authorities as a whole, the floor does not specifically relate to individual services – so decisions about where to spend the money are for councils to take.

The National Assembly for Wales is responsible for local government revenue finance in Wales. Funding is allocated to each authority for all local government services, including fire; there is no breakdown for individual services.

In Northern Ireland, the Fire Authority is funded by the Department of Health, Social Services and Public Safety.

Capital

Fire authorities' capital expenditure is met by a combination of borrowing, under credit approvals set by central government, receipts from sales of capital assets, and other income. Borrowing in any year is constrained by the level of credit approvals issued to the authority. These may take the form of basic credit approvals (BCAs) for general capital expenditure or supplementary credit approvals (SCAs) for specific purposes.

The ODPM allocates fire credit approvals for English and Welsh fire authorities from its overall financial provision. The majority of fire credit approvals are issued in the form of BCAs, which fire authorities are free to use as they see fit (and shire authorities are free to move BCAs between different service blocks).

The Central Fire Brigades Advisory Council

Section 29 of the Fire Services Act 1947 requires the Secretary of State to constitute a Council, to be called the Central Fire Brigades Advisory Council (CFBAC). The Council advises the Secretary of State on any matters on which he or she is required by the Act to consult the Council and on any other matters (except pay, conditions of service and discipline) arising in connection with the operation of the Act which the Council has taken into consideration, whether on reference from the Secretary of State or otherwise.

The main matters on which the Secretary of State is required to consult the Council are the making of Regulations (on matters such as criteria for Appointments and Promotion) and orders relating to the Firefighters' Pension Scheme. The Council, which covers England and Wales, was first established in 1947.

The Act provides that the Council is to consist of a chairman appointed by the Secretary of State – in practice this is the Minister with policy responsibility for the fire service – and of representatives of fire authorities and fire service employees appointed by the Secretary of State in numbers determined by him or her. The Secretary of State may, in addition, appoint other persons with special qualifications as members of the Council, either generally or for the consideration of any particular matter. Northern Ireland has observer status on the CFBAC.

Section 36(18) of the 1947 Act requires the Secretary of State for Scotland to set up a separate Central Fire Brigades Advisory Council for Scotland with a parallel remit in relation to the fire service in Scotland.

Much of the work of the two CFBACs is undertaken through joint committees, including five joint advisory boards. Membership of the joint committees reflects that of the Councils and many issues are resolved there without reference to the main Councils. However, representation on the joint committees is not restricted to those bodies represented on the Councils and other government departments and outside organisations can be represented if this is necessary.

It should be noted the Independent Review of the Fire Service report *The Future of the Fire Service: reducing risk, saving lives* published in December 2002 recommended that, 'The CFBAC should be replaced. The new body should draw on the widest range of expertise relevant to helping the Minister decide upon the strategic principles under which the fire service should be operated, and the future direction of policy.' The Government is proposing to respond to the recommendations in the review by means of a White Paper.

Office of the Deputy Prime Minister (ODPM) 999/112 Liaison Committee

The ODPM organises and supports meetings of the 999/112 Liaison Committee, a forum which brings together representatives of the Emergency Authorities (EAs) – (Police, Fire, Ambulance and Coastguard), the Public Telecommunication Operators (PTOs) (fixed and mobile) and other organisations with an interest in the 999 service. These include HM Fire Service Inspectorate, Chief and Assistant Chief Fire Officer's Association (CACFOA), Association of Chief Police Oficers (ACPO), the Ambulance Services Association, the Scottish Office, Oftel and the Department of Trade and Industry (DTI).

The Committee, which meets twice a year under the chairmanship of ODPM, discusses issues and matters arising from the provision of the 999/112 public emergency call service. The Committee encourages liaison between the EAs and PTOs at a more local level and considers what mechanism might be introduced to resolve disputes between them. It has also introduced Codes of Practice and Memorandums of Understanding, covering such issues as methods of handling 999 emergency calls on the fixed and mobile telephone networks.

The 999/112 Liaison Committee was responsible for producing the 'Strategic Framework for Combating Malicious Hoax 999 Calls' issued as DCOL 9/96 (in Scotland as DFM 8/1996).

Any problems which need to be resolved are progressed through a spirit of co-operation and goodwill between the relevant parties; the 999/112 Liaison Committee has no statutory powers or authority.

1.2   Oftel

The Office of Telecommunications, Oftel, is the regulator – or 'watchdog' – for the UK telecommunications industry. The Director General of Telecommunications heads Oftel. The Director General is appointed by the Secretary of State for Trade and Industry and the appointment usually runs for five years.

Oftel will be merged into a new regulatory body, The Office of Communications (Ofcom) later in 2003. This will have responsibility for all work currently carried out by Oftel.

Oftel was set up under the Telecommunications Act 1984. Oftel regulates through monitoring and enforcing the conditions in all telecommunications licences in the UK, and initiates modifications to these licence conditions.

All telecommunications operators – such as BT, Cable & Wireless, local cable companies, mobile network operators and the increasing number of new operators – must have an operating licence. Licences set out what the operators can – or must – do or not do.

Under the Telecommunications Act 1984, Oftel has a number of functions.

These include:

- ensuring that licensees comply with their licence conditions;

- advising the Secretary of State for Trade and Industry on telecommunications matters and the granting of new licences;

- obtaining information and arranging for publication where this would help users; and

- considering complaints and enquiries made about telecommunications services or apparatus.

Under the Act, the Director General has a duty to carry out these functions, some of these duties include:

- ensuring that telecommunications services are provided in the UK to meet all reasonable demands for them (this includes emergency services, public call boxes, directory information services and services in rural areas);

- promoting the interests of consumers;

- ensuring that those providing services are doing so efficiently; and

- promoting research and development.

The Director General has extensive powers under the Telecommunications Act, particularly when enforcing or modifying licence conditions. He or she can direct licence holders to comply with a certain condition – or conditions – in their licences. If they continue to breach the same condition/s the Director General can make orders which are enforceable through civil action.

Oftel is also responsible for administering the numbering scheme in the UK and allocates blocks of telephone numbers to operators. A separate Numbering Administration Unit within Oftel deals with this.

As well as being a regulator, Oftel monitors developments overseas. Nowadays UK operators are international businesses and so are their major customers. Oftel takes a global view and ensures that UK policies and decisions reflect international developments; it is also closely involved with telecommunications developments in the European Union.

Oftel is a non-ministerial government department, and is, therefore, independent of ministerial control.

Each year the Director General is required to submit an Annual Report on the department's activities and those of the Competition Commission in the telecommunications area, to the Secretary of State. This report is laid before Parliament.

Funding is provided by Parliament, but the cost of Oftel is offset almost entirely by the licence fees paid in by the operators.

Oftel staff are civil servants, and many are experts from consumer, business and industrial backgrounds. The Director General also has six Advisory Committees to advise him or her on telecommunications matters. Only one of these committees has a direct relevance to the Fire Service, this being the Advisory Committee on Telecommunications for Disabled and Elderly People (DIEL).

BT provides a service called 'In Contact Plus' which makes provision for 999/112 calls and also allows incoming calls for a low charge. The purchase of a pre-pay card allows outgoing calls to be made. Disconnection policy may include barring of outgoing calls as an alternative to disconnection and may allow emergency calls still to be made.

1.3 Radio Management

Radiocommunications Agency

The Radiocommunications Agency (RA) was established as an executive agency of the Department of Trade and Industry on 2 April 1990. Previously the RA operated as the Radiocommunications Division of the DTI.

The RA is responsible for most civil radio matters, other than those of telecommunications policy, broadcasting policy and the radio equipment market. Its main activities are:

- licensing the use of radio equipment under the Wireless Telegraphy Act 1949;

- investigating interference and enforcing the relevant legislation;

- representing United Kingdom interests in international meetings on radio spectrum management matters;

- seeking to ensure that all United Kingdom users, manufacturers and installers of radio equipment comply with the relevant European Union measures and with the relevant provisions of international agreements to which the United Kingdom is a party;

- developing policy for, and planning and regulating use of the radio frequency spectrum, the geostationary orbit and other orbits of telecommunications satellites by all non-government users of radio equipment in the United Kingdom except where otherwise agreed; and

- monitoring the radio frequency spectrum as an aid to its management, enforcement, and ensuring freedom from harmful interference.

Radio Investigation Service

The Radio Investigation Service (RIS) is the enforcement arm of the Radiocommunications Agency. Its aim is to ensure that authorised radio users can operate without undue interference. This is achieved by ensuring that licensed users adhere to the conditions under which they are authorised to operate and, if necessary, by taking legal enforcement action against those who operate radio equipment without regard to other authorised users.

The RIS has several roles:

- resolution of interference problems;

- inspection of installations at customers' premises; and

- help and advice with radio problems and provision of a paid diagnostic service to commercial and domestic radio users.

The RIS inspects all police and fire service radio installations as part of its work. This is to ensure compliance with the conditions of the radio licence. The RIS has indicated that it will contact users beforehand to arrange a convenient date and time for the inspection.

NB Both the Radiocommunications Agency and the Radio Investigation Service are going to become part of Ofcom.

(See website www.radio.gov.uk for further information.)

Radio Spectrum Regulations and Management

In 2001 the Department of Trade and Industry and HM Treasury asked Professor Martin Cave to independently review Radio Spectrum Management in the UK. His report recommended that:

- public safety users should continue to benefit from guaranteed access to the radio spectrum subject to full spectrum pricing applicable to comparable private mobile radio users; and,

- the Radiocommunications Agency should rationalise disparate assignments and widen the pool of spectrum reserved specifically for the delivery of public safety services, under the management of the Public Safety Spectrum Policy Group (PSSPG). Wherever possible, a technology neutral approach should be taken to the systems adopted for use to allow for competition.

The remit of the PSSPG should be broadened to encompass an expanded group of approved users, including: commercial and local government organisations with a public safety remit; and specialist users whose spectrum needs are currently met from within Home Office managed bands. Bands currently managed by the Home Office should be placed under the control of the PSSPG.

The Government has broadly accepted Professor Cave's recommendations and arrangements are being made to transfer Home Office managed spectrum to the PSSPG.

The Home Office and the Office of the Deputy Prime Minister, have agreed that the current management arrangements for existing fire service allocations will continue until the transfer to the PSSPG is completed. The PSSPG has assumed responsibility for evaluation of applications for new spectrum requirements.

Queries relating to management of existing allocations should in the first instance be referred to:

> Home Office Information and
> Communications Technology Unit,
> Radio Planning Section,
> Room 507 Horseferry House
> Dean Ryle Street London SW1P 2AW

Applications for fresh spectrum should be referred to the PSSPG.

Policy and Regulation

The Home Office Radio Planning Section assigns frequencies to its user services to meet specified operational requirements. Wherever possible, this takes account of national and international frequency management policies.

Policy changes and regulation will continue to be circulated in the form of 'Radio Frequency Policy Statements'. Current Home Office Policy Statements will remain in force until revoked or revised by the PSSPG.

These documents, which are classified as 'Confidential' under the Government Protective Marking Scheme, are sent to all Chief Officers of Police, and Fire Services, and to certain other interested parties such as the Radiocommunications Agency. The documents form the basis on which assignments are licensed and regulated.

Radio Frequency Policy Statements can include operational limitations on the use of channels where it is considered necessary to maintain the efficient use of the radio spectrum.

Type Approval and the R&TTE Directive

The introduction of the R&TTE Directive (99/5/EC) on 8 April 2000 has brought about a change in the way manufacturers of radio equipment and telecommunications terminal equipment (TTE) can gain access to the European marketplace for their products. This change means much greater flexibility and opportunity at the price of greater responsibility for the conformity of their products. Manufacturers will no longer have the comfort of a type approval regime but must shoulder the full consequences of product liability.

The R&TTE Directive was signed by the European Parliament on 9 March 1999 and published in the Official Journal of the European Communities (OJEC) on 7 April 1999. The Directive was transposed into national law via Statutory Instrument (SI) 730/2000.

The Directive aims to provide the European radio and TTE industry with a more deregulated environment than at present. The involvement of a third party in conformity assessment is not necessary in most cases. Those who place equipment on the market will, in general, be regarded as taking full responsibility for its conformity to essential requirements, and for properly informing purchasers of its suitability for its intended use. Only in the case of radio equipment for which harmonised standards are not available is it mandatory to consult a notified body. Further information on the role of notified bodies, harmonised standards and the conformance assessment process, which is beyond the scope of this statement, may be found in Home Office Guidance Note HGN(P) 38.

Compliance with the R&TTE Directive will be controlled by market surveillance. There will no longer be any obligation for a manufacturer to obtain type approval before placing equipment on the market. The manufacturer can simply declare conformity and attach the 'CE' mark. The Directive therefore offers a lighter regime, considerably reduced administration, and a faster route to the marketplace. Radio equipment, for which EU member states apply restrictions on service and operation, will display an additional identifier after the 'CE' mark.

This identifier is commonly referred to as the 'alert' mark, that is, (!)

The new directive replaces Directive 98/13/EC and all national type approval regimes for radio equipment and TTE. It also contains requirements on Health and Safety and Electromagnetic Compatibility (EMC) based on the Low Voltage

Directive (73/23/EEC) and EMC Directive (89/336/EEC). This means that manufacturers will only have to declare conformity to one directive to enable them to place their products on the market anywhere in the European Union.

Other Sources of Information

Radiocommunications Agency (RA) website:
 www.radio.gov.uk

Department of Trade and Industry (DTI) website:
 www.tapc.org.uk

European Telecommunications Standards Institute (ETSI) website:
 www.etsi.org

Radio call signs

The call signs for all police and fire brigade radio schemes start with 'M2' followed by two letters which identify the particular radio scheme, e.g., M2FH. The use of call signs, and radio operating procedures generally, are dealt with in the Fire Service Manual – Training. The basic call sign of a fire brigade is shown on the brigade radio licence.

The Police and Fire Comprehensive Radio Licence

In accordance with the Wireless Telegraphy Act 1949 all users of radio frequencies **must** be licensed by the Secretary of State. The organisation responsible for issuing radio licences or authority to use frequencies is the Radiocommunications Agency.

The 'Police and Fire Comprehensive Radio Licence' has been designed to cover all current Home Office assignments allocated to a particular user. Any assignments that a user holds which are in civil bands will need to be licensed separately. The only exception is 'Citizen Band' (CB) channels which are covered by the Police and Fire Licence.

For the fire service the licensee referred to in the licence document is normally the Chief Fire Officer. Under the terms of the licence, the licensee shall only use the Fixed Stations and Mobile Stations to send and receive wireless telegraphy relevant to the operation of the fire services.

Private Contractor Access to Fire Assignments in the Home Office Bands

Subject to the agreement of the licensee, private contractors are permitted to use brigade frequencies during the maintaining of brigade radio equipment. A test and development licence is required by the maintenance authorities wishing to maintain brigade equipment outside agreed operational areas. This licence will attract a separate charge.

Licences for other frequency bands

The Police and Fire Comprehensive Radio licence does not authorise use of any frequencies other than those in the Home Office frequency bands. It is, therefore, necessary to apply for a separate licence from the Radiocommunications Agency for each channel. **A separate licence fee is payable for each licence.**

Use of Radio Channels in an Emergency

No automatic right exists for any authority or person(s) to use any frequency not allocated to them. However, in specific circumstances, e.g., an emergency, or for carrying out tests associated with maintenance and repair activity, such authority may be prior issued in writing or verbally. If an emergency situation exists, such person(s) must, at all times, utilise correct voice procedures which specifically ensure that the call sign of the correct licensee is used with specific suffixes allocated to 'approved' external users.

Licence Schedule

The licence schedule consists of a number of pages relating to every base station site used (one per page), as well as all mobile and fixed equipment used within each brigade. The detail contained within the schedule relates to the technical parameters associated with every base station site used by the brigade and its mobile equipment, and the radio frequencies the equipment is authorised to use.

Examples are:

 Transmit power – The maximum transmitted power is normally that which enables the user's operational requirement to be met. This limits the risk of interference to other users and allows re-use of channels; and

Height above ground – The height of the aerial above ground may have to be limited to that required to give the required coverage.

Chief Fire Officers who require additional radio frequencies on any equipment, if access to a channel of a neighbouring brigade is required, **must**:

seek the permission of the relevant Chief Fire Officer, forwarding the approval response to Home Office the Information and Communications Technology Unit, for the frequencies and channel to be included on the schedule.

Local Authority Chief Fire Officers are authorised to allow access to VHF and UHF incident channels (used within their Authorities area) by any member, of any fire brigade, providing assistance with fires in accordance with Section 2 of the Fire Services Act 1947 or to secure the discharge of an authority's function under Section 12 of the Act, subject to the conditions set out in Radio Frequency Policy Statement FPS 16.

Interference to Home Office Frequency Assignments

The Home Office allocated radio spectrum is used exclusively for the assignment of frequencies for use by emergency services, and other Home Office user radio systems. Thus, co-channel interference on the mobile channels is likely to be from another police force or fire brigade, as applicable. The bands used for VHF/UHF links are shared by both police and fire brigades and interference could, therefore, be from either. Other users on a radio channel is one of the factors taken into account when assignments are made. However, during periods of high atmospheric pressure, co-channel interference from other users at a considerable distance may be experienced, due to enhanced radio propagation.

Emergency service radio schemes are often co-sited with other privately operated systems. Some of these prime sites are heavily used, and there is a

consequent high risk of interference between schemes due to the generation of intermodulation products. Although steps are taken to avoid assigning frequencies that may cause interference to existing channels, the probability of intermodulation interference depends largely upon the standard of engineering at the site.

If interference is suspected of being generated from electrical or telecommunications apparatus operated by another user, the Radio Investigation Service should be informed. The RIS have details of all users at each site, are highly experienced at solving interference problems, and do not normally levy a charge if the interference is caused by another user. However, if the investigation concludes that the interference was caused by a deficiency within a police or fire brigade's own equipment, then a charged may be levied accordingly.

Where interference is thought to involve another Home Office assigned service, the Home Office Radio Planning Section should be informed immediately. Where no suitable engineering solution is possible, consideration will be given to the reassignment of one of the services involved.

Air/Ground Communications

The police are making increasing use of aircraft, both rotary and fixed wing. The Home Office has access to two 25 kHz bandwidth air/ground assignments in the military band area of highband VHF, both of which are available for fire service co-ordination with the police.

Fire brigades are authorised to use one VHF simplex channel and three UHF simplex channels for air/ground use operating within the Home Office UHF band. Two of the UHF channels are contained within the six UHF 'Fire Incident' channels, namely Channel 1 or 6.

Brigades may select Channel 1 or 6 for air/ground use but **not both**. The choice of channel adopted by each brigade MUST BE notified to the Information and Communications Technology Unit for recording on the brigade's radio licence. The third available channel is one allocated to the police from the National UHF Channel Plan. This channel has been agreed by the Association of

Chief Police Officers (ACPO), primarily to facilitate the safe landing of other emergency services aircraft on roads.

Brigades may occasionally have a need to communicate from air-to-ground or vice versa utilising the police VHF or UHF channels. Before doing so, prior approval of the relevant Chief Constable must be obtained.

All equipment (regardless of channels used) in aircraft **must** comply with the technical parameters and approvals as laid down in Radio Frequency Policy Statement FPS 11.

1.4 Home Office Communications Advisory Panel (HOCAP)

The original 'HOCAP' arrangements were established to provide a Home Office Communications Advisory Panel to liaise with police forces and fire brigades after their post-WARC VHF mobile radio systems were installed to ensure that anything that they did harmonised with the developing strategies in the centre. Part of this process was the need to formalise guidance on common issues, and these became the HOCAP Guidance Notes.

Now that responsibility for the fire service has passed to the ODPM, the Home Office will no longer produce 'HOCAP' Guidance Notes for the fire service, although there are a number of existing documents that are still relevant and some that are no longer valid.

HOCAP Guidance Notes that were written for the fire service are no longer considered to be live Home Office documents, although they may continue to be used if applicable to brigades. Some need to be cancelled because they have been superseded by other documents or by changes in regulations. The following categories refer.

1.4.1 No longer valid and cancelled

- **HGN(F)7 Issue 2**
  The implications of the relaxation of the Duopoly Policy

- **HGN(F)8 Issue 2**
  Radio Spectrum Issues for the Fire Service
  19 April 2002
  (Has been superseded by PSSPG Guidance Note 1)

- **HGN(F)14 Issue 4**
  Type Approval of Radio Equipment
  10 January 1994
  (Has been superseded by RTTE Directive)

- **HGN(F)17 Issue 1**
  Radio Paging – Interference Potential
  11 February 1993

- **HGN(F)19 Issue 1**
  Type Approval of 25 kHz Bandwidth UHF Radio Equipment
  10 January 1994
  (Has been superseded by RTTE Directive)

- **HGN(F)21 Issue 1**
  The EMC Directive and Home Office Type Approval
  29 March 1996
  (Has been superseded by RTTE Directive)

- **HGN(F)22 Issue 1**
  Radio interference with Medical Devices
  7 April 1997
  (Corresponding police version HGNP33 Issue 2 has been updated and copied to ODPM)

1.4.2 Remain applicable to the Fire Service

- **HGN(F)1 Issue 4**
  UHF Handheld Radios for the Fire Service
  21 October 1992

- **HGN(F)3 Issue 3**
  UHF Band Reversal
  23 September 1992

- **HGN(F)11 Issue 3**
  CTCSS for UHF and VHF Radio Schemes
  23 September 1992

- **HGN(F)15 Issue 1**
  Hazards caused by Radio Transmissions
  16 February 1993
  (Police version currently being updated to take account of new regulations for use of radios on petrol forecourts)

- **HGN(F)16 Issue 1**
  Communications in Noisy Environments
  3 October 1996

- **HGN(F)20 Issue 1**
  Hazards associated with Microwave Transmitters
  5 August 2000
  (Issued with DCOL 6/2001 in October 2001)

- **HGN(F)24 Issue 1**
  The Automotive EMC Directive 95/54/EC
  8 May 2001

- **HGN(F)26 Issue 1**
  The Automotive EMC Directive 95/54/EC and application to Fire Service vehicles
  31 January 2003

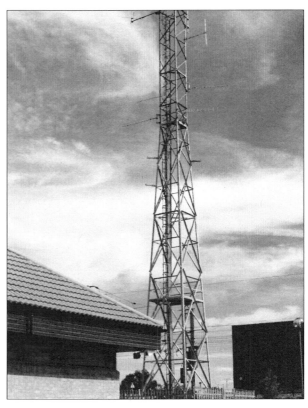

Figure 1.1 Shared police and fire service site.
(Photograph: Bedfordshire and Luton Fire and Rescue Service)

1.4.3   Site Sharing

The Home Office VHF bands are used to support wide-area coverage schemes using dominant radio sites. Such sites are often shared with other users. These may be other police forces or fire brigades as well as private users. Often the police and fire brigade will have several radio channels covering one part of the operational area. At each site, therefore, several transmit and receive frequencies from the same and different bands will be in operation.

The Home Office Radio Planning Section offers a free service which can advise police forces and fire brigades on the frequency compatibility of site sharing applications. FMG utilises specialist software to predict the spurious signals that may be generated when several transmitters operate on the same site. Some of the spurious signals may cause interference to co-sited base receivers or to mobiles which may be close to the site.

When considering site sharing applications, Communications Officers are strongly advised to

seek assistance from their engineering advisors on the likely wind loading of the additional aerials and the resultant overall wind load on the mast. The overall capacity of the power supply to the site also needs to be assessed. Further advice can be found in the relevant Policy Statement on site sharing.

1.4.4   Retained Firefighter Alerter Systems

Fire Alerter systems used by fire brigades operate on a 25 kHz bandwidth FM alerter channel in the VHF highband portion of spectrum.

The Home Office Information and Communications Technology Unit allocate the alerter tones to brigades. The UK has been divided into hexagonal cells 50 km across, with each cell being divided further into 127 smaller cells with each smaller cell being 5 km across. Seven codes are allocated to each smaller cell, making a total of 889 codes. Fire stations are allocated codes on the basis that the minimum re-use distance is 50 km.

All equipment must be type approved by the Information and Communications Technology Unit.

Licences allowing fire brigades to operate the above type of alerting system will be withdrawn after 31st December 1998. Thereafter, systems will comply with MG-4.

1.4.5 MG4 Specification Systems

In 1991 a new alerter system specification was introduced, produced to Home Office Specification MG-4 (Issue 2), which employs a recognised industry standard signalling system know as POC-SAG (Post Office Code Standardisation Advisory Group). The transmitters operate at a maximum output of 25 W Effective Radiated Power (ERP). The system architecture is structured to provide each brigade with a unique coded address, together with up to 2,000 separate address codes which may be allocated within the brigade to a station, a team or an individual as required. (Radio Frequency Policy Statement FPS 7 refers.)

All MG-4 base station transmitters must comply with the Radiocommunications Agency Specification MPT 1325 and Home Office Specification MG-4 (Issue 2). Base station aerial heights MUST NOT exceed 10 m above ground level without the prior approval from the Information and Communications Technology Unit.

Chapter 2 – Fire Control Centres

The Fire Service Act 1947 Section 1(i)(c) requires fire authorities to secure the provision of efficient arrangements for dealing with calls for the assistance of the fire brigade in case of fire and for summoning members of the brigade.

To meet this duty, fire authorities usually have a continuously staffed mobilising and communications centre, equipped with computer based Command and Control systems to deal with the receipt of emergency calls and the alerting and despatching of fire service resources within its mobilising area. Although these are considered to be the 'core' activities of a Control Centre, many additional 'non-core' duties are performed by control personnel as stipulated by the Chief Fire Officer/Fire Master.

All emergency communications for the fire service are channelled through the Control Centre which acts as a general communications and information resource for the fire brigade. It is usually housed in either a Control Suite at Brigade Headquarters or in a purpose built building within the county.

A Control Centre is staffed (in shifts to provide 24 hour cover) by uniformed professionals who, although employed under different conditions of service to firefighters, are an important part of the fire service.

Secondary and tertiary control systems are also maintained to ensure a continuity of service. There are no national standards of efficiency for handling fire calls but many Chief Officers have set their own standards which are set out in Brigade Orders/Service Instructions or their Citizens Charter.

In most cases the Control Suite comprises a Control Centre, training room, offices, equipment rooms, kitchen/rest-room, store rooms, locker-rooms/toilets, etc.

These rooms and facilities should be well designed and within easy access of the Control Centre room. Control personnel performing duties away from the Control Centre may need to be recalled if there is a sudden spate of calls, or if personnel become busy for other reasons. Easy access from anywhere within the suite will enable personnel to respond quickly.

Comprehensive guidance on the design of Control Centres was issued in DCOL 8/1997 (in Scotland as DFM 8/1997) (FRDG Publication 2/97). This is an updated version of Volume 5 of the Home Office Guidance usually referred to as 'Logica'.

The document includes advice on Control Centre design and ergonomics, procurement and legislation.

The recommended Control Centre rank structure is:

Fire Control Operator	FCOp
Leading Fire Control Operator	LFCOp
Senior Fire Control Operator	SFCOp
Fire Control Officer	FCO
Group Fire Control Officer	GFCO
Principal Fire Control Officer	PFCO

Not all these ranks are represented in every brigade.

2.1 Basic Call Handling Procedures

The primary function of a Control Centre is to provide the essential communication link which enables the provision of emergency firefighting, rescue and humanitarian services to the public when they call for assistance.

The basic principles of running a Control Centre have a common theme. However, the responsibilities and accountability of each rank may vary depending upon the size of the brigade.

The detailed procedures for handling an emergency call differ in each brigade according to its size and the type of communications and mobilising systems used.

Fire Control Operators are trained to elicit information from those calling for assistance. This activity requires the identification of the incident address and confirmation of the type of emergency for which assistance is required.

Difficulties in obtaining this information may result if the caller is unduly anxious or excited. A Fire Control Operator will still need to bear in mind that the primary purpose is to obtain information and will need to use effective call-handling skills to overcome these difficulties, possibly by calming and reassuring the caller. It may be necessary to give advice for dealing with the emergency whilst waiting for fire service attendance.

Techniques used to achieve this outcome could include a sympathetic approach or, perhaps, the adoption of an authoritative tone. The exact style is dependent upon the Operator's perception of what is appropriate in the circumstances.

It is possible that the caller may be in some personal danger. It is easy to understand that such circumstances might create a wide range of behavioural responses on the part of the caller.

Traditionally, Fire Control Operators are taught the appropriate inter-personal skills by a combination of initial training including simulation exercises and 'on the job' training by experienced personnel.

The first contact an emergency caller has with the Fire Service is with the Fire Control Operator. The way the operator handles the call is vital and to this end the operator must be immediately available to take control of the call. This will enable effective collation of call details to mobilise, and will indicate to the caller that he or she is being dealt with efficiently.

Further information on the training of Control Centre personnel is given in Appendix 1.

The responsibilities of each rank within Control Centres vary from brigade to brigade and many of them overlap.

The following list gives **examples** of skills and responsibilities within each rank.

Fire Control Operator (Core Skills)

- Receive emergency calls.

- Give advice to emergency callers as required.

- Identify and dispatch appropriate fire brigade resources to incidents (if necessary receiving guidance from senior ranks).

- Be familiar with the location of fire stations and their station ground.

- Keep officers informed of incidents/ occurrences as required.

- Liaise with other authorities and resources to keep them informed of incidents and request their assistance if necessary.

- Answer radio messages, relay radio messages to appliances and officers and act on information obtained.

- Deputise for Leading Fire Control Operators in their absence, subject to brigade requirements and competence of the Operator.

- Test and inspect equipment held in control, and the secondary control, carrying out such first line maintenance as appropriate.

- Answer non-emergency switchboard calls out of office hours and direct/advise callers.

- Answer non-emergency calls from station personnel and act on information received.

- Complete incident statistics.

- Work as part of a team and react appropriately as instructed and directed by officers.

- Ensure that levels of personal conduct are maintained in accordance with the standards prescribed in the Fire Service (Discipline) Regulations 1985 and by accepted Service Procedures.

- Ensure compliance with current Health and Safety Legislation, including Display Screen Regulations 1992.

- Comply with the brigade's Equal Opportunities Policy and other relevant legislation at all times.

- Undertake control/watch administration duties as required.

Leading Fire Control Operator

- Duties mirror those of a Fire Control Operator with the addition of supervisory duties.

- Assist and support other officers and be responsible to the Watch Officer in respect of the day-to-day management of the Control Centre and the development of personnel.

- Deputise for a Senior Fire Control Operator in their absence.

- Assume duties as Watch Officer in the absence of a Senior Fire Control Operator and/or Fire Control Officer, subject to brigade requirements and suitability of the Leading Fire Control Operator.

- Participate in the design, programming, running and monitoring of training programmes.

- Provide support and guidance to probationary Fire Control Operators and personnel preparing for examinations.

- Be familiar with the general command principles necessary to undertake the variety of other such tasks and duties as may be required, to meet the needs of the brigade.

Senior Fire Control Operator

The tasks listed below may be the responsibility of a Leading Fire Control Operator in those brigades that have Senior Fire Control Operators as Watch Officers.

In addition to a Leading Fire Control Operator's duties:

- Take charge of Command and Control activities during the absence of the Watch Officer.

- Assist and support the Watch Officer in respect of the day-to-day management and the development of personnel.

- Ensure that all resources have been dispatched correctly.

- Prepare and carry out the watch training programme and maintain training records as required by the Fire Control Officer.

- Undertake administrative/project work as required and assist in the supervision and completion of control/catch administrative workloads.

Fire Control Officer

In addition to the above:

- Monitor emergency calls and take command of the dispatch of all resources.

- Ensure that fire cover is maintained throughout the brigade area, utilising resources from neighbouring fire brigades if necessary.

- Ensure compliance with all Brigade Instructions, policies and guidelines.

- Identify training needs and manage the design, programming, running and monitoring of training.

- Management of control/watch administration duties including financial responsibilities as required.

- Assist and support other officers and be responsible to the Group Fire Control Officer (if applicable) in respect of the day-to-day management and development of personnel.

This may include conditions of service, sickness monitoring and welfare issues.

- Assist and support management in the development and planning of mobilising strategy.

Group Fire Control Officer

In some cases an FCO or GFCO may also hold other references within the brigade. These may include Personnel Officer, Communications Officer or, for example, in larger brigades the Watch Officers may hold the rank of GFCO.

The tasks listed below may be the responsibility of an FCO in a brigade which does not employ a GFCO:

- Responsible for the overall management of the Control Centre, its personnel, equipment and all other resources to ensure the effective, economic and efficient operation of the Control Centre, in line with brigade policies and procedures.

- Attend Control Centre during a major incident or spate conditions, and take strategic command and provide support as appropriate.

- Keep Control Centre personnel informed of brigade policies, procedures and standards.

- Monitor the welfare and motivation of personnel whilst constantly seeking to promote and improve teamwork and efficiency.

- Establish an effective working relationship with Control Centre personnel.

- Monitor all Control Centre personnel in respect of performance, conditions of service and training where appropriate.

- Development and planning of mobilising procedures.

- Development and planning of control/station communication systems.

- Maintain an efficient and effective Command and Control centre within allocated budgets provided.

Principal Fire Control Officer

The Principal Fire Control Officer rank is usually associated with the larger metropolitan brigades and generally performs the same role as FCO/GFCO in managing the Control Centre. Other brigades may introduce the rank to lead special projects or to be a head of section for the Control Centre – that is, Command/Control/Communications and IT – or perhaps to perform the management function of a DO with responsibilities for Personnel and Development, Equal Opportunities or Health and Safety.

However, in some brigades the PFCO may be responsible for developing brigade mobilising policy as part of the Principal Management Team.

2.2 Control Centre Staffing Levels

Her Majesty's Fire Service Inspectorate (and the Scottish Office Fire Service Inspectorate) is charged with the duty of obtaining information on how fire authorities are performing their functions, with particular regard to efficiency and effectiveness. Included in these functions are the brigade Control Centre and the manner in which it is staffed and operates.

To assist HM Inspectors and brigades in setting staffing levels within the Control Centre, a Staffing Model has been developed. This model was issued to brigades as **DCOL 6/1996 (in Scotland as DFM 6/1996)**.

The model is designed to give an **indicator** of the number of operators required to handle and process a given workload to a given *Grade of Service*. The model is not intended to take into account levels of supervision, sickness, training or control personnel required for projects, etc. It is used as a means of determining the number of operators required, from which decisions regarding establishment and officer levels can be made.

HM Inspectors will also use the model to assess the adequacy of brigade staffing requirements.

Brigade managers are, of course, free to run the staffing model within their own brigades. However, the Home Office recommends liaison with HM Fire Service Inspectorate to ensure correct interpretation and to develop a common approach.

HM Fire Service Inspectorate does not currently recommend a Grade of Service but may do so in the future.

Additionally, Fire Service Circular October 1975 recommends rank levels for control personnel established by reference to the population within the brigade area.

(See also Appendix 1 – Control Staff – Training, Competence and Promotion.)

Chapter 3 – A brief history of the 'Fire Control Centre'

In 1997 the Fire Service as we know it was 50 years old. Over those years a new career has evolved; that of Fire Control Operator.

In the very early days strategic mobilising to fires was virtually non-existent. During the 1800s, numerous fire insurance companies formed their own brigades of 'watermen'. Following a call to a 'fire', sometimes several of these insurance brigades would send their 'engines' and, on arrival, would look for the 'fire mark' to establish whether the victim was insured and by which company. In the free-for-all that ensued, the brigades could find themselves working against each other instead of working for the common good, to the detriment of the public. There was little co-ordination of resources or direction of the overall situation.

Over the next century and a half all this was to change significantly.

One prime innovation which would start the long haul to a unified and well-organised service, was the Metropolitan Fire Brigades Act, 1865. The act covered the City of London and *'all other Parishes and Places for the Time being within the Jurisdiction of the Metropolitan Board of Works'*. The Act also stated the need *'for the establishment of Telegraphic Communication between the several Stations in which their Fire Engines or Firemen are placed, and between any such Stations and other parts of the Metropolis'*.

This enabled the receiving and transmitting of locations of fires to all stations connected by the telegraph system. It was the first indication of mobilising from a source remote from the location of the fire and, by necessity, carried out by a fireman at the fire station receiving what was called a 'running call'. A situation that exists to the present day.

Metropolitan brigades had an advantage over the smaller rural brigades by nature of their size and the population they served. They were far better equipped financially to exploit the new technology that appeared, such as street fire alarms and fire detectors in commercial premises.

In rural areas, private telephones were scarce and public telephones were not as plentiful or well situated as they are today, and there were no street fire alarms. The firemen in rural brigades were usually part-timers who would rely on being called by a 'knocker up' or by the sirens that were installed during the First World War. These sirens were still in use well into the 1970s.

With the Second World War imminent, the Government mounted a recruitment campaign to encourage men and women to join the Auxiliary Fire Service (AFS). Women were encouraged to join as drivers, or to work in fire stations doing office work or watchroom duties. Some women opted for motorcycle training and driving lessons, while the majority learned watchroom procedures and the vital process of mobilising appliances. They all had basic firefighting training. (The AFS became the National Fire Service approximately one year after the war started. It was reformed in 1947 to run until the mid 1960s.)

One of the difficulties of forming a large number of small brigades into a National Fire Service was that most of the equipment, hose couplings, pump deliveries and appliances, etc., were all different. This caused obvious problems when one brigade was called to assist another. There was a desperate need for standardisation.

All emergency calls were received at the local General Post Office (GPO) telephone exchange, (at this time telephone exchanges only covered a

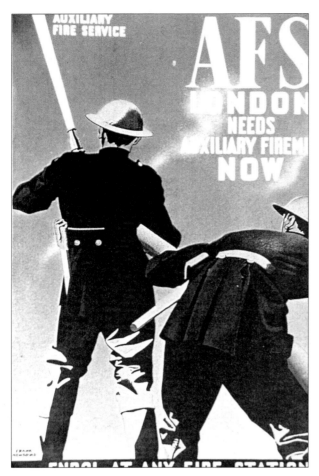

Figures 3.1 and 3.2 Posters used during a recruiting campaign in the Second World War.
(Photographs: HM Fire Service Inspectorate)

small area and there were a great number of them), and passed to the wholetime fire station in that area where a 'Watchroom' was continuously manned, either by a firemen, firewomen or a combination of both.

After the Second World War, communication was still a laborious and lengthy process. Watchrooms or Control Rooms in various brigades evolved differently, some were staffed by firemen who had a rota for 'Watchroom' duties while others were staffed by firewomen who had served in the National Fire Service. Many of these women stayed on after the war to become the forerunners of today's control operators.

In 1947 the Fire Services Act was passed to make further provision for fire services in Great Britain *'to transfer fire-fighting functions from the*

National Fire Service to fire brigades maintained by the councils of counties'. With brigades under the auspices of the County Councils the long process of standardisation of all equipment continued. This included the amalgamation of some fire station watchrooms into divisional control rooms which, because County Councils were also responsible for the ambulance service, were sometimes shared with ambulance personnel.

Unfortunately, whilst the Second World War had produced some well-managed and equipped fire control rooms up and down the country under the NFS, these were thought to be too elaborate for county brigades, and were dispensed with.

Mobilising was still carried out by the duty watchroom attendant who would take call details, dispatch the first attendance and, if necessary,

Figure 3.3 London Fire Brigade Control Room, 1937.
(Photograph: London Fire Brigade)

Figure 3.4 London Fire Brigade Wireless
Control Room at HQ.
(Photograph: London Fire Brigade)

Figure 3.5 AFS Fire Women in Watch Room.
(Photograph: Kent Fire Brigade)

Figure 3.6 GPO Telephone Exchange, late 1960s.
Note red lightbulb above unit for 999 calls.
(Photograph: Hertfordshire Fire and Rescue Service)

Figure 3.7 Kent Fire Brigade Control Room, 1960.
(Photograph: Kent Fire Brigade)

Figure 3.8 Control Room using VF System 'A', 1980.
(Photograph: Kent Fire Brigade)

pass the call to a divisional or district control. It was the duty of the watchroom attendant to record all fire calls, as well as officer and appliance movements, in the 'log book'. In fact everything was meticulously recorded, usually in beautiful handwriting.

In some cases, Kent for instance, brigade controls were responsible for plotting, logging of all calls with associated paper work, fire reports, accident reports and statistics, but not at any time talking to the originator of the call.

Improvements in the telephone network had revolutionised brigades. The introduction of a radio network was the next step towards improving brigade-wide communications.

The radio scheme was sometimes shared with the police (provided fire control asked 'nicely' and the police were not too busy, the scheme would be opened to allow for transmission) or, sometimes, the scheme was shared with another brigade. Police and fire brigade radio schemes were the responsibility of the Home Office Communications Branch, later the Directorate of Telecommunications, and remained so for many years.

These early systems, although now construed as relatively primitive, were to further enhance the capabilities of the service. Once each fire brigade had its own private mobile radio networks it became more practical to operate the radio from one location. It was one more step towards a single control.

Contact with fire stations was made by land line and 'part-timers' or retained firemen were called in by house bells or sirens; alerters for retained firefighters were not introduced until 1968.

Control rooms were now capable of reliable contact with stations by means of the 'K system' and subsequently, among others, the VF 'System A' and private wires, all of which used land lines. These mobilising systems were very reliable but rather slow, the method used to communicate with the station or stations required was the human voice, and all turn-out instructions, with additional information if necessary, were repeated. Mobilising was accomplished by checking the pre-

determined attendance (PDA) card for the parish or street to determine which appliance/s to send before alerting the station/s. These cards were kept in large 'bins' in the control room and if the brigade used street mobilising there were many hundreds of cards.

Operators prided themselves on their topographical knowledge and remembering the attendance for many areas or special risks, only using the PDA cards for confirmation. Fire calls were recorded by hand on individual incident forms and, in some cases, the old 'log book' was still running!

At this time, nationally, control staff personnel were a mishmash of backgrounds and experience. Control was thought to be the easy option and many operators were firemen who were on the run down to retirement, or sick and on light duties. Some brigades started to employ women because they couldn't get men to work shift work for the low rate of pay. Others, of course, had long established specialised personnel.

The developing use of computers generally in the 1970s inevitably led to thoughts of computerised mobilising. To have fingertip control of all brigade resources, PDAs, call logging, statistics and instant recall of information seemed very exciting. There was talk of 'paperless' control rooms! In fact, because of this belief, many of the consoles designed at that time had no 'working' space. This mistake was rectified next time around.

In 1972 two new courses were introduced at the Fire Service College. One was for Communications Officers, a post usually occupied by an operational fire officer, and the other was the very first course especially for Control Room staff, a Supervisory Officer's course. By 1974, in part due to local government reorganisation, the concept of a central control room for each County was well established. This was also the year in which Local Area Health Authorities were formed and fire brigades and the Ambulance Service went their separate ways.

1975 saw the standardisation of Control staff rank structure and markings, and recruiting was geared to the special skills required of an operator.

Grampian Fire Brigade was the first to use computer aided dispatch, closely followed by Greater Manchester Fire Brigade who went live with a fully computerised Ferranti Argos system in 1979. By the late 1980s almost all of the fire brigades in the United Kingdom had a computerised mobilising system although some were more sophisticated than others.

The number of emergency calls is increasing year by year, as is the type of emergency. To reflect the diverse nature of fires and special services they now attend, many fire brigades have changed their title to 'Fire and Rescue Service'.

The Home Office Guide to Fire Brigade Mobilising Systems, known as the 'Logica' report, was published in 1990 to help brigades with the specification, procurement and support of their second generation mobilising and communication systems.

Long gone are the days when all that was required of the watchroom attendant was to wind a handle for the station to turn out to a fire and hope that contact was made. The requirements and expectations of the control room have changed, and the improvements in communications have enabled a faster and more effective response.

Nowadays local knowledge is not enough; even the most experienced Control Centre personnel cannot rely on memory to process the large quantities of information required by a modern fire service. All incidents and relevant information are logged and stored on the database of modern computerised mobilising systems.

An operator now requires keyboard skills and a knowledge of computers, retrieval and statistical systems, chemical and hazardous materials, Management Information Systems (MIS), mapping systems and, most importantly, call-handling techniques.

An operator also requires a basic understanding of the many communication systems, be they voice or data, that are used in the fire service. In fact it is becoming increasingly difficult to distinguish between Communications, Mobilising and IT systems.

The Control Centre, as its name implies is, by its very nature, an essential part of any fire and rescue service. Instead of the free-for-all of the early days, firefighters can rely on being well informed about the incident they are attending, being kept up-to-date with all developments as they occur, and know that requests for help or assistance will be quickly and efficiently acted upon.

The skill of the operational firefighter together with the professionalism of their colleagues in the Control Centre, combine to provide an efficient and effective service to the public.

Communications and Mobilising

Chapter 4 – The 999/112 emergency service

4.1 British Telecom (BT)

The Public Telecommunications Operators (PTOs) are obliged, under the terms of their licences, to provide a public emergency call service by means of which any member of the public may, without charge, communicate as quickly as practicable with any of the appropriate local emergency authorities (EAs) to notify them of an emergency.

The 999 call service provides national coverage in respect of the four main emergency services – police, ambulance, fire and coastguards. Other services can also receive emergency calls via the police. These organisations are cave rescue, colliery rescue, mountain rescue, air/sea rescue, diver emergency and cardiac units.

BT answers approximately 34 million calls each year, including 18 million mobile 999s and almost 1 million calls from cable networks. Cable & Wireless take about 9.5 million – including mobile and cable company calls, whilst Kingston Communication handles 231,000 calls. Requests for police help account for some 60 per cent of emergency calls, while the ambulance service accounts for 30 per cent and the fire brigades for the majority of the remaining 10 per cent. Some 35 per cent of 999/112 calls from fixed networks and 75 per cent from mobile handsets are false calls, where callers make no request for an EA, and are safely filtered by BT and Cable & Wireless operators at the request of the EAs using agreed procedures.

4.1.1 The British Telecom fixed telephone system

There are currently nine BT Operator Assistance Centres (OAC); these are at Bangor, Blackburn, Grimsby, Hastings, Inverness, Milton Keynes,

Figure 4.1 A BT Operator Assistance Centre. *(Photograph: BT)*

Newport, Nottingham and Thanet. These Centres are served calls by three Operator Switches that queue calls to the first free operator across a number of centres, for resilience.

Circuit reservation facilities are used to ensure that there will always be a number of circuits reserved for 999/112 calls on routes between concentrator units and local switches and between local switches and trunk switches. All of the circuits in a route can be used for emergency calls if necessary, whereas it is not allowed that all circuits can be accessed by other sorts of call. This prevents unusually high levels of ordinary calls, e.g. on a television or radio 'phone-in', preventing a 999 call being made.

In addition each local switch is parented on at least two Trunk Switches, mostly three Trunk Switches, which gives access to a large number of alternative paths to an Operator Centre. These alternative paths can be continuously and automatically retried in the highly unusual circumstance of all being busy or unavailable.

Further 999 and 112 calls are given a protection marker which allows them to continue to be processed at any busy switches (when other call types may be rejected) and also to ignore any network protection measures. These measures are used in very busy network periods to limit the number of ordinary calls that can be progressed through the network, effectively rejecting some calls to maximise the number that can get through – 999 and 112 calls are not rejected.

Calls are spread evenly across the three Operator Switches mentioned above to help provide an even call distribution to centres and minimise the impact of call surges. This means that consecutive calls from the same location could be presented at any of the nine Operator Centres.

The connection of an emergency call involves four main phases:

(1) Connection of the caller to the operator via the 999/112 code;

(2) Selection by the operator of the required Emergency Authority Control Centre (EACC);

(3) Onward connection of the caller to the EACC; and

(4) Confirmation that the connection has been established with the appropriate EACC and ability to provide further assistance to the caller or EA when required.

The operator will monitor the call until the caller is clearly giving details of the incident. The operator normally holds the call in the system without listening (unless there are difficulties) until both the caller and EA have cleared the line.

It is the responsibility of the EA Control Centre staff to obtain adequate address information from the caller to enable the EACC to locate the incident being reported.

4.1.2 Operator call-handling procedures

The action of dialling 999/112 on BT's public telephone network in the UK automatically routes the caller through to one of its designated Operator Centres. Here, if it is not immediately answered, the call is visually and audibly signalled on all operating positions and, in addition, a special red light operates to ensure that a BT operator gives the call immediate attention.

To cater for unforeseen circumstances EAs have to provide three separate routes from the Operator Services Centres to the emergency service. The secondary and alternative routes would normally be used in sequence in the event of an unusually high level of traffic or a fault on the primary route.

BT will allow a 30 second delay with no reply before another route is attempted unless the EA control centre has a call queuing system, in which case additional time is allowed. For this reason it is vital to inform BT if any call queuing system is installed.

These routes are:

Primary
> This is the route that the operator will initially use to connect a caller to the EACC. The EA must provide sufficient capacity on this route to handle normal 999/112 traffic distribution.
>
> EAs will reserve primary routes exclusively for receiving 999/112 calls.

Secondary
> In circumstances where the PTO receives no reply on the primary number after 30 seconds, the operator will connect the call to a secondary number provided by the EA, except where call queuing is used. This procedure should only be necessary in instances when the EACC has an unusually high level of traffic or a fault in its switchboard or one of the PTOs' networks.

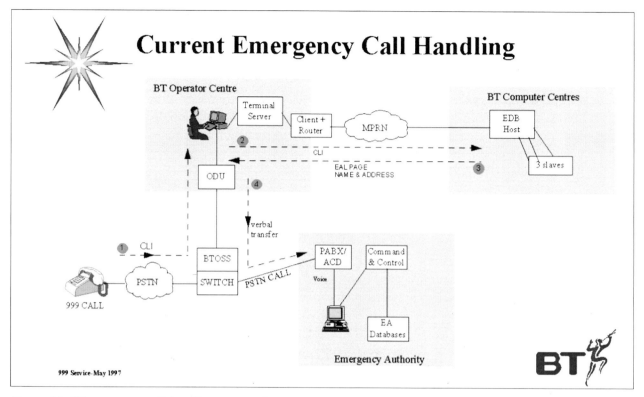

Current Emergency Call Handling

BT Operator Centre

Terminal Server

Client + Router

MPRN

BT Computer Centres

EDB Host

3 slaves

② CLI

EAL PAGE NAME & ADDRESS

③

ODU

④

verbal transfer

① CLI

PSTN

999 CALL

BTOSS

SWITCH

PSTN CALL

PABX/ ACD

Voice

Command & Control

EA Databases

Emergency Authority

BT

999 Service-May 1997

Figure 4.2 BT emergency call handling system. (Diagram: BT)

Alternative

In the event of a major problem that results in the primary and secondary routes to an EACC being unavailable to the PTO, the EA should provide the PTO with an alternative means of taking delivery of the call, ideally at a different EACC for maximum security.

To provide adequate security, this alternative number must be served by a different network route from that providing the primary and secondary routes. EAs would have to consider, where appropriate, which EA Control Centres are used as alternatives to each other.

These routes are agreed between the EA and the BT Emergencies Services Manager (ESM). Changes arranged by the EA have to be notified directly to the ESM giving at least two weeks notice.

All the routes have to be staffed on a 24 hour basis.

When a PTO operator answers an emergency the full national calling line identity (CLI) number will normally be automatically displayed on the operator's VDU. In the rare event of the CLI number not being present, the customer may therefore still need to be asked for their number so that the operator can route the call correctly.

The CLI number uses a different mechanism to BT's Caller Display and Call Return (1471) service. The number cannot be altered or withheld by the caller.

The originating calling information (CLI) will be used to automatically display details of the appropriate EACC connect-to numbers, and whether the brigade uses EISEC.

It is worth noting that an increasing number of business customers have direct dial in (DDI) systems and/or private networks. In these cases, the number automatically presented to the PTO operator is usually the outgoing number of the main switchboard. This will be the number passed to the EA and may be different to the number the caller may give if asked by either the PTO or EA operator.

There are also some private networks that extend over several areas. Their 999/112 calls could be fed into the PTO networks in only one of these areas. This will lead to inevitable problems as such calls will be presented with a telephone number applicable to this area and, therefore, routed accordingly. Problems will only become apparent at the EACC when the caller is questioned as to his or her location.

The operator establishes which EA is required by answering the call with **'Emergency, which service?'**

If the caller needs to be asked for their number this is entered onto the screen.

The operator hands the call over by passing the Operator Assistance Centre (OAC) name and the caller's number and listens while the caller passes their location. If this has been given clearly and the call is progressing, the call will be held in the operator's system (without listening) until the call is complete. Operators normally listen throughout in difficult cases – for example, panicking callers or where there is a language problem.

If the Enhanced Information Service for Emergency Calls (EISEC) is used, the caller's number and Operator Assistance Centre identity are automatically transferred to the EACR's switch and do not need to be verbally passed.

If it is necessary to speak to the BT operator again it should be possible to call him or her back onto the line while the call is being listened to. If a caller clears before all relevant information is obtained, and the EA operator does not end the call, the CHA will be alerted and an emergency operator will return to the line for rapid assistance. Calls are only released from the CHA when both parties clear.

If details of the call are required once the call has been released, it will be necessary to dial the Operator Assistance Centre number allocated to each brigade. **The allocated number should be used even if the call was passed via a different OAC.**

Every emergency call is recorded and stored on a call database that shows basic call details, such as time, telephone number and address of the call. BT's recording of emergency calls is from the time that the call is ringing in the queue at the BT Operator Assistance Centre to the time when the call to the EA is cleared by the caller and the EA. The speech of all parties is recorded. All 999/112 call records are held for three months.

4.1.3 Calls without service request from mobile phones

Very large numbers of accidental 999/112 calls are received from mobile phones. CHA operators try to obtain a response. In cases where nothing apart from general noise (no speech) is heard, or where the voice link is terminated during CHA questioning, it is recognised that there is a negligible chance that the call is genuine and the CHA operator can end the call.

Where there is no response, the voice channel remains open, and background voices are present, it is recognised that the CHA cannot decide whether an EA is needed. In this case the call is connected to a police voice response system at New Scotland Yard which asks the caller to press '5' twice if help is required. If 55 is pressed then immediate connection with the appropriate police authority is made.

For any cases where suspicious noises are heard the CHA can override the above procedure and connect to a police operator.

4.1.4 112 Calls with no service request from fixed phones

There is very little use made of 112 for calling EAs; the overwhelming majority of callers use 999. 112 calls can readily generate faults within networks or customer equipment and, despite the use of network filters to prevent such calls reaching the CHA operators, many are still received which appear as silent, open lines, or noisy lines with crackling and interference sounds.

Once CHA operators have asked their normal questions without response, noisy 112 calls are connected to the police voice response unit at New

Scotland Yard. This provides one final check to cover the rare case of a genuine 112 caller unable to speak being on the line.

4.1.5 Mismatches between EA and Fixed Network Operator Boundaries

BT has over 7,500 exchanges, each with its own code; it is these codes that determine how the call is routed. BT exchanges have grown up over the last 90 years as the cable network has evolved and each has a defined catchment area averaging 4 km in radius.

Exchange area boundaries do not always coincide with EA boundaries. Where 'mismatches' exist, careful planning and general agreement between neighbouring Chief Fire Officers/Fire Masters and the BT Emergency Services Manager has to be reached on which EA Control Centre will take the calls from the whole of the split exchange area in question.

The Fixed Network Operators will connect all directly connected customers to the appropriate EACC for the agreed geographical areas wherever possible. It will be the EA's responsibility to pass information if necessary to another EACC in these mismatches cases.

Increasingly, mismatches will be minimised as BT moves during 2003 to routing by map reference. This is being achieved in BT using a resilient database, known as Emergency Services Database, to find the installation address and postcode of the calling line. The postcode is converted to an Ordnance Survey map reference that is matched to EACC catchment areas and then to the numbers provided for those EACCs. If the caller's address is not available then the OS map reference of the centre of the exchange area is used.

4.1.6 Provision of ex-directory information

BT operators will only provide name and address information for numbers from which a recent 999/112 call has been made. All other routine requests for such information must be made by EAs through BT's Group Engineering Services Team.

In providing an XD/NC service BT undertakes not to give the number to anyone outside BT including EAs. BT has laid down procedures to enable urgent calls to be connected to XD/NC customers without revealing the number. EAs requiring such a connection must contact BT OACs using the 100 service where the operator will ask a number of questions to support the request before connecting.

4.1.7 Access to tape recordings of Emergency Calls

BT record all calls terminating on 999/112 circuits. Calls are recorded from the time the call is answered by a BT operator until the EA and caller clear the line, and the circuit is released.

Requests from EAs to listen to, to make notes about, or be given a copy of a recording of a 999/112 call must be referred to the BT Group Engineering Services Team.

These requests must be authorised by the agreed level according to the Code of Practice which states that access to emergency call records can be obtained in two forms:

Normal For investigatory purposes where it is required as evidence or for similar use. Arrangements for access will be agreed at the time of request.

Urgent Where instant access is required to respond to a 999/112 incident. This is dependent on the availability of suitable equipment within the CHAs (currently this is not available over BT).

For Cable & Wireless Communications and British Telecom the following levels of authority are required for access to be given:

Normal Assistant Chief Constable (Police)
Assistant Chief Officer (Fire)
Assistant Chief Officer/Director of Operations (Ambulance)
Regional Inspector (Coastguard).

Urgent Senior Duty Control Room Officer for all Emergency Authorities.

An authorised representative of the EA (not necessarily the Authorising Officer) must be present when the tapes are being played at an Operator Services Centre.

BT will only keep original 999/112 recording tapes for a period of three months. Evidential quality copies can be requested if necessary.

BT will apply to the Chief Officer of the relevant EA for similar recordings of calls made by the EAs.

PTOs inform all their customers to use the 999/112 code when making emergency calls. PTOs do not tape record emergency calls made on any other circuit. However, such calls are processed despite the use of the incorrect code.

4.1.8 Calling Line Identity (EISEC)

Since 1985 BT have been modernising their network and converting from analogue to digital exchanges. This gives 99 per cent CLI coverage, which means BT operators have instant access to the caller's address on over 40 million lines.

Once BT developed CLI for their own operators, the emergency services requested an enhancement to 999 services to reduce call-handling times and the number of hoax calls. Their requirement was an enhancement which allows the telephone number and address of the caller to be automatically displayed on the EAs own mobilising computer screen. BT have called it Enhanced Information Service for Emergency Calls (EISEC)

The advantages are:

- Caller's number and address automatically available on answer, no need for information to be passed verbally or the EA operator to type. Reduces typing errors.

- Name of BT OAC displayed, for call back if necessary.

- Caller can speak to the EA sooner, reducing frustration or panic. Address and telephone number is simply confirmed.

- CLI overcomes the problem of spelling, pronunciation and language difficulties.

- Early indication of hoax calls – the auto-address will reveal if the caller is giving a false address or at a payphone often used to make malicious calls, or perhaps a mental hospital.

BT have devised a system which requires an ISDN link from that operator centre which is used to forward CLI to the EACC. On receipt of a call the EA's mobilising system will dial into the BT database for address information which will be displayed on the EA mobilising screen; this will take approximately one second. This is technically a dial-up system (which will have a small cost implication for the EA) but will appear automatically to the EA operator.

It will only be possible to automatically obtain the telephone number and address of the caller if they have dialled 999 or 112. The system design dictates that these calls follow a certain technological route that safeguards the integrity of the BT system.

It will not be possible for any EA to type in a number and interrogate the BT database. By the act of dialling 999/112 it is deemed a caller has given consent for this information to be used and, therefore, complies with the Data Protection laws.

Although in most cases the emergency call is made from the address of the fire, it should not be assumed that this is so. Experienced EA operators will recognise the dangers and know how easy it is to get an affirmative answer to any question.

Training on call handling procedures will have to reflect this.

> Code of Practice for the Public Emergency Call Service (PECS) between Public Network Operators and the Emergency Services.

For more detailed information on the 999 system procedures please read the above document.

4.1.9 Network Resilience

In the event of a major failure to a part of a PTO's network, the PTO will notify the affected EAs as soon as possible after the failure is identified, or is anticipated.

The process for informing EAs of any BT exchange that fails to give customers 999 access commences when a regional Network Management Centre (NMC) detects a failure that causes loss of 999 access. Once the extent of the problem is known a report of the failure is faxed to each appropriate EA Control Centre (Police, Fire, Ambulance and Coastguard). In addition the Network Management Centre draws attention to the fax by telephoning the police control, which in turn telephones the other affected EAs. If necessary, progress reports are faxed at periodic intervals and finally a fault clearance report is sent once 999 access has been restored.

All reports of 999-affecting network failures are sent to BT's Lead 999 Operator Centre at Newport to provide a back up source of basic information.

The process includes the provision by BT to the EAs of maps showing the area covered by each of its exchanges.

These procedures are periodically reviewed.

EAs and relevant PTOs should prepare local contingency arrangements to cover the receipt of emergency calls during conditions of serious breakdown in the PTO network.

4.1.10 Priority Fault Repair Service

The conditions in the PTO's licence requires them to provide a free Priority Fault Repair Service to those emergency authorities who receive 999/112 calls on lines connected to the PTO's network. When notified of any fault or failure which causes interruption, suspension or restriction of the telecommunication services provided by the PTO, the PTO will restore those services as swiftly as practicable and with a priority, so far as is reasonably practical, over Fault Repair Services to other persons.

Where an EACC has connection directly to a PTO for an Emergency Call Service, the Priority Fault Repair Service will be extended to all 999/112 circuits in accordance with the relevant condition of the PTO's licence. This is Condition 9 of the Cable & Wireless licence and Condition 10 of the BT licence.

The BT Priority Fault Repair Service will apply where BT and Cable & Wireless use common terminations supplied by BT.

4.1.11 BT National Emergency Linkline

The National Emergency Linkline is a service designed to give nominated customers a quick and easy means of contacting BT to request assistance during emergency situations.

Nominated customers are primarily the Emergency Services and Local Government, as they would normally be responsible for co-ordinating emergency incidents. However, the service can be made available to Health Authorities, the Armed Forces and those public utilities that are likely to play a significant role in major emergencies.

This service has been specifically set up for use in the event of Civil Emergencies and major disasters. It is NOT for normal business enquiries.

BT's modern digital technology uses a flexible call routing tool known as Advanced Linkline Services. This facility can direct calls from a special national telephone number to one or more pre-selected answering points. The National Emergency Linkline number dialled from anywhere in the UK, connects the caller to the nearest Emergency Linkline reception point, normally the local BT Network Management Centre. The NMC will ensure that a request for assistance is handled promptly and that all necessary parts of BT are alerted.

The service is available 24 hours a day.

When requesting assistance a caller must identify themselves and the organisation that they represent and provide a telephone number on which they, or another representative, can be contacted. They should give as much detail about the incident as possible to enable BT to react quickly.

The information should include:

- Accurate location (e.g. address/grid reference).

- Casualty situation (e.g. is an enquiry bureau being set up?).

- Access problems (e.g. difficult terrain or parking restrictions).

- On-site security (e.g. will BT identity cards suffice?).

- Reporting instructions (e.g. who should BT people report to?).

- Communication needs (e.g. requirements at the scene/incident control?).

- Safety issues (e.g. are there any hazardous conditions to consider?).

The National Emergency Linkline number must not be disclosed outside the organisation. This will ensure unauthorised users do not abuse the service.

To get more information (including the National Emergency Linkline number) contact the local BT Zone Emergency Manager.

4.1.12 Government Telephone Preference Scheme

The Government Telephone Preference Scheme (GTPS) provides a contingency facility for the withdrawal of **outgoing** telephone services from the majority of customers on a telephone exchange. The scheme is designed for use in a serious crisis when increased use of the telephone network is causing severe congestion and preventing the emergency services and other essential users from making and receiving calls. At present the scheme only applies to BT and Cable & Wireless.

Lines that have their outgoing service withdrawn under the scheme will retain the capability of receiving incoming calls. Normal service will be restored to all customers as soon as possible.

Rules for the selection of lines for inclusion in the scheme have been set by the Government. These rules and other information about the scheme are contained in a Government Notice.

The GTPS can only be invoked by the Government in exceptional circumstances. However, the facilities GTPS provides can be used by BT or Cable & Wireless as part of their network management arrangements if their network is heavily overloaded or damaged.

The scheme only operates over PSTN lines. There is no charge for this service.

All exchange connections are placed in three categories.

Category 1 consists of those lines essential to the Government and the emergency authorities in a severe crisis or emergency which is affecting the public telephone network.

Category 2 includes lines additional to Category 1 that are required to maintain the life of a community during civil emergency.

Category 3 covers the remaining lines not entitled to special preference during an emergency.

All Government departments have a designated authority to nominate for inclusion in Categories 1 and 2; this process is known as 'sponsoring'. **Sponsors are required to notify BT or Cable & Wireless annually of the lines they wish to nominate for inclusion in GTPS.** The GTPS administration is handled by BT and Cable & Wireless Emergency Planning Managers.

4.1.13 Secondary Control

As well as providing PTOs with secondary and alternative numbers, fire brigades should have contingency arrangements to cover the receipt of emergency fire calls during conditions of serious breakdown, either in BT's network or their own brigade communications systems.

These arrangements usually involve a 'secondary control' set up either in a different building on the same site or at a different location. Consideration also has to be given to the receipt of calls during the interim period; EAs should make use of automatic call diversion facilities where possible.

Managers of Control Centres should regularly monitor the efficiency of their 999/112 emergency call arrangements and arrange regular liaison meetings at local level. Exercises should be run periodically to ensure that all staff are familiar with the contingency arrangements.

4.1.14 Publicity/Public Education

Both the PTOs and the four emergency services continue to be actively involved in various education programmes aimed at young school children.

Apart from the Strategic Framework for Combating Malicious Hoax 999 Calls (**DCOL 9/96**), many separate initiatives have been taken by the PTOs, Police, Fire and Ambulance services to educate the public as well as reduce the number of hoax calls.

It is interesting to note that the Coastguard service receives very few malicious or hoax calls.

Education and advice to the public is ongoing, and will become more important if the police and ambulance services introduce 'second priority' numbers for minor emergencies.

HM Fire Service Inspectorate and CACFOA advise against fire services using a 'minor emergency' number.

It is not generally recommended that persons should call the fire brigade by dialling the fire station or the fire control number direct. The reasons are:

- Directly dialled calls cannot be monitored by the BT operator.

- It is seldom possible to trace the origin of a directly dialled call.

- The call would be delayed if the fire brigade number were found to be engaged or out of order.

- Payphone users would need to insert coins which, in an emergency, might not be readily available.

Entries in telephone books

A standard page is included in the preface of all telephone books on the use of the 999/112 emergency service. Administrative telephone numbers of fire brigade headquarters and other departments or establishments should be inserted in telephone books under the heading of the local authority concerned.

4.2 Cable & Wireless 999 service

Cable & Wireless work within the Code of Practice for The Public Emergency Call Service between Fixed Network Operators and the Emergency Services.

In 2000 Cable & Wireless outsourced the management of their Operator Services to Vertex Customer Management. As well as handling all 999 calls, Operator Services handle Operator 100 and International Operator Services. Any query or problem with the 999 service should be referred to Cable & Wireless.

Cable & Wireless handles approximately 9.5 million 999 and 112 calls each year. It provides services to its directly connected customers and to the customers of a number of Other Licensed Operators including (but not exclusively) cable carriers and mobile networks. The Emergency Service operators are located in one call centre in Birmingham and two in Glasgow which handle all 999 and 112 traffic originating or connecting onto the Cable & Wireless network anywhere in the UK.

These call centres are fully resilient being on separate power supply lines with separate multiple connections to the Cable & Wireless trunk network. The sites have on-site emergency power generator provision. They share the same management and ancillary structure.

999 and 112 calls entering or originating on the Cable & Wireless network are routed by the shortest possible route to one of five dedicated switches for Operator Services traffic, located

around the country. These switches form a complete resilient, fully networked five-node system for routing traffic to the Operators. The system has full 24 hour support and queues have Real Time Management Information Systems to ensure all calls are answered immediately. The system is configured to give these calls priority over all other traffic on the Cable & Wireless network.

The emergency operators in both the Birmingham and Glasgow call centres connect callers to the Emergency Service. For reasons of cost and speed the call will route over the Cable & Wireless network emerging, if necessary, onto the BT network at the nearest Point Of Interconnect to the Emergency Authority. At no time are BT 999 operators involved in Cable & Wireless 999 procedures, only BT's local network where required by the Emergency Services own telephone network. On rare occasions BT may receive an emergency call for a Cable & Wireless customer, or vice versa. In these instances Cable & Wireless and BT will assist each other to ensure effective handling of the call.

4.2.1 Operator call-handling procedure

When a directly connected Cable & Wireless, T-Mobile, Virgin or cable customer dials 999/112, an Emergency Call attempt will be recorded within the call centre by means of an audible and visual signal. The call is immediately given the highest priority. The operator holds any existing call on the console and answers the emergency call 'Emergency, which service please?'.

Simultaneously, the operator will have received a display of the 'calling line identity' – originating caller's telephone number, in addition to which, automatic voice recording is activated. If the call has originated from a cellular caller a five-digit area zone code will also be presented.

At this time, the system will automatically initiate a search of the customer records database, Front Office Directory (FRNT), using the displayed calling line identity and/or zone code. If the search is completed successfully, the operator will have the following customer details displayed:

- Caller's telephone number.

- Name.

- Address.

- Primary connect – to numbers of each Emergency Service.

- Secondary connect – to numbers of each Emergency Service.

(Where zone code is used, the caller's name and telephone number will not be available.)

Where the caller's details cannot be retrieved from FRNT, then the operator will refer to a 'backup' screen on FRNT which will provide the relevant connect-to numbers in accordance with the caller's STD code presented in the calling line identity or cellular caller's given county location.

The operator will advise the caller that they are being connected to the requested Emergency Service. Once connected and an answer from the Emergency Service is gained, the operator will introduce the caller by announcing:

- Operator call centre identification;

- Customer's calling line identity; or

- Relevant telephone number if the call has been received from a call monitoring centre.

Once these details are given, the operator will hold the call on the console, and leave the call in progress, allowing the operator to become available to answer other incoming 999/112 calls. The operator will only remain on line if requested to do so by the Emergency Service, if the caller is particularly distressed and incoherent or if the caller requests more than one emergency service.

Once the operator has left a conversation in progress between the Emergency Service and the caller, the console will visually display the call status. When the call is complete, and all parties have cleared the call, the operator can relinquish the call. At this point, a call print-out will be generated, providing the following details of the call:

- Date.

- Relinquish time.

- Operator distribution cabinet number.

- Console number.

- Operator identity number.

- Call type (999/112).

- Calling line identity (and zone code where applicable).

- Number that the call was extended to.

- Any operator comments relating to the call.

There will be occasions where the caller is incoherent and unable to ask for the required Emergency Service. As before, the operator will initiate a search of the customer records database (FRNT), using the displayed calling line identity. The operator will then connect the call to the specified police control relevant to the area and advise of the caller's attempt and any other useful information, e.g., name, address, telephone number.

If during any 999/112 call attempt where a caller's details cannot be retrieved from FRNT, then the Operator will contact the Switch Network 'B' division (SNB) to obtain detailed information.

Where the caller is a subscriber of another Licensed Operator, dedicated 'hotline' numbers have been set up to each operator in order to provide speedy retrieval of information.

Cable & Wireless have no plans at present to automatically forward CLI information to brigade mobilising systems.

4.2.2 Enquiries and requests from Emergency Services

In some instances, an Emergency Service may find it necessary to request additional information or to seek clarification after a call has been released by the operator. The Emergency Service must call the designated numbers within the centres and advise what additional information is required. The request will be actioned by a Team Leader or operator immediately.

Access to emergency call records and recordings should be obtained in accordance with the Public Emergency Call Service Code of Practice.

4.3 Kingston Communications

Kingston Communications is a rapidly expanding UK communications group. It has run an operator services department in the Hull area since first established in 1904. One of these services is the handling of incoming emergency calls from anywhere in the Kingston Communications network, which is in and around the city of Kingston-upon-Hull and expanding the East Yorkshire area.

Kingston Communications operates under the Public Emergency Call Service Code of Practice, handles approximately 231,000 calls a year, and all operators are trained to handle emergency calls. Kingston Communications pass calls to Humberside Fire Brigade using the primary, secondary or alternative numbers.

4.4 Telephone Number Portability

Telephone number portability means that subscribers can keep their existing number when they change phone companies.

Portability was proposed by OFTEL to eliminate the problems and expense (mainly for business customers) of changing a phone number when changing from one licensed operator to another.

Agreed procedures between the Emergency Services and the PTOs (before number portability trials on the fixed networks took place), ensure that customer addresses are always available for use on emergency calls during the transfer.

Licence modifications are now proposed for the mobile networks, this means that from 1 January 1999 portability will be extended to mobile phones.

The ability to obtain customer record information on mobile networks will become almost impossible.

Portability will be between mobile and mobile or fixed and fixed networks. There will not be convergence between the two systems for the foreseeable future.

4.5 Emergency Text Telephone Service for the deaf

Typetalk – (DCOL 6/ 1995, DFM 5/1995)

The Director General's Oftel Advisory Committee on Telecommunications for Disabled and Elderly People (DIEL) advised that all involved in the 999 service should take account of the need to establish uniform access to the emergency services for people with severe speech and hearing difficulties.

To that end the Text Users' Emergency Service was launched in March 1995.

Run by Typetalk, which is part of the RNID, and funded by BT, it gives deaf, deaf and blind, deafened, hard of hearing and speech-impaired people access to the Emergency Services. A deaf or speech-impaired person who is unable to use an ordinary telephone uses a textphone, which is like an ordinary phone but has a keyboard and screen, to dial the Typetalk Text Users' Emergency Service on **18000**, although if the old number is used calls will still connect to Typetalk.

The operators employed by Typetalk are highly trained and fully familiar with the needs of deaf and other text phone users.

Typetalk procedure for dealing with incoming emergency calls

(1) Receive call on Text Users Emergency Service (TUES) terminal. Emergency calls take priority over all other switchroom activity.

(2) Establish the number from which the call is being made.

(3) Establish the service required by the caller.

(4) Attempt to obtain name, address and location of incident if different from caller's address.

Figure 4.3 Emergency text telephone for the deaf. (Photograph: Typetalk)

Figure 4.4 Using a text phone. (Photograph: BT)

(5) On obtaining minimum information (calling line number and service required) dial out to the required EA using BT 999 service and instruct BT to connect Typetalk to the EA for the calling number given.

(6) On connection with the EA, the Typetalk operator will relay the call between Text caller and EA by voice. BT operator will normally remain on line to monitor call and offer assistance with locations, etc.

Departures From Standard Procedure

In the event that Typetalk are not given the calling number by the caller, attempts will be made to establish their location. This allows for connection of the call to the appropriate EA using the county lists and direct connect-to numbers.

In the event of failed connections, or calls which go off-line mid-stream, the back up CLI (Calling

Line Identity) printer is used to try and establish the calling number. Attempts will be made to contact the calling number. If contact with the caller is not successful a call is made via the BT 999 service to the police control for the area of the calling number to report a failed Emergency Call.

It should be noted that the CLI is not always received. It may be suppressed by the caller, or be from a network which does not share CLI with the BT network.

The service is tested at regular and frequent intervals. This testing involves Typetalk and BT responses, EAs and BT are not informed of the times of any test calls.

4.6 Emergency calls from the Railway Industry Network

Since British Rail was fragmented into a number of different companies and franchises the collective term used is the 'Railway Industry'. The railway industry has its own telephone network – the Extension Trunk Dialling network (ETD) operated by Global Crossing Ltd.

The ETD network is almost exclusively used by railway personnel although, in some circumstances, possibly an emergency situation, it could be used by members of the public. Dialling 999/112 from this network connects the caller to a Global Crossing operator, not to a BT Operator Assistance Centre.

An Emergency Call is defined as a call from any source, concerning an incident, for which the caller requests the assistance of any of the Emergency Services. Emergency calls will only be answered by Global Crossing operators who are trained and certified as competent to do so. All emergency calls will take priority over any other call and be acted upon even if it is a repeat call.

Currently, the switchboard at Glasgow is designated to receive emergency calls.

Emergency calls from any part of the country could be received at any of the designated switchboards. The operator begins by asking which service the caller requires and the location of the incident. Location details are entered into the Telephone Operators Directory System (TODS) which will show the primary and secondary connect-to numbers of each Emergency Service Control Centre. Emergency calls on this network will almost always originate from a railway location, although it is possible for the switchboard to receive calls from non-railway locations. In this latter situation locational information is unlikely to be found on the TODS database and the Global Crossing operator will endeavour to obtain enough information (such as the nearest town, etc.) to correctly route the call.

If the EA primary number is unobtainable, or not answered within 30 seconds, the Global Crossing operator will try the secondary number. If this is not answered, or is unobtainable, then the call is passed to the civil police. The number given by the Global Crossing operator is an ex-directory emergency ringback number.

Global Crossing will remain on the line until the EA operator has all the required information and the call is complete.

All details of emergency calls are recorded in writing, as well as recorded on audio equipment.

4.6.1 Payphones

The card payphones situated on trains for public use are not part of the ETD network. These phones are GSM phones which accept prepaid 'smart' phonecard and credit cards. GSM public payphones are also installed on some domestic coaches, Scottish and cross-channel ferries and Eurostar trains. Phones on the cross-channel services have the added facility of 'roaming' onto the French and Belgium GSM networks.

It should be noted that it is not possible to make 999/112 calls from these payphones. The phones are clearly marked to inform the public that 999/112 calls are barred.

4.7 Cellular communications

Cellular radio is a telecommunications service which allows people with mobile phones to make and receive 'radio' telephone calls within the service area to and from almost all national, international and other mobile phone network numbers.

In the early 1980s the Government concluded that British business and industry would be handicapped without adequate mobile communications. To this end cellular licences have been granted by the Department of Trade and Industry (DTI) since June 1985.

The cellular licence issued by the DTI and monitored by Oftel initially prohibited some network operators from dealing directly with customers. This has created a complex multi-tiered market structure comprising of service providers, dealers and high street retailers which has implications when trying to trace an abandoned call on a cellular network.

The original cellular phones were analogue but now all UK operators provide a digital network (GSM).

The history of GSM started in 1978 at the World Administrative Radio Conference (WARC) where the radio frequency band for cellular mobile systems was agreed upon.

In 1982 a committee was set up to ensure that the frequencies allocated to cellular radio were being used correctly and to co-ordinate plans for a European standard. This committee was called 'Groupe Special Mobile', the European standard has taken its name from this committee, hence GSM.

In 1987 twelve countries agreed to sign a Memorandum of Understanding to design and implement GSM. Work on the technical development of GSM continues through the European Telecommunications Standard Institute (ETSI). In 1990 the GSM initials were changed to represent the new title 'Global Systems for Mobile Communications'.

GSM, now considered an international standard, was developed to ensure compatibility across cellular networks, allowing mobile phones to operate in different countries. However, it does mean that there is an increasing number of overseas customers using their mobiles on UK networks ('roamers'). 'Roaming' may cause problems in tracing silent or difficult emergency calls. Although the mobile number will inform the British PTO of the caller's country of origin, any trace can only be done by dialling the caller back. The call will be routed via the country of origin which has cost implications for EAs and PTOs. Additionally, some 'roamers' have incoming calls barred because the cost of the call from their country of origin is charged to their own number.

It is unlikely a call from a 'roamer' could be traced. Therefore, it is reasonable to consider these calls un-traceable.

4.8 Cellular 999 services

The cellular operators 999/112 call service is supported by Operator Assistance Centres (OACs). BT handles all 999/112 phone traffic for O_2 (UK) Ltd, Vodafone and Orange. T-mobile emergency traffic is handled by Cable & Wireless.

Cellular network companies found several problems that had not been experienced to such a degree with the fixed networks. To help overcome these problems a Code of Practice was devised under the auspices of the Home Office 999 liaison Committee – The Public Emergency Call Service (PECS) for Mobile Radio 999 Emergency Access.

When a 999/112 call is generated on the cellular network, it is received by the base station providing the strongest signal; generally this will be the nearest cell site to the location of the caller. However, because the transmission is radio, several conditions may influence where the signal travels to – local topography, poorly positioned aerials on vehicles, or weather conditions, and so on. Also the fact that cellular frequencies travel exceptionally well across water adds to the contributing factors.

The incoming call will be transferred to the fixed network providers (BT or Cable & Wireless) and from their OACs they will be passed to the Emergency Operator together with a four-digit zone code and the mobile telephone number. Zones roughly mirror county boundaries and are used to direct the call to the correct emergency authority. The zone code accesses the OAC's database and produces the connect-to number for each of the emergency services.

Mis-routing of cellular calls generally happens because of the reasons stated earlier. It is also not possible to impose the same boundaries as with a land line system, this is especially apparent in rural regions where one transmitter may service quite a large area. With the growth of the cellular phone industry and zones becoming smaller, the occurrence of long distance mis-routing is likely to become less common.

The mobile companies recognise the difficulties incurred by the Emergency Services when a call is mis-routed. To keep these calls to a minimum, they require the emergency authorities to inform the mobile companies when a mis-route has taken place to enable them to investigate the circumstances surrounding the call.

The final responsibility for the overall correct routing of 999/112 emergency calls rests with the cellular companies.

4.8.1 Name and Address Information of Mobile Callers

Previous licence agreements of O$_2$ (UK) Ltd (Formerly Cellnet) and Vodafone restricted these companies from dealing directly with subscribers, their services could only be bought through Service Providers.

Although EAs require 24 hour access to subscriber records, the records of O$_2$ (UK) Ltd and Vodafone customers, who purchased their mobile telephone from an independent Service Provider, are not always available on a 24 hour basis. The Service Providers are not obligated by their licence to provide 24 hour access to customer information. Therefore, it may be difficult to follow up a 999 call which has been interrupted or terminated in suspicious circumstances.

The Federation of Communication Services (a trade association of the mobile communications industry) have informed Oftel that it is not commercially viable for all Service Providers to provide 24 hour access given the small number of cases involved.

The licence agreements of T-mobile and Orange enable them to supply their customers directly. Therefore, most of their subscriber information is available on a 24 hour basis. Only a small percentage of their customers use high street service providers.

It should be noted that licence agreements are subject to change. However, brigades will be kept informed of all 999 issues through the 999 Liaison Committee.

Subscriber information for cable customers is held by BT and Cable & Wireless.

4.8.2 Release of Subscriber information

The principles and procedures applying to the recording of calls, and the release of subscriber information for emergency calls originating in any mobile network, is the same as those applying to emergency calls originating in the fixed network.

All the cellular 999/112 services work in much the same way. Therefore, a detailed description of one system may be helpful.

4.8.3 System Description – Orange

Orange Personal Communications System (PCS) – sometimes referred to as Personal Communications Network (PCN) – uses BT's facilities to connect 999/112 calls to Emergency Authorities Control Centres (EACCs). All 999/112 calls generated from the Orange digital network will be routed to one of six BT Operator Assistance Centres from one of 13 Orange switches, each of which has a minimum of two routes into BT's network. Routing design within Orange ensures that 999/112 calls are sent to BT at the originating switch first, followed by alternative routing throughout the total network. This guarantees delivery of a 999/112 call in all cases, barring a major network disaster within Orange.

The principle behind the PCN system is the multiple re-use of valuable radio channels. The country is divided up into a series of 'cells', each served by

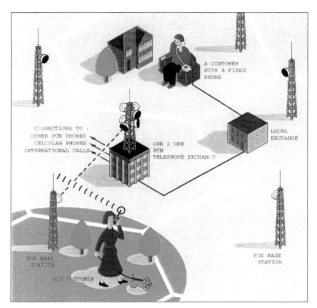

Figure 4.5a

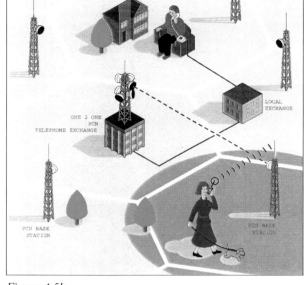

Figure 4.5b

Figure 4.5 As the caller moves around between cells, the PCN system automatically switches the signal between base stations without interrupting the call. (Graphic: One 2 One)

its own low powered transmitter/receiver (base station). Each of these base stations is assigned a set of frequencies differing from those assigned to adjacent cells. The resulting pattern can be repeated enabling radio channels to be used again but geographically far enough away to prevent interference.

The very nature of the provision of radio telephone communication means that users do not necessarily always know their exact location and the radio communication system cannot pinpoint the caller either. The resultant problem in a cellphone user having to detail this information when making a 999/112 call is likely to lead to delays in satisfactorily reporting particular incidents. However, in every case the caller should be asked to state their location.

4.8.4 Cell/EA Boundaries

Comparison of cell site boundaries and EA boundaries shows that cell boundaries are unlikely to overlap more than two adjacent EA coverage areas. Where a cell coverage area straddles two or more EA coverage areas, a particular EACC will be agreed and nominated, to which 999/112 calls from that cell are to be directed.

4.8.5 Routing 999/112 Calls to EACCs

To enable the operator to connect a cellphone 999/112 call to the correct EACC, a map of the UK with EA zone code areas has been created. The EA boundaries generally follow county boundaries. Each cell within the agreed EA boundaries is mapped to a four-digit zone code (Orange zone codes are prefixed with '3', O_2 (UK) Ltd zone codes are prefixed with '2', Vodafone zone codes are prefixed with 'O'). Each zone code will be mapped by BT to the four EAs within the boundary area, each of which will have advised the PTO of primary, secondary, alternative and evacuation EACC connect-to numbers.

On receiving a cellphone 999/112 call, the BT operator's console displays the caller's number (CLI) and a zone code which identifies the originating base station. A list of EACC connect-to numbers corresponding with the zone code is automatically displayed for selection by the operator.

4.8.6 Cell ID Look-Up Failure

Failure to display an EACC connect-to number on receipt of cell ID/zone code information is extremely unlikely. However, should this occur the mobile operator will tell the cellphone caller that there is a network fault and that some information checking will be necessary. Orange will locate the origin of the call by determining the cell ID, using its inherent network facilities. Once the cell ID is provided, the mobile operator will refer to a look-up table and forward the call to the corresponding EACC. This call tracing facility can only be carried out if the calling cellphone holds the connection. Post event traces are not possible but records are kept by Orange which include time of call, duration, originating cellphone number and the cell which received the call. These records are kept and are available for cross-checking for approximately three months.

4.8.7 EACC Connect-to Numbers

The EAs should aim to provide at least two weeks notice of changes to primary, secondary, alternative and evacuation connect-to numbers to the relevant PTO and mobile operator, who will follow their mutually agreed update procedure. The date and time that the new numbers become effective should also be stated.

4.8.8 Mis-routed Calls

Base station radio reception areas cannot be sharply defined or matched exactly to EA boundaries. Therefore, for some base stations, it is inevitable that a small proportion of calls will originate outside the boundaries of the agreed EACC. Also, there are several other reasons why the base station handling the call may not be the closest to the incident; that is:

- The caller is moving and delays reporting an incident.

- A distant base station across water can sometimes provide a stronger radio path than a closer one on land.

- If the caller is in the radio shadow of a hill or large building, a more distant base station may be selected.

- Cellular radio signals can travel long distances in certain weather conditions.

- The nearest base station is already fully occupied.

It is also possible that the cell ID could give an incorrect but apparently valid code to the operator due to a faulty console or a fault in transmitting the display information from the switch.

Once connection is established to the EACC, it is the responsibility of the EA operator to establish that the call is relevant to the EA area, or to instigate means of transfer if it is not.

This can be achieved in a number of different ways:

- In most cases the EA will take the details of the call and pass the information on to colleagues in the correct authority.

- The EA operator may recall the operator back into circuit and request that the call is passed to another EACC. This may be a different emergency service.

- The EA operator may advise the operator of the correct EA to handle the call. The Operator will then look-up the appropriate connect-to number.

If it is not possible for the EA to advise the correct connect-to number, or even the correct county, then the operator will instigate a call trace procedure as described, resulting in Orange providing the location where the cellphone accessed the Orange network. The operator will then re-route the cellphone customer to the applicable EA.

The ultimate responsibility for redirecting the call to the correct EA, however, will rest with the operating company who will take all reasonable steps to do so.

4.9 The satellite telephone

Worldwide communications cover is now readily available to all by way of satellite telephone technology and is capable of transmitting both voice and data. It would be true to say, however, that until recently many brigades felt that such systems were out of their reach due to the high cost factor in providing the equipment, together with the unit cost of calls. The situation has changed over the past few years: satellite communication is now considered an everyday form of transmitting speech and data. Today, equipment is both readily available and owing to competing commercial service providers, can be supplied and operated at a reasonable cost.

Satellite communication has been employed by some UK brigades in the past, most notably when a system was obtained on loan and used successfully by members of UK Fire Brigade Search and Rescue Teams whilst in Armenia, following the earthquake disaster. Some fire brigades in Europe regularly use satellite telephone systems as an acceptable form of communication and their specialist rescue units carry it as part of their normal equipment.

The modern range of satellite communications systems offers a light and compact package of equipment that can be set up and ready for use in a matter of minutes. It is also true to say that it is a reliable form of communications that is simple to use from anywhere in the world.

Figure 4.6 Satellite telephone.
(Photo: Inmarsat)

Communications and Mobilising

Chapter 5 – Control Centre equipment

All fire services now use computerised mobilising systems to support call taking and mobilising procedures. These systems comprise several main elements:

- Integrated Voice Communication Switch.

- Computerised Mobilising System.

- Communications Interface.

- Bearers.

- Station End Equipment.

Integrated within the mobilising system are other optional features which aid the control staff; for example, resource displays, mapping systems and automatic vehicle location systems.

5.1 Control Centre Design

A comprehensive guidance document for Fire Service Control Centres has been published by the Home Office in DCOL 8/1997; in Scotland as DFM 8/1997 (FRDG Publication 2/97). The document includes advice on Control Centre design and ergonomics, procurement and legislation.

The most common workstation configuration for a Control Centre Operator comprises two PC terminals, one connected to the mobilising system and the other to the communications switch. Through these two terminals the operator can carry out all mobilising and communications tasks. In some cases a third PC will be used to hold resource display maps and geographical information systems (GIS).

Typically, each operator position in a Control Centre will be fitted with the same workstation configuration and system facilities through which the basic tasks of message handling, logging and resource despatching are carried. Occasionally a specially configured supervisory position may be installed to provide additional facilities such as special monitoring functions or access control, etc.

5.2 Communications

5.2.1 Administrative Communications

In centralised mobilising schemes there are considerable advantages in segregating administrative communications systems from operational systems, though it is common for links to be provided between them to give flexibility of usage. The main advantages of this principle are firstly that the operational systems can be much simpler and more easily duplicated at a number of operating positions, and secondly that both operational and administrative systems can be operated simultaneously at maximum capacity at any time without causing mutual interference.

In the majority of Control Centres segregation applies. The administrative PABX telephone switchboard is usually in an ideal situation elsewhere than in a Control Centre and operated by non-uniformed staff during normal office hours. After office hours, however, when the switchboard is closed, incoming calls are switched on 'night service' extensions to terminate on Control Centre equipment.

5.2.2 Safeguards for Emergency Communications

The mobilising scheme should have been planned so that it is not seriously affected by congestion, either in the Control Centre or on its communications systems. It should also be able to function normally and without interruption in the event of a mains electricity supply failure, either locally at fire stations, or centrally in the main Control Centre. Either the central control should be safeguarded so that there is virtually no chance of a complete breakdown there or, alternatively, there should be arrangements made for a secondary control to take over in the event of a serious failure of the main control or its communications.

There are, however, technical and economic problems involved in choosing, equipping, staffing and keeping an alternative control centre in being, solely for the use in the event of a breakdown at the main control. The general practice has, therefore, been for the fire authority to invest available financial resources in a highly reliable communications system and in safeguarding the central control to the maximum possible extent. These safeguards normally include:

- Diversified communications bearers.

- Standby power facilities which automatically come into operation immediately in the event of failure of mains electricity supplies.

- Adequate fire precaution arrangements; e.g., smoke detectors in communications apparatus rooms and also in plant rooms and roof spaces which are normally unattended and are a potential risk.

Finally, there should always be predetermined and practised last-ditch arrangements, including the use of radio and of pre-arranged telephone contacts. At fire stations there should be an emergency uninterrupted power supply (UPS) to maintain computer and turn out equipment. It is essential for Control Centres to have an uninterrupted power supply and emergency power generator facilities.

5.2.3 Provision of Suitable Circuits

The mobilising 'scheme' should not rely totally upon access to the public telephone network, since this may become congested due either to peak normal usage or to the direct effect of a flood of emergency calls to a large incident.

Whilst, in the past, fairly widespread use of the public telephone network was made by fire services for remotely controlling fire station alerting systems, this practice was, generally speaking, only acceptable when mobilising was decentralised on fairly small units such as districts or divisions. Nowadays, due to the automation of the telephone network and the rapid expansion in the amount of telephone traffic, it is no longer regarded as satisfactory for emergency call-out purposes. Therefore, the primary bearer could be Kilostream circuits, ISDN, radio or a commercially available public data system such as RAM or Paknet. Whilst the secondary bearer should be independent of the primary bearer it could be ISDN, radio, data radio, or PSTN.

Some fire services also have a tertiary bearer and utilise a commercial paging network to operate the firefighters call-out system. Other brigades use an overlay paging scheme on the brigade main scheme radio as their tertiary bearer.

So far as the initial connection of 999/112 emergency callers is concerned, it is important that adequate facilities are provided to enable telephone operators to do this very quickly. As a standby against breakdown of these circuits and for use during peak periods, a number of ex-directory exchange lines are provided in the Control Centre where they appear at all operator positions together with the trunk-subscriber circuit terminations.

5.2.4 Alternative Routing of Cables

To minimise the effects of a possible breakdown, the scheme should always include what is known as 'true alternative routing' of the lines serving the central control building – that is the provision of at least two separate cables in different cable duct routes. This principle should extend, so far as is practical, to all fire stations particularly where important or 'key' stations are involved. At the central control end, essential operational circuits used for receiving incoming emergency calls and for remote control call-out facilities should be equally divided between the different duct routes so that, if one of them is interrupted, for example, due to flooding, at least half of the circuits remain in operation.

5.2.5 Monitoring of Remote Circuits

The circuits which carry remote control facilities, which are the essential backbone of any mobilising scheme, should always be of the monitored type. These give automatic indication to the Control Centre operators of faults as and when they develop on the network, enabling immediate action to be taken to get the faults rectified and to implement predetermined alternative arrangements for alerting the affected stations or personnel.

GD92 can be set up to generate traffic which tests the bearer at regular intervals. The timing of these test signals is set on installation and can be determined by the brigade. (See page 51.)

5.2.6 Exchange Telephone Lines

Exchange telephone lines are commonly used in Control Centres and terminate either on telephone instruments, telephone switchboards, line concentrator units or a digital switch linked to a touch sensitive screen. Some are earmarked for exclusive operational use whilst others are for administrative purposes.

The common tendency is for the operational circuits to be terminated on line concentrator units or a digital switch and the administrative circuits to be terminated on a PABX switch so that calls may be connected to extensions throughout the organization. A PABX or private branch exchange is a semi-automatic switchboard that allows the majority of connections via the PABX to be dialled direct by the extension users, and also may allow external callers to directly dial the extension. The function of the switchboard operator is then mainly confined to answering calls on the main switchboard numbers, where the caller is unaware of the extension number, or answering queries from callers on the internal extensions.

5.2.7 Operational Lines

Where a number of exchange lines serve a Control Centre it is usual for them to be 'ex-directory', under which arrangement the numbers are not disclosed to the public. It is advisable to keep at least one line free for outgoing calls or have these numbers allocated as incoming calls barred lines. It is

not uncommon to apply the facility of 'auxiliary working' to such groups of exchange lines, so that when the first numbers or lines of the group are in use, the caller is connected automatically to one of the free numbers or lines in that group.

However, one disadvantage of this arrangement, when applied to operational lines in a centralised mobilising scheme, is that certain fault conditions in the telephone exchange might put all the lines of the group out of action.

It is highly desirable, therefore, to split the emergency and the administrative lines between telephone exchanges so that they would not all be affected by one fault.

5.2.8 Line Concentrator Units and Digital Switches

Line concentrator units allow the operator to accept a call from whatever source with a single action, and are more convenient to use than a number of telephone instruments. These units can accommodate a variety of types of termination including exchange lines, private wires and telephone extensions as well as control terminations for the various facilities on fire service radio schemes.

These units can easily be repeated, each with identical terminations and facilities at any number of operating positions in a Control Centre, to facilitate the simultaneous handling of a number of different calls during busy periods. The termination of all operational circuits on concentrator units avoids the unacceptable bottleneck which would be created if a conventional type of telephone switchboard was used.

Where concentrator units are repeated, an incoming call is indicated at all positions with a flashing lamp signal. When the call is answered at one position, the flashing signal on all other positions changes to a slow wink on the lamp, which indicates that the call has been accepted and that the circuit is engaged until the lamp goes out.

The unit is also suitable for use when monitoring facilities are required, e.g., to enable a supervisor or Officer-in-Charge to listen in to calls and to

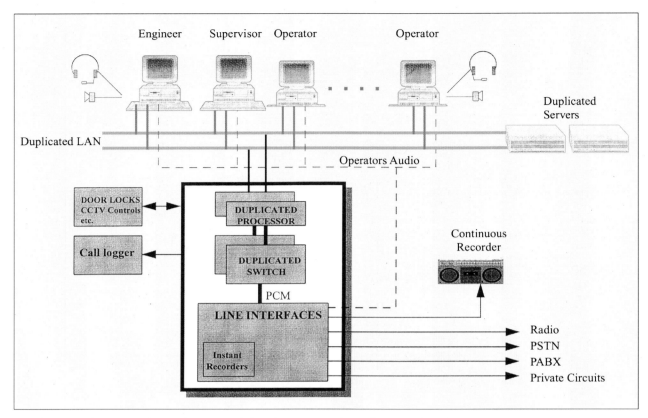

Figure 5.1 Architecture of typical Integrated Communications Control System (ICCS).
(Graphic: Securicor Information Systems)

break in and assist when necessary. The same terminations and facilities as on the operational position units appear also on the supervisory position units, and the circuitry can be so arranged that the action of 'listening in' does not degrade or in any way interfere with the call.

The line concentrator unit is being superseded by a digital switch that incorporates duplication of all key elements of the switch. Most switches use a PC and 'touch screen' which is linked to the switch via a high speed local area network or a serial link. Many are Integrated Communications Control Systems (ICCS) which incorporate both telephony and radio functions.

Different levels of access are available for operators, supervisors and maintenance staff.

Operator access to the system is by means of a touch screen colour which displays coloured representations of press-buttons. **When touched, the system responds and begins a series of operations related directly to the selected button to provide full control and status displays of all radio and telephone functions.** Most switches are capable of queuing incoming calls and, if necessary, present the operator with the oldest call first. This is especially useful during spate conditions when emergency calls may be waiting on the system.

The high speed local area network interconnects operator positions and the central switch. This network broadcasts simultaneous updates to all positions and whilst performing particular functions, the operator is able to call up information from the system's database including relevant help messages and telephone directories. In addition, each operator position has access to all the facilities provided by the line concentrator units such as call monitoring by any other position, intercom and indication of line state and extra facilities such as a system database, recording and playback facilities and configuration of both telephone and radio facilities.

Figure 5.2 A touch screen in use in a Control Centre.
(Photograph: Simoco)

Communications switches either have an integrated call-logging facility which can be accessed through an RS232 port to an engineering terminal, PC or separate printer, or the ability to connect an independent call logger. These PC-based systems provide flexible report generation and the ability to customise information presentation of all telephone and radio traffic.

Open interface capabilities give the flexibility to connect external devices including CCTV, lighting systems, door entry mechanisms and alarm systems.

5.2.9 Automatic Call Distribution

An additional function of call-handling equipment is Automatic Call Distribution (ACD). Incoming emergency and administrative calls are automatically presented to available operators on a highest priority basis. Each call in the queue is presented to the first operator who releases a line and is available to accept the next incoming call. An electronic tone is transmitted to alert the operator that he or she has been allocated a call.

When the call has been completed, the operator is allowed a predetermined amount of time in which to carry out other essential actions. The system will automatically present another call when the lapsed

time is reached. Alternatively in cases where a number of actions is required, a manual option to 'suspend' the operator from the system is available.

This system is more likely to be used in larger brigades; for example, it has been installed in London.

5.3 Computerising Mobilising System

Clearly the primary Control Centre tasks of incident logging and resource availability are team activities which require all operators to have access to the same information.

A variety of mobilising systems is available to brigades and may be known as Computer Aided Mobilising Systems, Command and Control Systems, or Mobilising and Communications Systems. All these systems are, as the names suggest, systems which incorporate computers to aid the reception and logging of calls and the despatch of the brigades' resources to incidents.

The incident and resource information recorded on the mobilising system is of interest to a number of departments outside of the Control Centre, e.g., statistics, Press Officer, etc. Much of the information required by control may be prepared or maintained by other departments and, hence, database update facilities must be made available to these departments.

An operator or supervisor's workstation will have a visual display unit (VDU), a keyboard and often a 'mouse' to provide access to the mobilising computer. The keyboard may be a standard typewriter (QWERTY) layout or a standard keyboard with some of the key functions changed to dedicated functions.

5.3.1 Mobilising System Functions

The main function of the mobilising system is to aid the recording of call information and the despatch of the selected resources. Secondary functions include displaying alarm conditions for the system and the generation of statistical information.

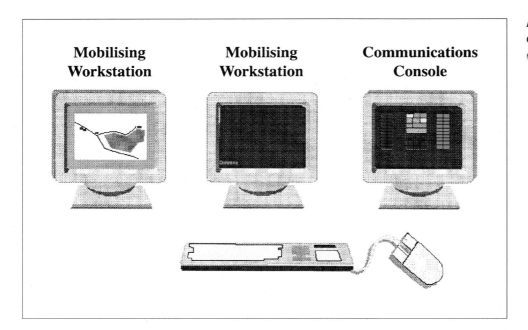

Figure 5.3 Typical Operator Workstation.
(Graphic: Fortek)

Mobilising Workstation	Mobilising Workstation	Communications Console

Upon entry of an incident type and address into a predefined format the system will interrogate its database to match the address information. If an exact match is not found the system may offer similar sounding addresses to the operator. It may also search for risks, duplicate incidents, telephone kiosks, map references, etc., so as to provide the operator with as much information as is possible. When an address match is made the operator is presented with a predetermined attendance and possibly a recommended attendance.

The operator is then able to accept the offered attendance, mobilise an alternative or defer the incident, placing it in an incident queue or merging it as in 'same as all calls'.

To undertake these tasks the mobilising system accesses various databases that are part of the system, such as an address gazetteer, predetermined attendances, risks, special procedures and CHEM-DATA information. It also records the status of appliances and officers and will not offer resources that are already committed to other incidents.

It must be stressed that the system only makes recommendations which can be overridden by the operator. The mobilising system also maintains a log for each incident recording all the actions associated with that incident. Other logs are also maintained recording other aspects of the system not related to incidents, such as communication failures, operators logging in and out of the system, tests, etc.

Once the mobilising system has been used to set up and verify the incident details and proposed resources to attend, the turn-out instructions must be conveyed to the appropriate fire stations and the crews alerted. The system will encode the data and deliver it to the communications network for onward transmission to the appropriate destination including automatic data recovery.

In most instances the communications network comprises a primary, secondary and possibly a tertiary back-up bearer. The primary bearer will be the most appropriate bearer for the station taking into account the number of calls and other facilities required, as well as available technology.

Secondary bearers should be independent of the primary bearer so that any failure will not affect both bearers. Examples of bearers used are Kilostream circuits, ISDN and PSTN telephone lines and radio links including dedicated data networks and brigade radio schemes.

The station end equipment must be able to receive and respond to Control Centre messages for turn-out instructions from both primary and the secondary bearers.

In 1992 the Home Office produced a specification (known as GD92) for a standard communications protocol to be used for all operational communications between the control centre and the station end equipment.

The main objectives of the specification were that products could be procured by fire authorities under a Framework Arrangement and would be interoperable with other products from the same, or different, contractors. In addition, brigades and contractors should be able to enhance the basic products without affecting interoperability.

The Framework Arrangement was able to meet the need for provision of equipment with the differing capacities and performance required by different brigades whilst providing the benefit of economies of scale and boundary independence.

The specification was also beneficial to brigades procuring equipment outside the Framework Agreement.

GD92 defines a standard protocol and message format for mobilising systems over commercially available bearers. This protocol is now used by the majority of fire brigades.

For each of these bearers the protocols and message formats have been designed to ensure that the mobilising system and, hence, the operator, is advised of the delivery or non-delivery of each 'turn out' instruction. The protocol supports administrative messages, equipment status messaging and other functions such as burglar and fire alarm activation, power failure and restoration messages, and tests of mobilising links to ensure their availability.

GD92 also supports two-way messaging and, hence, station personnel can prepare messages locally and send them into control, the most common example of this being the entering of staffing levels at a change of shift.

Station equipment has become increasingly sophisticated and is generally controlled by a microprocessor or computer. The system will check the incoming data to ensure that it is valid and then undertake a series of localised actions which may include:

● Control of mains powered equipment such as lights, doors and exhaust extract.

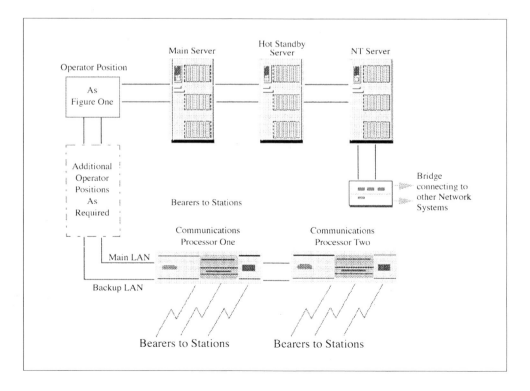

Figure 5.4 Mobilising System Configuration.
(Graphic: Fortek)

- Signalling to crew alarms including audible alarms, alerters, appliance indicator lights.

- Confirmation to the Control Centre that uncorrupted data has been received, peripheral equipment has operated and crews have acknowledged.

In addition to the above, the system may also run self-checking routines to ensure that they are functioning correctly, that the bearers are functioning correctly, and that other conditions such as mains power and battery status are monitored.

Mobilising system architecture generally falls into two basic categories:

(1)   Central Processing; or

(2)   Distributed Processing

The workstations and/or processors are linked together by means of a local area network thus providing a communications path to the various elements of the system. Redundancy is provided in the system so that failure of any particular element will not inhibit the mobilising process.

(1)   Central Processing
Various standard computer configurations can be implemented in a client/server system, consisting of PC workstations connected via a network to a server. The server requires an operating system that is not only able to support multiple work stations, but also has the processing capabilities of high level programming languages, systems such as 'Unix' or 'Microsoft NT' fulfil these requirements. In addition, processors are required to provide the communication interface from the mobilising system to the bearers and hence the station ends.

Resilience is provided by incorporating redundancy within the system. Dual servers are provided, one operating as a 'hot' standby, i.e., the secondary server is continually being updated in 'real time' by the master server, so that in the event of failure of the master server the secondary server is able to take over the function of the master server.

The secondary server may be in a different location within the same site – that is, a different building within a control complex. This builds in some additional resilience in the event of system failure to the main control room.

The communications processors are also duplicated but as these do not incorporate dynamic databases both are operating together but are able to mobilise the brigade independently in the event of failure of one of them. As the workstations are in effect 'dumb' terminals, failure of one workstation processor will not jeopardise the mobilising system but will only render that workstation inoperative. Again, it is possible that the communications processors are in different locations within the same site.

(2)   Distributed Processing
Workstation processors are connected by a local area network but in this configuration the workstation processors are high-level processors which hold all the database information such as incident logs and PDAs. One processor is deemed to be the master processor and co-ordinates the processes of the other workstations. In the event of failure of this master processor, then another workstation can be designated as the master. Communications with other peripherals are carried out by other processors on the network; for example, each fire station has a processor on the network which is connected to the bearer interfaces.

It is desirable that local area networks used for mobilising should not allow access from other networks as this could lead to congestion or failure of the mobilising system by corruption.

Consideration should be given to protecting the mobilising local area network and providing appropriate 'fire walls' where necessary. With current technology it is possible to provide more than one network connection on a workstation thus providing an operator with the presentation of information from different networks but not providing any interconnection of the networks.

Information required by other brigade departments, such as incident logs or statistical data, may be downloaded at predetermined times or in 'real time' to another computer system for interrogation and processing. Conversely, data may be retrieved from other brigade computer systems by the mobilising system for use in the processing of incidents.

5.4 Ancillary Control Facilities

5.4.1 Voice Recorders

Fire services record incoming emergency calls automatically by using various types of voice recording machines ranging from the relatively simple single and multi-track tape machines to PC-based digital recorders with automatic time injection. It is customary to devote an individual track on a multi-track machine to each workstation and derive the audio from the connections to the operator's headset enabling all land lines and radio channels in use at that particular workstation to be recorded.

Although voice recordings are not used to assist turn out, they are nevertheless sometimes useful for verifying the accuracy of an address or other information. They are also used at subsequent enquiries to prove what in fact was said on a particular occasion both by the caller and the fire service operator. Recordings of emergency calls are frequently used as training aids to help trainees appreciate the problems in extracting adequate information from agitated callers. They are also used to aid identification of callers who have made malicious fire calls; there have been instances where, when faced with a voice recording, the culprit has confessed to being the originator of a false alarm.

5.4.2 Availability and Fire Situation Display

Every Control Centre must have, in one form or another, an accurate and up-to-date record of the location and availability of appliances, equipment and officers on standby for immediate turn out. This display is used primarily for ensuring that the resources of a predetermined first attendance are in fact available for despatching when an emergency call is received. It is also used as an aid to

the Officer-in-Charge of the Control Centre when considering 'covering' moves to maintain an equal distribution of fire cover throughout the area during periods of intense activity.

Fire situation information throughout the mobilising area at any time must also be displayed, and this must be kept up-to-date and in step with the mobilising moves. The fire situation display would normally show the address of the incident, the appliances and officers attending, whether or not a stop message has been received and very brief information likely to be needed by Control Centre staff or senior officers.

The types of display used for these purposes vary a great deal in different brigades. In the past, it was usual to display a wall map showing general-purpose and appliances-availability, with separate boards marked out for recording fire situation details. In such cases the map indicates every available appliance by means of tallies or coloured lamps. It is also common to find separate 'officer-availability' boards which indicate where an officer is and whether he or she is available by radio, pager or telephone.

When an appliance or officer leaves their station, whether it be to an incident, for drill or other purposes, the appropriate tally or lamp is deleted from the availability board or map and shown either on the fire situation display board, in the case of an incident, or in an appropriate section of the mobilising board. Provision is made to record on the board those appliances and/or officers not immediately available for operational duties.

With the advent of computerised mobilising systems, the display of information from fire situation displays, appliance or officer lists, to PDAs and incident logs has become the norm at each operator position. The introduction of graphical information systems (GIS) has enabled displays similar in appearance to those presented by the traditional lamp system to be reproduced electronically. These resource availability displays, for both appliances and officers, are driven from the changes in status of the resources held in the mobilising system database. These displays can be presented at each operating position on a VDU and also projected on to a wall display, either from the front or the rear.

5.4.3 The Gazetteer

To receive and validate details of an emergency, an operator relies on a comprehensive gazetteer of streets and special locations and/or premises. All mobilising systems will utilise a gazetteer to validate the location of an incident, so clearly the speed and accuracy with which an operator can confirm the incident location will depend upon the quality and comprehensiveness of the gazetteer.

As a minimum the gazetteer holds a list of street names for the major towns and district, or parish names for rural areas. Additionally, special risk locations will be included. Associated with each entry in the gazetteer is a list of nearest pumps and special appliances (PDA) from which the operator can select the most suitable response to the incident.

Increasingly, the quality of the gazetteer data is being improved and extended to work with digital maps. In most brigades an extensive map database is held on the mobilising system, this graphical database providing an alternative method for validating addresses and also a more appropriate means for holding risk and general reference data.

5.4.4 Maps

In Control Centres, Ordnance Survey maps of the mobilising areas are available for reference purposes, giving such information as the boundaries of the area and the location of stations. In addition, larger scale maps may be held, together with plans and diagrams giving details of motorways (with their access and exit points), dock and harbour areas, new city development complexes, unusual special risks, etc. Street maps, often in a book format, are also held to assist with the location of streets and to enable directions to be passed to appliances, officers or other agencies.

As the amount of information available to a brigade and required by firefighters is added to, almost exponentially, the Control Centre is seen increasingly as the most suitable repository and distributor of this data. It is now an essential requirement that this information can be easily retrieved through the mobilising workstation either for review by control, or for dispatching to a fire station or incident ground.

Figure 5.5 Graphical Information System Display.
(Photograph: Bedfordshire and Luton Fire and Rescue Service)

This has led to the introduction of commercially available computer software to support these functions. Much of this information is held in graphical, or geographical form, Graphical Information Systems (GIS) allow mapping data to be manipulated and presented to the Control Centre operator. The maps can also be linked to the mobilising system so that when a database search matches the address criteria, the correct map showing the address location is displayed to the operator.

GIS can interact with other software such as word processors and graphical presentation technology so that composite packages of information can be developed. Maps of a fire service's area can be presented with facilities to zoom from small scale to large scale presentations. As these maps are composed as a series of layers it is possible to select the level at which certain features are displayed; for example, text can be displayed only when it is possible to read it. Overlays can be added to give details of hydrants and water mains or gas pipe lines or any other features that have a significance to fire service operations.

Using the facilities of word processing and other software, it is possible to display information from the inspection of premises as text, with drawings, photographs, diagrams and even video clips all linked and accessed from the map presentation. As with the resource display, this information may then be displayed at the operator's VDU or projected on to a wall display and, because the data is held electronically, may also be transmitted to other computer systems including those on appliances or special vehicles.

Brigades obtain maps from the Ordnance Survey through Service Level Agreements.

5.4.5 Automatic Fire Alarm (AFA) Terminations

The majority of manual AFA terminations were disconnected when Control Centres migrated to computer-aided mobilising systems. However, AFA activity can be properly supported by computer technology. Such a system is currently in use by some fire services which act as a collector station to commercial premises.

5.4.6 Secondary Control Facilities

Control mobilising systems incorporate a number of levels of resilience. Duplicated computer systems and fall-back bearers each add their own levels of security to the system, as does the ability to alert crews locally from the fire station.

These facilities do not, however, cater for the rare possibility of having to evacuate the main control centre. Systems have been developed which permit restoration of basic turn-out facilities from other locations.

Different mobilising systems provide different secondary control provisions. These range from a portable laptop computer containing the basic mobilising system and communications interface to a duplicate control on the same site as the main control or at a remote location.

Secondary control facilities should be provided with facilities for the reception of emergency and other incoming and outgoing calls, the despatch of resources and the operation of the main scheme radio at a location that would not be affected by any disruption to services provided at the main control. This may necessitate locating the secondary control with emergency telephone lines from a different exchange to those of the main control.

5.4.7 Control Centre Software

With the advent of mobilising equipment based on PCs and the proliferation of computer systems in the workplace, Control Centres now have commercially available software packages for their use. These systems may reside on PCs that are also used for mobilising or on stand alone machines.

These packages generally include:

- Word processors for producing text; for example, aides memoires, help files and specific instructions which can imported into other systems, including the mobilising system.

- Spreadsheets for manipulating data.

- Presentation software to produce lectures.

- Databases to produce statistical analysis and performance indicator criteria – required by the Home Office.

- Graphical information systems to produce maps and analyse statistical data in a mapping format.

- Fire reports (FDR1) – required by the Home Office.

5.5 Equipment at Fire Stations

5.5.1 Mobilising Computer

Some brigades utilise a computer on each station that acts as a station controller, controlling most of the equipment associated with the mobilising of crews. There are a number of data links to various items of equipment as defined below.

5.5.2 Printers

Printers in fire stations are primarily used for receiving turn-out instructions from controls following the operation of the 'turn out' alarm system. They can also be used for the receipt of other non-urgent operational information.

5.5.3 Alerter Base Station

Fire stations with retained crews have a base transmitter to activate the alerters. This equipment alerts crews with a number of different signals. A positive acknowledgement is transmitted via the data link to the mobilising computer, and onward to the brigade control room when the equipment is actuated.

5.5.4 Public Address System

Many fire stations and headquarters have public address systems of one kind or another, with loudspeakers sited strategically throughout the building. These broadcast routine and domestic announcements.

A number of brigades now use improved types of remotely controlled public address systems on whole-time stations for alerting crews and the broadcasting of turn-out instructions as well as routine announcements. A number of different tones can be sent over the systems that enable crews to distinguish the type of message being broadcast.

Station Bells/Alert Tones

A system of alarm bells/tones, usually referred to as 'turn-out bells' has, from time immemorial, been part of the normal equipment of fire stations. The system is used primarily for alerting personnel, turn-out instructions being passed by teleprinter or telephone.

On some stations, a simple system of one or more circuits of bells is used for alerting personnel in all parts of the premises.

5.5.5 Turn-out Lighting

Fire stations, both those continually staffed and retained, usually have automatic facilities for switching on selected lights to illuminate those parts of the station that are used by personnel responding to calls in the hours of darkness.

These lights are generally controlled via a relay box which, in turn, is connected to the mobilising computer. Following operation they may be on a

time switch and stay on for a fixed period of time, or may be reset manually.

5.5.6 Alternative Power Supply

Mobilising equipment should be provided with alternative power facilities for use when the normal power supply fails. The types of system can vary, but in most cases consist of an uninterrupted power supply (UPS), consisting of a bank of batteries that is continually charged to supply power in the event of a failure The system is connected to the mobilising computer by a data link and will inform the control room of both the failing of and restoration of normal power.

5.5.7 Exhaust Extraction Systems

A number of brigades utilise exhaust extraction systems to remove exhaust fumes from the appliance bay. These are normally actuated by the turn-out system and will remain on for a set period of time. They can also be reset manually.

5.5.8 Control of Traffic Signals

In large towns and cities, provision can sometimes be made for traffic signals in the vicinity of the fire station to be operated by the mobilising computer or from the watchroom, to stop traffic and give fire appliances a clear exit from the station.

5.5.9 Automatic Appliance Room Door

Electrically operated appliance room doors are provided on some fire stations. In addition to having manual controls and built in safeguards, these will be linked so that they can operate concurrently with the station alerting systems.

5.5.10 Running Call Facilities

Some fire stations provide facilities (with instructions on how to use them), to enable members of the public calling personally at the station to summon the brigade. This type of call is known as a 'running call'.

Where there is always someone available in the fire station premises, a switch, usually labelled 'Fire', is sometimes provided on the front of the

Figure 5.6 Station Equipment.
(Graphic: Fortek)

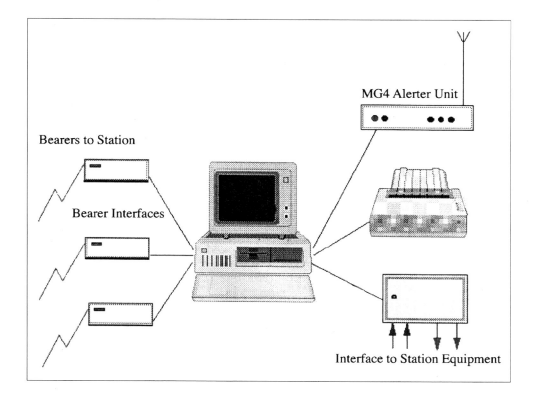

Bearers to Station

Bearer Interfaces

MG4 Alerter Unit

Interface to Station Equipment

fire station for use by the public. This actuates an alarm within the station, which alerts personnel for a turn-out and summons a firefighter to the front doors to obtain particulars from the caller.

At fire stations where there is not always someone available on the premises – for example, a day staffed or retained station where all personnel attached to the station turn out on the fire appliances – there is usually a special telephone at the front of the fire station for use by 'running callers'.

This telephone, suitably illuminated and labelled with instructions, may sometimes be an ordinary exchange line with limited dialling facilities to prevent misuse, on which the caller may dial 999/112. It could also be a telephone linked to a private wire communications network which, when the handset is lifted, connects the caller instantly to the appropriate Control Centre.

In all cases instructions should be displayed on how to use 'running call' facilities and include

directions as to what the caller should do if the system is out of order or if for any other reason there is no response. If facilities are not provided notices should be displayed informing the caller of the location of the nearest public telephone.

5.5.11 Enquiry Bell

It is common practice for continuously-staffed stations to have an enquiry bell circuit installed, with a press-button labelled 'Enquiries' at the main entrance to the station. Operation of the button actuates a bell or buzzer in the station. The enquiry bell is normally intended for non-urgent enquiries only.

5.5.12 Other Ancillary Equipment

With the introduction of modern systems virtually any piece of equipment can be operated on the actuation of the station alerting systems. Such examples are the switching off of cookers, kettles and other electrical equipment.

Chapter 6 – Automatic Fire Alarm Transmission Systems

The purpose of automatic fire detection equipment is to give early warning of fire to the occupants of buildings and to possibly alert the fire service. Where the equipment is designed to automatically call the fire service it is essential that automatic fire signals should be transmitted to Fire Service Control Centres as quickly and reliably as possible.

There are many different types of automatic fire detectors installed in buildings, either as part of a fire detection and alarm system or as self-contained detection/alarm devices. These detectors are designed to detect one or more of the characteristic phenomena of fire (heat, smoke, gas or flame) and actuate alarm devices or systems.

6.1 Arrangements for linking automatic fire detection systems with a brigade Control Centre or a commercial Central Alarm Station

The main elements are:

- A means for transmitting signals from the protected premises to a remote manned centre (RMC), such as a brigade Control Centre or a Central Alarm Station, sometimes referred to as an Alarm Receiving Centre (ARC);

- and, where the RMC is a Central Alarm Station, the processing of signals at the Central Alarm Station.

- A means of communication between the Central Alarm Station and the Brigade Control Centre.

No matter how comprehensive and efficient an automatic fire detection system may be, its task is not fully completed until it has informed those responsible for taking appropriate action that there is a fire in the building. Therefore, if the purpose of the system is to ensure prompt fire service attendance due to the potential life safety situation (as, for example, is the case in a hospital, or to protect property) there will need to be an efficient and reliable method of transmitting fire signals automatically to a remote manned centre unless there is a very reliable means of on-site monitoring, e.g., at a permanently manned security gate-house. If the fire detection system is intended to satisfy the fire insurer, this will normally be a requirement.

Where an automatic transmission system is provided, it should transmit a signal to the RMC as soon as the automatic fire alarm system operates. In exceptional circumstances, a time delay unit (TDU) may be provided to permit an investigation prior to the transmission to the RMC. A TDU should, however, only be used if there is a false alarm problem that cannot be addressed by other means. A TDU is not normally acceptable in life risk premises e.g. in hospitals or premises providing sleeping accommodation; in other premises, TDU should only be provided after consultation with the fire service and the insurer.

The performance of the alarm transmission link may be expressed by the probability of an alarm call being received at the brigade Control Centre within a specified time. The 'time of transmission' is the period, expressed in seconds, between the start of the transmission of the alarm signal from the premises and the point in time of connection to the fire brigade control. Ideally, this should not exceed 60 seconds.

6.2 Transmission Methods and Reliability Issues

As is always the case, economics is a relevant factor which directly affects system planning and, since highly reliable communication systems cost more than less reliable ones, there is a variety of systems in use throughout the UK.

Mention has been made of automatic fire detection systems (AFD) being connected via remote manned centres (RMC) to local authority fire brigade Control Centres. There are, in a few areas, facilities for AFD to be connected directly to fire brigade Control Centres.

One fire and rescue service collects signals from data transmitters direct into its command and control computer. Two simple key operations will display the predetermined attendance (PDA) to the premises on the operator's screen. Another station also monitors AFD within its county (and some outside its county) via a digital communicator.

Several small areas of the UK are covered by the 'Alarms By Carrier' (ABC) system. This uses the subscriber's normal exchange line onto which is superimposed an inaudible signal so that the line is continuously monitored by British Telecom. Fire signals are routed directly to, and monitored at, the brigade Control Centre where the ABC system interfaces with the mobilising system, enabling the alarm signals to be displayed directly onto the mobilising screens.

Other than in a small number of areas, examples of which are contained in the first paragraph above, and in the four counties served by ABC, there are generally no facilities for the transmission of fire signals from AFD direct to fire brigade Control Centres (except by 999 autodiallers, the use of which is now discouraged). Signals are normally routed to an alarm company Central Alarm Station.

A British Standard Code of Practice, BS 5979 (Code of Practice for Remote Centres for Alarm Systems) gives recommendations for the planning, construction, facilities and operation of Central Alarm Stations that monitor fire alarm, intruder alarm and/or social alarm systems.

The Code recommends that the date and time of origin of all incoming and outgoing signals, and incoming and outgoing communications are automatically recorded.

With regard to communications with the fire brigade, BS 5979 recommends that there be two independent means of outgoing communication between the Central Alarm Station and the Control Centre of the fire brigade appropriate to the geographical area from which alarm signals are received.

This latter caveat is particularly important. It is unacceptable for a Central Alarm Station to receive connections from AFD in areas for which there is no acceptable means of communication between the Central Alarm Station and the relevant fire brigade. It has been known for a Central Alarm Station to dial 999 in the hope that the local brigade will connect them to the appropriate brigade.

Use of the fire brigade administrative telephone number for passing fire calls is also unacceptable.

The Code also recommends that the two means of communication with the fire brigade be selected from the following:

- A dedicated voice transmission path.

- A supervised data transmission path.

- An ex-directory telephone number for the Control Centre **(this should be recognisable at the Control Centre as an emergency call from the Central Alarm Station).**

- The 999 system, provided this will result in the public telecommunications operator routing the call to the appropriate fire brigade (this is clearly only possible if the Central Alarm Station and the protected premises are located within the same fire brigade area).

A single ex-directory telephone number served by two or more lines on a hunting group at the fire brigade Control Centre is regarded as two independent means of communication.

On receipt of a fire alarm signal at the Central Alarm Station, action should immediately be taken to establish communications with the appropriate fire brigade Control Centre within:

(a) 30 seconds for 80 per cent of fire alarm signals received; and

(b) 60 seconds for 98.5 per cent of fire alarm signals received.

These times exclude delays in transmission of the signal from the protected premises to the Central Alarm Station, and any delays in answering calls at the fire brigade Control Centre; they represent a form of Central Alarm Station response time.

The Loss Prevention Certification Board (LPCB) operate an approvals scheme for Central Alarm Stations that monitor AFD. The Central Alarm Stations are approved to the LPCB Loss Prevention Standard LPS 1020: Requirements for Remote Centres for Fire Alarm Systems.

It is a requirement of LPS 1020 that the Central Alarm Station must be able to offer an LPCB approved system for the transmission of fire alarm signals from the protected premises to the Central Alarm Station. This is the case even if the approved Central Alarm Stations can also offer other methods of transmission that are not approved. Although the LPCB are responsible for the approval scheme, it is operated jointly by the LPCB and its sister organisation The National Approval Council for Security Systems (NACOSS). Inspection of Central Alarm Stations is carried out by NACOSS.

The LPCB publish a list of Central Alarm Stations that have been approved under LPS 1020. This list indicates, for each Central Alarm Station, the geographic areas from which the Central Alarm Station is approved to receive fire alarm signals. This provides confidence to the user that there is third party verification that the Central Alarm Station complies with good practice, that there is an agreement with the relevant fire brigades and that there are suitable means for passing fire calls to them.

CACFOA and the British Fire Protection Systems Association (BFPSA) have developed a Model Agreement relating to AFD connected to brigade Control Centres via commercial Central Alarm Stations. The Model Agreement would be between the Local Fire Authority and the Central Alarm Station.

Brigades should make use of the ODPM/ CACFOA/BFPSA document entitled 'Avoiding Unwanted False Alarms Generated by Automatic Fire Detection Systems', as issued in DCOL 6/96 (in Scotland as DFM 8/1996).

In general, the requirements of LPS 1020 are incorporated within BS 5979. One of these requirements is that LPS 1020 approved Central Alarm Stations must prepare a written report describing the circumstances, and action taken, in all cases where the time between receipt of a signal and transmission of information to the fire brigade exceeds three minutes. (This includes any delay in answering the incoming call at the fire brigade Control Centre.)

There are four distinct means for transmitting fire signals from protected premises to RMCs.

These are:

1 Digital Communicators which automatically dial the Central Alarm Station using PSTN, and transmit a coded signal to a receiver at the Central Alarm Station.

2 Private circuits, which provide a permanent monitored transmission path between the protected premises and the Central Alarm Station.

3 British Telecom 'CARE' system, which is similar in principle to ABC, in that it uses the subscriber's normal telephone line to carry alarm signals 'piggy back', but is used to route signals to a Central Alarm Station rather than the fire brigade.

4 'Paknet Radio Access' links a protected promises to a Central Alarm Station using Vodafone's public data network. Connecting an alarm panel to a Paknet Radio-Pad provides access to the network, enabling alarm signals to be sent to the Central Alarm Station.

Prior to the widespread introduction of the digital telephone network, research showed that digital communicators, which are probably the simplest and least expensive form of transmission system, were relatively slow and less reliable than methods involving private circuits. The speed and resilience of the digital telephone network have improved and should have enhanced the reliability of this method.

Private circuits may, over relatively short distances, be established on a 'point to point' basis, but over longer distances part of the path between protected premises and the Central Alarm Station is usually shared by many subscribers. In the latter case, the many subscribers are connected via the alarm companies' 'Satellites' (data concentrators), from where a large number of signals are multiplexed via private data circuits. This offers a reliable, fully monitored signal path at, possibly, the highest cost.

British Telecom's CARE system is quite economical because it uses an existing telephone line. Cost is, therefore, independent of distance between the Central Alarm Station and the protected premises. This system is now available in most areas of the UK.

Vodafone Data Network provides periodically monitored communications between a protected premises and a Central Alarm Station, using Paknet Radio Access. The cost is the same as a monitored telephone line and is independent of distance between the protected premises and the Central Alarm Station. With radio coverage approaching 95 per cent of the UK population, Vodafone Data Network is increasingly being adopted for alarm communications.

6.3 Social and Community Alarms Centres

Social (community) alarms are found in the homes of approximately one and a half million people in the UK. Most of these people are elderly and/or disabled but there are many other groups (including victims of domestic violence or racial harassment and individuals discharged early from hospital), who depend upon social alarms.

Many of the alarms are located in individual dwellings and are connected to a 24-hour Community Alarm Centre (CAC) via their normal telephone line. A call to a CAC can be triggered by pressing the button on a portable pendant or on the telephone unit. Calls can also be triggered automatically by a range of sensors (for example, smoke, gas or PIR detectors) installed as part of a telecare package.

Similar alarms located in sheltered housing schemes for older people are often activated by pulling a cord (similar to a corded light switch). Such alarms normally form part of an integrated alarm system for the whole of that development and, in these circumstances, the alarm system will also provide for control of door entry systems and onward transmission of building safety alarms (e.g. fire alarm panels). During the day alarms within sheltered housing schemes are often monitored by the resident warden and switched to a CAC when the warden is off duty.

When an alarm is actuated, a call is automatically placed to the CAC. Equipment at the CAC will auto-answer the call and information about the location and nature of the alarm call will be transferred to the CAC's equipment. This information is placed in the call queue display on the CAC operator's computer screen and used to provide the operator with detailed information about the caller from the CAC's customer database. Information within the database will include data on the caller's personal circumstances including any medical details provided by the customer.

Approximately 95 per cent of all calls received are not emergencies and are usually requests for reassurance or information. Out of the remaining 5 per cent of calls, approximately half can be appropriately addressed by sending out a mobile warden or alerting family, friends or neighbours. However, there will be times when the CAC operator needs to contact a health professional or an emergency service, including the fire service.

The sector is represented and regulated by the Association of Social Alarms Providers (ASAP). ASAP has published a Code of Practice for the management and delivery of social alarms services, which has been adopted by the Office of the Deputy Prime Minister. ASAP has contracted with the Security Systems and Alarms Inspection Board to provide an independent audit scheme for its Code of Practice. ASAP and CACFOA have agreed procedures for filtering and passing emergency calls from sheltered housing schemes to the fire service. Operation to these procedures is a requirement for conformance with the ASAP Code of Practice.

While many users of social alarm services are likely to ring 999 directly in the event of an obvious emergency, this being the quickest way of contacting the fire, police or ambulance services, there are times when a call will be routed via a CAC. This will occur because:

- many residents in sheltered housing do not have a telephone and their only means of calling for the emergency services is via their social alarm;

- the circumstances of some individuals mean that they are unable to summon help, other than by activating their pendant. This may be because of a medical emergency, an impaired ability to move (e.g. a fall) or because the alarm provides the quickest or only means to call for assistance in the circumstances;

- calls from automatic sensors will always be routed to the CAC;

- some social alarm users are confused and have difficulty communicating with others, including the BT operator, or do not realise that the situation requires one of the emergency services.

The information held by the CAC and the training and experience of their staff means that the service is able to identify the most appropriate response for the caller. As a result, the CACs not only ensure that customers receive the most appropriate service but filter out many calls which might otherwise have led to an unjustified 999 call.

There are approximately 350 CACs in the UK. Most are operated by public sector bodies; for example, a local council or housing association.

Between them, these centres receive approximately 27 million calls per annum and a number of these calls will require a response from the emergency services. Where calls are forwarded via a social alarm, CAC operators are in a unique position to help emergency service control staff because they:

- are trained and experienced in dealing with emergencies;

- have detailed information about the person in need of help;

- are trained and used to dealing with vulnerable people and can reassure the individual and liaise with them until assistance arrives;

- can alert and liaise with other agencies and carers who might need to be involved; and

- will have detailed information on the address of the emergency and on emergency access to sheltered housing schemes, which can be passed to crews attending the incident.

Although some CACs utilise direct (ex-directory) lines into the Fire Service Control Room, most calls are now being placed via the 999 system. To avoid mis-routing of calls within the 999 system, CACs operate an agreed procedure with BT's 999 services which results in the system disregarding the network CLI of the CAC and substituting this with the telephone number for the premises at which the incident has occurred. To ensure effective management of incidents, ASAP advises its members to provide emergency service controllers with the location details for the incident, as well as contact details for the CAC handling the call.

Further details of the Association of Social Alarms Providers (ASAP) are available from their website

HYPERLINK http://www.asap-uk.org www.asap-uk.org

or from their office (telephone 08700 43 40 52).

Copies of the ASAP Code of Practice are available free of charge from the publications page of the website.

Communications and Mobilising

Chapter 7 – Automatic Vehicle Location Systems

Automatic Vehicle Location (AVL) systems are not new. They have been used in the United States for some considerable time and by a number of security firms in the UK. Ambulance services and, more recently, some fire services in the UK have also taken advantage of the technology to enhance their vehicle availability and running times.

In the early 1990s many fire services were replacing their mobilising and communication systems and revamping or moving into new Control Centres.

During the upgrade many of the large and expensive resource display boards were replaced with screen-based resource displays incorporating simple mapping or full Graphical Information Systems (GIS).

As fire service personnel became aware of the operational potential of GIS, it became obvious that questions about AVL systems would soon be asked.

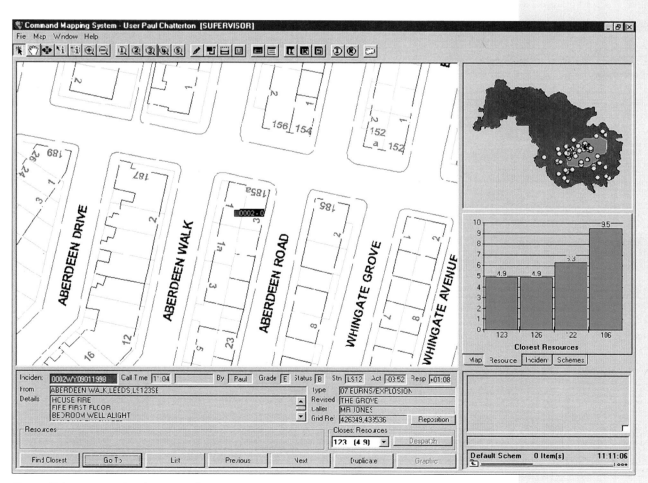

Figure 7.1 An example of a screen format using an Automatic Vehicle Location (AVL) system. (Graphic: Fortek)

HM Fire Service Inspectorate took the initiative, and a consultancy project was set up to **'investigate the applications and possible benefits of Automatic Vehicle Location (AVL) in the U.K. Fire Service'**.

The consultancy contract was awarded to Fortek Computers Ltd by the Fire and Emergency Planning Directorate as part of a Research and Devlopment programme managed by HM Fire Service Inspectorate. The contract was awarded in October 1994 and ran for two years.

The findings of this consultancy were published in DCOL 8/1997 (in Scotland as DFM 8/1997).

The following are extracts from the original consultancy report plus updates that were added in January 2003.

7.1 AVL Technology

An Automatic Vehicle Location (AVL) system has the capability to report vehicle positions to a central control centre either at regular intervals or on demand, or a combination of both. Several technologies are used worldwide although in the UK two systems predominate.

Systems based on the Global Positioning System (GPS) utilise time signals received from a constellation of 24 satellites moving through precisely defined orbits to calculate the position of GPS receiving equipment located in the vehicle. Timing signals are transmitted by the satellites on an almost continuous basis and hence the vehicle position is always known, providing sufficient satellites are in view of the GPS receiver. A minimum of three satellites (ideally four to eliminate certain minor inaccuracies) must be in view of the receiver for it to calculate its position.

A terrestrial-based system, as supplied by Securicor Datatrak Ltd, uses a series of low frequency radio base stations to distribute a matrix of radio signals from which a Datatrak receiver can calculate its position using a form of triangulation.

Recently, mobile phones have been developed which use triangulation techniques with their base station locations to determine the location of the

mobile phone, although at this time these systems are not widely available. This technology will have the benefit of working inside a building.

Today, the systems based on GPS can determine a vehicle's position to better than 10 m and frequently better than 5 m. There are, however, locations (such as built-up urban areas) where accuracy is compromised by physical or geographical phenomena (for example, high rise buildings obscuring the satellites from the GPS receiver).

Both the GPS and terrestrial based systems described here require a data network to deliver the positional data to the communications centre. The Datatrak system utilises a national radio network set up specifically for AVL reporting.

Figure 7.2 'Mobile Radio' data terminal.
(Photograph: Fortek)

Figure 7.3 Securicor Datatrak Ltd data terminal.
(Photo: Fortek)

Figure 7.4 GPS Antenna mounted in centre of appliance roof.
(Photograph: Fortek)

Figure 7.5 Data terminal installed in cab.
(Photograph: Fortek)

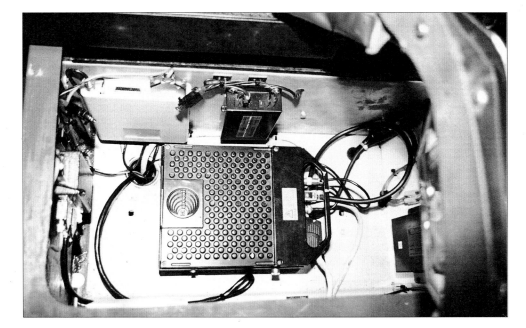

Figure 7.6 Mobile Data equipment in appliance under rear seat.
(Photograph: Fortek)

GPS-based systems require a mobile data network to be provided. This requirement can be met by using a public data network, a PMR channel with a data capability, or by utilising features of the new public safety radio schemes currently being introduced to the emergency services. These are based upon digital trunked radio schemes. Such schemes offer GPS originated AVL data as part of the radio infrastructure.

The final component of the AVL system involves the processing and presentation of the AVL data to support the control room and management task of a fire brigade. While the computation and delivery of a vehicle's position can be achieved through the use of commercially available components, the application of this data to benefit a fire brigade requires a degree of customisation.

7.2 Potential Benefits of AVL to the Fire Service

Potentially, AVL data can be used to assist in the deployment and mobilising of brigade resources and to improve the quality of data recorded against each incident.

Knowing the precise location of each brigade resource should enable the control room to optimise resource deployment and to ensure that the most appropriate (quickest suitable) resources are despatched to an incident.

Present mobilising policy seeks to achieve this objective by using Pre-Determined Attendance (PDA) which is compiled on the basis of appliances being at their home station. Hence, they are despatched from a known location to an incident, which in a typical brigade will be the case for approximately 80 per cent of incidents.

AVL data, therefore, has the potential to assist in the mobilising of resources to some 20 per cent of incidents by providing accurate positional fixes for the mobile appliances. This information can then be used by the mobilising system to compute the nearest/quickest appliances to the incident.

Generally speaking, the resource deployment strategy used by brigades necessitates standby moves to key stations to cover station areas when

appliances from that station area are unavailable. Without AVL, an appliance crew report their position as within a station ground which in many cases is a rather imprecise location.

Through the use of AVL systems, a far more precise location is available for each resource and, hence, the opportunity exists to deploy resources to more accurately reflect the needs of the risk areas and their corresponding standards of fire cover.

Until recently the only parameter available for defining location in the preparation of incident statistics and analysis has been station ground. This situation has in most cases now improved. Some brigades now include a grid reference in their traditional streets and places gazetteer. Other brigades are now using or combining their traditional streets and places gazetteer with commercially available data sets which effectively geocode every addressable premises within the brigade area. This information can be filed with the incident log and used in subsequent analysis.

AVL fitted appliances booked in attendance would also be reporting their exact position and, hence, the position of the incident. Further information, such as the position of the appliance when it booked mobile to incident could also be saved for future response time analysis.

7.3 AVL System Implementation

AVL systems have been implemented in many commercial organisations and other emergency services within and outside of the UK. An investigation into the performance of a number of these systems and the experience gained from the Pilot System installed in Avon Fire Brigade has highlighted features which, at the time of the pilot scheme, would compromise the effectiveness of AVL in the fire service, the more significant of which are discussed below.

Most fire brigades see the main benefit of an AVL system as being the ability to identify and despatch the nearest/quickest appliances to an incident, regardless of whether they are mobile or not.

In 1998 when this report was first written it was felt that whilst theoretically it was possible, the current AVL systems had not been designed to meet this requirement and it would not easily be achieved.

The difficulties in achieving this principal objective arose from the errors in the data that would be used in the calculation of the nearest appliance list. These errors were derived from:

(1) Inaccuracies in the incident location.
(2) Inaccuracies in the reported positions of each mobile appliance.
(3) Inaccuracies in computing, for each appliance, the running time from its present position to the incident location.

Today the inaccuracies of the first and third points have been overcome. This is because of:

● The availability of commercially prepared geocoded datasets that enhance / replace the traditional gazetteer.

● The availability of commercially prepared road networks and tools that allow the theoretical running time to be estimated with confidence.

It is still true to say that, as with the existing PDA system, the acceptable level of inaccuracy will vary according to risk and the associated standards of fire cover. In brigades where parish, or area, mobilising is used, the largest source of error will invariably come from the incident location. In urban and high-risk areas where street and premises locations are held in the gazetteer the inaccuracies in the reported vehicle positions will be the more significant.

It is difficult to set an expectation of what can be reasonably achieved. However, as a guide, it is reasonable to expect the computed running times in the list of nearest appliances to be accurate to within 90 seconds.

The Avon Pilot System showed that the required functionality can be met if AVL positional updates are transmitted with each resource status update, upon a request from the mobilising system or Control Centre operator and, periodically, at a rate dependant upon the resource status.

If a public data network is to be used then the interval for periodic updates could be several minutes without seriously compromising the system integrity. However, if updates can be delivered at no cost other than network loading then the interval set should be such that it does not impact upon the other data traffic.

Most mobile data terminals and portable PCs can now be fitted with a GPS transceiver and most transceivers will compute the vehicle position to the level of accuracy required for fire brigade applications. Certain units will perform better than others in difficult areas such as urban areas where satellites may be hidden from view by high buildings. Therefore, **the performance of the proposed GPS transceiver should be checked in various key locations throughout the brigade area.**

The same approach is recommended if a terrestrial solution is being considered.

7.4 Operational Considerations

Present mobilising procedures and PDAs reflect the principle that, if available, an appliance will be despatched to an incident in its own station ground. In the Avon Pilot site, where this policy applies, there were numerous occasions where other appliances, sometimes at home station and sometimes mobile, were calculated to be nearer than the appliance in whose station ground the incident had occurred.

Furthermore, in busy periods it is conceivable that appliances will get drawn across the brigade area, on the basis of being the nearest available appliance in a sequence of incidents, into areas with which they are not familiar and for which they may not carry appropriate information.

With current resource deployment and mobilising policies AVL data will be of relevance for approximately 20 per cent of emergency calls; that is when mobile appliances are considered for mobilising.

Through the use of AVL data, it is possible to move appliances to standby points other than fire stations while still being able to identify the nearest appliances to respond to new incidents. Such a policy, operated in a limited form in one brigade, has already shown savings by reducing the need to turn out retained fire stations.

7.5 Implementation Costs

The infrastructure requirements (that is, the need to provide a two-way mobile data network) of a GPS-based AVL system are such that it would be wholly uneconomic to consider setting up a system solely for AVL. **For both operational and economic reasons a brigade should consider the introduction of GPS-based AVL only as part of a programme to introduce mobile data.**

If mobile data can be justified in its own right, then the incremental cost of introducing AVL will be relatively small and should definitely be considered.

If the cost of introducing mobile data cannot be justified, then the benefits which could be provided by AVL at relatively minor additional cost may make the difference in justifying the introduction of the mobile data network.

The terrestrial solution offers a different approach. Since the required infrastructure has already been put in place by Datatrak, it is viable to introduce AVL with relatively little up-front investment. Such a system should provide two-way data which, for mobilising purposes, is considered essential.

As mentioned earlier, the new public safety radio schemes currently being introduced to the emergency services are based upon digital trunked radio schemes. Such schemes offer GPS originated AVL data as part of the radio infrastructure.

7.6 Conclusions

An AVL system operating as part of a two-way mobile data scheme will provide a brigade with the opportunity to simplify, and improve, its mobilising procedures by providing information which can be used in selecting the nearest/quickest appliances to an incident.

Clearly, busier brigades with a high proportion of wholetime crews stand to gain the greater benefits from AVL. However, even these brigades would need to consider changing a number of existing operational procedures.

There may also be a need to make a significant investment in upgrading the data, particularly the gazetteer, used by the mobilising system and possibly to upgrade the mobilising system itself.

Without such a commitment it will not be possible to realise the benefit of improved mobilising.

AVL data will also improve the quality of operational and management information, by accurately locating all incidents, and logging resource journeys to those incidents.

The technology exists to deliver quality data to firefighters and control staff alike, with AVL data being just one element. In formulating a strategy for the introduction of new technology it is inappropriate to view an AVL system as an independent item, since it will only be effective if it is introduced as part of a broader overall scheme to improve the quality of data brigade wide.

At the time of initial writing (1998), the incremental cost of including AVL technology as part of a mobile data scheme was relatively low. However, the investment required to create the environment in which it can be exploited will be significant.

This document was updated in 2003, by which time a relatively small number of fire brigades had taken the step of introducing the AVL into their mobilising and communications systems. The Fire Service Inspectorate Firelink project (see Chapter 9) expects to have a national public safety radio scheme for the fire service fully operational by 2007. This radio scheme will provide AVL data to the fire service, removing the significant investment requirement for each brigade that has been detailed above.

Communications and Mobilising

Chapter 8 – CCTV in the Fire Service

Since the text for this section was first prepared in 1998, a great deal of development work in the field of closed circuit television and associated systems has taken place and is still progressing. In due course, a report will be produced and provided on fire service applications. In the meantime the original text has been retained for information purposes.

Most fire service staff will be familiar with the type of Closed Circuit Television (CCTV) used for monitoring premises for security purposes, either in shops, car parks or used to survey headquarters, remote fire stations or even Control Centres – especially at night!

Some years ago the fire service recognised the benefits of capitalising on the technical developments of video cameras, and their ability to transmit images by a variety of means, having the potential to provide the Service with new tools to improve the efficiency of rescue and assist with the command role at incidents. These objectives, as well as the secondary, but important, benefits of improving debriefs, identifying training needs and informing the public accurately, led the Service to explore ways of using this equipment effectively on the fireground.

Because firefighters need to know as much as possible about the emergencies and the dangers they may be facing, the best substitute for seeing something directly is to have real time video of it.

Visual information of this kind does not add to information overload in the way that manuals, plans and procedural documents do. Irrespective of whether the incident involves a collapsed building with casualties, a dog lost in a warren or a huge fire that can only seen from one side, there are benefits in providing vision using available technology which would otherwise be difficult, dangerous or impossible to obtain. The use of video equipment allows fire service personnel to achieve this.

Information may need to be relayed to a Strategic, Tactical or Functional Command, or the Control Centre. Without doubt, officers want good quality information about major incidents for debrief, training, enquiries and public relations. Cameras can be provided in small robust units, and the transmission and recording methods available adapted to meet fire service needs.

A number of brigades have been awarded test and development licences by the Home Office to assess the operational potential to transmit audio and visual colour pictures within an incident environment.

Microwave spectrum is used because a wide bandwidth is required to transmit moving colour images.

For example, one brigade concentrated on getting visual information from confined spaces, such as collapsed buildings or sewers, and used video to improve the effectiveness of command at major incidents.

Another brigade uses two separate systems which enhance their CCTV applications. One has a microwave transmitter/camera, and the other is a Cellsend System. A Bodyworn microwave transmitter/camera system is also used.

These systems are described below.

The Modular Remote Control Rapid Deployment Camera System (MAVIS) is a remote-controlled camera sending high-quality

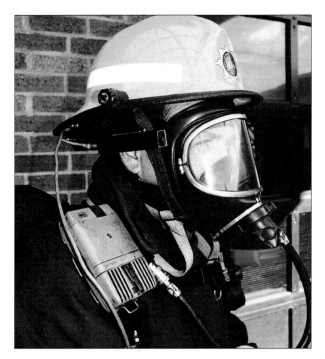

Figure 8.1 Detail showing camera mounting position. The transmitter can just be seen under the rim of the helmet. The camera sees approximately half the normal field of normal vision.
(Photograph: Bedfordshire Fire and Rescue Service)

Figure 8.3 One of the original tripod mounted hard wired cameras. These cameras are high-definition colour but do not have a zoom facility. Both new and not so new technology work well together.
(Photograph: Bedfordshire Fire and Rescue Service)

Figure 8.2 (left) BA wearer equipped with high-definition helmet mounted colour camera. The combined battery pack and control module (mounted on the waistbelt) permits changeover to thermal images from the TIC. The microwave antenna and transmitter are mounted on the helmet to ensure both protection and height for better image transmission.
(Photograph: Bedfordshire Fire and Rescue Service)

Figure 8.4 Video rack installed in an Incident Command Unit. At the top are the quad and master monitors. Below this is the real/lapse time video recorder. Below this is the Cellsend modem that permits transmission of pictures to Brigade Control via GSM telephone. Below this is the video processor for manipulating up to eight inputs, freezing images and electronically enlarging if required. Other facilities are also available. The unit below this controls the mast mounted camera. The bottom (lighter) unit is the control case for MAVIS and the receiver for the microwave transmissions from both MAVIS and the helmet camera which normally operates within the ICU. (Photograph: Bedfordshire Fire and Rescue Service)

vision and sound to a control unit operating either in the ICU or, being fully portable, from a forward point or any remote location.

The CCTV system consists of tripod mounted colour cameras, a maglight camera, bodyworn camera and ISG thermal imaging camera as well as a remote control decoder/microwave transmitter and a control case. The control case is mounted in the ICU video rack to receive and process the signals from the microwave cameras. It can operate independently of the ICU.

The bodyworn high-resolution colour camera (about 50 cm long and 1 cm square) is mounted under the right rim of a helmet. A microwave transmitter is mounted under the rear rim of the helmet. This transmitter takes its power from a harness mounted battery pack, the antenna is also mounted on the rear rim protruding upwards.

The harness also has a lapel viewer in the form of a small LCD screen which allows the wearer to monitor the video picture being transmitted.

The camera is wired to the transmitter through a connector to allow connection of the ISG thermal/video overlay camera if necessary.

The 'Cellsend' system uses digital technology to send video images from the ICU to the Control Centre.

Two modems are installed, one in the ICU and the other in the Control Centre. The modem in the ICU is linked to a mobile GSM phone, while the other is linked to an extension of the Meridian telephone system. The GSM link can be established from either end.

Control Centre operators can choose the resolution best suited to the image. In high resolution the image is updated a few seconds behind real time, at the lower resolution the image is updated more slowly but technology will continue to improve these times. This update rate is achieved by only updating those parts of the picture that move and the images can be recorded at both ends, it is also possible to incorporate an audio facility. 'Cellsend' does not require a PC to operate the system.

Results from tests in these brigades indicated the technology worthy of further research.

The general public are more aware of the fire service than they were in the past. Unusual incidents are of widespread interest, especially where rescue is involved and these pictures are in heavy demand from news media.

The opportunities provided by developments in communications, computing and video technology enable the fire service to provide a more effective, efficient and safe front line service.

Communications and Mobilising

Chapter 9 – Radio

Introduction

All fire brigades in the UK are planning to replace wide area radio systems as a result of changes to radio frequency spectrum allocation to the fire service in England and Wales and because existing systems are obsolete and are increasingly difficult to maintain.

The need for emergency services in general and the fire service in particular to deal with major catastrophes and chemical, biological, radiation and nuclear incidents has brought into sharp focus some of the complex issues associated with the radio systems used by the emergency services. The Presidents of the professional bodies representing the three primary emergency services (Chief & Assistant Chief Fire Officers Association, CACFOA; Association of Chief Police Officers, ACPO; Ambulance Services Association, ASA) agreed a joint statement on same service and inter-agency interoperability, which was subsequently ratified by the Government's Civil Contingencies Committee. In the light of this agreement, the Minister responsible for the fire service announced in May 2002 that the Government was going to procure and fund a national radio communications system for the fire service and that work should commence immediately. This national procurement project, called 'Firelink', is now under way. Completion of the national roll-out is scheduled for December 2007.

The procurement is based on a functional output specification and is anticipated to result in a national radio network. In order to maintain an open and competitive procurement, and to comply with EU procurement rules, it will be left to industry to propose the technical solution to the fire service user requirement. Fire brigades will therefore not be required to specify technical matters relating to the wide area radio infrastructure, other than in national performance terms but will be provided with a set of core speech and data services to meet the national user requirement. Brigades are likely to have the option of specifying additional features from a national framework agreement, to meet local requirements.

Firelink is in direct liaison with individual brigades to identify and agree requirements, interim arrangements and implementation plans.

Firelink is also planning and implementing changes to existing fire service radio systems to both enhance capability, in particular to improve same-service interoperability and to improve the reliability of existing systems until the national radio system becomes available.

The Manual of Firemanship first contained, in 1954, a section dealing with Radio. At that time, the majority of fire brigades shared the radio scheme (system) used by the local police force. The situation now is completely different; for many years every fire brigade has had its own radio scheme with a high percentage of fire appliances and other vehicles, equipped with modern 'transceivers' (radio sets capable of transmitting and receiving).

In addition to the standard radio sets fitted in vehicles which communicate primarily with the Brigade Control Room, there is specialised vehicle radio equipment which can communicate with personal radio sets. These provide on-the-spot fireground communications and can have the added facility of being able to link personal radio sets into the main brigade radio scheme.

The primary objective of this chapter is to provide a basic knowledge of how radio schemes work, their capabilities and their limitations, sufficient to enable fire service personnel to get the best possible use from what is a highly sophisticated technical resource.

There is a continuous demand for improvements and expansion to radio schemes; for additional radio 'channels', which permit appliances at different incidents to be dealt with independently; and for new facilities of various kinds. Unfortunately, the unlimited expansion of radio as a medium of communication is not possible. It is a finite resource with clear limits and, as a result, the extent and purposes for which radio may be used are strictly controlled.

9.1 Frequency Spectrum characteristics, selection and allocation

9.1.1 The Frequency Spectrum

Radio signals travel through space as a 'wave' which, for the purpose of this explanation, can be likened to a wave on the surface of water. Every such wave consists of alternative crests and troughs to which the following terms apply:

CYCLE – the portion of the wave between successive crests or troughs, which is repeated over and over again to form the continuous wave.

WAVELENGTH – the distance between successive crests, or successive troughs.

FREQUENCY – the number of cycles of wavelengths, which appear to pass a given point in a specified time, usually one second. Wavelength, frequency and velocity are related in a very simple way:

$$\text{Velocity} = \text{Frequency} \times \text{Wavelength.}$$

However, this formula does not show the relationship very clearly. Normally the velocity is a constant for a particular type of wave in given conditions so more specifically:

$$\text{Frequency} = \frac{\text{Velocity (constant)}}{\text{Wavelength}}$$

or,

$$\text{Wavelength} = \frac{\text{Velocity (constant)}}{\text{Frequency}}$$

Radio waves are just one form of what is known as 'electromagnetic radiation', other forms being 'microwaves', infra-red (heat), visible light, ultra-violet and X-rays.

These all have one very important common characteristic which is that they all travel through space with the same very high velocity. This velocity measures 300 million metres, or 186,000 miles per second. For all earthly distances this is virtually instantaneously.

The only difference between the various forms of electromagnetic radiation is that they each occupy different ranges of frequency and, hence, different ranges of wavelengths. Radio waves occupy the lowest range of frequencies (and, hence, the longest range of wavelengths) followed by infra-red, visible light, ultra-violet, and X-rays. Even though they occupy the lowest part of the spectrum, the frequencies of radio waves are quite high in numerical terms. The lowest usable frequency for radio communication is about 10,000 Hertz, corresponding to a wavelength of 30,000 m. The highest frequency currently in use for radio communication within the fire service is about 2,300,000,000 Hertz, corresponding to a wavelength of 0.13 m. The police use higher frequency bands, up to 50 GHz for very short (5 km) links.

1,000 Hertz (Hz) is called 1 kiloHertz (kHz)
1,0,000 Hz = 10 kHz
1,000 kHz = 1 MegaHertz (MHz)
1,000 MHz = 1 GigaHertz (GHz)

Thus 2,000,000,000 Hertz is more compactly called either 2,000 MHz or 2 GHz.

The following are two worked examples using the above formulae:

If a transmission has a wavelength of 4 metres, what is the frequency?

$$\text{Frequency} = \frac{\text{Velocity (constant)}}{\text{Wavelength}} \text{ Hertz}$$

where

Velocity   300,000,000 metres per second
Wavelength   4 metres

Thus

$$\text{Frequency} = \frac{300,000,000}{4} = 75,000,000 \text{ Hz or 75 MHz}$$

If a transmission has a frequency of 450 MHz, which is a similar frequency to the Fireground channels, what is the wavelength?

$$\text{Wavelength} = \frac{\text{Velocity (constant)}}{\text{Frequency}}$$

where

Velocity   300,000,000 metres per second
Frequency   450 MHz or 450,000,000 Hz

Thus

$$\text{Wavelength} = \frac{300,000,000}{450,000,000} \text{ metres} = 0.88 \text{ metres}$$

The result is that the higher the frequency the shorter the wavelength. For radio waves, 'wavelength' is measured in 'metres' and 'frequency' is measured in 'Cycles per second' for which a special name 'Hertz' is used.

Figure 9.1 shows how the various forms of electromagnetic wave occupy different parts of the range of frequencies which are known as the 'electromagnetic spectrum'. With the exception of visible light the boundaries of the various forms are not sharp and there is considerable overlap.

Our interest is with the radio wave portion, extending slightly into the microwave portion, and that is expanded in Figure 9.3 with the corresponding wavelengths added.

Radio waves occupy a wide range of frequencies with the maximum being several million times larger than the minimum. This contrasts with the very narrow range occupied by visible light in which the maximum is only about twice the minimum.

The result is that, whereas the various colour components of white light normally all behave in the same way, the lowest range of radio frequencies, e.g., below 100 kHz, will behave quite differently from the highest range, e.g., above 1 GHz. This leads to the 'radio frequency spectrum' being divided into relatively small frequency bands, within each of which all frequencies behave in much the same way and are, therefore, suited to a particular purpose. Since every frequency has a unique corresponding wavelength the different frequency bands correspond to different 'wavebands'.

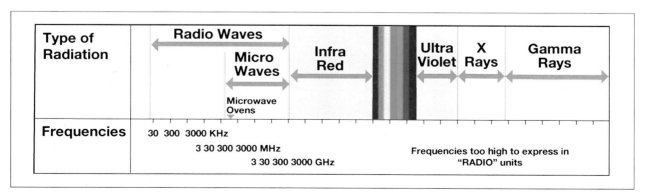

Figure 9.1 The electromagnetic spectrum.

9.1.2 Characteristics of the different Frequency Bands

Figure 9.2 shows, in very broad terms, how the different frequency bands (wavebands) differ with particular reference to the way they travel through space – their 'propagation characteristics', the size of the aerials and the power required.

Consider the size of the aerial. It is common knowledge that, for receiving, the size of the aerial is not very important, indeed the vast majority of transistor radio receivers operate very well with no visible aerial at all. The aerial is a coil wound round a magnetic rod (ferrite aerial). However, for transmitters the position is totally different; for effective transmission an external aerial is essential and its length must be carefully matched to the wavelength being transmitted. For the type of aerial fitted on vehicles the correct length is almost precisely one-quarter of the wavelength; e.g., at 30 MHz, wavelength 10 m, this would be 2.5 m. For higher frequencies it is shorter, but for lower frequencies it is longer.

From that, and Figure 9.2, it can be seen that, currently, only two parts of the radio frequency spectrum are suitable for land-based, mobile and personal radio schemes: the VHF and UHF parts. Unfortunately these parts are also eminently suitable for many other uses, notably the entertainment side, i.e., broadcast radio and television.

There are also allocations to marine, aeronautical, armed services, public utilities, and others to commercial user requirements. There is, therefore, only a limited allocation available to the emergency services, of which the fire service is only one.

9.1.3 Frequency Selection and Allocation

From the spectrum characteristics in Figure 9.3 it is clear that the allocation of radio frequencies is not a matter which can be handled in isolation by any one service, by any one government department, or even by any one country. Agreement has to be reached on an international basis as to how the different parts of the spectrum are to be shared between the different types of service for which they are best suited. For broadcasting, civil aviation and the mercantile marine, operation in the same bands of frequencies may be either by regional cover or world-wide.

Block allocations of frequencies, by function, are agreed from time to time at conferences of the International Telecommunications Union, of which practically all countries are members. These block allocations by broad function are then divided nationally among the various users of each type of service.

In the United Kingdom, control of the frequency spectrum is vested in an inter-departmental committee comprising representatives of all Government departments with responsibility for frequency-using services. These include the Radiocommunications Agency of the Department of Trade and Industry, the Home Office and the Ministry of Defence. **(See Chapter 1, Regulatory Issues.)**

Frequencies	KHz 30 100 300 1000 3000 KHz			GHz 1 3 10 30			
			MHz 1 3 10 30 100 300 1000 3000				
Band Names	Low Frequencies Long Waves (LW)	Medium Frequencies Medium Waves (MW)	High Frequencies Short Waves (SW)	Very High Frequencies Short Waves (VSW)	Ultra High Frequencies (UHF)	Super High Frequencies (SHF)	
Wavelength (metres)	1000 3000	1000 300	100 30	10 3	1 0.03	0.1 0.03	0.01
				CMS 100 30	10 3		1cm

Figure 9.2 Divisions of the radio spectrum.

Low Frequencies (LF) or Long Waves (LW)
30 – 300 kHz   10,000 – 1,000 metres

Follow earth's curvature. Not screened by mountains etc. Consistent long range both by day and by night. Requires very high transmitter powers and very big aerials. The top end of the band is widely used for broadcasting.

Medium Frequencies (MF) or Medium Waves (MW)
300 – 3,000 kHz   1,000 – 100 metres

Longer ranges by night than by day. Rapidly varying effects at sunrise and sunset. Requires high transmitter powers and big aerials. Widely used for broadcasting, ship-shore radio, marine navigational aids, etc.

High Frequencies (HF) or Short Waves (SW)
3 – 30 MHz   100 – 10 metres

Short range over ground, but reflection from upper atmosphere gives very long range both by day and by night with very little power. Vulnerable to atmospheric disturbances, sunspots, etc. Frequency changes needed every few hours to maintain continuous communication. Widely used for long range communication.

Very High Frequencies (VHF) or Very Short Waves
30 – 300 MHz   10 – 1 metre

Screening and reflection by hills, large buildings, etc., becomes noticeable, gradually approaching visible light characteristics giving significance to line-of-sight. Generally short range over ground, 20 miles (30 km) average, almost wholly dependent on upon line-of-sight, i.e., height of aerial. Fairly constant results both by day and by night but vulnerable to long-range interference during abnormal weather conditions. Ideal for two-way land mobile schemes due to relatively short aerials and moderate power requirements.

Ultra High Frequencies (UHF) or Ultra Short Waves
300 – 3,000 MHz   1 – 0.1 metre

Broadly similar to VHF but closer still to visible light characteristics. Screening and reflection more noticeable, but less long-range interference. Shorter range over ground, but line-of-sight even more significant. Lower part of the band is ideal for two-way, hand-held personal radio schemes due to very short, but efficient aerials and low power requirements.

Figure 9.3 Characteristics of different frequency bands.

9.1.4 Channel Spacing

It is not possible to convey information by using just a single frequency. A narrow band of frequencies is required which is known as a 'Channel'. Different channel widths are required for different services: for example, a television video channel must be many times wider than a speech channel. Channels are normally known by their centre frequencies and the centre frequencies of adjacent channels must be separated by at least the required channel width in order that there shall be no overlap which would result in unacceptable interference.

In fact the centre frequency spacing of adjacent channels are slightly greater than the 'bandwidths' occupied. Several technical factors, including the design, build standard and achievable frequency stability all determine allowable channel spacing. Technical advances have made it practicable to reduce channel spacing progressively from 50 kHz to 25 kHz and, currently, to 12.5 kHz. Reduction of the channel spacing specification to which all users and, hence, all manufacturers must comply has the effect of increasing the number of channels which can be made available within a given frequency bandwidth. A 100 kHz allocation will take two 50 kHz channels or eight 12.5 kHz channels. This is shown in Figure 9.4.

Nevertheless, there are still not nearly enough radio channels available to meet the growing demands from would-be mobile and personal radio users. Further reductions in channel widths and channel spacing will inevitably be sought as technology continues to improve. The alternative, using digital technology, is to place multiple speech channels onto one radio carrier by giving each one a time slot. The TETRA system which is proposed for the Public Safety Radio Communications Project (PSRCP) has four speech channels in a 25 kHz bandwidth channel, whereas GSM, which is the system used for digital cellular radio, currently has eight speech channels in a 200 kHz bandwidth channel. Advances in technology will soon increase this to 16 speech channels in a 200 kHz bandwidth channel.

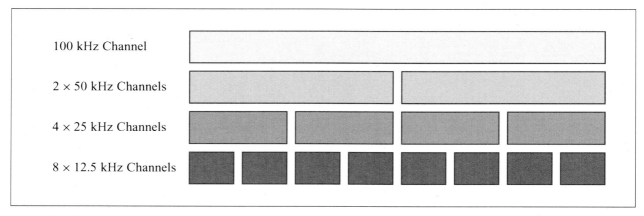

100 kHz Channel	
2 × 50 kHz Channels	
4 × 25 kHz Channels	
8 × 12.5 kHz Channels	

Figure 9.4 Channel spacing.

9.2　Radio Scheme Engineering

9.2.1 Modulation methods

The technique of superimposing a speech signal on a radio wave is called 'MODULATION'. The radio wave then becomes the 'carrier' for the speech and it is often referred to as the 'carrier wave', or simply, the 'carrier'. Basically the radio wave is a single frequency of constant 'amplitude' which means that all the peaks in the wave have the same height and all the troughs have the same depth.

Modulation can be superimposed by varying either the frequency or the amplitude of the basic radio carrier. Thus, there are two techniques currently in use in the fire service:

(1)　Frequency Modulation (FM)

(2)　Amplitude Modulation (AM)

Whichever method of modulation is used, the result is to produce 'side frequencies' just below and above the carrier frequency. It is the presence of these side frequencies which causes the radio signal to require a small band of frequencies, and they determine the 'bandwidth' of the signal.

Within the narrow channels used for mobile radio (12.5 kHz) there is little difference between AM and FM in terms of 'user-noticeable' performance. In schemes originally provided by the Home Office prior to 1989, the main and mobile

transmitters use amplitude modulation (AM). Whereas in schemes provided and maintained by commercial suppliers they may use both AM and FM depending upon a brigade's stated need to communicate with any adjacent AM brigades.

Simplex and Duplex

The two terms can be taken as a pair. Within the context of emergency services' radio, 'simplex' working is that, while transmitting (sending), it is not possible to receive, so the person receiving cannot interrupt. Any attempt to do so means neither person hears anything. A vital part of 'simplex' operating procedure is the use of the word 'over'. The speaker must say the word before switching from 'transmit' to 'receive', and the listener must hear the word before switching from 'receive' to 'transmit'.

All equipment normally rests in the 'receive' mode, and operation of a 'press-to-speak' key, sometimes known as a 'pressel switch', switches the equipment from 'receive' to 'send'. The key or switch must be released before transmissions from other stations can be received.

'Simplex' working makes it impossible to speak and listen simultaneously, but it has the advantages of encouraging a concise and efficient operating procedure and an economy in the use of words, and of discouraging lengthy conversations. Further, the equipment required is simpler than that needed for 'duplex' working.

Single-frequency

Single frequency radio equipment is designed to transmit and receive on the same frequency. Clearly such equipment can only operate in the 'simplex' mode and, in such equipment, the receiving portion is always effectively switched off when the transmitter is activated.

Single frequency working is not used in main VHF radio schemes between brigade control rooms and mobiles, but single frequency personal radio equipment is commonly used by fire brigades for direct person-to-person working over short distances both with VHF and UHF (Figure 9.5).

Home Office supplied VHF vehicle-fitted radios are capable of operating on the two VHF channels allocated to manpack working.

Double-frequency or two-frequency

Double or two-frequency equipment is radio equipment which is designed to receive on one frequency and transmit on another and **all** fire brigade main radio schemes operate on this principle (Figure 9.6). The need to occupy two channels of the limited available spectrum is a disadvantage but that is outweighed by the advantages it affords.

'Two-frequency' working permits 'duplex' operation but, in practice, all fire brigade mobiles are 'two-frequency simplex', mainly because of the advantages of 'simplex' already given.

The advantages of 'two-frequency' working are that it permits the control station to operate in the 'duplex' mode, which in turn allows a mobile to 'break-in' to a control station transmission when urgent attention is required due to a priority message. It also permits the engineering of multi-station, wide area coverage schemes.

9.2.2 Talk-through

An important difference between 'single frequency' working and 'two-frequency' working is that 'single-frequency' provides an 'all-hear-all' system, whereas 'two-frequency' does not. In 'two-frequency' working, all the mobiles can hear control, and control can hear all the mobiles, but the mobiles cannot normally hear each other.

A pip-tone 'busy' signal (short 'beeps' about one second apart) is, therefore, transmitted by control whenever it is receiving from a mobile. It is an important aspect of radio scheme discipline that no mobile transmits when the 'pips' are on except in urgent, high-priority circumstances.

Figure 9.5 Principles of single frequency simplex working.

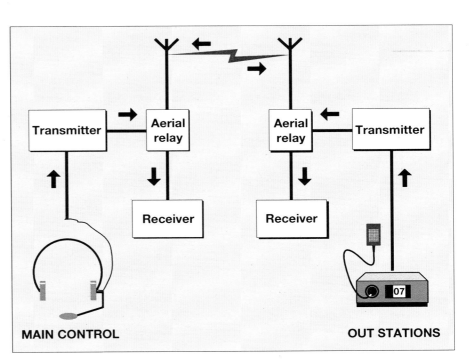

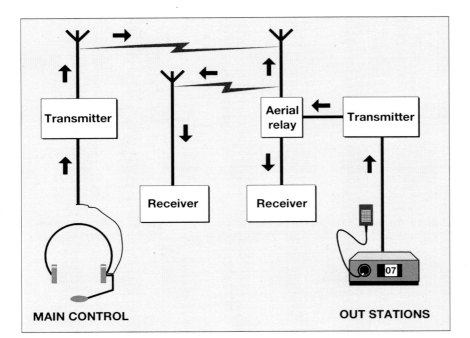

Figure 9.6 Simplex working – out stations only (double frequency).

MAIN CONTROL

OUT STATIONS

Although the mobiles in a 'two-frequency' system cannot normally hear each other, there are occasions when it is more convenient for them to communicate directly rather than requiring the control operator to relay a message. To make this possible, 'two-frequency' systems are provided with a facility known as **talk-through**. 'Talk-through' is selected by the control operator and, when it is selected, the incoming speech from any mobile is 'turned round' and re-transmitted. It is, therefore, received by all other mobiles in exactly the same way as speech from the control operator.

The control operator can, of course, still hear all the mobile transmissions, and retains full control of the scheme. When 'talk-through' is selected, 'pip-tone' is automatically inhibited, either completely or whenever speech is received from a mobile, and there may be an increase in the level of background noise and some degradation of speech quality, which may be noticed because of the link-up between the incoming and outgoing channels.

'Two-frequency' working provides a measure of security because unauthorised listeners can only hear one way, normally the 'outgoing' transmissions from control to mobiles. However, in fire brigade communications security is a lower priority than speed and it is usually more important for two mobiles to talk to one another. 'Talk-through'

provides that speed, and some fire brigades choose to operate their schemes permanently on 'talk-through'.

9.2.3 Wide Area Coverage

This implies that radio communication is required over an area greater than that which can be served by a single base station, no matter how favourable its location may be. All county fire brigade radio schemes fall into this category with the result that at least two, and in some cases more, base stations are required. Hence these schemes are known as **'multi-station schemes'**.

One approach would be for the individual main stations (hill-top sites) to operate on different channels, in other words a number of single station systems, and not an integrated scheme. However, mobiles receiving from one main station would not benefit from the 'fill-in' effects of other main stations as they moved into difficult areas.

The system adopted must **appear** to be a single station system even though two or more stations are involved. It might be thought easy to set all the main transmitters at all the hill-top sites on exactly the same frequency, so that their signals merge into one in the mobile receivers but it is, in practice, virtually impossible.

9.2.4 The Spaced Carrier System

The original, classic solution to the problem was to deliberately offset the frequencies of the main transmitters within the allocated channel width. The 25kHz channel spacings used at the time permitted at least three slightly offset main transmitter frequencies, which a mobile would receive as a single integrated signal without any noticeable interaction.

However, with the compulsion to reduce channel spacing to 12.5 kHz, the spaced carrier system has had to be abandoned because the narrower channel width does not permit sufficiently large offsets to prevent noticeable interaction.

9.2.5 The 'Quasi'-Synchronous or 'Common Frequency' System

Fortunately, technical advances have improved the stability of the frequency generators used for scheme main transmitters to the stage where they can maintain an almost constant frequency over long periods of time in spite of changing temperatures, etc. The main transmitters at different sites are not exactly synchronous, but they are almost or **'Quasi'-synchronous**.

For all practical purposes they all have the same, common frequency.

As an example, fire brigade scheme main transmitters currently operate at about 70 Mhz. The stability of the quasi-synchronous frequency generators is such that the individual main transmitters in a scheme keep to within less than 0.5 cycle per second of each other.

In conjunction with the method of modulation used, a scheme can be described as 'quasi-sychronous amplitude modulation' or 'quasi-synchronous frequency modulation', with the alternative term 'common frequency' in place of 'quasi-synchronous'.

Doppler Effect

An effect which may be apparent in quasi-synchronous (common frequency) systems, which was never apparent in spaced carrier systems is the 'Doppler' effect. This is the effect where there is an apparent change of frequency whenever there is relative motion between a transmitter and receiver. If the vehicle is moving towards a fixed transmitter the frequency appears to increase slightly, but if the vehicle is moving away the frequency appears to decrease slightly.

'Doppler' effect is of no consequence in a single station scheme because the change is so small compared with the channel width. Likewise, it was unnoticeable in spaced carrier systems because the changes were so small compared with the deliberate offsets. However, Doppler effect may well be noticeable in quasi-synchronous (common frequency) systems when a mobile is in an area where it receives more or less equal signal strengths from two hill-top sites and is travelling towards one but away from the other. One frequency appears to increase while the other appears to decrease with the result that the difference superimposes a 'warble' or flutter on the received speech which varies with vehicle speed.

Every effort is made to engineer schemes, by location of hill-top sites, by adjustment of transmitter power, by use of directional transmitter aerials, etc., so as to minimise the effects, but because its cause is a natural phenomenon it can never be completely avoided.

When a mobile is stationary in a position where it receives more or less equal signals from two stations, the small difference between the two frequencies may be noticeable as a slow 'whooshing'. Normally it is only noticeable when the transmitter is on without any speech and it does not impair speech intelligibility. If it is intrusive, a small change of position, to take advantage of local screening from one station, can be advantageous.

9.2.6 Scheme Engineering

A number of carefully sited main stations (hill-top sites) are required to give brigade-wide radio communication coverage. Figure 9.7 illustrates the way in which such schemes are engineered. There are a number of variations, particularly as far as the linking arrangements are concerned.

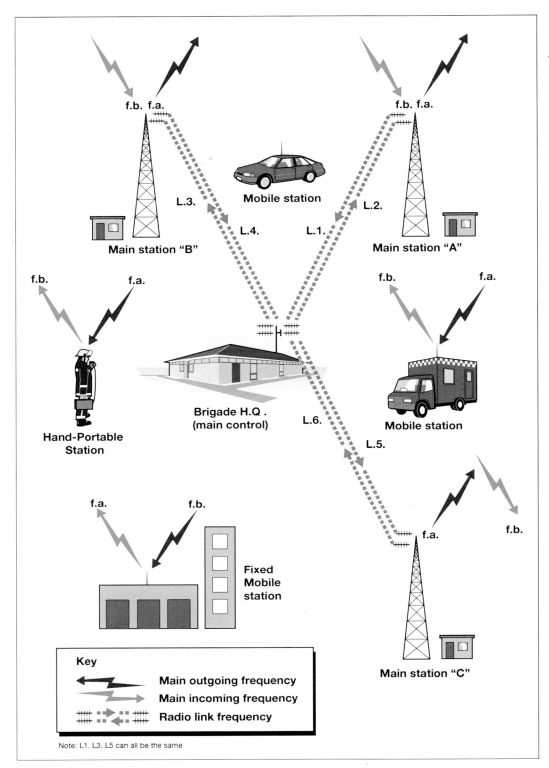

Figure 9.7 Multi-station double frequency area coverage scheme.

Under 'two-frequency working' there are two main frequencies:

(1) **Outgoing** – main station transmit and mobile receive.

(2) **Incoming** – mobile transmit and main station receive. These main frequencies use aerials at, or near, the tops of the masts, and the aerials are almost always omni-directional to cover the largest possible area.

9.2.7 Links

The links between the control station and the main radio stations can be by land line but the majority use radio links (see Figure 9.7). Each radio link has its own dedicated pair of frequencies so that there is no mutual interference. Directional aerials are used – commonly known as 'yagis' – similar, apart from size, to TV or FM sound broadcast aerials. These aerials 'look at each other' from opposite ends of the link to 'beam' the signals and provide the 'point-to-point' mode.

Although, in general, each link has its own transmitter and receiver and its own pair of dedicated frequencies, the outgoing links from control are identical. Channels can be saved and made otherwise available if a single outgoing link transmitter is used and its output is split between two or more aerials, each pointing at a main station. There is then only one outgoing link frequency but there must always be independent incoming link frequencies and link equipment.

At each main station, the link transmitters and receivers are interfaced with the main transmitters so that, for example, a signal from control is:

(1) Transmitted by the link transmitter at the control station.
(2) Received by the link receivers at all the main stations.
(3) Re-transmitted by all the main transmitters.
(4) Received by the mobile receivers.

A similar sequence, in the opposite direction, occurs when a mobile transmits to control.

In general the above descriptions refer to the use of VHF High-Band linking. Some brigades also deploy onward linking at UHF. However, in recent years there has been a move towards vacating both VHF and UHF linking in favour of microwave, or land line links. The Information and Communication Technology Unit has issued policy statements addressing this subject (see Section 9.2.14).

9.2.8 Frequencies

It is of interest to add up the number of frequencies which are permanently needed in multi-station,

wide-area coverage radio schemes. In the example shown (Figure 9.7), a permanent assignment of eight frequencies is required with independent outgoing links to all stations, and this can only be reduced to six if they all share a common outgoing link – that is, two main frequencies and either six or four link frequencies to provide just one operational radio channel.

9.2.9 Equipment

It is normal practice in fire service radio schemes to provide two sets of equipment at every station, known as the 'main' and the 'standby' equipment. Basically, only one is operational at any one time and their purpose is to ensure continuity of service in the event of failure. Change-over from 'main' to 'standby' equipment and vice-versa is normally under the control of the control station operator or supervisor. In addition, every station will have several items of ancillary equipment.

At the main control station of a radio scheme, facilities are provided to enable control room staff to isolate the main stations individually when for one reason or another they are troublesome (e.g., when a temporary, very high noise level is caused by the effects of static electricity during a severe storm in the vicinity of a site).

Remote control facilities are also provided to enable control room staff to switch main station equipment from 'main' to 'standby', such equipment can be changed either individually (i.e., just one faulty piece of equipment) or collectively (i.e., all equipmen)t. In addition to the duplication of equipment, a further safeguard is normally provided against complete link failure, perhaps due to aerial damage. The equipment at all main stations is arranged so that in the event of a link failure the station changes to **'automatic talk-through'**.

This means that any signals received from mobiles by the main receiver are automatically re-transmitted to the mobiles by the main transmitter instead of, or in addition to, being transmitted via the link transmitter to control. This system provides some measure of service, which is better than none, until the faulty link can be repaired.

Under the automatic talk-through system the mobiles can at least talk to each other but the control is isolated if the link is completely severed. Even if the link still works one way the control will either hear what is going on without being able to participate or will be able to speak out without knowing whether anyone is receiving.

Voice Infrastructure

Figure 9.8 shows a typical voice communication system infrastructure which comprises a number of control operating positions, each of which is provided with a headset and microphone, a loud-speaker and means of making a radio channel selection. The operating positions provide the Control Operator interfaces to the Integrated Communication Control System (ICCS) which in turn provides access to individual radio channels and also to telephone circuits.

The Main Fire Service voice channel is broadcast on low band VHF from a number of hill-top radio sites, at each of which is located either a single or duplicated Base Station. Each Base Station is connected to the control site by means of either a microwave link network or by a private wire circuit. At the control site the receive and transmit

signals are brought together in a voting unit. The main purpose of the voter is to accept all of the receive signals from the hill-top sites and select (vote) the best signal to pass to the ICCS.

A local base station at headquarters is connected to the ICCS to provide for fall-back control of the main VHF channel when it is operating in talk-through mode, by operating on the mobile frequencies. This base station may also provide communication on fireground Channels 21 and 22 in the locality of Fire HQ. Further levels of fall-back protection are also provided by means of a desk mounted mobile radio located at Fire HQ or at an alternative location.

A local base station at headquarters is also connected to the ICCS. This provides for inter-brigade communications with adjacent county fire services.

9.2.10 Fixed Mobiles

A unit can either be fixed or mobile but it cannot be both. However, in the radio sense, a 'fixed mobile' is a radio transmitter/receiver which has all the attributes of a mobile radio (it might even be physically identical) except that it is installed in a

Figure 9.8 Voice infrastructure.
(Graphic: Simoco)

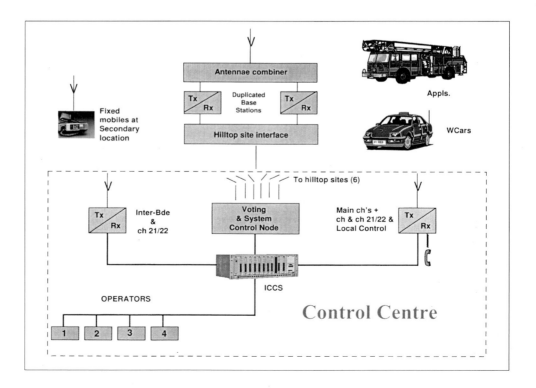

fixed location, within a building, instead of in a vehicle. Such a unit transmits on the 'mobile' transmit frequency and receives on the 'main station' transmit frequency. Transmissions from such a unit are received by control in exactly the same way as transmissions from true mobiles and, apart from an identifying call sign, are totally indistinguishable from them. Within the context of a mobile radio scheme the expression 'fixed mobile' is, therefore, quite logical and understandable.

If a 'fixed mobile' is installed in, or can be operated from, the control room then, in the event of total link failure, a control room operator will be able to fully participate in whatever remains of the radio scheme through the 'auto-talk-through' facility. In this context, the 'fixed mobile' may alternatively be described as 'reverse frequency' equipment because its transmit and receive frequencies are the reverse of those for the normal outgoing and incoming control channels. It can also be used as a realistic way of checking hill-top site performance within radio range of control.

Although individual fire brigade systems are totally independent, it is very useful, for the control rooms at least, to have access to the schemes of neighbouring brigades. This facility is useful when incidents occur over brigade boundaries or when assistance is sought at large incidents. Such access is also provided by a 'fixed mobile', each control room having a radio which operates on the mobile transmit and receive frequencies of the neighbouring brigade(s).

The fixed mobile originally supplied by the Home Office can also be programmed with the two-man-pack frequencies (Channels 21 and 22) allowing, when radio range permits, direct radio communications in an emergency when main scheme failures occur, between control and vehicles.

9.2.11 Main Control

A radio scheme with a considerable number of users all operating on the same channel is almost unworkable unless one station is made responsible for its overall control. That station is known as the 'Main Control', or simply 'Control'. The 'two-frequency' system automatically gives the control station the ability to 'dominate' all other radio

scheme users. Normally they can only hear control and not each other unless talk-through is a permanent arrangement.

In the fire service the main radio control is invariably in the centralised mobilising control room for the brigade, whilst the radio equipment is located in an adjacent room or building known as the 'link room'. Outside is a tower or mast on which the directional link aerials, each pointing at a distant hill-top site, are fitted, along with simpler aerials for any 'fixed mobile' equipment in the control complex.

The control equipment will be duplicated at two or more operating positions, the number of such positions depending upon the size of the brigade and the number of separate radio channels it uses.

9.2.12 Transportable Equipment

'Mobile' equipment in 'hand-portable' form can either be in a briefcase, haversack or 'backpack'. It operates on the mobile transmit and receive frequencies and contains its own (usually) rechargeable batteries. It permits direct contact with control whilst away from a parent vehicle and is an alternative to the personal radios. Because of the lower transmitter power, imposed by the limited weight for the batteries, and the less effective aerial, this type of equipment is not as good as a vehicle radio particularly for transmitting back to control.

9.2.13 Power Supply Arrangements

The control and main scheme radio equipment at the control station, and all the radio equipment at hill-top sites, are operated from the normal domestic electricity supply of 230 volts, 50 Hz, AC. Fire service personnel should be aware of the potential danger arising from the presence of such voltages and should **never attempt to go inside any equipment**.

At all key stations there will usually be a standby power supply in the form of a diesel driven generator with automatic start-up and change-over to ensure the scheme is never put out of action by a mains supply failure.

'Fixed mobile' equipment may be designed to operate direct from the AC mains, i.e., genuinely

fixed equipment made to operate on 'mobile' frequencies, or it may be a 'mobile radio' made to
operate from a vehicle battery, 12 volts, DC, with
a 'mains power unit' made to operate from a vehicle battery, 12 volts, DC, with a 'mains power unit'
which converts 240 volts, AC, direct to 12 volts,
DC, without the need for a battery.

9.2.14 Microwave

Earlier in this chapter (see Section 9.2.6, Scheme
Engineering) it was explained how as many as
eight separate frequency channels are required in a
three-station scheme to support just one operational channel for a brigade. Additional stations
require at least one additional frequency (possibly
two) whilst an additional operational channel will
require a complete additional set of frequencies.
Only two of the frequency channels supporting
each operational channel are used to actually communicate with the mobiles – the main outgoing
and incoming frequencies shown as 'f.a' and 'f.b'
in Figure 9.7. The remainder are 'link' frequencies,
shown as L1 to L6 in Figure 9.7 and they serve to
carry speech, on a point-to-point basis, between
the control station and the hill-top sites.

Until very recently the frequencies used for link
channels have always been in either the VHF band
or the lower part of the UHF 2 band. An obvious
disadvantage is that frequency channels which are
ideally suited for mobile communication on a
broadcast basis are being used for point-to-point
links and, with the intense competition and
demand for additional mobile channels, that
'waste' of mobile channels can no longer be tolerated.

To release currently used VHF and UHF 'link'
channels for 'mobile' use, regulations now require
that all new point-to-point links shall immediately
operate in the **microwave** part of the frequency
spectrum and that all currently used VHF and UHF
'link' channels shall be moved to microwave as
part of a 'rolling plan'.

Definition

The term 'microwave' is one which has no precise
and universally accepted definition which fits in
with the generally accepted frequency and wave

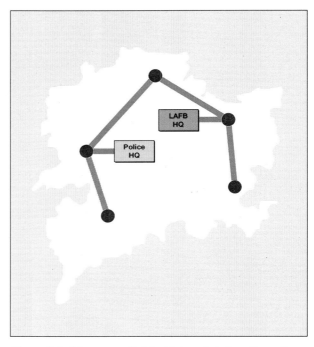

Figure 9.9 Microwave link in 'daisy-chain' configuration.

bands given in Figure 9.2. However, for our
purposes, 'microwaves' means frequencies above
1000 Mhz (1 Ghz) – that is, wavelengths shorter
than 0.3 m. 'Microwave' ovens operate at 2.45 Ghz
so the frequencies which will be used for emergency services links – in the range 1.79 to 2.3 Ghz
– can be legitimately described as 'microwaves'
although Figure 9.2 clearly shows such frequencies
to be in the upper part of the UHF band.

As far as the operational user of a radio scheme is
concerned, its linking arrangements – its 'scheme
engineering' – should be completely transparent,
in other words operationally 'invisible'. The user
may work on the assumption that radio signals
pass directly between the vehicle aerial and the
mast at brigade HQ, although it is better to appreciate the limitations of radio communication over
the ground, and the need for linked multi-station
schemes.

Limitations

At microwave frequencies, an unobstructed line-
of-sight path between aerials at opposite ends of a
link, is essential. This is in contrast to VHF links
for which a degree of obstruction from hills, trees,
or buildings, was acceptable. As a result it is not

always possible to replace a VHF link path with an identical microwave link path. Some reconfiguration, either re-routing between existing stations or additional stations, may be necessary. In VHF linked schemes every effort was made to link every main station (hill-top site) direct from control in a radial 'cartwheel spoke' configuration.

That was not always possible and in some cases a very remote main station is linked to control through another main station which is then known as a 'master' or 'repeater' station. With increasing congestion of the VHF band, the use of 'master' or 'repeater' stations was becoming progressively unworkable, but the move to microwave has removed that particular problem, and microwave links are equally likely to be arranged in a 'daisy-chain' configuration (see Figure 9.9).

Microwave links are wide-band in contrast to VHF links which were narrow-band. In this context narrow-band is 12.5 kHz, just one channel width, whilst wide-band is several hundred kilohertz which is many channel widths. This means that it is possible for a single microwave link to carry many separate speech channels using a technique known as 'Multiplexing'.

9.2.15 Multiplexing

'Multiplexing' on a wide-band microwave link does not save on frequency spectrum occupied because the total width of a number of multiplexed channels in a single wide–band channel is greater than the sum of the widths of the same number of channels in individual narrow bands. The big saving is on equipment – one analogue microwave link can, for example, carry up to 36 separate speech channels – and on the number of aerials required on the masts.

The multi-channel capability results in another potential change in linking philosophy. Whereas in the past all the individual emergency services have had independent radio systems, with perhaps their equipment in different rooms or even different buildings at hill-top sites, microwave linked systems are planned on a 'combined user service area' basis to meet all the operational requirements of all sharers. Each system will be designed individually taking into consideration:

- Topography of coverage area, e.g. a county or counties.

- Disposition of existing, and possibly future, hill-top sites.

- Disposition of operational controls, e.g., fire and police headquarters, etc.

- The number of channels and the routing required by each user. Figure 9.9 illustrates a possible linking arrangement with fire and police headquarters in different parts of the county.

9.3 Mobile, Transportable and Personal Radio Equipment

9.3.1 Conventions

Although the actual use of individual channels is not subject to regulations there are obvious dangers if there is not disciplined use. **DCOL 4/88 (in Scotland DFM 5/1988)** recommends that use of channels should be identified.

DCOL 6/1992 Item 12 Appendix 1 (in Scotland DFM 4/1992) recommends the primary and secondary use for each channel.

Since 1 January 1993, six UHF 'at incident' channels and two VHF channels have been available, in addition an inter-agency channel is provided. Other users, such as airport fire brigades and works fire brigades may be permitted to use one channel if the local authority Chief Fire Officer/ Fire Master considers this could improve operational efficiency and subject to the approval of the Information and Communications Technology Unit.

This, for instance, enables an airport fire officer instant radio contact with responding local authority appliances equipped with UHF facilities.

9.3.2 Mobile Equipment

The World Administrative Radio Conference in 1979 directed that all emergency services in the UK still operating in the 88–108 Mhz VHF Broadcast Band must move to alternative bands by

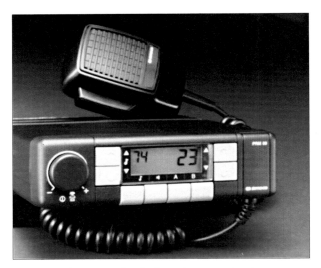

Figure 9.10 A typical vehicle radio control unit.
(Photograph: Simoco)

the end of 1989. Conversion of old equipment to operate in the new bands was not considered worthwhile and the opportunity was taken to re-equip and standardise.

Current mobile equipment for the fire service normally receives in the 70–72 Mhz band and normally transmits in the 80–82 Mhz band.

The standard 'mobile radio' consists of two main parts:

(1)   the transmitter/receiver unit; and

(2)   the control unit.

The transmitter/receiver unit is the larger of the two and is placed within the vehicle. The smaller, control unit is mounted in a convenient position for operation by the driver and/or the front seat passenger. A multi-core control cable with multi-way plugs or sockets connects the two units together. Connected to the control unit are the handset and the loudspeaker. Provision is made for two loudspeakers so that one can be fitted in the cab and one at the rear of an appliance if required. Connected to the main unit are the aerial and the battery.

Installation

The installation of radio equipment in all vehicles (motor cars in particular), is controlled by the

requirements of health and safety and compatibility (non-interference) with other sophisticated vehicle electronic systems which are now fitted as standard equipment.

'Standard fits', in which the precise location of every part of the radio equipment is defined, should be agreed by the vehicle manufacturers, the fire service and the service provider. The staff who actually install the equipment have no authority to deviate from the 'standard fits' because, in attempting to meet the wishes of vehicle owners, a physically or electrically dangerous situation may result.

Further guidance on installation of mobile radio equipment in fire appliances is available from the Information and Communications Technology Unit.

9.3.3   The Aerial

It is a truism that any mobile radio is only as good as its aerial. Hence, the design of the aerial, and its location on the vehicle, largely determine the overall performance obtained. The type of VHF aerial currently fitted on fire mobiles is known as a 'quarter-wave rod', its length being almost exactly one quarter of the transmitted wavelength.

For a transmitter frequency of 80–82 Mhz, the wavelength is about 3.66 m, so a quarter wavelength is just under 1 m.

Ideally the aerial should be mounted in the centre of a flat electrically conducting (i.e., metal) surface, such as the roof of a car or van.

Many modern fire appliances have fibre-glass bodies and it is customary for an area of metal foil or mesh to be moulded into the roof of the cab during manufacture to which the aerial must be fitted. If other roof-mounted equipment, such as ladders, are fitted first, care is needed to avoid encroaching on the critical 'aerial space'. Metal close to the aerial will absorb the radio energy resulting in inferior performance.

The aerial is connected to the transmitter/receiver unit by a coaxial cable, similar to that used to connect a TV aerial to a TV set. The performance of such cable is impaired if it is sharply bent or squashed even though there may be no visible sign of damage.

9.3.4 Channel Selection

The mobile radio (as originally supplied by the Home Office) has the capability of accessing up to 255 channels but only a limited number have been used. Channels 1 to 20 are 'brigade allocated' and each brigade has made its own selection. The brigade's own channel or channels will normally be on channels 1, 2, 3, etc., as required, followed by the channels of neighbouring brigades by mutual agreement. Channels are selected using a numerical key pad and the illuminated display will show the channel number entered.

Channel numbers 21 and 22 are allocated for working both with man-pack VHF radio equipment and directly between vehicles fitted with suitable radio equipment.

9.3.5 Squelch

All mobile radios operating at VHF (and UHF) have an automatic 'squelch' or 'mute' which completely switches off the receiver output to the loudspeaker and earpiece when no transmission is being received. The 'squelch' is necessary to suppress the noise which would otherwise be heard in the absence of a signal. The receiver automatically 'opens up' when a signal of sufficient strength to over-ride the noise and give an intelligible output is received. An incorrect setting of squelch levels aimed at reducing unwanted noise could mean that very weak operational signals will not open up the receiver.

9.3.6 Transmission Timer

To avoid the risk of the transmitter being permanently locked on transmit due to a faulty handset pressel switch or the handset falling into a position where the switch is jammed on, the transmitter is fitted with an automatic transmission timer. A jammed-on transmitter would block the complete radio scheme, and, being 'simplex', the receiver would be inactive so that no signals could be received. The transmission timer automatically switches off the transmitter after about 30 seconds continuous operation. Normal transmissions rarely last that long but, if necessary, the pressel switch is simply released and pressed again to continue transmitting. The 30 seconds time-out applies to Home Office supplied mobile radio equipment. However, similar principles, but possibly with actual different time-outs, will normally apply to all transceiver equipment.

9.3.7 Power Supplies

The standard mobile radio is designed to operate from the standard 12-volt DC vehicle battery source. Connection is made between the transmitter/receiver unit and the vehicle battery via a heavy duty two-wire cable with a suitable fuse in the 'non-chassis' (usually positive) wire, the fuseholder being as close to the battery terminal as possible. The use of two wires, avoiding 'earth return' through the vehicle chassis, helps with equipment 'compatibility' by reducing the risk of mutual interference with vehicle electronic systems.

No problem arises in standard 12-volt vehicles but many larger vehicles and fire appliances have 24-volt electrical systems. Radios can be built to work off 24 volts, but it is not economic to have two standards so 24-volt vehicles are fitted with 12-volt radios.

There are two ways in which this can be done:

(1) **By battery tapping.** The 24 volts is normally provided by two 12-volt batteries 'in series' so the radio can be connected across the 'lower' one (the one with one terminal to chassis). This works reasonably well, although the battery supplying the radio will be discharged more than the other one and this can cause battery maintenance problems. The main disadvantage of this method is the risk of a vehicle mechanic, unfamiliar with the unusual arrangement, re-connecting the radio across the full 24 volts when replacing the batteries. This method is, of course, not possible if the 24-volt battery is a single unit with no access to the intermediate 12-volt point.

(2) **By using 24-volt to 12-volt converters.** These units, which are readily available, are easily fitted to 24-volt vehicles and are far more

satisfactory. However, there are cost, installation and maintenance overheads to consider.

9.3.8 Fixed Mobile Version

A fixed mobile version of the standard mobile radio is available, designed to be fitted in a rack or cabinet in a building and powered from the normal 230-volt domestic AC mains supply via a power unit with an output of 12 volts DC.

The radio can be controlled either 'locally' at the rack or cabinet, or 'remotely' by a control unit designed to fit in a console or be free-standing on a desktop. Emergency power can be made available either via the building backup generator, from the uninterrupted power supply, or a direct 12-volt DC switchcable or plugged battery supply.

9.3.9 Special Features

(1) Single frequency working

The standard mobile radio normally works in the 'two-frequency simplex' mode communicating with 'Control' over the main VHF radio scheme via the hill-top sites and the linking system. The outstanding feature of 'two-frequency' working is that the mobiles can only hear 'Control'; they cannot hear each other unless the control operator has engaged 'talk-through'. Direct 'mobile-to-mobile' communication is possible with 'talk-through' engaged, but that ties up the whole of the 'main-scheme'. The control operator may wish to monitor the messages, but all other mobiles are unnecessarily involved.

To provide greater flexibility the 'standard' mobile radio will have one or two 'single frequency' channels (usually Channels 21/22) and any two or more mobiles switched to one of those channels will be able to communicate directly **and totally independently of the main scheme**, within a very limited geographical area. The size of the area will be almost entirely determined by the intervening terrain and is likely to be severely restricted in heavily built-up areas.

It is of course necessary to prearrange the switch to the single frequency channel. It is even more important to switch back to the normal two-frequency channel because there is no way in which

'Control' can contact a mobile switched to the single frequency channel. Under normal circumstances permission will be requested from 'Control' before a mobile switches to Channels 21 and 22. The fixed mobile can also operate on Channels 21 and 22 for direct emergency communications with mobiles when no other normal channel is available.

(2) CTCSS

CTCSS stands for 'continuous tone controlled signalling system'. It is an optional feature, already fitted in a small number of brigade hill-top receivers. Mobile radios may be similarly fitted.

In the normal way, the squelch or mute of a receiver is opened by the reception of a 'carrier' signal of adequate strength. The audio output of the receiver is then fed to the loudspeaker or earpiece to reproduce any speech modulation superimposed on the carrier.

This normal system works reasonably well but it has two disadvantages:

(1)   There is no way in which individual mobiles, or groups of mobiles, can be called independently so that only those for whom a particular message is intended will hear it; and

(2)   There are circumstances in which a radio receiver can be 'fooled' by natural or man-made 'radio noise' so that its squelch opens when no real signal is present resulting in a 'noise' output. This particularly affects main VHF scheme hill-top receivers which control rooms need to maintain a constant listening watch.

CTCSS overcomes these disadvantages by superimposing a continuous low-pitched tone upon the radio 'carrier' at the transmitter, in addition to the speech. The continuous tone is also used to open the squelch at the receiver, after which it is 'filtered out' so that it is not heard at the audio output. In the outgoing direction (hill-top to mobile) different tones can be selected by the control operator and different mobiles, or groups of mobiles, will respond to different tones. This provides what is known as 'selective calling' in which only selected mobiles will receive the transmission.

When CTCSS is fitted to inhibit hill-top radio receivers, remote technical arrangements must be fitted to allow Control to switch off the brigade's CTCSS. This arrangement is necessary to allow non-CTCSS fitted mobiles to access the brigade's radio scheme upon such a request from the non-CTCSS fitted brigade Control.

9.3.10 Transportable Equipment

'Transportable' in this context, as distinct from 'mobile' or 'personal', means equipment which is completely self-contained with its own batteries and aerial, which can thus be transported from place to place and used anywhere, but which is usually set down, rather than operated whilst being carried, as is the 'norm' for personal equipment. The distinction is, however, somewhat vague and some transportable equipment is certainly capable of being used 'on the move', for example, a hand-held radio (see Figure 9.11).

The standard equipment has a 99 channel capability and the first 20 channels are 'brigade allocated' in the same way as a standard mobile. It also has the same 'single frequency' capability using

Figure 9.11 Hand-held radio in use.
(Photo: Hertfordshire Fire & Rescue Service)

Channel 21 or 22. It has all the features and facilities of a standard mobile except 'public address'.

The biggest demand on the battery is during 'transmit' and a compromise must be made between the transmitter power and the acceptable size and weight of the battery. Of necessity, the transmitter power is about half that of a standard mobile but in all other respects the performance is identical.

9.3.11 Personal Equipment

'Personal' equipment is small enough to be carried in the hand or pocket, or in a suitable lightweight body-harness. Its small size means miniature construction techniques which can create difficulties in the transmitter. The battery size is severely limited and those two factors restrict the transmitter power to a fraction of what is obtainable from 'transportable' equipment (normally approximately 1 W in the majority of equipment). This, coupled with the restricted aerial dimensions and efficiency, limits the range of the transmitter section. The receiver performance will be comparable to that of a transportable under similar conditions.

Personal equipment can operate in either the VHF or the UHF band, but VHF equipment, other than in the single frequency mode, would normally be expected to transmit into the main scheme hill-top sites. VHF equipment is perfectly satisfactory on a single frequency basis to other personal, transportable or mobile units over short ranges, but in general UHF offers better performance for personal radios. The use of FM offers advantages for personal radios, if only because it permits greater transmitter power to be obtained from a given size of battery, and all UHF personal radios used within the fire service operate on FM.

9.3.12 Methods of using Personal Radios

Personal set communication can be organised in a number of different ways to meet various operational needs which, in broad terms, break down into the following categories:

● Direct person-to-person communication on an exclusive single frequency channel over very short distances e.g., between individuals at an incident, or when carrying out dry riser tests, or other duties, in high-rise buildings.

• Similar communication but in which one of the units is mounted in a vehicle.

• Two-frequency communication between personal sets via a vehicle-mounted or portable VHF repeater.

• Two-frequency communication between personal sets and the brigade control room via a vehicle-mounted UHF/VHF repeater.

Single Frequency Operation

Figure 9.12 (1) involves personal sets only and, although only two are shown, any number can be used on an 'all-hear-all' basis subject to the limitations of range imposed by location and environment. With more than two units it may well be the case that unit 'B' can communicate perfectly with both unit 'A' and unit 'C' whereas units 'A' and 'C' cannot communicate directly at all.

When switched to a single frequency channel a personal radio transmits and receives on the same radio frequency and, when used without the aid of any control station equipment, has the following limitations:

(i) Its effective direct range between individuals is seriously affected by the screening phenomena. Therefore, the general range and performance must be expected to vary constantly as the individuals move about.

(ii) It is not possible to forecast accurately what the performance will be in any particular building or other environment, and it does not follow that because good results are obtained in one building, similar results will necessarily be achieved in a nearby and similar building.

To summarise, performance can be expected to vary from one extreme, where screening is severe and when communication even over very short distances is unreliable, to the other extreme where there is little or no screening, good communication over several kilometres is not uncommon.

Figure 9.12 (2) illustrates the use of a control set using single-frequency equipment. Provided the control point equipment is well sited and has an efficient aerial, this arrangement has the advantage that working range between the control point and individual personal set users is greatly improved. However, with single-frequency working, effective range between individuals depends upon them being within direct range of each other, this

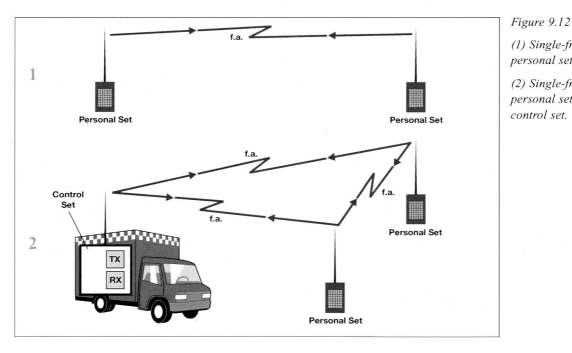

Figure 9.12

(1) Single-frequency personal set usage.

(2) Single-frequency personal sets with control set.

arrangement will not improve person-to-person communication. The control set operator could, however, personally relay messages from one personal set user to another where they are not within direct working range.

It should be noted that when a user is transmitting, they will not be able to hear calls from other users.

Two-Frequency UHF Personal Set Channels

UHF multi-channel personal equipment with three or more channels has been adopted by most fire brigades for fireground purposes. Four of the six channels utilise single-frequency working and two utilise a two-frequency channel. The two-frequency channels cannot be used for direct person-to-person communications without a suitable control set. Figure 9.13 illustrates the arrangement which is adopted when using two-frequency personal set channels. All outgoing transmissions from the 'control' set are on one frequency (f.a.) and all incoming transmission from 'personal set' users are on another frequency (f.b.). Therefore, since all 'personal set' receivers' two-frequency channels are tuned to frequency (f.a.) they cannot hear transmissions direct from other 'personal set' transmitters, which are tuned to frequency (f.b.).

Nevertheless, this arrangement has advantages over single-frequency working especially when there is a need to increase working ranges between individuals. This is achieved by a 'talk-through'

facility on the 'control set' for use when it is necessary to automatically re-transmit on the outgoing frequency (f.a.) all incoming signals received from 'personal set' transmitters on frequency (f.b.). When the 'talk-through' facility is off, the 'control set' operator will hear and be able to communicate with all 'personal set' users within range, but 'personal set' users will not be able to hear each other.

There is no reason why the 'control set' should not be switched to 'talk-through' on the two-frequency channel and left unattended when the requirement is for good communication between 'personal set' users. All six channels may be used simultaneously at the same incident without mutual interference.

The 'talk-through' facility provides considerably enhanced range between 'personal sets'. This is above that obtainable with single-frequency working because of the greater performance of the vehicle-mounted or 'portable set' and its aerials.

It is customary to designate the direction from control as 'outgoing' and the direction to control as 'incoming'. The equipment thus has a true control function, exactly the same as that which the brigade control room has over the main VHF scheme. Hence, the operator at the fireground can control the miniature UHF scheme in just the same way. Vehicle-mounted sets are normally fully duplex, usually with separate transmit and receive aerials although it is possible to use a single aerial with an additional unit, known as a 'duplexer'

Figure 9.13 Double frequency personal set usage with control set on talk-through.

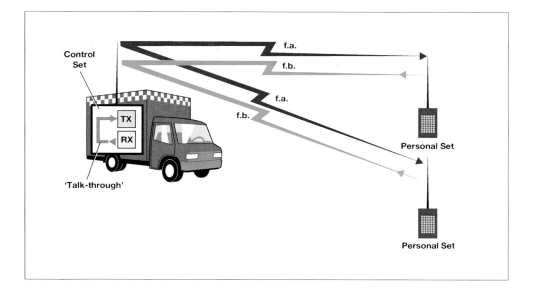

which enables the transmitter and receiver to operate independently and simultaneously with a single aerial.

The conditions of licence, under which frequencies are allocated and radio communications authorised, restrict the use of fire service personal sets to low power, short range communications. In consequence, the setting up of fixed base stations on 'personal set' frequencies, VHF or UHF, to give greater working ranges (for example, throughout a town or city) is not permitted because of the risk of causing interference to other brigades in neighbouring areas.

Normally there is no operational requirement for permanently engineered fire service 'personal set' schemes similar to those which are an operational necessity for the police. The normal fire service requirement is the need for completely portable short range systems which can be set up and brought into use at very short notice anywhere at incidents.

At specific locations fixed UHF base stations (repeaters) may be authorised by the Information and Communications Technology Unit for use in road tunnels or airports, etc. However such use is strictly limited according to the stated operational requirement when the licence was granted.

9.3.13 Composite Units

A vehicle with a mobile UHF control unit will usually have a VHF mobile radio fitted, and an operator in the vehicle can then communicate with both the 'personal set' users and the brigade control room. An added facility is an interface, usually in the form of a combined UHF/VHF control unit which connects to both transmitter/receiver units.

The equipment collectively is now known as a 'VHF/UHF Repeater Unit', and it can be used in three distinct ways:

(1)   Local control of a two-frequency UHF network and VHF communication with brigade headquarters but not simultaneously because although the repeater control unit has two loud-speakers, it has only one handset which is switched to VHF or UHF as required.

(2)   Talk-through between 'personal set' users with the vehicle set unattended or, with the vehicle operator solely involved with VHF communication to the brigade headquarters.

(3)   With the vehicle control unit switched to 'repeat', all signals received on VHF are re-transmitted on UHF and vice versa so that the 'personal sets' all hear brigade headquarters just like a mobile or transportable, and individual personal set users can speak directly back to brigade headquarters.

In (3) above the vehicle will usually be unattended or at least without a designated operator, and it can be arranged that when on 'repeat' the two loud-speakers are switched off to prevent unauthorised 'listening-in'. One minor drawback of the repeater is that personal sets are not able to directly communicate with each other; the vehicle VHF radio is 'simplex' and its receiver switches off when transmitting. When a personal radio transmits, the UHF receiver in the vehicle switches on the VHF transmitter back to control, and the VHF receiver is switched off. The UHF transmitter to the personal sets therefore also switches off and they appear to go 'dead'.

9.3.14 Personal Hand-Held Radio Sets

A limited number of frequencies are specially allocated on a national basis for use by fire brigades, some in the VHF band and others in the UHF band. Fireground communication is presently carried out using the UHF band of frequencies. Personal hand-held equipment is normally designed to accommodate a minimum of three channels. However, with the allocation of six UK-wide fire service UHF frequencies for their exclusive use and possible additional channels for other purposes, there is a need for synthesised multi-channel equipment which, for a number of years, will be used in addition to existing three-channel equipment.

9.3.15 Intrinsically Safe Personal Radios

Ordinary personal radio equipment is capable, in flammable atmospheres, of causing explosions or fire. Intrinsically safe equipment is designed, when correctly used and maintained, to operate safely even if it develops a fault. Personal radios certified to BS 5501 Part 7 or European Standard EN50 020 for Category ib, Group IIC and Temperature Class T4, provide the minimum standard that should be used. The harmonised standard CENELEC Eex ib IIC T4, is often referred to and equally valid. (See DCOL 8/95 Item A, in Scotland DFM 6/1995 item A.)

Each user should satisfy themselves that this equipment is suitable for use at incidents in their area. If in any doubt about the suitability or use of their equipment then HM Fire Service Inspectorate should be consulted.

9.3.16 BA Radio Communications Interfaces

Fire Service Circular 3/75 recommended that all future purchases of breathing apparatus should comply with BS 4667 and should be covered by a Certificate of Assurance (C. of A.) issued in accordance with the Joint Testing memorandum.

From 1 January 1990, under COSHH Regulations, BA equipment had to be suitable for its intended use and approved by the HSE. It is certificated under HSE Testing Memorandum No. 3 (TM3). New CEN standards will apply as they become available.

Any fitting must have the prior approval of the manufacturer who, as the holder of the C. of A., can ensure that, if any amendment is required, it will be HSE approved.

9.3.17 Disadvantages of use of radio with BA

Firefighters should particularly bear in mind that there are disadvantages to the use of radios with BA.

● Radio signal penetration in some types of buildings can be limited.

● Some atmospheres are so potentially hazardous that only communications equipment with the highest standard of explosion protection should be used.

● Radio systems can operate explosive devices designed to be operated remotely.

● Radio transmitters may interfere with building control systems.

9.3.18 User Discipline

The increased general use of BA fireground radio requires good radio discipline. A very complicated radio call-sign system could interfere with operational flexibility and command at an incident and, therefore, self-evident call signs are recommended. Call signs, however, should be such that the brigade can be identified from them, especially in a multi-brigade incident.

The possibility of cross-incident and inter-brigade interference from the use of over-powerful transmitters should be guarded against.

The transmit/receive ratio of the use of radio should always be considered. Transmissions should, wherever possible, be of short duration with an adequate pause to allow other users of the frequency, with perhaps a higher priority message, to transmit.

It is not possible for two UHF repeaters or mobile base stations on the same channel to operate simultaneously within range of each other. Therefore, it is essential that, where this happens, the repeater/base station in the least advantageous location is switched off.

Regulatory approval must be obtained from the Information and Communications Technology Unit before specifying the installation of fixed UHF equipment which will be left permanently switched on.

Radio communications could be required to operate deep within an underground railway system, railway tunnels, building sub-basements and other complex constructions. **The potential of deploying 'throw-out' and 'inbuilt' leaky-feeder systems should be considered even during the early stages of an incident.**

A great deal of research is currently underway to improve underground radio communications.

9.3.19 Security

Modern hand-held radio equipment is of significant financial value and can also be of great value to others, outside the service, if used unlawfully. Accordingly, radio equipment should never be left unattended on appliances unless it is suitably secured against theft.

Hand-held or portable radios should never be left exposed to public view in unattended cars, even if the car is secured.

Arrangements for securing hand-held radios on appliances will vary between brigades. Suitable arrangements could, for example, include a locked container secured in the appliance from which the radio can be taken when required.

9.3.20 Care Of Hand-Held Radio Equipment

The initial purchase of any hand-held radio equipment should include suitable protective carrying cases. The design chosen must take account of local requirements. Even if the equipment procured is water-resistant the protective case should be designed so that it minimises the chances of the battery terminals/connections, aerials or controls coming into contact with water or spray. This is particularly important where the design of the radio equipment is such that water can collect near any of these fittings.

Virtually any electrical or electronic equipment will fail if subjected continually to heat in excess of that in which it was designed to operate.

Battery compartments of radio equipment should be kept closed (and locked in the case of intrinsically safe equipment) except when batteries are being changed. Batteries of explosion protected equipment must never be changed within the hazard area.

Radio equipment should be carried in such a way that it cannot easily be dropped, strike another solid object, become exposed to water, water spray, corrosive chemicals, or be subject to any unnecessary or abnormal mechanical stress.

Radios should not be carried in containers with other metallic objects which could make accidental connection with battery charging or other external radio connections. When external equipment is not connected to any socket, a protective cover should always be in place over the exposed connectors.

9.4 Trunked mobile radio systems

The growth in mobile radio systems over the years, and the subsequent demand for frequencies has placed an ever increasing load on the spectrum managers. The concept of 'Trunked' radio schemes goes far in addressing this problem.

'Trunking' makes greater use of the available channels, but leaves users less aware of the congestion on that channel. Users share a pool of channels and are only allocated a channel when they need to make a call. In practice, not all users wish to make a call at the same time and, therefore, 'trunking' theory is based on the probability that there will be free a channel when required.

Telephone networks have used 'trunking' theory for a great many years, but it has only recently become economically possible in radio systems with the advent of microprocessor circuitry.

Fundamentally, 'trunked' radio systems are engineered in a similar way to cellular telephony systems, with coverage being modelled in polygon shaped cells, although it is of course possible that area coverage is satisfied by a single trunked base station. A mobile radio will be constantly 'speaking' to its local base station via a control channel.

When the mobile wishes to send a message the control channel will allocate a speech channel dynamically, and communications will be available. At the end of the transmission, and at each subsequent transmission, different channels may be used. Once the transaction is complete these channels would then be available for other users.

Unlike cellular systems, which are generally designed to be a one-to-one service, 'trunked' radio schemes can set up user groups in which multiple users will be able to talk.

Figure 9.14 Trunking concept.

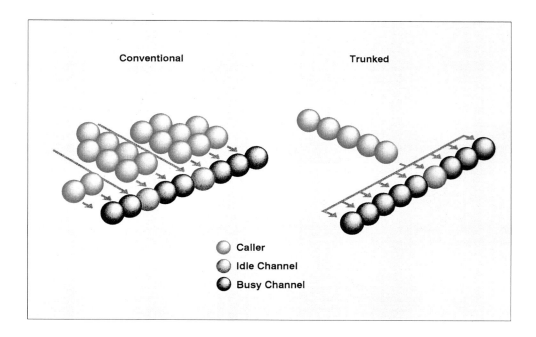

'Trunked' radios permit 'roaming' throughout the area required, with handover between radio cells as a mobile passes from one base station area to another. The mobile receiver will be constantly hunting for a control channel during this exercise. Once a signalling channel is identified the signalling information is examined and checked by the mobile and, if validated, locks the mobile to this channel. The process happens automatically, and transparently to the user.

In the UK, the MPT 1327 signalling standard is used to facilitate analogue 'trunked' private mobile radio services in 'Band III' (174–225MHz) although there is no reason why 'trunked' systems could not operate in different frequency bands.

Future development in trunked systems is currently being addressed by ETSI under the remit of the Trans European Trunked RAdio project (TETRA) which will be a digital TDMA product, with an effective bandwidth per voice channel of 6.25 KHz, giving four time slots possible within the 25 KHz bandwidth.

TETRA will operate for the Emergency Services in the band 380–400 MHz, and commercially by PAMR service providers in 410–430 MHz.

Communications
and Mobilising

Chapter 10 – Radio Alerting System

Fire brigades in the UK currently use alerting systems to the Home Office MG4 specification. This forms the final link in the overall mobilising system. The simple schematic shown in Figure 10.1 indicates the system elements involved.

Mobilising and communications components of the overall system are dealt with elsewhere in this publication. But in the interest of understanding the alerter system itself, a brief outline may be of help.

The mobilising system contains a large amount of detail covering the whole of the brigade and this is available to Fire Control Officers when assessing the operational needs of any particular incident. Having determined a station or stations to be turned out, the detail concerning the incident location, the appliances to attend and, if **Wholetime**, the operation of sounders or, if **Retained**, the operation of alerters, is passed to the Communications Processor.

The Communications Processor makes use of the further Home Office Protocol, GD92, which specifies the way the processor works and the facilities it must provide. The processor will normally have a number of communication links, known as

'bearers', between Control and the stations and will use these on the basis of laid down preference and availability. The types of 'bearer' available to such systems can change as technology makes them available and cost effective.

The station end GD92 is similar but on a smaller scale and serves to interpret incoming instructions and also operate printers, sounders, lights, doors, appliance bay indicators, etc. In the case of stations with a retained element instructions from the GD92 unit are passed to the **Alerter** using the MG4 protocol.

10.1 Alerter – General Description

The requirements of an alerter system have not changed in essence since radio alerting was introduced. The fundamental need is still to call a retained crew to a station in the case of an incident or to send a test call to the alerters. What has changed is the means and the method of operation resulting in greater detail concerning a calls progress being available at Control.

It is not always the case that an MG4 alerter is installed at the same time as the communications

Figure 10.1 Radio alerting system to the Home Office MG4 specification.
(Graphic: Multitone)

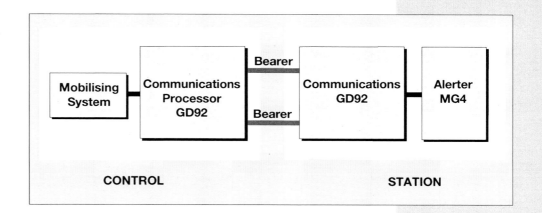

system is upgraded to GD92. In general, older Mobs/Comms systems presented simple relay contact closures to the alerter and expected a simple relay contact closure in return to indicate a successful or failed call. MG4 systems have to be able to operate in this mode leaving the more advanced MG4 signalling protocol to be implemented at a later date. This gives a brigade the flexibility to upgrade the overall system on a staged basis.

There are basically two component parts to the system, an encoder and a transmitter. The encoder generates the call required and the transmitter sends it to the alerters. It was common practice with the previous alerter bays to use two transmitters in a main and standby configuration. This has all but disappeared with the new MG4 alerter systems. Modern transmitters are far more reliable, but the facility is still available giving brigades the opportunity to take financial advantage of the improved technology and still use dual bay at certain locations if operational needs dictate.

Standby power is still required to cover for the eventuality of mains supply failure and normally this would take the form of batteries designed to give the brigades a stated period of operation. However, various options are available and a choice depends on the period of back up and whether or not other devices share the back up source; for example, the GD92 Comms unit. The charging of such batteries would be by stand alone equipment or perhaps via one of the units already part of the system. This is dependant in part on individual suppliers and their particular approach. Although there is flexibility in the type of power supply that can be provided, and brigades may request back up periods less than the MG4 recommendation, suppliers must be in a position to achieve the 24 hours if requested.

10.2 Encoder

This component of the alerter performs the following basic functions:

● Provides two-way communication, using the MG4 protocol, with the station GD92 Comms equipment.

● Generates required team fire or test call using

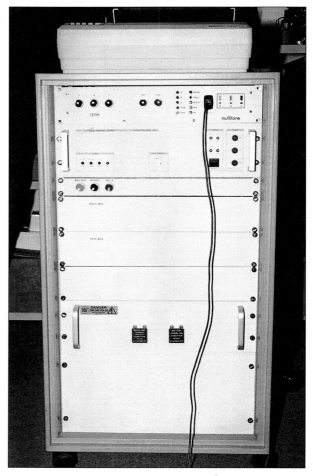

Figure 10.2 MG4 Alerter Transmitter combined with a GD/92 compliant 'Station end' mobilising terminal. (Photograph: Multitone)

POCSAG (Post Office Code Standardisation Advisory Group) paging protocol, as described earlier (see Section 10.1, Alerters). Up to three teams and the combinations of such are catered for.

● Generate paging calls with alpha numeric messaging if required for individual calls.

● Control both the sending of calls via the transmitter and the monitoring of transmitter parameters during calls. (See Section 10.3, Transmitter.)

● Record the operation of team 'off air' monitoring receivers to determine the transmission of correct call data.

- Assemble the monitored transmitter parameters and the 'off air' receiver status and produce a message indicating a successful or failed call and send this to the station GD92 Comms unit for onward transmission to control. If the call is a failure the type of failure is also returned.

- Provide a considerable degree of configuration in order to be able to replace a faulty encoder at a station with the minimum of delay.

- Where stations may overlap from the point of alerting, coverage provision is made for coping with simultaneous mobilisation of the two stations.

It should be noted that although individual manufactures have to comply with the requirements of the MG4 specification they are not restricted from providing additional features, either, of their own idea or at the request of brigades, subject to these not compromising the prime requirements. As these vary from manufacturer to manufacturer it is not intended to include them here.

10.3 Transmitter

The transmitter performs the following basic functions:

- As the paging code specified in MG4 is **POCSAG**, this requires the transmitter to use Frequency Fast Shift Keying (FFSK) modulation. The POCSAG code is referred to in a little more detail under Alerters (Section 10.4).

- MG4 requires that a minimum of two transmitter parameters be monitored during a call; namely, forward power and reverse power.

- The transmitter must be capable of 25 W output with the ability to set alarm trigger levels. As a guide, a level of 12 W (3 db down) would set the alarm. The reverse power alarm indicates the efficiency of the aerial and would normally trigger at what is termed a 'voltage standing wave ratio' of 2:1 or approximately 10 per cent reduced power transmitted.

- The allocated frequency for firefighters alerting is 147.8 Mhz and this frequency is used throughout the UK. A separate frequency allocated to Emergency Services is often used to provide wide area officer paging and, on a few occasions, has been used for mobilising purposes. This frequency is 153.05 Mhz

10.4 Alerters

MG4 calls for the use of alerters working to the POCSAG format which are produced by virtually all manufacturers, and uses a seven-digit numeric address or **RIC** (**R**eceiver **I**dentity **C**ode) code.

The Home Office instituted a numbering scheme whereby the last three digits are fixed for each brigade. The POCSAG code allows the first four digits to range from 0000 to 1999, a total of 2,000 codes per brigade. In the event of a brigade requiring more RIC codes, if for instance multi-RIC code alerters are used for officer paging, then these are available on application to the Home Office. Two RIC codes are normally required for firefighter alerters but versions with four codes are available.

Alerters need to be robust and have protection against ingress of moisture and dust. POCSAG pager design is driven by the large area wide paging market demands of companies such as BT and Vodapage together with other international service providers. Final design is, therefore, a compromise between design requirements, ready availability, and competitive pricing. This results in low cost units which are often cheaper to replace than repair.

Firefighter alerters are normally of the 'tone only' type although some have limited display options to highlight Fire or Test calls. A flashing LED operates on receipt of a call. The use of rechargeable batteries has virtually disappeared although they are still available. A limited need for intrinsically safe alerters exists, more from the point of view of equipping firefighters with such units because of the hazardous areas in which they normally work rather than operational reasons. A vibrate option is available where the normal workplace is subject to high noise levels.

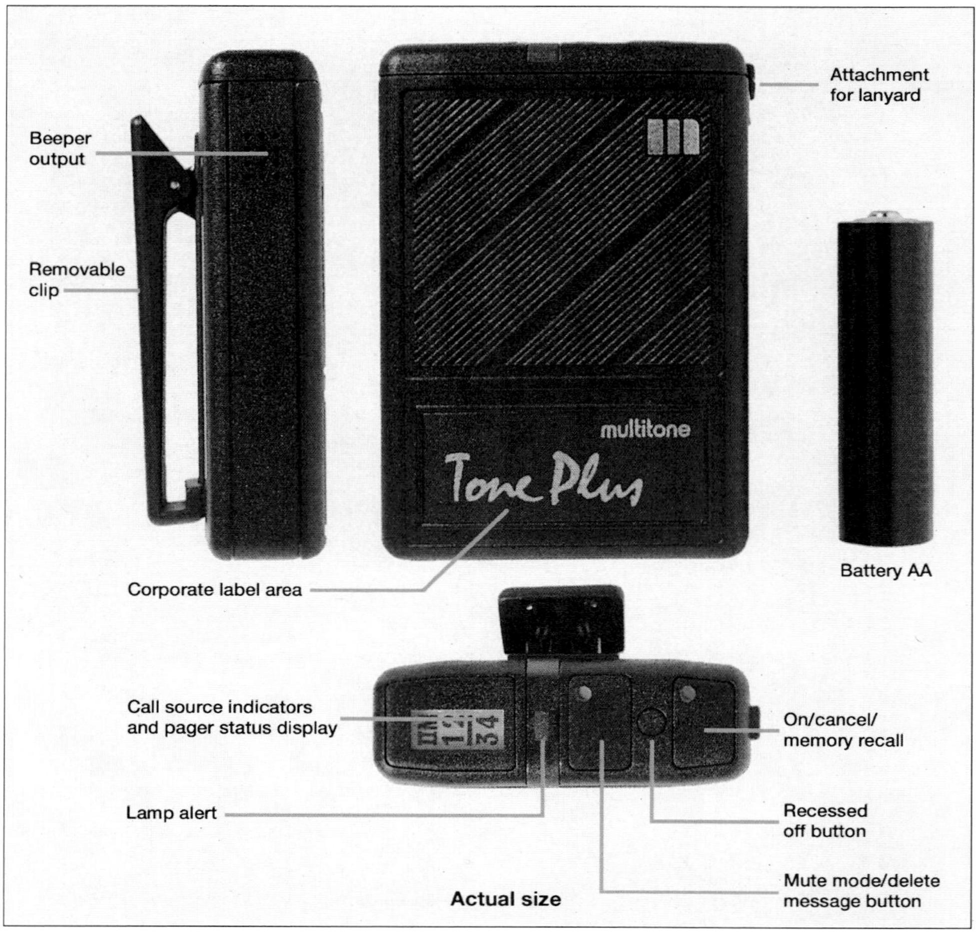
Beeper
output
Removable
clip
Attachment
for lanyard
multitone
Tone Plus
Battery AA
Corporate label area
Call source indicators
and pager status display
On/cancel/
memory recall
Lamp alert
Recessed
off button
Mute mode/delete
message button
Actual size

Figure 10.3 Alerter.
(Photograph: Multitone)

The alerter is required to sound for a minimum of 30 seconds. POCSAG pagers vary in the period of call generated by receipt of a single call from an MG4 unit and multiple calls are often used to achieve the overall required alerting period. This has the advantage of increasing the chance of receiving a call as, say, four calls separated by 12 seconds improve the chances of receiving at least one good call in areas of weak coverage. Alerters are equipped with a call cancel button.

Chapter 11 – Mobile Data

11.1 What is Data?

In data communications, information is transmitted in the form of characters, namely letters, figures, and symbols. The information is represented by binary signals, which are characterised by different states. When considered electrically, these signals correspond to, for example, tone ON, and tone OFF.

In digital message transmission over radio circuits, the signal elements of the characters are transmitted in turn (serially). Figure 11.1 shows the relationship between DC keying and VF keying.

The fire service has been using data at incidents for many years in various formats. 'Data' can be interpreted as telemetry, resource updates, risk information, and can be deployed as direct links to command and control systems.

Data transmission capabilities and speeds (rates) are dependant largely on bandwidth. The bandwidth and, hence, capacity, has risen over the years and hence typical data rates now are around 9,600 bits per second (bps). Developing technology and compression techniques will lead to an expected rise in these rates in the next few years.

Data can be sent both to and from vehicles and used to supplement information held onboard. The development of personal computers and associated software/hardware now allow more data to be stored onboard vehicles.

11.2 History

Resource Availability Status (RAS)

The use of mobile data in the fire service began in the late 1970s when Resource Availability Status

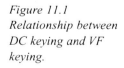

*Figure 11.1
Relationship between
DC keying and VF
keying.*

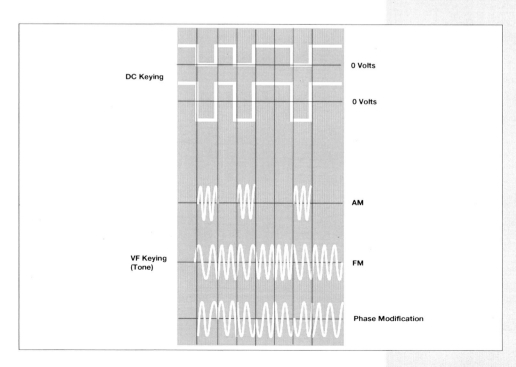

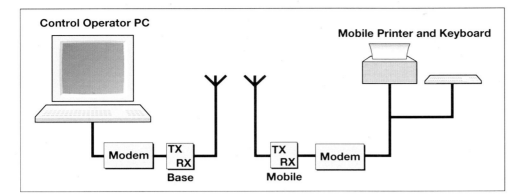

Figure 11.2 Typical data transmission arrangement.

Control Operator PC

Mobile Printer and Keyboard

Modem | TX RX
Base

TX RX | Modem
Mobile

(RAS) was incorporated into some existing radio systems. The first systems operated at 100 baud and were literally one way only with the acknowledgement being a single tone. Further development enabled these systems to operate at 300 baud with two-way communications.

All systems operated over the speech radio network. One of the major problems associated with this shared voice/data radio channel was that data tones would block speech traffic using the radio channel. This had the effect of data being received but speech having to be re-transmitted.

Mobile Data

Depending upon the level of traffic, it is sometimes better to provide a radio channel specifically for mobile data. This allows a number of different types of messages to be passed. The first systems went live in 1989 providing a series of services to the incident. For example:

- Mobilising messages Command and Control (C&C) to mobile.

- Administrative messages (C&C to mobile).

- RAS messages in both directions.

- Access to Management Information Systems.

- Access to Chemdata central information.

- Incident messages (mobile to C&C).

This was the first time it was possible for mobile information to be directly input into C&C systems

(RAS had been possible earlier) without any Fire Control personnel action and requiring the Control Room staff to change working procedures.

Figure 11.2 shows a typical data transmission system. The Dispatcher computer interrogates the mobile to establish contact and, on receipt of an acknowledgement, sends the data as a burst transmission. The mobile is able to display and/or print out the received message. The process can also work in reverse, with the mobile initiating the call.

11.3   Current Technology

The term 'mobile data' encompasses data sent when an appliance is mobilised, available en route to an incident, or available at an incident. Requirements for the provision of data vary from brigade to brigade leading to a multiplicity of system configuration. The main elements of these vehicle mounted systems are communications processors, visual displays, printers, keyboards, radio modems, etc.

11.4   Radio Communications

Various options are available for the transmission of data between mobiles, or from mobiles and/or fixed locations.

Existing Brigade Radio Schemes – It is possible to transmit data over existing analogue radio schemes; some brigades use this bearer as part of their mobilising arrangements. Typical transmission rates are 1,200 bps Frequency Fast Shift Keying (FFSK).

Figure 11.3 Data infrastructure.
(Graphic: Simoco)

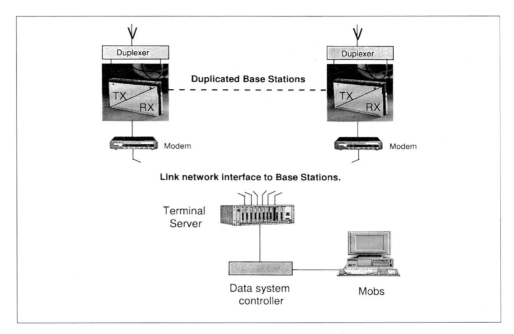

Figure 11.3 shows a typical data system infrastructure developed for use by UK fire services. Such a system is able to make use of the fire station mobilising system that already forms part of the Command and Control facility at any Fire Service Communications Control Centre. The mobilising system is connected to a Data System Controller the purpose of which is to control the operation of the radio data system via the Terminal Server. One function of a Data System Controller is to transparently convert data into a form suitable for transmission to the hill-top site Radio Modems. The Terminal Server distributes data to and collects data from the Radio Modems. 'Best hill-top site' information is stored and continually updated for all mobiles and the appropriate site used for any communication with a mobile.

It is possible to send general text messages between Control and mobiles as well as status messages from mobiles to Control.

GSM Cellular Telephone – The Global Systems for Mobile Communications (GSM) networks incorporate both voice and data modes of operation. These networks consist of individual radio base stations that communicate with the users, each base station forming a cell. In the data mode the system offers a circuit switched, end-to-end communications service and, at present, transmission speeds of up to 9.6 K bits per second.

Packet Radio Service – There are several commercial packet radio data networks. These networks deliver data in the form of bursts or packets. Each packet contains address information, information data and some form of error correction.

11.5   Data on Vehicles

The data available on mobile resources is as diverse as that held in an office environment. Individual brigade requirements vary, from providing limited information held on mobile computers, to being able to access personnel records, building plans, status messaging, global positioning, updating the central mobilising system, receiving turn-out information and chemical and risk data.

Data must be current for it to be of value. The provision of many geographically scattered mobile data terminals leads to difficulties in maintaining

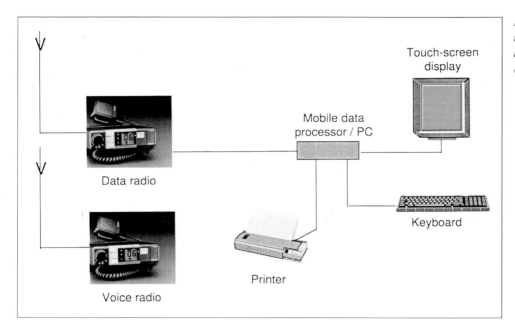

In Cab Equipment

Figure 11.4 shows equipment used in a typical vehicle installation in a fire appliance. This comprises two separate mobile radios, including a Voice Radio which can be used for voice transmissions at any time and a separate Data Radio operating on its own radio channel.

The Data Radio, which would incorporate a modem, would be used to pass status information, various data, and free text information from a Mobile Data Terminal (MDT) which comprised an in-vehicle PC, Touchscreen Display, Printer and Keyboard. The PC might be equipped with large capacity hard drive memory which can hold map data, chemical data, risk files, etc., which can be triggered by the incoming data to give information specific to the incident. The Touchscreen gives the mobile operator the facility of manually accessing data, maps, etc., or of inputting status or text messages, as required.

The printer makes it possible to produce hard copy of diplayed data, while the keyboard may be used for the inputting of text or for maintenance access.

the data, especially if a large percentage is held on the terminal. It is essential that a robust and effective system is established so that updating the stored information is carried out on all the terminals within a minimum period of time.

Part of this process should include an audit trail so that it is possible at a later stage to verify when and who amended any of the data files. Various methods have been adopted to carry out this procedure including updates by floppy disk, CD–ROM, radio or wire connection to each data terminal when the vehicle is in the fire station.

11.6 Typical Data Requirements

The following data packages are available for use on mobile processors, whether held on the mobile, retrieved from a central source or a combination of both. Software licensing issues, along with other factors, may influence whether data is held centrally or dispersed amongst the mobile terminals.

1 Status Messaging

The ability of the mobile resource to update the central mobilising computer of any change of status instead of using a voice radio or cellular telephone scheme. This facility should also incorporate the ability to send other standard messages – for example, assistance and stop messages – and should have a free text option to cater for any non-standard messages.

2   Risk Information

Information gained from the inspection of premises, under the relevant section of the Fire Services Act 1947, shows the layout of the premises, the utility supply inlets and isolating points, the location of water supplies and any risks to firefighting. This data, which has traditionally been held in paper form, lends itself to being held electronically, thus making it available to all mobile data terminals and centrally on the brigade's own network. Building plans and maps may also be linked to this risk information.

3   Brigade Information

Brigade orders, firefighting information, operational and technical procedures, any information produced by brigades or from other sources may be held in an easily retrievable format so that the Officer-in-Charge of an operational incident has all the information available.

4   Hazardous Information

Information relating to hazardous substances may either be held on the mobile or centrally. Chemdata, for example, when held centrally, can be distributed by radio and only comprises of a relatively small amount of data.

5   Graphical Information Systems (GIS)

GIS software, which requires mapping data, gives the operational crews access to maps of the brigade area, ranging from 1:50,000 raster-based maps to vector-based maps, which enable the operator to zoom in to display individual buildings. These maps can then be linked to building plans, street maps are also available to replace the map books carried on vehicles.

Hydrant and water main information may also be superimposed on the maps so that the information resources of the brigade are available to all mobile terminals.

Clearly, mapping data files are quite large. Therefore, storing the maps centrally and transmitting the data on demand would require high transmission rates, cause congestion on the radio network and be expensive. Currently, it would be better to store this type of information on the mobile data terminal hard drive (if available).

6   Automatic Vehicle Locating Systems (AVLS)

AVL systems have been available commercially for some years. It has only been more recently that the fire service has investigated the technology for its own use (see DCOL 8/1997, in Scotland DFM 8/1997).

There are two basic systems in use; land based and satellite. The vehicle is fitted with a suitable AVLS receiver which, following the reception of signals, allows the geographic position to be computed and then transmitted to a central mobilising system. The position of the vehicle is then displayed on a map at the central control.

These systems have varying degrees of accuracy but care must be taken when attempting to predict the precise location and direction of movement of a vehicle using this system. It is possible for example, that the AVL system could indicate that a vehicle is located close to the scene but it transpires that it is on the wrong side of a river or motorway to attend the incident. (See AVLS, Chapter 7.)

7   Vehicle Telemetry

With the provision of a processor on a vehicle and a wireless connection with a central point it is possible to send telemetry information. For example, information on the vehicle engine systems could be routed to the Brigade's transport department or show quantities of water, foam or other operational

consumables to the Control Centre or mobile
control unit.

11.7 Mobile Control Units

With the development of reliable data transmission
technology and vehicle-based computer systems,
mobile control vehicles used for major incident
command and control are now being equipped
with IT systems linked to brigade computer networks and mobilising systems.

These vehicles include complex computer systems
and voice/data message handling facilities.
Bespoke software packages have been developed
specifically for this purpose.

Chapter 12 – Breathing Apparatus Telemetry

There is an increasing need to provide firefighters, particularly those protected with breathing apparatus, with enhanced information in order to improve both their safety and operational effectiveness. This could include, for example, information on remaining cylinder contents and respiration rates, ambient and body core temperature, heart-rates, and so on.

Data can be displayed to the wearer in full, or more practicably in an abbreviated form, perhaps by means of a display in the firefighter's breathing apparatus face mask. They can also be recorded in an electronic database and downloaded at the conclusion of an incident, to provide a record which can be added to personnel records and used in the investigation of any injury or malfunction of the apparatus.

It is also possible to transmit some of this data by radio to those personnel controlling the incident, including the Incident Commander or the breathing apparatus Entry Control Officer. Here, data provides information which can be used to facilitate better control of the incident and to improve firefighter safety.

Such provision of a radio data link between firefighters and those controlling the incident will also permit the remote signalling of other safety signals. These include the transmission of information to the breathing apparatus Entry Control Officer of a message in a data format indicating the automatic or manual operation of a breathing apparatus Distress Signal Unit, and the transmission of a message in a data format causing the operation of an Evacuation Signal either to all in the risk area or selectively.

It also facilitates the signalling of a radio message in a data format indicating that the operator is withdrawing from the risk area for reasons of personal safety. This last information, particularly if more than one team signals it, will assist the

Figure 12.1 Radio Distress Signalling Unit. (Graphic: Marconi)

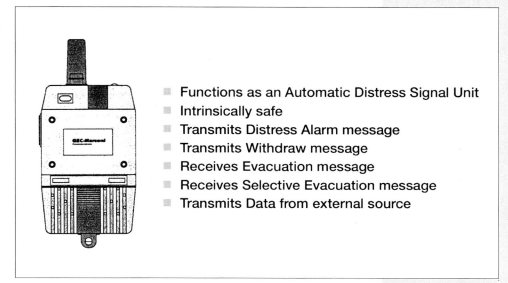

- Functions as an Automatic Distress Signal Unit
- Intrinsically safe
- Transmits Distress Alarm message
- Transmits Withdraw message
- Receives Evacuation message
- Receives Selective Evacuation message
- Transmits Data from external source

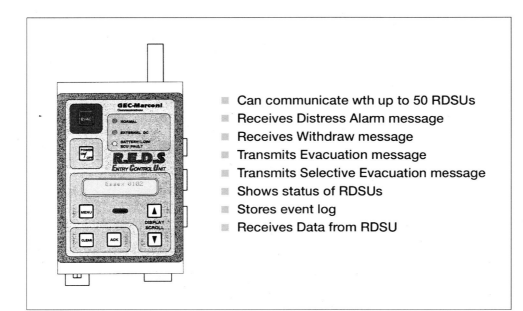

Figure 12.2 Entry Control Unit.
(Graphic: Marconi)

- Can communicate wth up to 50 RDSUs
- Receives Distress Alarm message
- Receives Withdraw message
- Transmits Evacuation message
- Transmits Selective Evacuation message
- Shows status of RDSUs
- Stores event log
- Receives Data from RDSU

Figure 12.3 Telemetry in use during trials. (Photograph: Essex Fire and Rescue Service)

Incident Commander in deciding whether emergency evacuation of the risk area is justified.

The Home Office has developed a User Requirement (**JCDD/40**) for fire service telemetry based on the use of a dedicated radio frequency in Home Office regulated radio frequency bands, supported by a Type Approval (**MG41**) specification and a common-air interface.

Figure 12.4 Telemetry in use by a Breathing Apparatus Entry Control Officer. (Photograph: Essex Fire and Rescue Service)

This User Requirement specifies the minimum functionality for such equipment. It includes remote signalling of Distress, remote signalling of Evacuation, Selective Evacuation, and signalling that the operator is withdrawing for reasons of personal safety. It allows the equipment to be combined with, and provide the functionality of, an Automatic Distress Signal Unit. It also allows connection by means of a standard interface and transmission of data to and from operator-worn equipment providing enhanced functionality such as cylinder contents, temperature and so on. A similar connexion is specified for the remote element of the system to allow the use of equipment for managing incoming data.

Chapter 13 – Potential hazards of using radio equipment

13.1 General

There is continuing work and development taking place in this field, in particular, new EU guidelines being produced regarding Intrinsically Safe Radios and other equipment (ATEX Equipment Directive). Details will be issued to the service in due course but, in the meantime, the text from the 1998 edition of the Manual has been retained for this chapter.

FLAMMABLE ATMOSPHERES NEAR EXPLOSIVES OR IN HOSPITALS

Special precautions are necessary when working in areas where potential ignition or explosion hazards exist. These could be due to the presence of flammable dusts, gases or vapours, such as in oil refineries, petrol storage depots, some factories and commercial premises, coal mines, etc. The introduction of electrical or electronic equipment, such as a radio, by firefighters to such environments may constitute an ignition hazard.

Where such environments are known to exist, or are suspected, then the electrical equipment needs to be safe for such use. Equipment must be designed in such a way that it does not present a hazard and should be certified accordingly. A number of design concepts for achieving certification exist, the most common for radio terminals being that of '**Intrinsic Safety**'. Any equipment designed to be safe in such environments is generically described as 'explosion protected'.

Additional precautions are also necessary when using radio transmitters in the vicinity of explosives, ignition hazards or other devices that may be adversely affected by radio transmissions. This chapter deals in some detail with the technical aspects of terminal equipment to be used in potentially flammable dusts, gases or vapours. It concludes by giving operational advice and guidance on precautions to be taken by firefighters when using radio transmitters in potentially flammable or explosive atmospheres and use of radio in the vicinity of explosives, petrol stations and medical devices.

13.2 Explosion Protection – Standards

In the EU, standards for electrical equipment designed for use in environments containing flammable gases and vapours are those approved by the European Committee for Electrotechnical Standardisation (**CENELEC**). Such equipment may be certified as meeting the relevant standard by an EU approved Certification Body. The relevant EU standards for Intrinsic Safety are drawn from EN50 014, EN50 020 and EN50 039.

Terminal equipment in current use in the fire service may have been manufactured to comply with an earlier standard, e.g., British Standard 1259, 1958 or a BASEEFA standard SFA 3012 1972. This equipment can continue to be used until replaced.

Outside the EU other standards exist. In particular, in the USA and those parts of the world where US standards prevail, equipment is certified to standards specified by either Underwriters' Labaratories or the Factory Mutual Research Corporation.

For the UK fire service the recommended standard for general applications where a potential ignition hazard exists is for equipment that is certified for use in Zone 1 with gas group IIC and a maximum temperature rating of T4. Such equipment would be indelibly marked EEx ib IIC T4 (or T5 or T6). It is also recommended that such equipment should satisfy a degree of ingress protection of at least IP54 to EN60 529.

It is recommended that radio equipment purchased for use with breathing apparatus should, as a minimum, conform to this standard. The equivalent US standards are Underwriters' Laboratories UL913 or Factory Mutual Class No 316.

13.3 Ignition Sources

The presence of radio terminal equipment in a potentially explosive, or flammable, dust, gas or vapour environment can give rise to a number of possible hazards from which ignition might result. These include overheating of the radio terminal during use or, more likely, during fault conditions. In modern mobile radio terminals from reputable manufacturers this is most unlikely to occur under any working conditions, as long as the equipment is fully serviceable and operated correctly.

Other potential dangers from the use of radio terminal equipment can arise where there is a possibility of sparking. Sparks of sufficient energy to cause ignition may be produced by two quite distinct mechanisms, as follows:

(1) Sparking may occur when contacts are made or broken in circuits carrying electric currents, or containing sources of electrical energy; and

(2) Whenever a radio transmission is made the electromagnetic field radiates radio frequency voltages in all conducting materials in that field. The induced voltages in adjacent conducting elements, or between conducting elements and 'earthy' conductors, may be sufficient to break down intervening insulating layers of oxidation, grease, air, etc., and cause dangerous sparking. This hazard is directly related to the nature of the environment, the characteristics of the transmission (power, type of modulation, etc.,) and the distance of the antenna from the hazard.

13.4 Protective Measures

Fire service mobile radio terminals (with an expected transmit power between 5 W and 25 W) potentially introduce all of the hazards described in Section 13.3 above into the risk environment. There is also a potential risk that the antenna of a vehicle mounted radio may directly touch a conductor during transmission causing sparks.

> **The design of such equipment precludes the adoption of any satisfactory explosion protection strategies and, therefore, such radio terminals must be regarded as a potential ignition risk when used in a potentially flammable or explosive dust, gas, or vapour environment.**

When transmitting, they may also introduce a risk of inducing a current in a conductor, causing remote sparking and ignition or some other unwanted consequence.

> **The only practicable safeguards are to exclude the radio terminal (or the vehicle on which it is mounted) from the hazardous environment.**

The risk from any 'fixed' mobile radio terminal equipment should be minimal since it is expected that any potential hazard should have already been taken into consideration before siting is decided. However, some brigades employ transportable terminal equipment which may be used, for example, as temporary controls or talk-through stations for special purposes. Such terminal equipment generally has transmitter power of 5–25 W; therefore, the potential risk is comparable to that of a vehicle installation but it may be used in locations inaccessible to vehicles. This category of equipment should not be overlooked in any hazard assessment or the drafting of relevant orders.

For relatively low-powered hand-held radio terminal equipment (typically less than 1 W) the maximum radiated power is usually insufficient to create induced currents in adjacent conductors. Potential ignition hazards with hand-held radio terminals are, therefore, limited to the development of dangerously high temperatures, and sparking caused by making or breaking electrical circuits. High temperatures are only likely to exist in fault conditions – for example, by a component failure or breakdown of insulation and steps can be taken to prevent internal sparking that has sufficient energy to ignite a flammable or explosive dust, gas or vapour.

Thus, it is a practical proposition to design an hand-held radio terminal which can be used with safety in a potentially hazardous environment – that is, explosion protected equipment. Explosion protected terminal equipment nearly always exists as well in a normal, unprotected form. In comparison with the non-protected terminal equipment the protected equipment will often have a reduced maximum transmitter power, be more expensive to procure and maintain, will require 'special' batteries and may have reduced functionality.

The selection of protected types of hand-held radio is also likely to be much less than unprotected types and purchasers may have a limited choice of equipment from which to select equipment for procurement.

13.5 Intrinsically Safe Design Criteria

The requirement for explosion protected equipment certified for use in flammable, or explosive dust, gas or vapour environments, means that the equipment must be incapable of causing ignition, even under fault conditions or when subjected to gross mishandling. This necessitates design features which often have performance penalties in normal conditions.

It is usually necessary, for example, to make it impossible for batteries to be fitted or removed within the hazard area, because of the potential danger of sparking during this process. Therefore, it is usual for the battery compartment of such equipment to be fitted with a key-operated lock so that the compartment can be locked and the key retained outside the hazard area. In ordinary fire brigade practice using such a fitting may be an inconvenience.

To obviate the possibility of components overheating, current limited devices (often resistors) have to be fitted which may reduce performance. Extra thermal insulation may have to be provided, making the equipment more bulky than it would otherwise have been. The mandatory distance separations of components, and conductors on printed circuit boards, may also affect equipment size.

Requirements for special materials or plating, necessary to withstand long-term exposure to certain chemicals, involve considerable additional cost; as does the incorporation of other non-standard requirements already mentioned. These are some of the factors which combine to make the idea of using this equipment for all purposes quite unattractive from both a size and costs viewpoint.

13.6 Selection of Explosion Protected Equipment

The current recommendation to the fire service is that radio equipment purchased for use with breathing apparatus should be certified by an EU approved Certification Body for use in Zone 1 with gas group IIC and a maximum temperature rating of T4. Such equipment will be indelibly marked as follows:

EEx ib IIC T4 (or T5 or T6)

Certified equipment must cater for worst-case conditions for the whole of its working life under continuous operation in a hazardous environment. It must also take into account carelessness, clumsiness and ignorance on the part of the operator.

In perspective, the few occasions when faults will develop in modern personal radio sets are considered and the fact that fire service personnel are trained to comply with instructions regarding care of this equipment, the occasions when all the above special design features would be needed is small.

13.7 Radio Use in the Vicinity of Explosives, etc.

Radio transmissions impose a potential ignition or initiation hazard near commercial explosives, military ordnance (including nuclear weapons) and terrorist devices.

> **Current guidance to the fire and police services is that no radio transmitting equipment should be used within 10 m of the risk, that only hand-held terminal equipment (less than 5 W) should be used within 10 m and 50 m, and that vehicles fitted with mobile radio terminals should not be taken within 50 m of the risk unless the radio is switched off.**

In this context, many modern radio terminals, for example, data capable radios, 'trunked' radios and radios using public cellular or public data services are capable of auto-transmission. Unless the transmission function can be inhibited by the user such equipment should be switched off if it is necessary to take it into the protected area appropriate for the type. For example, where it is necessary to take a public cellular radio terminal to within 10 m of the hazard.

13.8 Radio Use in the Vicinity of Retail Petrol Stations, etc.

> Current guidance to the fire and police services is that similar restrictions to those applicable to the use of radio terminals in the vicinity of explosives should be applied in respect of retail petrol stations, petroleum transfer stations and oil depots.

13.9 Radio Use in the Vicinity of Air Bags

> Because of the potential, but remote, danger of actuating an air bag in a vehicle which has been involved in a road traffic accident in which an air bag has not actuated, no radio terminal should be used to transmit a message within 10 m of the vehicle.
>
> Additionally, to avoid the remote possibility of unwanted actuation, no hand-held radio terminal or hand-held cellular radio terminal should be used inside a vehicle equipped with an airbag unless it is connected to an aerial system external to the vehicle.

13.10 Radio Use in the Vicinity of Medical Devices

There is a potential hazard that radio transmissions may have unwanted effects on medical devices.

> No fire service hand-held radio can be considered as being safe to use in radio sensitive areas of hospitals, nor can any 'safe-distance' be recommended. Accordingly, hand-held radios should only be used for transmission in hospital buildings in exceptional circumstances and where the circumstances are unavoidable. If a hand-portable radio has been used then this should be reported locally to the hospital/medical staff so that they can initiate whatever checks they might think necessary to detect and rectify any effect that the transmission might have had.

In this context, many modern radio terminals, for example, data capable radios, 'trunked' radios and radios using public cellular or public data services are capable of auto-transmission. Unless the transmission function can be inhibited by the user, such equipment should be switched off if it is necessary to take it into a hospital premises.

Where a hospital has placed a restriction on the public use of cellular radios then these restrictions should also be taken as applying to hand-portable radio terminals or any public cellular radio terminals that may be used by firefighters.

13.11 Radio Use within Silos

The presence of a radio may cause ignition of any flammable dust, gas or vapour that exists in a silo. There is also a remote possibility that a transmission from a hand-held explosion protected (intrinsically safe) radio may result in a spark caused by an induced current.

Accordingly, similar restrictions to those applicable to the use of radio terminals in the vicinity of explosives should apply to the use of radios near or within silos until it has been established by monitoring that there is no trace of a potentially flammable dust, gas or vapour within the silo.

Notwithstanding the foregoing, the Officer-in-Charge may decide to permit the **limited use of explosion protected (intrinsically safe) hand-held radios or telemetry equipment within a silo, provided that a risk assessment has been carried out and it is considered that the operational and safety benefits of so doing exceed the remote risk of ignition.**

Communications
and Mobilising

Glossary of terms and abbreviations

Address Point Ordnance Survey digitally co-ordinated postal address data.

Alerter system A call-out system utilising pocket-alerters, carried by retained firefighters, which are triggered by a radio signal transmitted by a remotely controlled alerter transmitter usually located at a fire station.

Algorithm A procedural model used when computing complicated calculations (e.g., routes and drive times).

Analogue An analogue signal is one which can vary continuously, taking any value between certain limits. The human voice, for which the public telephone network was designed, is an analogue signal varying in frequency and volume.

ACD Automatic Call Distribution.

AFA Automatic Fire Alarm.

AVL Automatic Vehicle Location.

BA Interface An interface designed to permit a handheld radio set to be used in conjunction with breathing apparatus.

Bandwidth The range of signal frequencies which can be carried by a communications channel subject to specified conditions of signal loss or distortion.

Base Station The transmitter/receiver and associated equipment at a fixed location.

CACFOA Chief and Assistant Chief Fire Officer's Association.

Call sign An identifier, normally comprising a name, numbers or letters, by which an appliance or officer is identified when being called by radio.

CCTV Closed Circuit Television.

Cellular A technique used in mobile radio telephony to use the same radio spectrum many times in one network. Low power radio transmitters are used to cover a limited area or "cell" so that frequencies in use can be re-used in other parts of the network.

CHEMET	Chemical Meteorology.
CIMAH	Chemical Incident Major Accident Hazard.
CLI	Calling Line Identity.
Concentrator	Any communications device that allows a shared transmission medium to accommodate more data sources than there are channels currently available within the transmission medium.
COSHH	Control of Substances Hazardous to Health.
CTCSS	Continuous Tone Controlled Signalling System. In PMR, a method of using sub-audio tones to effect selective transmissions to a mobile or group of mobiles.
Cycle	The portion of the radio wave between successive crests or troughs, which is repeated over and over again to form the continuous wave.
DCOL	Dear Chief Officer's Letter.
DDI	Direct Dial In.
DIEL	OFTEL's advisory committee on telecommunications for Disabled and Elderly People.
Digital	Communications procedures, techniques and equipment where information is encoded as either a binary '1' or '0'.
Digital data network	A network specifically designed for the transmission of data, wherever possible, in digital form.
DTI	Department of Trade and Industry.
Duplex working	A communications technique in which it is possible to transmit and receive simultaneously, e.g., as in an ordinary telephone conversation.
EAs	Emergency Authorities.
EACC	Emergency Authority Control Centre.
ERP	Effective Radiated Power.
ESM	Emergency Services Manager.
ETD	Extension Trunk Dialling Network.
ETSI	European Telecommunication Standards Institute.
Fire alarm call point	A device to operate the fire alarm system manually.

Fire alarm system	A fire alarm system comprising components for automatically detecting fire initiating an alarm of fire and taking other action as arranged. The system may also include manual call points.
Frequency	The number of cycles of wavelengths, which appear to pass a given point in a specified time, usually one second.
FFSK	Frequency Fast Shift Keying.
FRNT	Front Office Directory.
Geocode	Assignment of a specific grid reference to an incident, address or rendezvous point, etc.
GIS	Geographical Information Systems.
GPO	General Post Office.
GPS	Global Positioning System. Navigation system developed by the United States Defence Department as a worldwide navigation and position resource for both military and civilian use. It is based on a constellation of 24 satellites orbiting the earth at a height of over 20,000 km. These satellites provide accurate three-dimensional position and velocity as well as precise time, and act as reference points from which receivers on the ground triangulate their position.
GSM	Global Systems for Mobile communications. European standard for digital cellular networks operating at 900 MHz worldwide and supporting data transmission.
GTPS	Government Telephone Preference Scheme.
Handshake	A predefined exchange of signals or control characters between two devices that sets up the conditions for data transfer or transmission.
Hertz (Hz)	Measurement of frequency where one Hertz equals one cycle per second.
Hill-top Sites	Or **Main Stations** are normally on high, open ground (hence the alternative name) from which it is possible to 'see', in the radio context, a considerable portion of the brigade area. 'Main' equipment operates in an 'omni-directional' mode to cover the largest possible geographical area.
ICCS	Integrated Communications Control System.
ICTU	Information and Communications Technology Unit
ICU	Incident Control Unit.
Interface	A shared boundary, a physical point of demarcation, between two devices where the electrical signals, connectors, timing and 'handshaking' are

defined. The procedures, codes and protocols that enable two entities to interact for the meaningful exchange of information.

IS
Intrinsically Safe Equipment designed to be operated safely in an environment consisting of flammable or explosive dusts, gases or vapours.

ISDN
Integrated Services Digital Network. An internationally agreed public network offering switched end-to-end digital services for voice and data.

KiloStream
The registered trademark for BT's digital network services, used for connecting a variety of high-speed applications including computers, LAN interconnect and switchboards.

LAN
Local Area Network is one which spans a limited geographical area, usually within one building or site, and interconnects a variety of computers and terminals, usually at very high data rates.

Leaky feeder
A linear aerial which radiates radio signals throughout its length. Such an aerial is particularly suited to facilitating radio communications in sub-surface premises in conjunction with a UHF base station.

Link Transmitters
And **Link Receivers** provide communication between the control station and the main stations. 'Link' equipment operates in a "Point-To-Point" mode in which every effort is made to send signals only in the intended direction and only so far as necessary.

Main scheme radio
A radio system giving wide area radio coverage throughout the area covered by the mobilising control.

Main Control
'Control Station' or simply 'Control'. This is the place where the operators who control the scheme, and the main transmitting and receiving equipment of a scheme are located.

Main Transmitters
And **Main Receivers** send radio signals to, and receive radio signals from 'mobiles'.

'Mobiles'
Are the transmitter/receivers fitted in fire appliances and other vehicles.

MIS
Management Information Systems.

MMC
Monopolies and Mergers Commission.

Modem
Modulator/Demodulator. A device for converting analogue signals into digital signals and vice versa.

Multi-station scheme
A scheme served by several main stations, e.g., a large country scheme.

NOU
Network Operations Unit.

OAC
Operator Assistance Centre.

OFTEL	Office of Fair Trading for Telecommunications.
Out-stations	All radio stations in a scheme, including two-way fixed and mobile sets and fixed receivers but excluding main stations, main and sub-controls.
PABX	Private Automatic Branch Exchange.
PCNs	Personal Communications Networks.
PCS	Personal Communications Systems.
PDA	Pre-Determined Attendance.
PECS	Public Emergency Call Service.
PMR	Private Mobile Radio. A network developed for one particular organisation, usually an emergency service.
POCSAG	Post Office Code Standardisation Advisory Group.
Private wire circuit	A dedicated telephone circuit permanently connected between two or more points for transmission and reception of speech and/or data.
Protocol	A set of rules governing information flow in a communication system.
PSTN	Public Switched Telephone Network.
PTO	Public Telecommunications Operator.
Public Address	A loudspeaker system which may be operated by remote control from a central control room or locally for both operational and administrative purposes.
RA	Radiocommunications Agency
RIC	Receiver Idendity Code.
RIDDOR	Reporting of Injuries, Diseases and Dangerous Occurrences.
Roamer	The term used to describe a person who takes their mobile phone abroad with the specific purpose of making or receiving calls.
Roaming	The term used to describe the ability provided to a person and which allows them to take their mobile phone abroad and be able to make and receive calls in a country with which their own network operator has signed a roaming agreement.
Running call facility	A facility at a fire station which enables a running caller to give an alarm of fire.

Secondary Control	A mobilising control (possibly in another fire brigade area) to which, in an emergency, the functions of receiving emergency calls and mobilising appliances are passed, in the event of an evacuation of the normal mobilising control.
Simplex working	A communication technique in which it is not possible to transmit and receive simultaneously.
Single frequency scheme	A scheme using one common frequency for transmitting and receiving by all stations.
SMS	Short Message Service.
Switch	A switch is the core element of a radio or telephone system. It provides control, management and the routing of voice and/or data calls between radio system infrastructure, mobiles and portables, telephones, controllers and computer terminals.
Talk-through	A facility on two-frequency radio schemes which interconnects incoming and outgoing channels. Used to enable out-stations on a scheme to hear and talk to each other.
Telemetry	A means of establishing measurement remotely.
Terminal	A device for sending and/or receiving data on a communication channel.
TODS	Telephone Operator's Directory System.
TOPS	Total Operations Processing System.
Transportable Radio	A portable transmitter/receiver of roughly the same power as a mobile set.
TUES	Text User's Emergency Service.
Two-frequency operation	A means of operation whereby radios receive on one frequency and transmit on a different frequency (also known as double-frequency peration).
UHF base station	A radio installation which allows boosted signals of double frequency operation with UHF equipment. This equipment is usually provided as a mobile version but, exceptionally – for example, at major airports – there are authorised fixed installations.
WAN	Wide Area Network. Interconnects geographically remote sites.
WARC	World Administration Radio Conference.
Wavelength	The distance between successive crests, or successive troughs.

Control staff – Training, Competence and Promotion

Background

Fire Authorities have a legal duty to ensure personnel are adequately trained. The Fire Services Act 1947, Section 1 (1), states 'Every Fire Authority shall secure ... the efficient training of the members of the Fire Brigade'.

More recently the Integrated Personal Development System (IPDS) has been introduced to provide an overarching strategy for the development of all fire service staff at every stage of their career.

The IPDS recognises that learning and development should be timely, appropriate to roles as defined in the appropriate National Occupational Standards and should consolidate the knowledge, skills, attitudes and behaviours required to enable people to deal with a wide range of situations efficiently and safely.

There is no one best way of delivering development because:

- People learn in different ways and at different speeds

- Certain types of training can only be delivered in one way

- Some training can be delivered in a variety of ways

- There may be financial and resource constraints

It may therefore be necessary to provide different ways of developing people so that they are able to achieve competence. This will meet the needs of the individual and their circumstances.

Development can be broken down into four skill areas as first identified in Fire Service Circular 15/97 A Competence Framework for the Fire Service. The skill areas are:

- **Task Skills**
  The routine and largely technical components of a role

- **Task Management Skills**
  The skills needed to manage a group of tasks and prioritise between them

- **Contingency Management Skills**
  The skills required to recognise and deal with things that go wrong and deal with the unexpected

- **Role/Job Environmental Skills**
  Ensuring safety, interacting with people and the ability to cope with the environmental factors required in fulfilling the wider role

The combination of these four skill areas provides the framework for achieving competence within any given role.

Additionally, when learning and development opportunities have been made available to people, the outcome of the process must be measured. This should not only look at how well the development has 'worked' but should also assess the benefit, not only to the individual but to the organisation as well.

Learning and development is a continuous process. Environmental and resource constraints within Emergency Control Rooms often means that the majority of learning and development opportunities occur within 'the real work place'. However, central, regional and local training provisions should also be utilised where it is considered appropriate.

The role of the 'Fire Control Operator'

The competences required by Fire Control Operators are clearly defined within the National Occupational Standards and are supported by the Emergency Fire Service Modular Development Objectives. This latter consists of a CD–ROM database, which provides detailed information on knowledge, skills and understanding and the attitudes and behaviours required to underpin competence in the role. There are three phases to any form of personal development.

Phase 1 – Acquisition of knowledge and skills required for the role.

Phase 2 – Application of the knowledge and skills within the work place

Phase 3 – Maintenance of competence

These phases of development will be applied each time individuals undertake to progress to new roles within their careers, i.e., progression to supervisory management, change of department and so on.

The processes of continuous work place assessment and personal development review should be used throughout to ensure learning is taking place.

A key function of fire control is to provide advice and support to callers. This may sometimes require the operator to provide survival guidance to those who are in danger.

In respect to fire survival the following guidance is offered. **At all times the operator must:**

- Communicate with the caller clearly to obtain and provide relevant information and advice.

- Analyse and dynamically risk assess the information gained to determine the potential impact of the advice given on the caller's situation.

- Act in an assured and unhesitating manner, adopting a communication style to reflect the needs of the caller.

Fire survival guidance should only be given when it has been clearly established that intervention by the operator is required to enable survival. At all other times the standard fire safety advice **'GET OUT AND STAY OUT'** should be given.

Full guidance on 'Training in Emergency Call Handling Techniques and Fire Survival Guidance' is contained in the Fire Control Personnel Training package issued to complement Fire Service Circular 10/1993.

Promotion and advancement

Under the IPDS, advancement will be achieved through the demonstration of competence in the present role coupled with the demonstration of potential to advance. These two qualities will be identified through continuous work place assessment and participation in the appropriate assessment centres.

People selected for advancement may be given temporary promotion, but will not be confirmed in the role until role critical training and the demonstration of competence in the key safety and leadership aspects of the role have been achieved. People will need timely access to relevant training courses and development opportunities.

The Fire Service College (Learning and Development Opportunities)

The introduction of the IPDS as a role-based development structure for the fire service has necessitated a review of the approach to learning and development adopted by the Fire Service College (FSC).

Development opportunities are now clearly linked to the essential information required to satisfy the specific roles as defined in the National Occupational Standards. These roles occupy four levels:

- Control Room Operations
- Supervisory Management
- Middle Management
- Strategic Management

(Photograph: HM Fire Service Inspectorate)

Each role is described by a number of centrally designed development modules, which were derived from the National Occupational Standards. The FSC offers development programmes for each level of management based on the appropriate combination of modules, which can be accessed singly or in clusters. This modular approach ensures that the FSC can provide development activities that meets both the needs of individuals and their organisations.

When applied in the context of Control Staff Development, the following example can be applied.

Supervisory Management
Provides development up to Watch Manager level and encompasses the role of Control Supervisor.

Middle Management
Provides development up to Group Manager level and encompasses the role of Control Manager.

Strategic Management
Provides development up to Brigade Manager level.

It is recognised that the roles within Fire Controls vary significantly from brigade to brigade. Control staff could, therefore, access any or all of the development opportunities according to their organisational role and their individual development needs.

National Vocational Qualifications

The Fire Service National Occupational Standards have been accredited up to Watch Management level as National Vocational Qualifications (Level 3). These are the only national qualifications specifically designed for fire control personnel. Brigades not wishing to pursue accredited vocational qualifications must still use the development framework provided by the National Occupational Standards, linked to an objective assessment process, to consistently measure the performance of their people.

Further guidance on the Competence Framework and The Integrated Personal Development System can be found in:

FSC 10/97 (revised)

FSC 9/2002

List of relevant DCOLs/DFMs (in Scotland) and FSCs

DCOL		DFM
4/1988	=	5/1998
6/1992	=	4/1992
4/1995	=	4/1995
6/1995	=	5/1995
8/1995	=	6/1995
6/1996	=	6/1996
9/1996	=	8/1996
8/1997	=	8/1997
1/1998	=	2/1998

FSC3/1975

FSC15/1997

Acknowledgements

HM Fire Service Inspectorate is indebted to all who helped with the provision of information and expertise to assist the revision of this volume, in particular:

Bedfordshire and Luton Fire & Rescue Service
Buckinghamshire Fire & Rescue Service
Cheshire Fire Brigade
Cornwall County Fire Brigade
Devon Fire and Rescue Service
GEC Marconi
Hertfordshire Fire & Rescue Service
Kent Fire Brigade
London Fire Brigade
Surrey Fire & Rescue Service
West Sussex Fire Brigade
Information and Communications Technology Unit – Home Office
ASAP
BT
BT Tallis Consultancy
Cable and Wireless
Cellnett
Fortek
Kingston Communications
Marconi
Multitone
One 2 One
Orange
Simoco
Racal BRT
C.S. Todd & Associates
Typetalk
Vodafone

UK POLITICS

To Complete Your Study for Component 1 you will need to study the core political ideas in addition to the chapters in this book.

1 Democracy and participation

The concept of democracy is fundamental to an understanding of politics as it underpins all the concepts, ideas and topics you will be studying, yet it is a concept that sparks fierce passions and debates and conflicting attitudes. Take, for example, the debates over Brexit. On the one hand you had people arguing that the democratic 'will of the people' needed to be respected, as the result of the 2016 referendum indicated the people wanted to leave the EU. On the other hand, you had people and groups arguing that it was up to Parliament to decide and that MPs should vote in the 'national interest' and choose to remain in the EU. Western democracies like the UK boast of their democratic institutions and accuse totalitarian regimes like North Korea of being undemocratic dictatorships, yet North Korea, like many Communist states, calls itself a 'Democratic People's Republic'. How can there be such disagreement over a term that is so central to politics?

Democracy is simply a term that means 'rule by the people'. It comes from the Ancient Greek *demos* (the people) and *kratia* (rule or power), but who the people are and how the will of the people should be translated into action is a matter of fierce debate. In fact, Plato, one the great Greek philosophers, saw democracy as undesirable and worried that mob rule by the uneducated masses would be damaging and could lead to anarchy and chaos, a view that persists even today in many debates about democracy.

Democracy is just a word, but how that word is interpreted and how it is applied to modern politics determines much of the decision-making and political systems that affect all our lives. Democracy today is largely seen as a 'good thing', but people still debate its meaning and how it should be applied, as evidenced by the heated debates over Brexit. At its heart, democracy is about the process and means of translating the will of the people into coherent plans and action, but how this is brought about and how it operates in practice is fluid and ever-changing, which is why it is important to start by getting to grips with what democracy means in practice.

Anti-Brexit protester Steve Bray (left) and a pro-Brexit protester argue as they demonstrate outside the Houses of Parliament

Current systems of democracy

Direct democracy

We normally divide the concept of democracy into two main types. These are **direct democracy** and **representative democracy**. Direct democracy was how the idea was first conceived in ancient Greece, mainly in the city state of Athens in the fifth century BC. Hence it is sometimes described as 'Athenian democracy'.

What made it a democracy was the idea that every tax-paying citizen would have one vote of equal value to all others and all citizens were able to contribute to a decision. Thus, the assembled free citizens would make important decisions directly, fairly and equally, such as whether the state should go to war or whether a prominent citizen who had committed anti-state acts should be exiled. After Athenian democracy declined in the fourth century BC, direct democracy, with a few exceptions, disappeared as a democratic form until the nineteenth century.

Today, direct democracy has returned in the form of the referendum, now relatively common in Europe and some states of the USA (referendums and their use will be covered in more detail in Chapter 3). However, direct democracy today should be seen as an addition to representative democracy rather than a separate system, one that can add great **legitimacy** to the decisions made by politicians. Some decisions are considered so vital, and also so unsuitable for representatives to make them, that they are left to the people. However, the size and nature of modern politics would make the regular use of direct democracy impracticable and so it cannot be considered as an alternative to representative democracy in the twenty-first century.

Direct democracy has its critics as well as its supporters. Table 1.1 summarises the main advantages and disadvantages of direct democracy.

Key terms

Direct democracy All individuals express their opinions themselves and not through representatives acting on their behalf. This type of democracy emerged in Athens in classical times and direct democracy can be seen today in referendums.

Representative democracy A more modern form of democracy, through which an individual selects a person (and/or a political party) to act on their behalf to exercise political choice.

Legitimacy The rightful use of power in accordance with pre-set criteria or widely held agreements, such as a government's right to rule following an election or a monarch's succession based on the agreed rules.

Table 1.1 Direct democracy — is it desirable?

Advantages	Disadvantages
It is the purest form of democracy. The people's voice is clearly heard	It can lead to the 'tyranny of the majority', whereby the winning majority simply ignores the interests of the minority and imposes something detrimental on them
It can avoid delay and deadlock within the political system	The people may be too easily swayed by short-term, emotional appeals by charismatic individuals
The fact that the people are making a decision gives it great legitimacy	Some issues may be too complex for the ordinary citizen to understand

Representative democracy

Representative democracy is the most common model found in the democratic world today.

The basis of this type of democracy is that the people do not make political decisions directly; instead, they choose representatives to make decisions on their behalf. The most common way of choosing representatives is to elect them through a formal, competitive election process. Indeed, if representatives are not elected in a vote with some degree of choice, it calls democracy into question. Elections are therefore what we first think of when we consider representation.

In addition to choosing representatives, representative democracy ensures that those elected to positions of power and responsibility have to be held to account by the people. **Accountability** is essential if representatives are to act responsibly and in the interests of the people. At election time both individual representatives, such as MPs in the UK, and the government as a whole are held accountable when the people go to the **polls**. During the election campaign, opposition parties will highlight the shortcomings of the government and will offer their own alternatives. At the same time, the government will seek to explain and justify what it has done in an effort to be re-elected. Similarly, individual representatives will be held to account for their performance: how well they have represented their **constituents** and whether their voting record in the legislature meets the approval of those same constituents. Of course, MPs are often faced with a dilemma of how best to represent their constituents, either by voting according to their conscience or to vote for their constituents wishes, or as the 18th century Conservative thinker Edmund Burke wrote, 'your representative owes you, not his industry only, but his judgment; and he betrays instead of serving you if he sacrifices it to your opinion'. We shall explore this dilemma in greater detail later in this chapter.

Useful terms

Accountability Where those who have been elected in a representative democracy must be made responsible for their policies, actions, decisions and general conduct. Without such accountability, representation becomes largely meaningless.

Polls Another term for elections; polls simply establish the number of people who support a particular person, party or issue.

Constituents The ordinary voters who elect a particular representative, usually based on residence in a particular geographical area.

Synoptic link

The nature of voting behaviour and the role of opinion polls will be considered further in Chapter 4.

Voters queuing to vote at a polling station

Accountability is less certain between elections, but those in power can be held to account regularly through investigations, media scrutiny and individual representatives asking questions on behalf of their constituents. However, the individual representatives are normally safe until the next election.

Having said that representatives in a democracy are elected and are accountable, we need to explain the concept of representation in general. It can have different forms and meanings.

Different types of representation

When people consider representation, they usually think of someone who will express the concerns and needs of the local community, acting as a sort of 'spokesperson' or champion for the local area that elected them. Representation can take different forms, however, which can have implications for interpreting and evaluating the strength of representative democracy in the UK.

Social representation

Social representation implies that the characteristics of members of representative bodies — whether national parliaments, regional assemblies or local councils — should be broadly in line with the characteristics of the population as a whole. In other words, they should be close to a *microcosm* of society as a whole and 'look like' that society. For example, just over half should be women, a representative proportion should be drawn from ethnic or religious minorities, and there should be a good range of ages and class backgrounds in representative bodies. Of course, this is difficult to achieve and the UK Parliament certainly falls short. This is explored further below when we discuss the state of representative democracy in the UK specifically, as well as in Chapter 6.

Study tip

Democracy underpins everything in UK politics, so it is vital that you make sure you are comfortable with the functions and different types of democracy as these will help you evaluate all other aspects of the course.

Synoptic link

The main representative body in the UK is Parliament, so understanding how different types of representation impact on Parliament will help explain how democratic and effective Parliament is at carrying out its representative role, as outlined in Chapter 6.

Representing the national interest

Though representatives may be elected locally or regionally, if they sit in the national Parliament they are expected to represent the interests of the nation as a whole and do what they believe is right, rather than what the people may want. Sometimes this may clash with the local constituency they represent, so they have to resolve the issue in their own way. For example, an MP representing a constituency near a major airport may be under pressure to oppose further expansion on the grounds of noise and pollution, but they may see it as in the **national interest** to expand that airport.

Activity

Carry out some research into how your local MP (or their predecessor) voted on the issues of HS2, Heathrow expansion and Brexit. On what basis do you feel they cast their vote (national interest, constituency interest or party interest)?

Constituency representation

The locality that elects a representative in UK national politics is known as a constituency. The idea is that a geographical area will have similar social and economic concerns that a representative will speak about in the elected body. The main focus, therefore, is on local issues. Such representation can imply three things:

1 It can mean representing the interests of the constituency *as a whole*, such as funding for local services, or whether a new railway or airport should be built in the area.
2 It can also mean representing the interests of *individual* constituents. This is often described as the **redress of grievances**. In this case, a representative will champion a constituent who feels they have been treated unfairly by the tax office or local hospital for example, or who needs help with an overseas issue.
3 Finally, it can simply mean that a representative listens to the views of their constituents when deciding about a national issue. This can lead to another dilemma. What happens if the elected representative does not personally agree with the majority of the constituents? This becomes a matter of conscience that has to be resolved by the individual concerned.

Activity

Find out which parliamentary constituency you live in. Access the website of the local MP. What local issues are currently prominent in your area? What type of representation do you feel your local MP is delivering?

Party representation

All modern democracies are characterised by the existence of political parties. Furthermore, the vast majority of those seeking and winning election are members of a political party. It is unusual in modern democracies to find many examples of *independent* representatives who do not belong to a party. Parties have stated policies. At election time these are contained in a list of party promises called a manifesto. It follows that members of a party who are seeking to be elected will campaign on the basis of the party's manifesto. This means that they are representing their party and the voters understand this.

Synoptic link

Manifestos are a list of policy promises made by political parties at election time to persuade voters to vote for them. This will be covered in greater detail in Chapter 4.

Occupational or social representation

Some elected representatives will represent not only their constituency or region, but also a particular occupational or social group. For example, those who support and are supported by trade unions will often pursue the cause of groups of workers; others may represent professions such as doctors or teachers. This function can also apply to social groups such as the elderly, those with disabilities, members of the LGBT+ community or low-income groups.

Causal representation

Where representative bodies are not representing people so much as ideas, principles and causes, this is called causal representation. In a sense this represents the *whole* community, in that the beliefs and demands involved are claimed to benefit everyone, not just a particular group in society. Typical causes concern environmental protection, individual rights and freedoms, greater equality and animal rights. Though elected representatives often support such causes and principles, most causal representation is carried out by pressure groups.

As we can see from the points above, the type of representation being followed by an MP is often down to a combination of factors and may depend on the nature of the issue being presented. In order to evaluate the nature of representation in the UK, you need to consider the advantages and disadvantages of representative democracy as outlined in Table 1.2.

Table 1.2 The advantages and disadvantages of representative democracy

Advantages	Disadvantages
Representatives can develop expertise to deal with matters the public does not have the time or knowledge to deal with	Representatives may not act in the best interests of their constituents
Representatives can be held to account for their actions at election time	It can be difficult to hold a representative to account between elections
Representatives have the time to deal with a variety of complex matters, leaving the public free to get on with their own lives	Allowing voters to delegate responsibility to representatives can lead to the public disengaging from social issues and other responsibilities
In a large modern country, it is the only practical way to translate public opinion into political action	Representative bodies can be unrepresentative and may ignore the concerns and needs of minorities

Debate

Is direct democracy a better form of democracy?

Advantages of direct democracy

- It is the purest form of democracy. It is the voice of the people.
- Decisions made directly by the people have more authority.
- Decisions made by the people are more difficult for future governments to change or cancel.
- Direct democracy can help educate the people about political issues.

Advantages of representative democracy

- Elected representatives may have better judgement than the mass of the people.
- Elected representatives may be more rational and not swayed by emotion.
- Representatives can protect the interests of minorities.
- Elected representatives may be better informed than the general public.

Look over the points for both sides of the debate and consider which side of the debate you believe to be stronger by comparing the relative advantages of the two forms of democracy and deciding why one form, overall, would be better.

The nature of representative democracy in the UK

Having explored the concept of representative democracy, we can now consider how representative democracy operates in the UK and evaluate how effective it is.

The whole administration of representative democracy is regulated by the Electoral Commission. This body ensures that representation is fair, that all those entitled to vote can register to vote and that the political parties do not have any undue influence through spending. It can be said that representation in the UK today is broadly uncorrupted, fair and honest, at least when compared with the past. However, there have been some notable exceptions to this, with some peers and MPs breaking the rules and acting dishonestly. These ideas will be explored in greater detail in Chapters 2 and 4.

Levels of representation in the UK

First, we can see that the people are represented at different *levels* of government. Table 1.3 demonstrates how this works in the UK.

Table 1.3 Levels of representation

Level	Jurisdiction
Parish or town councils*	The lowest level of government. They deal with local issues such as parks and gardens, parking restrictions, public amenities and small planning issues
Local councils	These may be county councils, district councils or metropolitan councils, depending on the area. They deal with local services such as education, public transport, roads, social services and public health
Combined authorities	Where groups of two or more local councils in England join together to share resources and increased powers devolved to them from central government. These may be presided over by an elected mayor, such as in Greater Manchester, or not have a mayor, such as the combined autority in West Yorkshire.
Metropolitan authorities*	This is big city government, such as in London. These bodies deal with strategic city issues such as policing, public transport, arts funding, environment, large planning issues and emergency services. They normally have an elected mayor and strategic authority
Devolved government	The governments of Wales, Scotland and Northern Ireland. They have varying powers, but all deal with health, social services, education, policing and transport. All three have elected representative bodies (an assembly in Northern Ireland, Parliaments in Scotland and Wales)
National government	This is the jurisdiction of the UK Parliament at Westminster and the UK government

* In England and Wales only

So we can see that all citizens of the UK are represented at three levels at least and that many enjoy four or five levels of representation. It is also clear that representation has become increasingly **decentralised** with the advent of devolution, and the delegating of increasing powers to city administrations.

Forms of representation in the UK

Having established at what levels of government we are represented, we can now examine what forms of representation flourish in the UK.

Constituencies

It is a cornerstone and an acknowledged strength of representative democracy in the UK that every elected representative should have a constituency to which they are accountable and whose interests they should pursue. These constituencies may be quite small, such as a parish or a local ward, or they may be very large, like those for the Northern Ireland Parliament or the Greater Manchester area (see Figure 1.1), but the same principle applies to all. This principle is that individuals in the constituency should have their grievances considered, that the interests of the *whole* constituency should be given a hearing in a representative assembly, and that the elected representative is regularly made accountable to their constituency. The levels of constituency in the UK, from smallest to largest, are shown in Table 1.4.

Figure 1.1 The parliamentary constituencies in the Greater Manchester area as of the 2019 general election

Table 1.4 Levels of constituency in the UK

Level	Representatives
Ward or parish	Parish and local councillors
Parliamentary constituency	MPs
City region	Assembly members
Metropolitan authority	Elected mayors
Devolved assembly constituency	Members of the Scottish Parliament (MSPs), Members of the Senedd in Wales and Members of the Legislative Assembly (MSLs) in Northern Ireland

Knowledge check

Who represents you? Who is your local MP? What other bodies represent your household?

Parties

The UK is unusual in that political parties play a much more central role in representation than in some other democracies. This is for two reasons:

● First, political parties have evolved out of ideological principles (usually expressed in their manifestos) and are therefore united by a set of core beliefs and principles at the heart of the party, such as conservatism for the Conservative Party, socialism for the Labour Party and liberalism for the Liberal Democrats. This means that, at their heart, members of UK parties have a shared ideology and set of beliefs, whereas in some other countries, such as the USA, parties arose in reaction to particular events or conflicts, so they are looser confederations with a shared label but large differences in principles.

- Second, it is usually the case that one single party governs in the UK, which is rare compared with many of the democracies across Europe. There have been exceptions: between 2010 and 2015, when a coalition ruled; and following 2017, when the Conservatives formed a minority government with support from (but not coalition with) the Democratic Unionist Party (DUP), a small party of 10 MPs that represents the unionist side of the political debate in Northern Ireland. However, the *norm* is for single-party government. Since the 80-seat majority secured by the Conservatives in December 2019, the UK has returned to its more 'normal' position of single-party government.

Government representation

The people as a whole are also represented by the elected government. As we shall see again below, it is a mark of a true democracy that the winning party or parties should govern on behalf of the *whole* community and not just those sections of society that typically support it. While it is true that there will be a *tendency* to support some groups more than others, this does not alter the fact that the elected government represents the whole nation.

Pressure groups

Pressure groups in the UK (and indeed in other democracies) are representative bodies in two main ways:

1 Some groups will have a formal membership and will represent their 'section' of society by promoting policies that will benefit them. This applies to sectional pressure groups such as the British Medical Association (BMA) and the National Farmers' Union (NFU).
2 Other groups are engaged in causal representation. Here they represent a set of beliefs, principles or demands that they believe will benefit the whole community, such as Friends of the Earth (environmental causes) and Liberty (human rights campaigning).

All pressure groups represent us in various ways. Whatever we believe, whatever we do and whatever our occupation, there is probably a pressure group working in our interests. It is all part of a **pluralist democracy** and a healthy **civil society**. The role of pressure groups in the UK is explored more fully later in this chapter.

How democratic is the UK?

If we are to attempt an assessment of democracy in the UK, we need to establish what we mean by the term 'democracy'. More precisely, we should ask two questions:

1 What constitutes a democratic *political system*? A word of caution is needed before this assessment. Democracy is a contested term. There is no single, perfect definition. Therefore, the elements described below add up to a guide, a collection of the most commonly accepted features of a democracy by western, liberal standards.
2 What constitutes a democratic *society*? This is a broader question and is explored below.

The UK's system of liberal democracy

When we talk about democracy in the UK, we are often referring to the concept of a 'liberal' democracy. This goes back to the seventeenth century and thinkers like John Locke, who believed that governments ruled by the consent of the governed and that a social contract existed between the people and those in power. This was a radical idea for the time as it rejected the idea of absolute monarchy and the divine right of kings, which suggested leaders only answered to God. Instead, leaders should answer to the people. In addition to this, to help to ensure the people would be free to live their lives and to prevent the government from becoming too powerful, a series of limitations should be introduced to restrict the power of the government in order to create a free society. This liberal form of democracy provides the key features of the UK's democratic system today, given below (summarised in Figure 1.2).

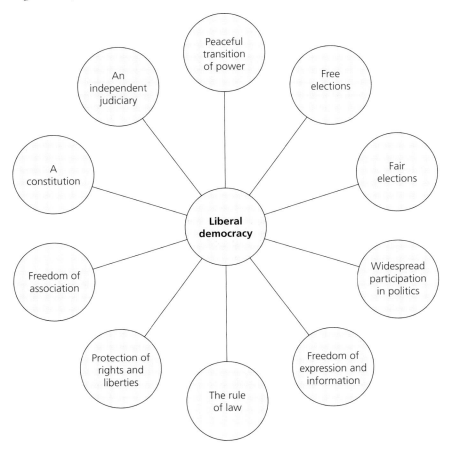

Figure 1.2 **The features of a liberal democracy**

The peaceful transition of power

This is a feature that is often taken for granted in democracies, but it is not guaranteed in many societies. It means that those who lose power by democratic means accept the authority of those who have won. If they do not, politics breaks down and non-peaceful conflict is likely to ensue. This helps to ensure that democracy can hold governments to account and ensures the legitimacy of those who have won an election.

Free elections

Elections are a cornerstone of democracy. Without them it is impossible to imagine democracy working in any meaningful way. Indeed, it is probably the first thing we look for when assessing whether or not a system is democratic. The description 'free' means that all adults (however that is defined) are free to vote and to stand for office. This is described as '**universal suffrage**'. If significant groups are excluded then elections are not truly free and democracy is flawed. Elections also need to be free to ensure that everyone can exercise their right to vote without fear, threats or intimidation. One way of achieving this is through the **secret ballot**, while rights to vote must be strictly enforced by the courts in order to ensure people are not unfairly denied the right to vote. If a secret ballot and a strict adherence to these rights is not in place, votes can be bought and sold and voters can be coerced into voting a certain way, or not voting at all.

Fair elections

This is a more difficult criterion. In the strictest sense, this means that everyone has one vote and all votes are of equal value. It also suggests that there are safeguards in place to avoid electoral fraud and **ballot rigging**. However, what fairness means is open to some debate; what may appear fair to some will appear unfair to others. A candidate who wins the most votes can be said to have won the election fairly, but if they gained only 25 per cent of the total votes cast, then it could be seen as unfair as 75 per cent of voters did not choose that candidate. Such debates surround the UK's various electoral systems and whether or not they are fair systems. These different interpretations are explored more fully in Chapter 3, but it is worth remembering that they underpin the very concept of democracy in the UK and why there are so many debates over how democracy works in the UK.

Widespread participation in politics

It is important for the health of a democracy that a large proportion of the population participates in politics. A well-informed and active population can prevent government becoming too dictatorial, and without the people participating in the political process there is a breakdown in communication between the government and the governed. This is why the issue of political participation is so important and will be explored later in this chapter.

Freedom of expression and information

One of the fundamental features of a democracy is the right of the people to express their opinions and criticise the government. This is known as a civil liberty and means that people cannot be arrested or persecuted for expressing negative opinions of those in power, their policies, or their competence. There should also be free access to public information to enable the people to check the government and consider how well it is governing. Few governments enjoy being criticised or scrutinised, but this is what marks out a democracy compared with a dictatorship, where public discussion and evaluation of the government are banned or restricted. This requirement implies a free media and no government censorship or interference. The development of the internet has helped as it allows free access for all, though whether or not the information provided is accurate leads to questions about its validity. This issue has become more stark in recent years with the rise of fake news and growing popularity of conspiracy theories, which have made it harder for people to take publicly expressed views as being based on fact or truth.

Freedom of association

Linked to freedom of expression is freedom of association. In terms of politics, this means the freedom to form parties or pressure groups, provided their aims and methods are legal. Parties and pressure groups are such vital vehicles for representation that if they did not exist, or were suppressed, democracy would be almost impossible to sustain.

Protection of rights and liberties

Linked to freedom of expression and association is the idea that the rights and liberties of citizens should be firmly safeguarded. This implies that there should be some kind of enforceable 'Bill of Rights' or 'Basic Laws' to protect rights and liberties in such a way that the state cannot erode them. The European Convention on Human Rights (ECHR) is just such an example, as is the US Bill of Rights, the first ten amendments of America's Constitution. In the UK, the Human Rights Act performs this role, while the Equalities and Human Rights Commission operates in England and Wales to promote and protect human rights.

The rule of law

The rule of law is the basic principle that all citizens should be treated equally under the law and that the government itself should be subject to the same laws as its citizens. It is linked to the concept of limited government and ensures that no one, even those in power, can break the law and if they do, they will be held to account on the same basis as anyone else.

Independent judiciary

The existence of the rule of law implies one other feature: an independent judiciary. It is a key role of the judiciary in a democracy to ensure that the rule of law is upheld. For this to happen, the members of the judiciary (the judges) must be independent from government and the whole process of politics. In this way they will ensure that all individuals and groups in society are treated equally under the law and that the government does not exceed its authority. It also means, of course, that the rights and liberties of citizens are more likely to be upheld.

A constitution

Democracy is at risk if there are not firm limits to the power of government. Without these, there is a possibility that government will set aside democratic principles for its own purposes. We expect this to happen sometimes in times of warfare and emergency, but not normally. The usual way to set the limits of government power is to define them in a constitution that will be enforced by the forces of law. This is known as constitutionalism and all democracies have a constitution.

How democratic is the UK political system?

Having established the features that make the UK a democracy, we are now able to assess the extent to which the UK political system is democratic and then to consider how it might be reformed.

Table 1.5 shows a 'balance sheet' considering whether the UK has a healthy democratic political system. However, there remain a few serious flaws. Collectively, these are described as a '**democratic deficit**'. The main examples of the UK's democratic deficit can be summarised thus:

- The first-past-the-post (FPTP) electoral system for general elections produces disproportional results, renders many votes wasted and elects governments with a relatively small proportion of the popular vote. It discriminates against small parties with dispersed support.

Study tip

Be careful not to confuse the European Court of Human Rights (ECHR), which is *not* an EU institution, with the European Court of Justice, which is and which enforces or interprets EU law.

Synoptic link

Constitutions, discussed in Chapter 5, are inherently bound up with democracy. Any democratic reforms would also be constitutional reforms and most constitutional reforms will have an impact, for better or worse, on how democracy operates in the UK.

Study tip

There is no right or wrong answer to the question of how democratic the UK political system is, but you will need to look at the arguments and consider what your judgement might be and why, as this is what you will need to explain and do in producing exam-style answers.

Key term

Democratic deficit A flaw in the democratic process where decisions are taken by people who lack legitimacy, due to not having been appointed with sufficient democratic input or not being subject to accountability.

- The House of Lords has considerable influence but is an unelected body.
- The sovereignty of Parliament, in theory, gives unlimited potential power to the government.
- The powers of the Prime Minister are partly based on the authority of the unelected monarch.
- The European Convention on Human Rights is not binding on Parliament, so individual rights and liberties remain under threat.

Table 1.5 How democratic is the UK political system?

Democratic feature	Positives	Negatives
Peaceful transition of power	The UK is remarkably conflict-free	Short-lived disputes have arisen when the results were not clear, in 2010 and in 2017, leading to some claims of a lack of legitimacy
Free elections	Nearly everyone over 18 can vote There is little electoral fraud and there exist strong legal safeguards	Some groups, such as prisoners and effectively the homeless, are denied their right to vote The House of Lords is not elected at all, nor is the head of state (monarch)
Fair elections	There are proportional systems in place in Scotland, Wales and Northern Ireland and other devolved and local bodies	The first-past-the-post system for general elections leads to disproportionate results and many wasted votes Governments are often elected on a modest proportion of the popular vote
Widespread participation	There is extensive membership of pressure groups, which are free and active. There is also a growing level of participation in e-democracy	Since 2001 voter turnout in general elections has been, on average, lower than in previous elections, while party membership, especially among the young, has generally been in decline. Despite some increases in party membership after 2015, it is still below levels experienced in the 1950s
Freedom of expression	The press and broadcast media are free of government interference. Broadcast media maintains political neutrality. There is free access to the internet	Much ownership of the press is in the hands of a few large, powerful companies such as News International, the owners of which tend to have their own political preferences Some information available on the internet is false and detrimental
Freedom of association	There are no restrictions on legal organisations People may organise and instigate public protests	The government has the power to ban some associations because they are seen as based on terrorism or racial hatred Public meetings and demonstrations can be restricted on the grounds of 'public safety'
Protection of rights and liberties	Strong in the UK. The country is signed up to the ECHR and the courts enforce it. The House of Lords protects rights, as does the judiciary	Parliament is sovereign, which means rights are at the mercy of a government with a strong majority in the House of Commons. The ECHR is not binding on the UK Parliament
The rule of law	Upheld strictly by the judiciary. The right to judicial review underpins this. The judiciary is independent and non-political	The monarch is exempt from legal restrictions There is statistical evidence to suggest that those of higher social and economic standing are treated more leniently than those at the lower end
A constitution	Parliament and the courts ensure the government acts within the law The Human Rights Act (see below in this chapter and in Chapter 8) acts as a restraint on the actions of the government, and constitutional checks exist to limit the power of the government	There is no codified UK Constitution so the limits to government power are vague. Parliamentary sovereignty means the government's powers could be increased without a constitutional safeguard. The prerogative powers of the Prime Minister are extensive and arbitrary

Is the UK in need of democratic reform?

Generally, the UK system of democracy is working, but there are arguments that it could be made to work better and that some of the traditional elements should be updated to reflect a more modern, diverse society. Individual reforms relating to Parliament, the judiciary, devolution, elections and parties will be considered in detail in the relevant chapters, but reforms in all these areas will have an impact on democracy, so Table 1.6 provides a summary of some of the potential reforms that could be considered. These are explored in more detail below.

Table 1.6 Potential democratic reforms for the UK

Potential reform	Advantages	Disadvantages
Replace the House of Lords with an elected chamber	It would remove an unelected, unaccountable body from the UK's democratic process	What replaced it might cause greater rivalry with the House of Commons, leading to gridlock in the political process The expertise in the Lords could potentially be replaced by career politicians
Replace the FPTP electoral system with a more proportional one	It would remove the negative features of FPTP, such as safe seats, minority constituencies, unfair representation and governments with a minority of support	Proportional systems make coalitions more likely and harder to hold to account. The systems are more complex and risk losing the close MP–constituency link that currently exists
Codify the UK Constitution	It would clarify the processes of the UK political system and provide a higher law that would be entrenched, rather than the flexibility of the current uncodified constitution	A codified constitution might prove too rigid and there are questions about who would write it and how it would be implemented. It would raise questions over the location of sovereignty It would give more power to unelected judges
Create a devolved English Parliament to equalise devolution	It would solve the West Lothian question (where MPs from devolved areas can vote in measures that no longer affect their constituents, covered more fully in Chapter 5) and create a more equal level of representation across the UK (see Chapter 5)	England is too large a single entity to work within a devolved system, but regional devolution has been rejected by voters
Introduce state party funding	It would allow politicians to focus on their main job rather than fundraising It would, potentially, remove the need to acquire money from powerful groups and vested interests that donate for their own ends, not the national interest	The process of fundraising helps to keep politicians and parties connected to voters Questions would be raised over how funding would be allocated and whether taxpayer money should be given to parties that some may find objectionable
Introduce compulsory voting	It would increase turnout in all elections, helping improve the legitimacy of elected officials	Forcing people to vote may not improve public engagement in politics The right to vote also includes the right *not* to vote
Replace the monarch with an elected head of state (president)	It would remove an unelected figurehead and replace them with an elected and accountable figure	The monarchy is popular and, being neutral, can act as a unifying figure in a way an elected politician cannot

Study tip

Although each reform may improve democracy in some way, it would also raise other questions and issues that might make the reform less desirable. Make sure you are able to explain whether the benefits of reform outweigh the negatives.

Knowledge check

Identify four things that make the UK a representative democracy.

Discussion point

Evaluate the view that the UK is in need of democratic reform.

Three questions you may wish to consider are:
1. Although democratic reform might be desirable, is it essential?
2. Is there a significant demand for such reforms?
3. Would these reforms potentially create more problems than they solve?

Knowledge check

Identify three problems or issues with the UK's system of representative democracy.

Political participation in the UK

The term 'participation' covers a variety of forms of political activity. Most citizens participate in politics in one way or another. However, there are two variables involved:

- What kind of participation?
- How intensive is that participation?

The first question can be answered by detailing the various ways in which it is possible to participate in political processes. The second can be answered by placing these forms of participation into some kind of order that expresses the degree to which they require intense activity. They are described below in order of intensity.

1 **Standing for public office** This is the most intensive. Many local councillors are part-time, but they do have to give up a great deal of their lives to attending meetings, campaigning, meeting constituents, reading information and making decisions. It goes without saying that full-time politicians have to immerse themselves in the job. Even those who stand for office unsuccessfully have to devote a considerable amount of time to the effort of trying to get elected.

2 **Active party membership** Many people join political parties, but only a minority of these are active members, also called 'activists'. Activists are fully engaged with the party they support. This may mean attending local meetings of the party, voting for officers, campaigning in the community and canvassing at election time to try to ensure as many party supporters vote as possible.

3 **Active pressure group membership** Like party activists, these pressure group supporters may be full members, helping to raise both money and awareness of the cause they support. Often this means attending or even organising demonstrations and other forms of direct action.

4 **Passive party or group membership** This means being enough of a supporter to join the party or pressure group, but taking relatively little active part. Such members often confine their activities to helping at election times or maybe signing a petition.

5 **Digital activists** Since the growth of social media and the internet, this has become a common form of participation. It requires only that the individual takes part in campaigns and movements that happen online. In other words, participation is possible without leaving one's home. It normally involves such activities as signing e-petitions, joining social media campaigns, expressing support for a cause on social media, etc.

6 **Voting** Voting is the most fundamental and yet the least taxing form of political participation. It has become especially convenient with the growing use of postal voting. Even with local, regional and national elections, plus referendums, most citizens have to vote only once a year at most.

We have seen above that high levels of participation in political processes are essential to a healthy democracy. If citizens are passive and do not concern themselves with politics, the system becomes open to the abuse of power. In other words, popular political participation helps to call decision-makers to account and to ensure that they carry out their representative functions.

The ways in which people participate are changing. Some also claim that political participation is in decline, especially among the young. Both these phenomena are examined below. The way in which participation has changed has consequences for how democracy operates in the UK, while a decline in participation can undermine the very practice of democracy itself.

Changing forms and levels of participation

The 2001 general election saw a turnout of only 59.4 per cent, a historic low, 12 per cent below what it had been in 1997 and 18 per cent below that of 1992. Coupled with declining membership of political parties, this led to a concern that Britain was experiencing a '**participation crisis**'. This may have reflected a situation where New Labour was so dominant that there was little real competition, or it may have been an early indication that methods of participation were changing. Nevertheless, as widespread participation is so integral to the functioning of a healthy democracy, any sense that there might be a crisis could lead to a democratic deficit where the legitimacy of those in power and the ability of the public to hold them to account are seriously undermined, leading to accusations of an **elective dictatorship**. As such, issues with participation need to be carefully considered.

Political parties

In the 1940s and 1950s membership of all political parties rose to over 3 million, mostly Conservatives. If one were to add trade union members affiliated to the Labour Party, this figure would be several million higher. Since then there has been a steady decline. Of course, those high figures did not mean that the mass memberships were politically very *active*, but they gave an indication of mass engagement with politics at some level. Figure 1.3 demonstrates this decline.

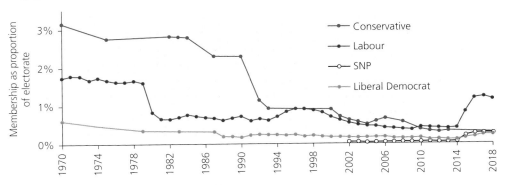

Figure 1.3 The decline in party membership

Source: House of Commons Library briefing SN05125

Party	Membership size	As a percentage of the electorate*
Conservative	180,000	0.38%
Labour	485,000	1.03%
Liberal Democrats	115,000	0.24%
SNP	125,000	0.27% (3.00% of the Scottish electorate)
Green	49,000	0.10%
Plaid Cymru	11,500**	0.02% (0.50% of the Welsh electorate)
UKIP***	29,000	0.06%
Total	994,500	2.11%

Figure 1.4 Party memberships in 2019

*Based on a December 2019 registered electorate size of 47,074,800 according to the Office for National Statistics

** Figures for Plaid Cymru as based on 2018 data, the most recent available

*** In 2020 former UKIP leader Nigel Farage created The Brexit Party and many UKIP members transferred their support to this new party, which had a membership of 85,000 by June 2020. UKIP membership declined as a result but it still exists as a party, though much smaller

Evaluate the extent to which the UK is suffering from a 'participation crisis'.

Three points you may wish to consider are:

1 What is the meaning of the word 'crisis'?
2 Is what is described in this section a crisis or an issue that would be nice to reform?
3 Would any potential reforms resolve all elements of the 'crisis'?

It is clear that parties are no longer the main vehicle by which most people wish to participate in politics. There are, though, exceptions:

- There was a surge in Labour Party membership in 2015 when, under new rules established by the then leader, Ed Miliband, it was possible to join the party for just £3 (normal subscriptions to a party are much higher). This was to enable a wider section of Labour supporters to vote in leadership elections.
- Following the 2014 referendum on Scottish independence, membership of the Scottish National Party (SNP) surged, and it claimed to have over 100,000 members in a population of only just over 5 million.
- There was an increase in membership of UKIP in the run-up to the 2015 general election. Nearly 50,000 had signed up to the party by the time of the election, making UKIP the fourth-largest party in the UK in terms of membership. After the 2015 election and in the run-up to the 2017 and 2019 elections, membership of the Liberal Democrat and Green parties also rose.

These three examples that buck the trend of declining party membership suggest that people still see parties as a vehicle for political action if they are proposing some kind of *radical* change. When it comes to more conventional politics and established parties, however, membership is continuing to decline.

Voting

The act of voting, in an election or a referendum, is the least intensive form of participation and also the most infrequent, yet it is also the most important for most citizens. The level of turnout (what proportion of registered voters actually votes) is therefore a good indicator of participation and engagement with politics. If we look at general elections, the trend has been mixed in recent years. Figure 1.5 shows the turnout at general elections since 1979.

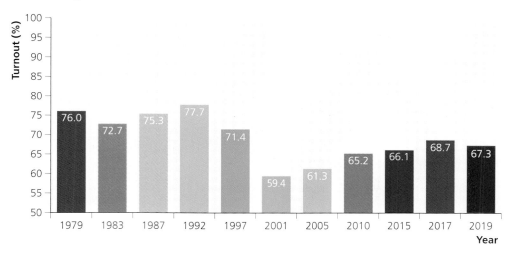

Figure 1.5 Turnout at UK general elections, 1979–2019

We can see that there is a general trend of falling turnout, though there has been a recovery since the historic low of 2001, a trend that was extended into the 2017 general election, but fell back slightly in 2019. The figure of two-thirds could be viewed as disappointing, but also not serious in terms of democratic legitimacy. It is useful to compare turnout in the UK with that of other democracies. Figure 1.6 shows comparative figures for the most recent general elections in other countries.

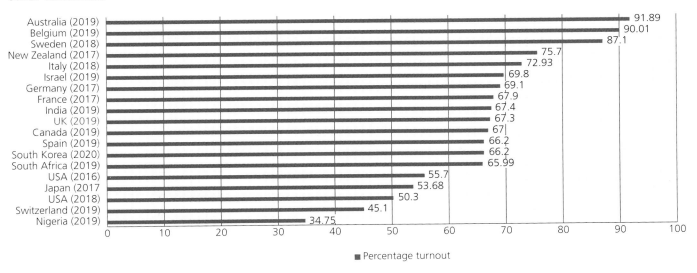

Figure 1.6 Comparative general election turnouts

Figure 1.6 shows that the UK stands in the middle of the 'league table'. This reflects the wider picture. It is interesting that the USA, often described as the purest democracy in the Western world, displays the lowest turnout figures in this selection, and one of the lowest in the democratic world, but it is worth remembering that in its figures the USA includes all possible voters, while in the UK turnout is based only on those who have registered to vote.

Turnout in referendums is rather more volatile in the UK. Table 1.7 shows turnout in a number of key referendums.

Table 1.7 Referendum turnouts in the UK

Year	Subject of referendum	Turnout (%)
1997–98	Devolution to:	
	• Scotland	60.4
	• Wales	50.1
	• Northern Ireland	81.0
1998	Should London have an elected mayor?	34.1
2011	The introduction of the AV electoral system	42.2
2014	Scottish independence	84.6
2016	British membership of the EU	72.2

We can see that referendum turnouts vary from 34.1 per cent concerning local government in London, up to 84.6 per cent in the Scottish independence referendum. Turnout is, of course, a reflection of how important voters consider an issue to be. Voters are certainly becoming more used to having a say on single issues and it is noteworthy that in the two most crucial referendums, EU membership and Scottish independence, turnout was higher than in recent general elections.

Should compulsory voting be introduced?

One potential reform to the democratic process would be to make voting compulsory. Compulsory voting exists in about a dozen countries, though in many it is possible to 'opt out' of voting before the election and so avoid a fine. In some countries the government does not enforce compulsory voting, though it exists in law. In Australia, compulsory voting *is* enforced and a fine can be levied. Voters there do not have to vote for any candidate(s) but must attend the polling booth and mark a ballot paper in some way. Some 'spoil' the ballot paper to avoid a fine. The turnout in Australia, not surprisingly, is above 90 per cent, and it is 90 per cent in Belgium for similar reasons. In Italy, voting was compulsory until 1998 when turnout was typically close to 90 per cent, but since voting has been no longer compulsory turnout has fallen (72.93 per cent in 2018). So, there can be no doubt that compulsory voting has a dramatic effect on turnout. The relatively low turnouts at UK elections, especially at local and regional levels, have led to calls for compulsory voting. The arguments for and against are well balanced.

Should the UK introduce compulsory voting?

Arguments for

- It may force more voters, especially the young, to make themselves more informed about political issues.
- By increasing turnout, it would give greater democratic legitimacy to the party or individual(s) who win an election.
- By ensuring that more sections of society are involved, decision-makers would have to ensure that policies address the concerns of all parts of society, not just those who typically vote in larger numbers.
- It can be argued that voting is a civic duty so citizens should be obliged to carry out that duty.

Arguments against

- It is a civil liberties violation. Many argue it is a basic right to *not* take part.
- Many voters are not well informed and yet they would be voting, so there would be ill-informed participation.
- It would involve large amounts of public expenditure to administer and enforce the system.
- It would favour larger parties against small parties. This is because less-informed citizens would vote and they may have heard only of better-known parties and candidates.

 Consider the idea of 'should' and why compulsory voting 'should' be introduced to help steer you to your overall evaluative judgement about which side is more convincing. Remember that in making your judgement you should address the question asked, not just an assessment of good and bad ideas from each side.

Attention tends to centre on young voters in the UK because they typically vote in smaller numbers than older people. Turnout figures at UK general elections among the 18–24 age group are typically about 35 per cent, while over 80 per cent of the over-60s tend to vote. This may result in governments favouring the older generation against the young when setting policy. However, civil rights campaigners are against compulsory voting, while the Conservative Party is unlikely to support it as younger people tend to be more left-wing than older people, so forcing the young to vote would favour Labour and other left-of-centre parties.

Digital democracy

E-petitions are a fast-growing form of participation, gaining greatly in popularity since official government petitions were introduced with the requirement that any petition gaining 10,000 signatures would receive a government response and any receiving 100,000 signatures would be considered for a parliamentary debate. Indeed, they have become so common that the term '**e-democracy**' has come into use. Such petitions are part of the wider spread of digital democracy, where campaign groups use social media and the internet to promote their causes.

E-petitions have the advantage of requiring little effort and it is immediately apparent how much support a particular issue may have. Combined with the use of social media, they can very rapidly build interest in an issue, causing a bandwagon effect. They are often criticised as a form of participation as it requires so little effort to take part and there is no guarantee that participants know much about the issue. Nevertheless, they are becoming an established part of modern democracy and do, from time to time, have some influence, perhaps most notably when they led to the re-opening of the investigation into the Hillsborough football stadium tragedy and the much-hyped televised debate about whether or not Donald Trump should be allowed into the UK.

Table 1.8 includes some of the most important e-petitions of recent times, and demonstrates how much impact they have had.

Useful term

e-democracy A name used to describe the growing tendency for democracy to be carried out online in the form of e-petitions and other online campaigns.

Table 1.8 E-petitions in the UK

Year	Subject	Signatures	Outcome	Platform
2007	Against a plan to introduce charges for using roads	1.8 million	The government dropped the plan	Downing Street site
2011	Calling for the release of all documents relating to the Hillsborough football disaster of 1989	139,000	Following a parliamentary debate, the papers were released, and a new inquest was launched	Downing Street site
2016	Should there be a second EU referendum?	3.8 million	A parliamentary debate was held on the issue but no second referendum was allowed	Parliamentary site
2019	'Don't put our NHS up for negotiation'	169,836	The government responded by saying 'The Government has been clear: the National Health Service (NHS) is not, and never will be, for sale to the private sector. The Government will ensure no trade agreements will ever be able to alter this fundamental fact'	Parliamentary site
2020	Offer more support to the arts (particularly theatres and music) amidst the Covid-19 pandemic	175,654	Debated in Parliament in June 2020. In response to this debate the government announced more funding to protect the arts	Parliamentary site

The importance of blogging, tweeting and general social media campaigning is also growing. A campaign on a current issue can be mounted in just a few hours or days. Information about various injustices or demands for immediate action over some kind of social evil can circulate quickly, putting pressure on decision-makers and elected representatives. Sites such as 38 Degrees and Change.org help to facilitate such social movements. Typical campaigns concern proposed hospital closures, opposition to road-building projects, claims of miscarriages of justice in the courts and demands for inquiries into the behaviour of large companies.

In party politics, social media has become a particularly important tool for campaigning in elections. While political adverts appearing on radio and television are prohibited in the UK, there is no such regulation on social media platforms that operate internationally, so there has been a rise of campaign videos and adverts that the parties can create and share on social media to influence voters, circumventing the controls in place in the UK and allowing parties with more resources to advertise more freely. Parties will also use data gathered from social media accounts to help target specific voters with specific issues that will resonate with them. They focus on key voters in key constituencies, without wasting resources on voters who are unlikely to vote or who will not be persuaded, though also ignoring large sections of the population. Social media is therefore changing the way in which political parties campaign and speak to voters, in ways that might be more democratic, as it allows more personalised campaigning that is relevant to key voters, but also in ways that are likely to benefit the wealthier parties and avoid the scrutiny of the Electoral Commission in trying to ensure elections are fair (covered in more detail in Chapter 3).

Pressure groups

As membership of and activism in political parties have declined, they have been partly replaced by participation in pressure groups. Many millions of people and organisations have formed themselves into pressure groups. Groups like trade unions and professional associations have been particularly prominent. For many, such participation may be minimal, but some are activists in these organisations and help

Activity

Access the 38 Degrees and/or the Change.org site and select two local and two national campaigns included on the site.
- Describe the nature of the campaigns.
- Describe the methods being used to further those campaigns.

with political campaigning. The position with promotional groups, on the other hand, is changing. These groups rely on mass memberships, seeking mass activism. In other words, they rely on mass active support rather than a large membership. This kind of participation is growing in the UK. The range and activities of pressure groups are explored later in this chapter.

The conclusion we are likely to reach is that political action is more widespread than ever before. It may be less intensive and it may place less of a burden on people's time, but the fall in voting turnout and party membership has been largely overtaken by the growth of alternative forms of political participation. Therefore, far from being in 'crisis', participation is simply evolving and adapting to modern society.

Suffrage

The term '**suffrage**' refers to the right to vote in free elections, also referred to as the '**franchise**'. The question of how people without the right to vote are able to persuade those in power to give them the right to vote is a fascinating one, most famously embodied by the campaigns to secure equal voting rights for women. Over the nineteenth century, fearing the possible violence that had erupted in the French Revolution of 1789, British governments gradually extended the franchise to more groups, from property owners, to skilled men, to most men, to all men and some women, until finally in 1928 all men and women aged over 21 got the right to vote on the same basis, or universal suffrage was achieved. In 1969 the age requirement was lowered to 18 to reflect changing expectations of adulthood in the UK. The main stages in the extension of the franchise in the UK are shown in Figure 1.7.

> **Key term**
>
> **Franchise/suffrage**
> Franchise and suffrage both refer to the ability/right to vote in public elections. Suffragettes were women campaigning for the right to vote on the same terms as men.

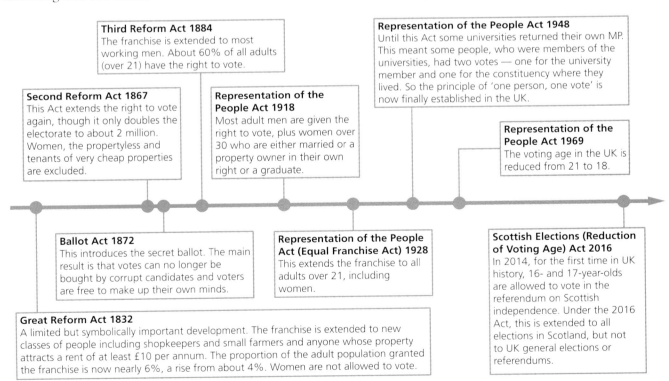

Third Reform Act 1884
The franchise is extended to most working men. About 60% of all adults (over 21) have the right to vote.

Representation of the People Act 1948
Until this Act some universities returned their own MP. This meant some people, who were members of the universities, had two votes — one for the university member and one for the constituency where they lived. So the principle of 'one person, one vote' is now finally established in the UK.

Second Reform Act 1867
This Act extends the right to vote again, though it only doubles the electorate to about 2 million. Women, the propertyless and tenants of very cheap properties are excluded.

Representation of the People Act 1918
Most adult men are given the right to vote, plus women over 30 who are either married or a property owner in their own right or a graduate.

Representation of the People Act 1969
The voting age in the UK is reduced from 21 to 18.

Ballot Act 1872
This introduces the secret ballot. The main result is that votes can no longer be bought by corrupt candidates and voters are free to make up their own minds.

Representation of the People Act (Equal Franchise Act) 1928
This extends the franchise to all adults over 21, including women.

Scottish Elections (Reduction of Voting Age) Act 2016
In 2014, for the first time in UK history, 16- and 17-year-olds are allowed to vote in the referendum on Scottish independence. Under the 2016 Act, this is extended to all elections in Scotland, but not to UK general elections or referendums.

Great Reform Act 1832
A limited but symbolically important development. The franchise is extended to new classes of people including shopkeepers and small farmers and anyone whose property attracts a rent of at least £10 per annum. The proportion of the adult population granted the franchise is now nearly 6%, a rise from about 4%. Women are not allowed to vote.

Figure 1.7 Timeline of the expansion of suffrage in the UK

The statue of Millicent Fawcett erected in Parliament Square in 2018 to mark the centenary of women gaining the right to vote

The last great struggle over suffrage was to give women an equal right to vote with men. The first petition to give women the right to vote was presented to Parliament in 1866 but was largely ignored. At the same time, the Manchester Society for Women's Suffrage was established, which inspired other local societies to form across the UK. These local movements would unite in 1897 under the leadership of Millicent Fawcett as the National Union of Women's Suffrage Societies (NUWSS), also known as the Suffragists.

The NUWSS was open to all and was internally democratic, practising peaceful campaigning to put pressure on those in power through letter-writing, producing material for publication, organising petitions and holding peaceful marches and protests. By 1914 the NUWSS had 100,000 members across 400 branches. Even after women aged over 35 got the right to vote in 1918, the NUWSS, renamed as the Fawcett Society, continued to campaign for equal rights to vote between men and women, which was achieved in 1928.

Despite the work of the NUWSS, some women felt the pace of change was too slow, leading to the creation of the Women's Social and Political Union (WSPU), or the **Suffragettes**, in 1903 by Emmeline Pankhurst and her daughters Christabel and Sylvia. Initially it was based in Manchester, but in 1906 it moved to London. Unlike the NUWSS, the WSPU was only open to women, was not internally democratic, and was focused on 'deeds, not words' using violence and illegal methods to publicise the issue of female suffrage and to put external pressure on those in power (see Table 1.9 for a comparison of the two groups). Methods would include members disrupting political party meetings, chaining themselves to railings, attempting to blow up buildings, destroying letters in post-boxes and going on hunger strike while in prison. They also sold badges, games and posters to help draw attention to their cause and adopted the three colours of purple, white and green to create an early form of branding for the movement.

The violence adopted by the WSPU certainly drew attention to the cause, leading people to talk about the issue, and the government response to supress them did win some public sympathy. However, the violence also cost them support, with some believing that giving women the right to vote would suggest the government had given in to terrorist actions, and that the violence somehow proved women were incapable of sensible thought. This alienated many moderate supporters, both men and women.

It was the work of women during the First World War and the fear of a resumption of the violence of the WSPU that ultimately persuaded Parliament in 1918 to give women over the age of 35 the right to vote. This was enough for the Pankhursts and they disbanded the organisation soon after that.

Useful term

Suffragettes Campaigners in the early part of the twentieth century advocating votes for women, who used both parliamentary lobbying and civil disobedience as their methods.

A gender protest, showing the legacy of the Suffragettes, with their green, white and purple colour scheme still being used in modern protests for women's rights

Knowledge check

What were the key differences between the Suffragettes and the Suffragists? What methods did each group use?

Table 1.9 Suffragists and Suffragettes compared

Suffragists	Suffragettes
Membership open to all	Membership for women only
Internally democratic	Run by the Pankhursts only
Peaceful methods of protest	Violent and illegal methods
Attempted to work with the government	Attempted to intimidate the government
A national organisation of committees	London-centred (after 1906)

Modern campaigns for suffrage

Although the UK has had universal suffrage since 1969 for everyone over the age of 18, there are still some groups that are excluded from voting:

- Those under 18 (although 16 and 17 year olds can now vote in some elections in Wales and Scotland)
- Prisoners (although Scotland now allows some prisoners to vote in Scottish elections)
- Those sectioned under the Mental Health Act
- Peers currently serving in the House of Lords (Peers not sitting in the Lords are permitted to vote).

In addition, the homeless are effectively prevented from voting as they lack a permanent address. It is also unclear whether those who live abroad in the EU will still be eligible to vote in the UK since the UK left the EU.

Activity

The Fawcett Society continues today to champion women's rights. Visit its website, and compare the aims and methods it uses today with those of the original NUWSS. Make a list of its current aims to compare with its aims in 1918.

Members of the royal family do not vote, but there is no legal or constitutional restraint on them voting, so they could if they choose to.

Votes at 16

Although 16- and 17-year-olds were given the right to vote in Scottish elections after 2014 and to vote in elections to the Welsh Senedd in 2020, the issue has not been settled in the UK overall. It seems inevitable that 16- and 17-year-olds will one day gain the right to vote. However, this may have to wait until a party comes to power that feels it will benefit from younger people having the vote. This is likely to be a more radical party of the left.

Debate

Should 16- and 17-year-olds be given the right to vote?

Arguments in favour

- With the spread of citizenship education, young people are now better informed about politics than ever before.
- Voting turnout among the 18–24-year-old age group is very low. This may encourage more people to vote and become engaged with politics.
- The internet and social media now enable young people to be better informed about politics.
- If one is old enough to serve in the army, get married or pay tax, one should be old enough to vote.
- The radicalism of the very young could act as a useful balance to the extreme conservatism of elderly voters.

Arguments against

- People of 16 and 17 years old are too young to be able to make rational judgements.
- Many issues are too complex for younger people to understand.
- Few people in this age group pay tax so they have a lower stake in society.
- It is argued by some that the very young tend to be excessively radical as they have not had enough experience to consider issues carefully.

 When considering this debate, really focus on the idea of 'should' and why 16- and 17-year-olds really must be given the right to vote (or not). While there may be many good reasons for giving 16- and 17-year-olds the right to vote, this is not quite the same as judging whether or not it needs to happen.

Case study

A modern campaign to extend the franchise: Votes at 16

Votes at 16 is a coalition of a number of different groups that believe the franchise should be extended to 16- and 17-year-olds across the whole UK in all elections. It was officially founded in 2001 under the direction and coordination of the British Youth Council. As of June 2020, the group had 4290 registered supporters and worked with organisations such as the British Youth Parliament, the Electoral Reform Society and the National Union of Students (NUS) to lower the voting age.

The campaign uses a variety of methods, including:
- producing and publishing information through its website

- providing templates and advice on how to email local MPs to raise the issue in Parliament
- providing information and advice on how to raise awareness and campaign locally, in schools and universities
- providing advice on how to lobby MPs
- organising an initiative called 'adopt a peer' to encourage members to contact and lobby specific members of the House of Lords.

Although the overall aim of lowering the voting age to 16 has not yet been achieved for general elections, the campaign has seen some success in moving the opinion of some political supporters and gaining wider support. This includes:

- Lowering the voting age to 16 was official party policy in the 2019 manifestos for Labour, the Liberal Democrats, the SNP, Plaid Cymru and the Green Party.
- Before the 2019 election, seven Conservative MPs publicly endorsed lowering the voting age to 16 (though five of those no longer sit in the Commons).
- The voting age for elections to the Scottish and Welsh parliaments has now been lowered to 16.
- In 2014, the voting age was lowered to 16 for the Scottish independence referendum.
- In 2012, a debate on the issue was held in Westminster Hall.

The history of the campaign shows gradual but steady progress moving towards lowering the voting age to 16. As a modern campaign, it has benefited greatly from the wider range of elected bodies in the UK and the greater use of referendums, certainly when compared with the campaign for women's suffrage in the early twentieth century. This has allowed the campaign to persuade different parties which hold real power and has seen a lower voting age implemented in parts of the UK, which has helped prove the ability of 16- and 17-year-olds to vote appropriately.

The campaign has also been hindered by politics; with age now the main dividing line in UK elections (see Chapter 4), the addition of about 1.5 million young voters who are overwhelmingly anti-Conservative could provide a significant boost to the more left-leaning parties that endorse the campaign. However, this makes it less likely that a Conservative government will favour such reform, so with an 80-seat majority it is unlikely national reform will be achieved until after the next general election at the earliest.

In many ways, the work of the campaign is in facilitating and enabling those who wish to campaign, by providing the necessary advice and strategy guidance to run a successful, individual campaign, rather than running a large national campaign itself. In this way, the campaign relies on an active membership to achieve its goals.

Study tip

You are required to know the work of a modern campaign to extend the franchise, and you may be required to reference this as an example in an exam answer.

Group activity

Pressure groups

A pressure group can be defined as a membership-based association whose aim is to influence policy-making without seeking power. Pressure groups have a variety of aims and employ different methods, but they all have in common a desire to influence government without becoming government itself. If a pressure group decides it wishes to exercise power, it becomes a political party. This happened when the trade union movement helped to form the Labour Party in the early twentieth century and when the UK Independence Party (UKIP) began to put up candidates at parliamentary elections after 1993.

The functions of pressure groups are as follows:

- To represent and promote the interests of certain sections of the community who feel they are not fully represented by parties and Parliament.
- To protect the interests of minority groups.
- To promote certain causes that have not been adequately taken up by political parties.
- To inform and educate the public about key political issues.
- To call government to account over its performance in particular areas of policy.

- On occasions to pass key information to government to inform and influence policy.
- To give opportunities to citizens to participate in politics other than through party membership or voting.

In addition, pressure groups are a vital part of democratic and pluralist society, ensuring an active and informed citizenry, offering the public choices and options that may not be recognised by the political parties, and raising awareness of issues to ensure all sections of society are heard and considered in the political process.

Classifying pressure groups

It is usual to classify pressure groups into two main types in order to help us understand how they operate. These types are **causal groups** and **sectional groups**. The main characteristics of each are as follows.

Causal groups

As their name suggests, causal groups seek to promote a particular cause, to convert the ideas behind the cause into government action or parliamentary legislation. The cause may be broad, as with groups campaigning on environmental or human rights issues, or narrow, as with groups promoting local issues such as the protection of green spaces or opposition to supermarket openings in high streets. Here are some prominent examples of causal groups operating in the UK:

- Greenpeace
- Friends of the Earth
- Liberty
- Unlock Democracy
- People for the Ethical Treatment of Animals (PETA)
- Campaign for Nuclear Disarmament (CND).

Sectional groups

These groups represent a particular section of the community in the UK. Sectional groups are self-interested in that they hope to pursue the interests specifically of their own membership or of those whom they represent.

Some sectional groups may be hybrid in that they believe that by serving the interests of their own members and supporters, the wider community will also benefit. For example, unions representing teachers or doctors will argue that the interests of their members are also the interests of all of us. Better-treated and better-paid teachers and doctors and medical staff will mean better education and health for all, they argue.

Prominent examples of sectional groups are:

- Age UK
- British Medical Association (BMA)
- Muslim Council of Britain
- Taxpayers' Alliance
- Confederation of British Industry (CBI)
- The MS [Multiple Sclerosis] Society

Useful terms

Causal group An association whose goal is to promote a particular cause or set of beliefs or values. Such groups seek to promote favourable legislation, prevent unfavourable legislation or simply bring an issue on to the political agenda.

Sectional group An association that has an identifiable membership or supporting group. Such groups represent a section of society and are mainly concerned with their own interests.

The features of causal and sectional groups are summarised in Table 1.10.

Table 1.10 **Features of pressure groups**

Causal groups	**Sectional groups**
They are altruistic in that they serve the whole community, not just their own members and supporters	They are largely (not always) self-interested in that they serve the interests of their own members and supporters
They tend to concentrate on mobilising public opinion and putting pressure on government in that way	Although they seek public support, they tend to seek direct links with decision-makers (insider status)
They often use 'direct action' in the form of public demonstrations, internet campaigns and sometimes civil disobedience	Their methods tend to be more 'responsible' and they often take the parliamentary route to influence
They seek widespread support	They usually have a formal membership

Insiders and outsiders

We can also classify pressure groups as 'insiders' and 'outsiders'. This distinction tells us a good deal about their methods and status. Insider groups are so called because they have especially close links with decision-makers at all levels. The main ways in which insider groups operate include the following:

- They seek to become involved in the early stages of policy- and law-making. This means that they are often consulted by decision-makers and sometimes can offer expert advice and information.
- Some such groups employ professional lobbyists whose job it is to gain access to decision-makers and make high-quality presentations of their case.
- Government at different levels uses special committees to make decisions about policy. Some groups may find themselves represented on such bodies and so have a specially privileged position. The National Farmers' Union (NFU) and the Institute of Directors (IOD) have advised government on these committees, as have trade unions and professional bodies representing groups of workers and members of the professions.
- Sectional groups may be called to testify before parliamentary committees, both select and legislative. Although they attend mainly to give advice and information, it is also an opportunity to have some long-term influence.

Outsiders are those groups that do not enjoy a special position within governing circles. This may be because decision-makers do not wish to be seen to be too close to them or because a group itself wants to maintain its independence from government. More radical groups, such as the Animal Liberation Front, which have a history of using illegal or violent protests to raise awareness of their aims, may find governments do not wish to be associated with them. The typical characteristics of outsider groups are listed below.

- They are usually, but not always, promotional groups. Sectional groups with identifiable memberships and support groups are a useful ally in policy-making, but promotional groups have less certain legitimacy.
- Their typical methods include public campaigning, in recent times often using new media to reach large parts of the population very quickly. They seek to influence not through direct lobbying or ministerial contacts, but by demonstrating to government that public opinion is on their side.

- Outsiders do not need to follow standards that the government will find acceptable, so have greater freedom in the choice of methods they use and are more likely than insider groups to use measures like civil disobedience, mass strikes and publicity 'stunts'.

Useful term

Lobbying An activity, commonly used by pressure groups, to promote causes and interests. Lobbying takes various forms, including organising large gatherings at Parliament or council offices, seeking direct meetings with decision-makers including ministers and councillors, and employing professional organisations to run campaigns.

Methods used by pressure groups

Access points and lobbying

The ways in which groups seek to promote their cause or interests depend to some extent on the access points they have available to them. Insiders who are regularly listened to by decision-makers will sit on policy committees at local, regional, national and even international level, such as through the United Nations (UN). Even at local level, groups will seek to foster special relationships with councillors or with the mayoral office to help provide them with opportunities for **lobbying** those with power. Of course, if groups do not have such access points available to them, they must look elsewhere for their methods.

Public campaigning

Groups without direct access to government tend to mobilise public demonstrations of support to convince the government to listen to them. Public campaigning ranges from organising mass demonstrations, to creating and publicising e-petitions, to using celebrities to gain publicity, to acts of civil disobedience. Some examples of such campaigns are described in Table 1.11.

Table 1.11 Campaigning methods

Group	Aims	Methods
Plane Stupid	To prevent airport expansions	• Invading airports and blocking flights • Occupying airport terminals • Blocking entrances to airports • Delaying Heathrow expansion with a judicial review case • Organising e-petitions
British Medical Association	To force government to withdraw a new contract for junior hospital doctors in 2016–17 after the failure of negotiations and lobbying and other insider tactics	• Regular withdrawal of labour for routine operations and treatments
The Stop the War Coalition (2003)	To demonstrate public hostility towards the proposed invasion of Iraq	• A large-scale public walk through London and speeches held in Hyde Park denouncing the proposed action
The Animal Liberation Front	To end mink farming for fur	• Breaking into mink farms around the world (including the UK) and releasing the captive minks into the wild
Extinction Rebellion	To persuade governments to take immediate action on climate change	• Organising public demonstrations that blocked major roads (including Oxford Circus in London)

Other methods

These can include the following:

- It is common for groups to make financial grants to political parties as a means of finding favour for their cause or interest. Trade unions have long financed the Labour Party. Many business groups and large companies send donations to all parties, but mostly to the Conservative Party. In this way they hope to influence policy.

- Some groups gain personal support from a member of Parliament. Most MPs and peers promote the interests of one group or another, raising issues in debate or lobbying ministers directly. They are sometimes able to influence the content of legislation, proposing or opposing amendments, if they sit on legislative committees.

- Media campaigns can be important. Groups may hope that the press, TV or radio will publicise their concerns. Although the broadcast media in the UK is politically neutral, some programming may publicise an issue to the benefit of the cause. Groups may help to finance such programmes. Press advertising can also be used.

- Groups may resort to direct action, such as public demonstrations or strikes that are officially organised but cause mass disruption. This may put pressure on the government to resolve, such as the threat of strikes by the National Union of Rail, Maritime and Transport Workers (RMT) over the proposed use of driver-only trains.

- Some groups have resorted to illegal methods. This is often a last resort when all else has failed, but they are also useful as a means of gaining publicity. Greenpeace, for example, has destroyed genetically modified (GM) crops to publicise the dangers, while members of Plane Stupid, wishing to demonstrate the dangers of airport expansion, have trespassed at Heathrow and disrupted flights.

- On some occasions a pressure group can pursue an issue through the courts by requesting a judicial review if it feels government or a state body has acted contrary to the rule of law and has discriminated against a group in society.

Factors in the success and failure of pressure groups

Why are some pressure groups more successful than others? This is an important question because it can go some way to explaining the direction of policy. To some extent the fortunes of pressure groups will change with time and with changes in government, but there are also several permanent factors, which are considered in Table 1.12.

Table 1.12 Factors affecting the success and failure of pressure groups

Factor	Success	Failure
Size of membership	The more supporters a group has, the more pressure it can place on decision-makers. Politicians do not like to fly in the face of public opinion because they will regularly face the need for re-election	Groups with smaller sizes can be overlooked or 'drowned out' by the campaigns of larger groups. With fewer people to participate and raise funds, such groups find it difficult to achieve their goals
Finance	Wealthy groups can afford expensive campaigns, employ lobbyists, sponsor political parties and purchase favourable publicity	Groups with less funding struggle to organise effective campaigns, hire lobbyists, fund the production of leaflets and websites and other research, and therefore struggle to make their voices heard
The strategic position of a particular sectional group	A group that is seen as important to the economy or a key service can put greater pressure on the government. Companies and industrial groups have a great deal of leverage because they are vital to the economy, as do NHS workers	Groups that are not seen as important can easily be ignored, especially if they are competing against a strategically important group; the Occupy movement failed in part because it was up against the strategically important finance sector
Public mood	The combination of public sentiment and strong campaigning can be successful in bringing an issue to the attention of decision-makers as politicians will be more likely to support a popular cause	Public mood can turn politicians against certain groups, either for the issue they champion, such as prisoner or terrorist rights, or because the methods they use alienate public opinion
The attitude of the government	Groups are far more likely to achieve success if the government of the day is sympathetic to their cause and position	If the government of the day is determined to follow a particular course of action that a group opposes, it is very unlikely that they will be able to change the government's decision, notably seen in the failure of the 2003 Stop the War Coalition

Case study

ASH

Name of group

Action on Smoking and Health (ASH)

Founding and objectives

- Founded in 1967 by academics and interested parties.
- Its objectives include the spreading of knowledge about the harmful effects of tobacco use and pressing governments to adopt policies and laws to reduce tobacco use.

Methods

ASH conducts research and publicises existing research into the effects of tobacco. It shares this with governments and the public. For example, it has sponsored research into the effects of passive smoking and on the effects of e-cigarettes. It is largely an insider group, concentrating on lobbying lawmakers and governments, mainly using scientific data to underpin its case.

Successes

There are many examples of success, including:
- restrictions on advertising tobacco products and tobacco sponsorship
- health warnings on cigarette packs
- persuading government to increase tax on tobacco to deter consumers
- restricting point-of-sale advertising and promotion
- campaigning for the law banning smoking in public places
- persuading government to develop a law banning smoking in cars carrying children.

Failures

ASH would like to go further on smoking bans and is now concerned that e-cigarettes may be harmful. As yet it has not succeeded in changing government policy in these areas.

Why is it successful?

It helps government to make policy by providing evidence and information. It acts responsibly and has built up a network of supporters within government and Parliament.

The RMT union

Name of group

The National Union of Rail, Maritime and Transport Workers (RMT)

Founding and objectives

- Founded in 1990 through the merger of two groups: the National Union of Railwaymen (NUR) and the National Union of Seamen (NUS), to create a single transport-industry trade union.
- Its objectives include the promotion of better pay and conditions for its members, including shorter hours and safer working environments.

Methods

The RMT negotiates contracts with major transport companies, including Transport for London, negotiating on behalf of its members for better terms and safety. It lobbies governments for better legislative protections and the safety of its workers. It also organises and conducts strikes to pressure employers to meet its demands, which can be effective due to the dominance of the RMT in key sectors, such as Tube drivers in London. The RMT also runs its own credit union to help support its members financially. Mostly affiliated with the Labour Party, the RMT has, in the past, put up candidates for election and endorsed other parties that it feels better represent its members' interests.

Successes

There are many examples of successes, including:
- In 2016 the RMT secured a £500 consolidation payment to all operational staff following the introduction of the Night Tube service.
- The RMT has helped members bring legal cases following injury and wrongful termination, including a £55,000 payment to a member who lost the tip of an index finger in an industrial accident.
- Securing bonus payments for workers during the 2012 Olympics.
- Improving safety standards on offshore oil platforms, as well as on trains and ferries.
- Creating a credit union to help members with their finances.

Failures

Despite long-running campaigns against the closure of many ticket offices on the Tube network, the offices have been closed. The government remains committed to removing guards from trains, which the RMT has vehemently opposed on the grounds of safety. In addition, not all its campaigns for higher wages and better conditions meet with similar levels of success, particularly in areas away from London where the transport network is not such an integral part of the local economy.

Why is it successful?

The RMT is the main union representing workers on the London Underground, meaning it represents workers on a strategically important transport network. Strikes can have a direct impact on the economy as well as possibly embarrassing the government of the day. The threat of strikes during the 2012 Olympics was particularly important as the UK had to avoid international embarrassment and the potential consequences of the transport network failing during the Games.

Pressure groups, society and democracy

The UK is a representative democracy. Political parties, social media and pressure group activity are the main components of a *pluralist democracy*. This term refers to the idea that there are multiple means by which different groups and sections of society can have their voices heard and that they have opportunities to influence government at all levels.

Pressure groups also form an important channel of communication between government and the governed. Citizens often feel that their influence through elections, referendums and political parties is too weak. Parties cannot represent a wide enough range of interests and causes, while elections and referendums are relatively infrequent. It is, therefore, important that there are alternative means by which citizens can constantly communicate with government. Pressure groups supply that link. Without them, citizens might feel powerless and ignored, which is a dangerous situation for a democracy.

Do pressure groups enhance or threaten democracy?

Ways in which they enhance democracy

- Pressure groups help to disperse power and influence more widely.
- Pressure groups educate the public about important political issues.
- Pressure groups give people more opportunities to participate in politics without having to sacrifice too much of their time and attention.
- Pressure groups can promote and protect the interests and rights of minorities.
- Pressure groups help to call government to account by publicising the effects of policy.

Ways in which they may threaten democracy

- Some pressure groups are elitist and tend to concentrate power in too few hands.
- Influential pressure groups may distort information in their own interests.
- Pressure groups that are *internally undemocratic* may not accurately represent the views of their members and supporters.
- Finance is a key factor in political influence, so groups that are wealthy may wield a disproportionate amount of influence.
- The use of civil disobedience, particularly illegal actions, can undermine the freedoms and rights of other citizens.

 Make a clear judgement about which side of the argument has the most weight to it. Pressure groups can both enhance and threaten democracy, but you must consider which side of the debate is more convincing and why.

Other collective organisations

Think tank A body of experts brought together to collectively focus on a certain topic(s): to investigate and offer solutions to often complicated and seemingly intractable economic, social or political issues.

Lobbyist A lobbyist is paid by clients to try to influence the government and/or MPs and members of the House of Lords to act in their clients' interests, particularly when legislation is under consideration.

Pressure groups are not the only external influence on decision-makers. There are also organisations commonly known as '**think tanks**', **lobbyists** and corporations, which seek to influence policy and decisions. Although they may adopt some of the methods of regular pressure groups, these other organisations tend to act in slightly different ways and operate in ways that mark them out as separate from ordinary pressure groups.

Think tanks

The term 'think tank' originated during the Second World War as a military term to describe bodies that developed strategy and ideas. Today they are considered public policy research organisations. Their main role is to carry out research and develop policy ideas that can then be adopted by political parties and governments. In this sense, think tanks carry out one specific role of a pressure group in order to influence those in politics. Think tanks are usually founded to research and develop ideas in specific areas, such as education, healthcare, social justice or economic matters. Usually they are funded by endowments from wealthy patrons or businesses, but they may also be funded by public donations or be affiliated to an academic institution, such as a university.

In carrying out the work of policy research, think tanks have replaced one of the traditional roles carried out by political parties. This is advantageous as it means policies can be considered and developed away from public scrutiny and can be tested before a party might adopt them as official policy. It also saves the party time and resources as it can 'delegate' the role of policy formulation. Of course, many think tanks are founded with a clear aim or objective in mind, and so may produce research to support a particular point of view that may not be in the public interest. Indeed, think tanks will often produce research to help support the demands of their donors.

One example of this is the Institute of Economic Affairs (IEA), a free-market think tank with close links to the Conservative Party – Dominic Raab spoke at its sixtieth anniversary event and launched an annual essay competition funded by the group. The IEA is critical of government measures to reduce or restrict harmful activities, such as smoking bans, sugar taxes and restrictions on fast food advertising, and has also called for the NHS to be replaced by a private, insurance-based system. One of its main donors is British American Tobacco, which raises questions about whether or not the public policy research being carried out is in the interests of the public or of the donors.

However, think tanks can play an important role in a democratic society. In February 2020, the National Institute of Economic and Social Research scrutinised the government's Budget, raising questions about the viability of its growth targets that were widely reported in the media. The competing views and range of ideas and opinions publicised by think tanks help to promote a pluralist and well-educated society.

Some prominent examples of think tanks are listed below.

Neutral think tanks:

- ResPublica — general policy issues
- Chatham House — international affairs
- Centre for Social Justice — policy on welfare issues
- Demos — current political issues
- The National Institute of Economic and Social Research — economic issues

'Left-wing' think tanks:

- Fabian Society — issues concerning social justice and equality
- Institute for Public Policy Research — various left-wing policy ideas

'Right-wing' think tanks:

- Adam Smith Institute — promoting free-market solutions to economic issues
- The Institute of Economic Affairs — another free-market group with close ties to the current Conservative Party
- Centre for Policy Studies — promoting ideas popular in the premiership of Margaret Thatcher

'Liberal' think tanks:

- The Centre for Reform — dedicated to promoting the values of the Liberal Democrats
- Reform — concerned with policies on welfare, public services and economic management

Lobbyists

'Lobbying' is the act of trying to persuade those in power to follow a particular course of action. In a sense, anyone in the UK can lobby, by writing to their MP, signing a petition or demonstrating, to try to persuade those in power of the validity of their views. In this way the act of lobbying is fundamental to a democratic society.

'Lobbyists', however, are distinct organisations or individuals that sell expert knowledge of the political process to those who can afford to hire them. Lobbyists and lobbying companies, sometimes referred to as public relations groups, usually employ people with close relationships with those in power (usually former advisers or staff for particular politicians) and with expertise in which bodies, committees and groups they target. They create strategies for their clients to access the political process. In this sense, they are selling insider status.

At a basic level, lobbyists provide clients with a 'map' giving them advice to follow in order to achieve their goals, effectively giving clients a political blueprint to help put pressure on those in power. This could be anything from a charity seeking additional government funding or trying to persuade the government to adopt a new strategy, to businesses trying to secure exemptions from certain laws or taxes that might impact on them. For example, in 2012 lobbyists developed a 'save our shops' campaign in conjunction with the National Federation of Retail Newsagents and the Association of Convenience Stores to persuade MPs to exempt local newsagents from new laws to keep tobacco products behind closed shutters. The campaign persuaded 80 MPs to back the exemption as well as gaining some public support through media reports. However, the campaign was a creation by lobbyists to persuade public opinion. It was funded by British American Tobacco, allowing it in this way to bypass the law that prevents tobacco companies lobbying MPs. Although this campaign ultimately failed, the fact that it was still able to persuade 80 MPs and to bypass the law that prevents tobacco companies directly trying to influence elected officials shows the potential danger of such activities.

At a more advanced level, lobbyists will arrange events for their clients to have an opportunity to meet with those in power, often through corporate hospitality by offering political figures free tickets to sporting or cultural events, at which they will be sat next to clients who have paid for the privilege. At the highest level, lobbyists will meet on behalf of their clients to try to directly persuade those in power in private meetings. This is why people with direct personal contact or high status are often hired by lobbyists to help gain this access. Former Foreign Secretary Jack Straw claimed, in 2015, that he had used his contacts in the EU to change sugar regulations on behalf of ED & F Man Holdings, which paid him £60,000 per year, while former special advisers (SpAds) like Denzil Davidson have gone to work for lobbying firms like Global Counsel (a group set up by former adviser, MP and current member of the Lords, Peter Mandelson).

Lobbyists gain access and help their clients achieve their goals. While they try to persuade, they are not always successful and politicians consider many factors before making decisions. In fact, sometimes by helping organisations access those in power, lobbyists can improve legislation by offering advice and perspectives that may otherwise have been missed. However, the perception remains that lobbyists, in selling their services, benefit those with money, often at the expense of the public interest, which undermines the concept of a pluralist society. It also undermines confidence in politics in general and raises questions about whom politicians serve, especially as an estimated £2 billion a year is spent by organisations on lobbying in the UK.

Prominent examples of lobbyists in the UK and the areas they represent include:

- The Cicero Group — financial services, infrastructure companies, energy and transport
- Frédéric Michel — News International
- Adam Smith — former SpAd who lobbies for Paddy Power
- PLMR — specialises in political lobbying and media relations
- Hanbury Strategy — specialises in political communication for anyone who faces a current political risk or issue, though the client list is not made public

Corporations

Large corporations such as Google, Starbucks, Virgin, Facebook and Amazon are so big and influential that they qualify as a kind of sectional pressure group on their own. They resist proposed legislation that might hinder their operations and seek to emphasise the positive role they play in the national economy. As they employ high numbers of people and account for a large proportion of economic activity, they have a *strategically* important place in the economy. This gives them great insider influence and they effectively have their own 'in-house' think tanks and lobbyists (usually their public relations department), thereby avoiding the need to hire lobbying companies.

One example of the success of these corporations in lobbying the government is successful resistance to calls for such companies to pay more in UK taxes on their profits (Starbucks, Google and Microsoft being prominent examples). Firms and industries such as alcoholic drinks manufacturers have campaigned against price controls proposed to reduce excessive drinking. In a similar way, the confectionery industry has resisted and toned down attempts by the government to reduce the sugar content of its products in an anti-obesity drive.

Rights in context

Human rights, civil rights and civil liberties

'Human rights' is a term used for the fundamental rights that apply to all people and that cannot be abridged or removed, at least in theory. Human rights developed in response to the horrors of the Second World War and combine two different concepts of rights: '**civil liberties**' and '**civil rights**'. These terms are often used interchangeably, but they mean slightly different things, focused on the role of the state.

Civil liberties is a term used to refer to the protections citizens have against government and the state. Civil rights is a term that refers to those rights that are *guaranteed* by the state. In other words, they are rights and freedoms in relation to the state itself. Prominent examples of each are listed in Table 1.13.

Table 1.13 Civil rights and civil liberties in the UK

Civil liberties	Civil rights
Freedom of speech	Right to life
Freedom of assembly	Freedom from discrimination
Freedom of the press	Right to exercise your vote
Right to trial by jury	Right to equal treatment
Freedom of religious worship	Right to an education

Human rights combine both civil rights and civil liberties and should be:

- absolute — meaning they cannot be compromised or diminished in any way
- universal — meaning they are applied to everyone
- fundamental — meaning they are an essential part of life and cannot be removed for any reason.

Useful terms

Civil liberties The rights and freedoms enjoyed by citizens that protect them from unfair and arbitrary treatment by the state and government. They are also those freedoms that are guaranteed by the state and the constitution. Civil liberties are sometimes referred to as 'civil rights', especially in the USA.

Civil rights Those rights and freedoms that are protected by the government, meaning the state must take an active role in ensuring people are protected and allowed to carry out these rights freely and equally.

The development of rights within the UK has also led to conflict between the judiciary and the legislative and executive branches of government. This is an important constitutional check and has altered the relationship between these branches of government covered in Chapter 8.

The development of rights and formal equality in the UK

Early rights

The first set of civil liberties introduced to the UK was in 1215 in Magna Carta. This was the first attempt to limit the power of the monarch (the government of its day) and ensure protections against arbitrary rule. Magna Carta included the right to trial by jury and that the monarch could only impose taxes with the consent of the people.

In 1689, under the influence of key thinker John Locke, Parliament drafted a Bill of Rights, another set of civil liberties designed to protect the people of England from a potential military dictatorship when they offered the throne to William of Orange. It ensured that the monarch could not take England into a foreign war without its agreement and that the people were free from 'cruel and unusual punishments'.

Common law rights

The traditional status of rights in the UK has been that every citizen was assumed to have rights unless they were prohibited by law. These rights were sometimes referred to as residual rights or negative rights. For example, it was assumed that people had freedom of movement unless there was some legal obstruction, such as if a person was convicted of a crime and sentenced to custody.

In addition, rights were sometimes specifically *stated* as a result of a court case when rights were in dispute. In these cases a judge would decide what was the *normal* or *traditional* way in which such disputes would be settled. Having made his or her decision, the judge would declare what he or she understood people's rights to be. In doing so the judge was declaring **common law**.

Let's take the example of a married or cohabiting couple. If they were to split up, there might be a dispute as to how to divide their possessions, in other words what *rights* the couple had against each other. If there were no statute law to cover the situation, a judge would have to state what the common law was. Once a judge had declared what the common law was under a particular circumstance, he or she had created a judicial precedent. In all similar cases, judges had to follow the existing judicial precedent. A great body of common law and common law rights was created over the centuries.

The Human Rights Act 1998

The main terms and status of the Human Rights Act (HRA) are described in Chapter 5; here we offer a brief description. The HRA brought into effect the European Convention on Human Rights, which was established by the Council of Europe in 1950. The UK helped to draft the Convention but did not accept it as binding on its government until 1998.

Useful term

Common law Traditional conceptions of how disputes should be settled and what rights individuals have. Common law is established by judges through judicial precedents when they declare what traditional, common law should be. It is sometimes described as 'judge-made law', although common law judgements can be found in law books or in digital form.

Traditionally, the UK relied on a series of negative rights, meaning people were allowed to do anything as long as it was not expressly forbidden by law. This meant these rights existed in the absence of law and were therefore very difficult to enforce and people's protections were limited. With the introduction of the Human Rights Act, which came into force in 2000 by making the European Convention on Human Rights a statute law, these negative rights were replaced with positive rights, that had to be protected and respected by law, giving the courts an important means of protecting the rights of citizens and the ability to act as a check on the government. This marked, perhaps, the most significant development in the long history of the development of rights in the UK.

The HRA establishes a wide range of rights to replace the patchwork of statute and common law rights in the UK. It is binding on all public bodies other than the UK Parliament (and it is politically binding on Parliament even if not legally binding; Parliament will rarely ignore it). It is also enforced by all courts in the UK, so that laws passed at any level should conform to its requirements.

Study tip

Remember that the European Convention on Human Rights (ECHR) has nothing to do with the European Union. It is a product of the Council of Europe. Therefore, the ECHR will continue to apply in the UK, whether or not it is a member of the EU.

The Freedom of Information Act 2000

Historically, citizens in the UK had no right to see information held by public bodies, whether it related personally to them or not. By the end of the twentieth century, however, it was clear that the UK was out of step with much of the modern democratic world in this respect. In many countries, including the USA, legislation had been passed, first to allow citizens to view information held about them — for example, by the tax authorities, or social security or schools — and then to view information held by these bodies that it would be in the public interest to see. Governments were too secretive, it was widely contended, and this was a barrier to making them accountable. The Labour government that came to power in the UK in 1997 therefore decided to redress this situation through the Freedom of Information Act in 2000.

Since the Act was passed it has proved an invaluable tool for social and political campaigners, for MPs and for the media, allowing them to discover information that was never available in the past. It has helped to improve such services as the health service, the police, the civil service and educational establishments by shedding light on their activities and helping to promote reform. Perhaps most famously, it was through a Freedom of Information Act request that the *Daily Telegraph* was able to reveal and publicise the MPs' expenses scandal of 2009.

The Equality Act 2010

There had been two parliamentary statutes prior to the Equality Act that established **formal equality** in the UK. The **Race Relations Act 1965** outlawed discrimination of most kinds on the grounds of a person's race or ethnicity. In 1970 the **Equal Pay Act** required employers to offer equal pay to men and women doing the same job. Important though these developments were, they failed to establish equality in the full sense of the word and missed out important groups in society who have suffered discrimination, notably those with disabilities and members of the LGBT+ community. Therefore, under the management of Harriet Harman, a Labour minister at the time, the **Equality Act** was passed in 2010.

Useful term

Formal equality Simply means legally established equality.

The Equality Act requires that all legislation and all decision-making by government, at any level, must take into account formal equality for different sections of society. Put another way, the Act outlaws any discrimination against any group. Equality is required and discrimination is outlawed on the following grounds:

- Age
- Disability
- Gender reassignment
- Marriage or civil partnership
- Race
- Religion or belief
- Sex
- Sexual orientation

In theory, *any* kind of discrimination is unlawful under the Act, but in practice it tends to apply to the following circumstances:

- Employment and pay
- Government services (local, regional, national)
- Healthcare (physical and mental)
- Housing (sales or renting)
- Education
- Financial services
- Policing and law enforcement

Equality of the kind described above is especially important in relation to group politics and a healthy pluralist democracy. By establishing equality, both formal and informal, between different groups and sections of society, it is more likely that their demands and interests can be taken into account.

Rights and responsibilities

Rights in the UK

In law, within the UK, all citizens have equal rights. This was a principle of UK law long before the 2010 Equality Act, but the Act finally confirmed it. This means that no individual and no group can be discriminated against as far as the law is concerned. As a result of the Act, people can now go through the courts to bring a case if they feel they have been discriminated against in any way, providing greater access to rights protections in the UK, thereby helping to develop the UK's democratic system.

Having these rights only matters if they are effectively protected, however. There is no doubt that these rights are more protected today than at any time in the past. The passage of such legislation as the Human Rights Act, the Equality Act and the Freedom of Information Act has ensured that rights are enforceable. Nevertheless, there are also weaknesses. The main issue is that the UK Parliament remains sovereign. In practice this means that Parliament has the ultimate power to create rights or to take them away. In other words, it is not possible in the UK to create a codified set of rights that is binding on successive Parliaments. Furthermore, the rights pressure group Liberty has pointed out that legislation alone does not guarantee rights. It is ultimately up to Parliament to ensure they are protected.

Activity

Research the European Convention on Human Rights. Outline what it says about the following issues:

- Family life
- Privacy

Knowledge check

What rights were established by the Freedom of Information Act 2000?

The passage of the Human Rights Act did *appear* to establish binding rights in the UK, but this was an illusion. The UK Parliament can, and occasionally has (for example, over anti-terrorism laws), ignored the European Convention on Human Rights. That said, Parliament remains reluctant to contradict the ECHR and all other public bodies must abide by its terms. It must also be said that the UK retains an international reputation for respecting human rights. It is one of the reasons so many migrants, asylum seekers and refugees are attracted to come here. Compared with many countries in the world, the UK is a haven for citizens' rights.

It is also true that rights in the UK can be suspended under special circumstances. All countries have such a provision, as it is necessary in times of crisis or emergency. Perhaps the best example occurred in the 1970s when the UK government introduced internment in Northern Ireland. Internment is the imprisonment, *without trial*, of suspected terrorists (people of German origin were interned in the UK during the Second World War for fear they might provide intelligence for the Third Reich or become subversives). This was done in Northern Ireland in response to the Troubles. In the early part of the twentieth-first century, too, Parliament allowed the government to hold suspected terrorists for long periods without trial (though not indefinitely) as a result of the Islamic terrorist threat after 9/11. More recently, in 2020, for public health reasons, various rights relating to associating with other people, meeting with family members and freedom to move around the country were suspended and restricted by law to help combat the Covid-19 pandemic.

Table 1.14 compares the strengths and weaknesses of rights protection in the UK.

Table 1.14 **Rights in the UK: strengths and weaknesses**

Strengths	Weaknesses
There is a strong common law tradition	Common law can be vague and disputed. It can also be set aside by parliamentary statutes
The UK is subject to the European Convention on Human Rights	Parliament remains sovereign and so can ignore the ECHR or can even repeal the Human Rights Act
The judiciary has a reputation for being independent and upholding the rule of law, even against the expressed wishes of government and Parliament	There is increasing pressure on government, as a result of international terrorism, to curtail rights in the interests of national security. The right to privacy, the right of association and expression, as well as freedom from imprisonment without trial, are all threatened
The principle of equal rights is clearly established	What equality means can be subject to interpretation and see some groups coming into conflict over the enforcement of their rights, such as religious groups and LGBT+ groups

Study tip

It is vital to understand the relationship between rights and the sovereignty of the UK Parliament. Full and equal rights can never be permanently guaranteed in the UK because Parliament is sovereign and can amend or remove them.

Activity

Research the following two cases:
- *A* v *Secretary of State for the Home Department* (2004) (also known as the Belmarsh case)
- *Steinfeld and Keiden* v *Secretary of State for International Development* (2018)

What do these two cases reveal about the strength of rights protection in Britain?

Civil liberties groups

The UK also has a variety of civil liberties groups that seek to champion and defend civil rights and liberties in the UK and internationally. In many ways, such groups act like pressure groups, allowing members to join and participate in demonstrations and activities in which they believe. Such groups tend to go beyond the traditional pressure group model and also work as think tanks and lobbyists, conducting research into rights issues, producing evidence and reports about rights abuses, trying to persuade those in power to champion a particular case or amend legislation, or even speak up in support of an issue on the international stage and bringing legal challenges on behalf of those who have had their rights denied.

Such groups have existed for a long time in the UK, but the introduction of the Human Rights Act and other key pieces of legislation, have given them important tools with which to promote and defend civil rights and liberties in the UK. The growth of judicial review in the twenty-first century has allowed these groups to become even more influential as well helping to promote a wider rights culture that they desire. Sometimes, the decisions, actions and organisation of these groups can become controversial, but generally they are seen as a positive force for promoting and defending civil rights and liberties, as the following case studies will show.

Case study

Liberty

Name of group

The National Council for Civil Liberties (NCCL) rebranded as Liberty in 1989.

Founding and objectives

- Founded in 1934 with the aim of challenging government measures to restrict freedoms in the UK and combat the rising threat of fascism.
- Its objectives are to fight to protect and uphold civil rights and liberties across the UK and to develop a wider 'rights culture' across society'.

Methods

Liberty has a number of methods it utilises. It carries out research and investigations into rights abuses and restrictions and seeks to publicise these through media campaigns. It utilises both mainstream traditional media and social media to spread awareness and develop support. Liberty also support and bring legal challenges against rights abuses, challenging what they regard as unfair or unjust laws that restrict civil liberties, such as a recent legal challenge to lockdown restrictions, as well as providing legal advice and support to cases of discrimination against gay rights, women's rights and disability rights.

Liberty will also work with the government and Parliament to advise on legislation and ensure it complies with the Human Rights Act. As well as media campaigns, Liberty regularly organises petitions (increasingly online) and offers pledges (to help demonstrate public support and develop a rights-based society). It also organises protests and public demonstrations to raise awareness of issues and demonstrate public support for their cause.

Successes

There are many examples of success, including:
- In 2020, Liberty used legal actions by bringing a case under the Human Rights Act to successfully pressurise Bournemouth, Christchurch and Poole (BCP) Council into removing parts of the Public Spaces Protection Order that had been used to criminalise rough sleepers and beggars.
- In August 2020, Liberty won a Court of Appeal ruling against the legal framework used by South Wales Police when using facial recognition technology.
- In 2017, Liberty brought a successful legal challenge to the Supreme Court against a loophole in the Equality Act which had allowed employers not to provide equal spousal provisions for same sex couples. The Supreme Court ruled the loophole was unlawful under EU law.
- Following the general election of 2015, Liberty launched a campaign to 'Save the Human Rights Act' to publicise and oppose the Conservative Party's manifesto commitment to repeal and replace the Human Rights Act with a British Bill of Rights. Although the Conservative government still talks about this step, it has not appeared in subsequent manifestos and, as of early 2021, the Human Rights Act remains in place.
- In 2012, Liberty campaigned against the Justice and Security Bill which would allow non-disclosable evidence (meaning secret evidence). The then leader, Shami Chakrabarti, attended the Liberal Democrat Party conference to successfully persuade the party (then in government) to pass a motion against the bill.

Failures

- Liberty has failed, so far, in its attempts to ban the use of facial recognition across the UK, which is still used by many security organisations including the Metropolitan Police.
- Despite organising online petitions and campaigns against lockdown restrictions in 2020 and 2021, the government has continued to impose such restrictions on peoples' freedoms in the name of public health.
- In October 2019 the Court of Appeal rejected an application by Liberty to bring a legal case that would have prevented a no-deal Brexit from the European Union.
- Despite persuading the Liberal Democrats to pass a motion against the Justice and Security Bill of 2012, the Bill became law in 2013, with the non-disclosure elements still in place.

Why is it successful?

Liberty is both a non-profit organisation that operates like a think tank, and a membership association, like a causal pressure group. That, and the fact that it is well established, gives it a large membership base which it can call on organising campaigns and demonstrations, whilst also employing legal expertise and experience to bring legal challenges to the courts. Since 2000, it has been able to use the Human Rights Act as well as EU law to bring legal challenges in UK courts. From 2003–2016 it also benefitted from a charismatic leader, Shami Chakrabarti, who was able to raise the public profile of the group as well as having close ties to the Labour Party (she would eventually go on to become a Labour Peer having conducted an investigation into Labour anti-Semitism claims at the request of Jeremy Corbyn).

Case study

Amnesty International

Name of group

Amnesty International

Founding and objectives

- Founded in London in 1961 by an English barrister, Peter Benenson, who claimed he was inspired by an account of two Portuguese students who had been sentenced to seven years in prison for drinking a toast to liberty.
- Amnesty International operates as a global campaign group or an International Non-Governmental Organisation (INGO) that aims to protest people wherever they believe justice, freedom, truth and liberty have been denied. In addition to exposing and ending abuses, they aim to educate society and mobilise the public to create a safer society.

Methods

Amnesty International's main work has been on raising public awareness of human rights abuses and other infringements of civil liberties, whilst also mobilising public support to put pressure on government to act and support reform. Traditionally this would be done by a letter-writing campaign — local branches and smaller groups are tasked with writing to an 'at-risk-individual' to show support as well as writing letters to the government concerned or to other governments in the hope that they will pressurise the offending government into taking action.

Today, email and Twitter are also used, such as the Twitter handle #FreeNazanin to pressurise the Iranian government into releasing British-Iranian Nazanin Zaghari-Ratcliffe. Amnesty International will give advice and produce proforma letters and emails for people to fill in themselves to add their voice. They also organise petitions, public demonstrations and vigils to raise awareness and put pressure on governments. They also carry out extensive research and publicise reports to highlight issues and educate the public as well as public officials. Perhaps their most important method is in co-ordinating their seven million members to ensure their campaigns are focused and targeted to help add more pressure.

Successes

There are many examples of success, including:

- The campaign 'Write for Rights' has seen a number of people released from prison or had their rights restored following pressure from Amnesty International, including the release of Yecenia Armenta (Mexico, June 2016), who had been jailed on the basis of a 'confession' extracted after 15 hours of torture; the release of Fred Bauma and Yves Makwambala (Democratic Republic of Congo, August 2016) who had been detained because of their pro-democracy work; and having Magai Matiop Ngong's death sentence commuted (South Sudan, July 2020) as he was only 15 at the time of his initial sentence.

- Following two major oil spills in Bodo, Nigeria, in 2008 and 2009, Amnesty International campaigned for compensation for the 15,600 farmers and villagers who were directly impacted by the spill which was alleged to have resulted from negligence by oil company Shell. In 2015 Shell agreed to pay £84 million in compensation.
- In 2013, Amnesty International used satellites to capture images of human rights abuses in Sudan, North Korea, and Syria which helped to raise awareness and could be used in future court cases.

Amongst other achievements in 2020, Amnesty International contributed to and played a role in the following success stories:

- The release of Teodora del Carmen Vasquez from prison in El Salvador, after her 30-year sentence for having an abortion (when in fact she had had a stillbirth) was reduced.
- 8000 prisoners of conscience were released in Myanmar.
- A referendum result in Ireland that overturned the ban on abortions, Amnesty International having released a 2015 report entitled 'Ireland: She is not a Criminal – the Impact of Ireland's Abortion Law'.
- India decriminalised same-sex relationships.
- The European Parliament passed a resolution calling for an international ban on fully autonomous weapons systems.
- The government of Malaysia and the US state of Washington announced plans to abolish the death penalty.

Failures

- Despite arranging 38,000 members in an online action and repeated calls for the international community to do something, China continues to detain Uighurs, Kazakhs and other Muslim groups living in Chinese territory.
- A number of national governments have publicly criticised Amnesty International for one-sided reporting and failure to recognise security threats, making it harder to persuade these governments to take the desired action. These countries include Australia, China, India, Israel and the USA.
- Despite their ongoing campaign to have Nazanin Zaghari-Ratcliffe freed, she remains imprisoned in Iran as of early 2021.
- Amnesty International has failed in its aim to persuade the Vatican to remove its objections to abortion.
- Amnesty International has received criticism for reportedly high salaries for its top officials and for excessive pay-outs for former senior figures in the organisation. Former Secretary General Irene Khan receiving a pay out of £533,103 following her resignation, despite being on an annual salary of £132,490.
- Following two suicides in 2018 by Amnesty International employees, both citing work-related issues, a 2019 report revealed that Amnesty International had a toxic workplace culture, with racism, sexism, bullying and harassment. By October 2019, five of the seven board members had resigned with 'generous' redundancy packages, further undermining the integrity of the organisation.

Why is it successful?

Amnesty International is successful due to its large international membership base and close relationship with other international organisations and governments. Its role in co-ordinating targeted efforts helps facilitate those who wish to fight for human rights in a way that is more likely to achieve success. The organisation also has strong financial resources which enable it to carry out detailed research and produce resources that are respected and recognised as well as employing full time staff across the globe to investigate and run their campaigns. Despite recent controversies, Amnesty International also benefits from strong global recognition, having developed strong ties to many western governments, including the UK and as a result of winning the Nobel Peace Prize in 1977. It also enjoys considerable celebrity support.

Responsibilities of citizens

With rights come responsibilities. The responsibilities of citizens have never been codified in the UK, but there is no doubt that they exist. With the increasing amount of immigration into the UK, the issue of what responsibilities or obligations citizens should have, especially new or aspiring citizens, has become more acute. It has been argued that rights can only be earned if they are matched by responsibilities, though this principle has never been firmly established. We can, however, identify a number of citizens' responsibilities that are widely accepted, and responsibilities that *may* exist but could be disputed. These are shown in Table 1.15.

Table 1.15 Citizens' responsibilities

Clear citizens' responsibilities	Disputed citizens' responsibilities
To obey the laws	To serve in the armed forces when the country is under attack
To pay taxes	To vote in elections and referendums
To undertake jury service when required	To respect the rights of all other citizens
To care for their children	To respect the dominant values of the society

It should be noted that the clear responsibilities are enshrined in law. If a citizen does not accept those responsibilities, they run the danger of prosecution. The responsibilities that are in dispute may well be enforceable, but many citizens will question them.

Collective versus individual rights

Although it is widely acknowledged today that the establishment and protection of individual rights are vital in a modern democracy, it also has to be accepted that the community *as a whole* has rights too, as do various *sections* of society. Problems can arise where the rights of individuals clash with the collective rights of the community or sections of the community. Very often there is no solution to these conflicts, but politicians are called upon to adjudicate. Occasionally, too, such conflicts may end up in the courts for resolution. Table 1.16 shows some examples of these kinds of clashes.

Table 1.16 Individual rights vs collective rights

Individual rights	Conflicting collective rights
Freedom of expression	The rights of religious groups not to have their beliefs satirised or questioned
The right to privacy	The right of the community to be protected from terrorism by security services that may listen in to private communications
The right to press freedom	The right of public figures to keep their private lives private
The right to demonstrate in public places (rights of association and free movement) and thus cause disruption	The right of the community to their own freedom of movement
The right to strike in pursuit of pay and employment rights	The right of the community to expect good service from public servants who are paid from taxation

Synoptic link

The role of the Supreme Court in protecting rights, as well as its relationship with other branches of government, is explained and analysed in Chapter 8. There are also important examples of key rights cases in Chapter 8.

Case study

Campbell v Mirror Group News Ltd, 2004

In 2001, the Mirror newspaper published pictures of supermodel Naomi Campbell leaving a clinic that dealt with narcotic addictions. This would trigger a legal case which would decide on whether, in this case, the right to privacy outweighed the newspaper's right to freedom of expression.

Naomi Campbell did not deny the allegations made by the paper that she was a drug addict or that she was seeking treatment, but chose to sue the owners of the Mirror for publishing the photographs that showed her leaving the clinic on the grounds that these breached her right to privacy. She stated it drew attention to the location of the clinic and would act as a deterrent to her, and others, using the clinic for future treatment.

The Mirror group claimed that it had the right, under freedom of expression, to publish the pictures as they helped to illustrate the published article and that, as Naomi Campbell had previously denied taking drugs and was a public figure, it was in the public interest to publish the supporting evidence.

Initially the High Court ruled in favour of Naomi Campbell, but the Court of Appeals overturned this decision, ruling that the photographs did not breach the right to privacy. This would lead to the case being settled in 2004 by the Law Lords, at that time sitting in the House of Lords.

The Law Lords had to balance the demands of Article 8 of the Human Rights Act, which covered the right to privacy with Article 10, which dealt with freedom of expression. As such, the Law Lords had to consider both aspects and the potential impact of the ruling. First, they had to determine whether or not the right to privacy had been breached by the publication of the photographs and, if the right to privacy had been breached, whether or not ruling against their publication would have made a detrimental impact to the Mirror's freedom of expression. In a divided opinion, the Law Lords ruled, 3:2, that in this case, the right of Naomi Campbell to privacy outweighed the Mirror Groups' right to freedom of expression, thus resolving a conflict that had arisen between differing rights contained within the Human Rights Act.

Summary

Having read this chapter, you should have knowledge and understanding of the following:
- → What democracy is and how democracy in its various forms works in the UK
- → The distinctions between direct and representative democracy and their relative advantages and disadvantages
- → The extent to which the UK is truly democratic and whether or not it is in need of reform
- → The nature of political participation in the UK, whether there is a participation crisis and possible ways in which it could be combated
- → The nature of representation in the UK
- → The nature of suffrage, how it developed in the UK and the issues surrounding changes to the franchise
- → The nature and activities of pressure groups and other organised groups in the UK
- → Issues concerning the operation of pressure groups and other organised groups in the UK
- → The general nature of political influence in the UK
- → The nature of rights in the UK and how they are protected
- → The conflicts between collective and individual rights

Key terms in this chapter

Democratic deficit A flaw in the democratic process where decisions are taken by people who lack legitimacy, due to not having been appointed with sufficient democratic input or being subject to accountability.

Direct democracy All individuals express their opinions themselves and not through representatives acting on their behalf. This type of democracy emerged in Athens in classical times and direct democracy can be seen today in referendums.

Elective dictatorship A government that dominates Parliament, usually due to a large majority, and therefore has few limits on its power.

Franchise/suffrage Franchise and suffrage both refer to the ability/right to vote in public elections. Suffragettes were women campaigning for the right to vote on the same terms as men.

Legitimacy The rightful use of power in accordance with pre-set criteria or widely held agreements, such as a government's right to rule following an election or a monarch's succession based on the agreed rules.

Lobbyist A lobbyist is paid by clients to try to influence the government and/or MPs and members of the House of Lords to act in their clients' interests, particularly when legislation is under consideration.

Participation crisis A lack of engagement by a significant number of citizens with the political process, either by choosing not to vote or to join or become members of political parties or to offer themselves for public office.

Pluralist democracy A type of democracy in which a government makes decisions as a result of the interplay of various ideas and contrasting arguments from competing groups and organisations.

Representative democracy A more modern form of democracy, through which an individual selects a person (and/or a political party) to act on their behalf to exercise political choice.

Think tank A body of experts brought together to collectively focus on a certain topic(s): to investigate and offer solutions to often complicated and seemingly intractable economic, social or political issues.

Further reading

Websites

The most important think tank and campaign organisation concerning democracy in the UK is possibly Unlock Democracy. Its site contains discussion of many issues concerning democracy: **www.unlockdemocracy.org**
The main rights pressure group is Liberty. Its site discusses current issues concerning rights: **www. libertyhumanrights.org.uk**
The two main sites for campaigning on local and national issues are:
- 38 Degrees: **http://home.38degrees.org.uk**
- Change.org: **www.change.org**

A selection of interesting pressure groups is as follows. Websites can be easily accessed through a search engine.
- Age UK
- British Medical Association
- Friends of the Earth
- National Union of Students
- Action on Smoking and Health (ASH)
- Frack Off

Civil liberties groups to research include:
- Amnesty International
- Freedom Association Ltd
- Liberty
- The Prison Reform Trust (prisoner rights)
- The Society for Individual Freedom (SIF)
- Stonewall (gay rights)
- Stop the Traffik

On the issue of rights in the UK, other than the groups listed above, it is worth looking at the website of the Supreme Court: **www.supremecourt.uk**

Books

Cartledge, P. (2016) *Democracy: A life*, Oxford University Press
Cole, M. (2006) *Democracy in Britain*, Edinburgh University Press
Crick, B. (2002) *Democracy: A very short introduction*, Oxford University Press
Eatwell, R. and Goodwin, M. (2018) *National Populism: The revolt against liberal democracy*, Pelican
Flinders, M. (2012) *Why Democracy Matters in the 21st Century*, Oxford University Press
Runciman, D. (2017) *The Confidence Trap*, Princeton University Press
Runciman, D. (2019) *How Democracy Ends*, Profile Books

Practice questions

1

> ## Source 1
>
> Low turnouts in UK elections and referendums have become a serious cause for concern. Many argue that democracy will decline if people do not participate in large numbers. One proposed solution is to introduce compulsory voting. This has been done in Australia and turnouts there are now above 90 per cent. Compulsory voting reflects the idea that voting is a civic duty, so we can justify forcing people to vote. It is also probably true that larger turnouts will produce a more representative electorate. As things stand in the UK, it is the elderly who vote in large numbers, while the young tend to stay at home.
>
> Falling turnout has accompanied a significant reduction in party membership and increasing disillusionment with party politics.
>
> However, it can also be said that low turnouts are not as important as we think. Those who do not vote, it could be said, have voluntarily opted out of the democratic process. It may also be said that non-voters are likely to be ignorant about political issues. It is also true that wider political activity is actually on the increase. What is happening is that increasingly large numbers of people see pressure group activity and participation in social media campaigns as more meaningful forms of activity.

Using the source, evaluate the view that the UK is suffering from a participation crisis.

In your response you must:
- *compare and contrast different opinions in the source*
- *examine and debate these views in a balanced way*
- *analyse and evaluate only the information presented in the source.* (30)

2

> ## Source 2
>
> The UK remains an outdated, ineffective democracy. Presided over by a hereditary monarchy with more than 800 unelected, unaccountable peers sat in the House of Lords, voting on issues that impact people who have no say in their membership. Is the Commons much better? The FPTP voting system meant Cameron could win a majority of seats with just 37 per cent of the vote, while many seats see no real choice and a range of wasted votes. This is our great democratic institution, which remains sovereign meaning there are no effective checks, such as a strong judiciary or codified constitution. Are our rights protected? Certainly not! They may be suspended or ignored as the government sees fit. Furthermore, what is the point of having these institutions if we are just going to pass the most important decisions of our lifetimes to the public in the form of a referendum? No wonder party membership and electoral turnout are in such poor shape!
>
> People may argue that things work, that the flexibility of the system makes it more rather than less democratic, that it is better to have elected officials determining rights cases than unelected judges and that democratic engagement has simply moved to other forms, but I say that without urgent reform, the UK is heading to a crisis, one that may spell the end of the UK as a democratic nation.

Using the source, evaluate the view that the UK is in the midst of a democratic crisis.

In your response you must:
- *compare and contrast different opinions in the source*
- *examine and debate these views in a balanced way*
- *analyse and evaluate only the information presented in the source.* (30)

3 Evaluate the view that the UK is in urgent need of democratic reform.
You must consider this view and the alternative to this view in a balanced way. (30)

4 Evaluate the extent to which organised groups enhance democracy in the UK. In your answer you must refer to at least three different types of organised groups.
You must consider this view and the alternative to this view in a balanced way. (30)

5 Evaluate the extent to which it is accurate to describe the UK as a nation with universal suffrage. In your answer you must refer to at least **one** campaign to extend the franchise.
You must consider this view and the alternative to this view in a balanced way. (30)

6 Evaluate how effectively rights are protected in the UK.
You must consider this view and the alternative to this view in a balanced way. (30)

In January 2019, a new political party was launched, the Brexit Party. Its aim was to ensure that Brexit would be delivered by the government. It organised members, provided an organisation for a membership that felt the political parties in Westminster were not delivering on the referendum result of 2016, and fielded candidates in all levels of elections, winning the most seats in the 2019 European Parliamentary election, and fielded candidates in 273 seats in the 2019 general election. The Brexit Party was the latest in a long list of political parties that have arisen in UK politics over the centuries to organise supporters, represent different groups within society, recruit members, contest elections and field candidates. As such, the Brexit Party, like all political parties, was helping to ensure that the UK continues as a functioning democracy. All parties in the UK rely on the strength of their leadership to develop policy ideas, seek office in council elections and devolved bodies, as well as trying to secure MPs to Parliament and these ideas will be explored throughout this chapter.

Parties are integral to the function of democracy in the UK, yet they are also controversial, seen as undermining the national interest as competing groups seek to promote the interests of party politics above reasoned and democratic debate. Rivalries between parties can lead to democratic debate and discussion, with competing ideas presented to the public to offer genuine choice based on ideological commitments to make the

Nigel Farage campaigning for the Brexit Party

nation all that it can be. However, partisan rivalries can also descend into something like a sporting fixture, with the goal becoming about winning for the team rather than presenting a vision for the nation. As John Adams, the second president of the United States, commented at the creation of American democracy, 'there is nothing I dread so much as a division of the republic into two great parties arranged under its leader, and concerting measures in opposition to each other. This, in my humble apprehension, is to be dreaded as the greatest political evil.' The failure of new parties, like the Brexit Party, to break the dominance of the Labour and Conservative parties perhaps suggests that Adams' fears have been realised, but it may be more accurate to say that party politics is what enables politics to happen in the modern world.

Objectives

In this chapter you will learn about the following:

→ The nature of political parties and what their functions and features are
→ How to distinguish between the terms 'left-wing' and 'right-wing'
→ How parties are funded and the nature of the political controversy over party funding, including proposals for reform
→ The origins of the Conservative, Labour and Liberal Democrat parties and the core beliefs of the three parties
→ The current policies of the three parties
→ The nature and impact of emerging and minor parties in the UK
→ The nature of the term 'party system' and other party systems that exist within the UK
→ Factors that affect party success

Principles of political parties

Features of parties

At its most basic level, a political party is an organisation of people with similar political values and views, which develops a set of goals and policies that it seeks to convert into political action by obtaining government office, or a share in government, or by influencing the government currently in power. It may pursue its goals by mobilising public opinion in its favour, selecting candidates for office, competing at elections and identifying suitable leaders.

Although we might argue about this definition, or add to it, it is a good enough summary that describes most organisations that we consider to be parties.

This definition also helps us to identify the features of political parties in the UK (see Figure 2.1):

- The members of parties share similar political values and views.
- Parties seek either to secure the election of their candidates as representatives or to form the government at various levels (local, regional, national).
- They have some kind of organisation that develops policy, recruits candidates and identifies leaders.

This is not a long list as the nature of political parties can vary in different parts of the democratic world. Typical variations in the features of parties include these:

- Some are mass membership parties with many members (UK Labour Party); others may have a small leadership group who seek supporters rather than members (the main US parties).

- Some parties may be highly organised with a formal permanent organisation (German Christian Democrats), while others have a loose, less permanent organisation (US parties that only organise fully during elections).
- Some parties may have a very narrow range of values and views, and are intensely united around those views (Brexit Party); others have a very broad range of views and values, and so may be divided into factions (UK Conservative Party).
- Some parties are very focused on gaining power (main parties in the UK and the USA), while others recognise they will not gain power but seek merely to influence the political system (Green parties).

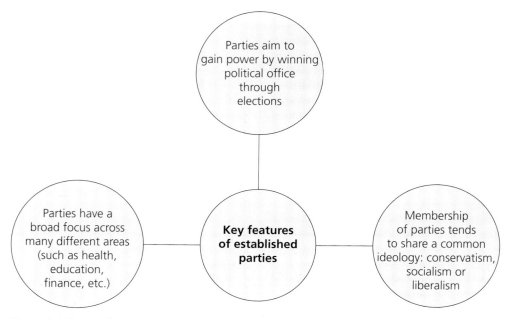

Figure 2.1 The key features of established political parties in the UK

Functions of parties

Synoptic links

Parties used to create policy themselves, but increasingly they have delegated this role to think tanks, as outlined in Chapter 1.

Making policy

Perhaps the most recognisable function of a political party is the development of **policy** and political programmes. This is a role that becomes especially important when a party is in opposition and is seeking to replace the government of the day. Opposition parties are, therefore, in a fundamentally different position from the party in power.

When a ruling party controls the government, its leadership *is* the government; there is virtually no distinction between the two. Therefore, the policy-making function of the ruling party *is the same as* the policy-making function of the government. It involves not only political leaders but also civil servants, advisory units and committees, and private advisers. Of course, the rest of the party, backbench MPs and peers, local activists and ordinary members, have some say through policy conferences and committees, but their role remains very much in the background. Most policy in the ruling party is made by ministers and their advisers.

Boris Johnson promoting the 2019 Conservative manifesto

In opposition, the leadership of a party is not in such a pre-eminent policy-making position. True, the leadership group will have most influence, the leader especially, but it is when in opposition that the general membership of the party can have most input into policy-making. Through various conferences and party committees, the membership can communicate to the leadership which ideas and demands they would like to see as 'official' policy and that therefore could become government policy one day. This kind of influence occurs at local, regional and national level.

Jeremy Corbyn launches the 2019 general election Labour manifesto

The policy-formulating function is also sometimes known as **aggregation**. This involves identifying the wide range of demands made on the political system, from the party membership, from the mass of individuals in society as well as from many different groups, and then converting these into programmes of action that are consistent and compatible. Aggregation tends to be undertaken by the party leadership group as these are the people who may one day become ministers and will have to put the policies of the party into practical government.

Representation

Parties claim to have a representative function. Many parties have, in the past, claimed to represent a *specific* section of society. For example, the UK Labour Party was developed in the early twentieth century to represent the working classes and especially trade union members. The Conservative Party of the nineteenth century largely existed to protect the interests of the landed gentry and aristocracy. This has, however, changed in the contemporary UK because all the main parties argue that they represent the *national* interest and not just the interests of specific classes or groups. So, when we suggest that parties have a representative function, we mean today that they seek to ensure all groups in society have their interests and demands at least considered by government. Of course, in reality, we understand that parties will tend to be prejudiced towards the interests of one section of society or another, based on their core values and ideologies.

One new phenomenon that has emerged, and which needs to be taken into account as far as representation is concerned, is the emergence of **populist** parties. Populist parties tend to emerge rapidly (and often disappear equally quickly). Typically they represent people who feel they have been ignored by conventional parties, in other words that they are *not* represented *at all*. The appeal of populist parties is usually emotional or visceral and plays on people's fears and dissatisfactions. They generally take root among the poor, who feel they are left behind. Generally, these populist movements tend to be defined by what they are opposed to (rather than what they favour) and are often anti big government, anti taxation, anti big business and anti established politics.

We are also seeing the rise of 'issue parties' that represent a particular cause. Green parties are the best example, but increasingly we are also seeing new parties dedicated to advancing women's rights in parts of Europe. Having said this, most contemporary parties in modern democracies still lay claim to representing the national interest.

Selecting candidates

Parties spend a great deal of their time and effort selecting candidates for office at all levels. They need to find prospective local councillors, elected mayors in those localities where such a position exists (notably London and other major cities), members of the devolved assemblies and the Welsh and Scottish Parliaments, and, most prominently of all, for the UK Parliament. This is mostly done at local and regional level, through party committees staffed by activists. The national party leaderships do have some say in which candidates should be chosen, but it is in this role that local constituency parties have the greatest part to play.

Identifying leaders

Parties need leaders and, in the case of the main parties, this means potential government ministers. They therefore have procedures for identifying political leaders. It is in this area that the established party leaders play a key role. For the ruling party, the Prime Minister completely controls the appointment of ministers. In opposition parties, the leader chooses a smaller group of 'frontbench' spokespersons who form the leadership. But despite the dominance of party leaders in this field, potential leaders cut their teeth to some extent in internal party organisations and committees. The formal organisations of parties give opportunities for members to become 'trained' as leaders.

The issue of political leadership was thrown into focus within the Labour Party in 2015–16. Following the party's 2015 election defeat, the former leader, Ed Miliband, resigned. This left behind a power vacuum. In finding a successor, the party ran into a huge controversy. The party membership voted overwhelmingly to elect Jeremy Corbyn. However, Corbyn's political views were far to the left of most of the Labour MPs and peers. He was the party leader until 2020, but many of the Labour MPs in Parliament refused to acknowledge him as *their* leader.

In contrast, the Conservative Party, which lost faith in its then leader, Theresa May, in 2019 after her failure to unify the party to deliver an agreed version of Brexit, had no such problems finding a successor. Boris Johnson was overwhelmingly the favourite among Conservative MPs and it became clear that most ordinary members of the party (who have the power to elect their leader) agreed with the MPs, leading to an emphatic appointment of him as party leader, which in turn saw him appointed Prime Minister on 24 July 2019.

A summary of some examples of key features with examples is provided in Table 2.1 with some of these features being dealt with in more detail later in the chapter.

Activity

Research the Conservative leadership election of 2019 and consider:
- Why did the leadership election take place?
- Were there any controversies concerning Johnson's election?

Table 2.1 Key features of the three main parties as of early 2021

Feature	Conservative Party	Labour Party	Liberal Democrats
Leader	Boris Johnson	Sir Kier Starmer	Sir Ed. Davey
Current policies	Get Brexit done Increase nurses by 50,000 Pensions to rise by at least 2.5% per year	Increase health budget by 4.3% Raise minimum wage to £10 per hour Stop state pension age rises	A 1 penny rise in income tax for the NHS Generate 80% of energy from renewables Free childcare
Number of MPs	365	200	11
Number of Peers	258	177	87
Number of MSPs (Scotland)	30	23	5
Number of MSs (Wales)	11	29	1
Number of London Assembly Members	8	12	1
Number of local councillors*	7445	6291	2527
Party membership in 1979**	1,120,000	666,000	145,000
Party membership in 1992	500,000	280,000	101,000
Party membership in 1997	400,000	405,000	87,000
Party membership in 2005	258,000	198,000	73,000
Party membership in 2019	180,000	485,000	115,000

*Figures cover the total number of County, District, Metro, London, Unitary Authority councillors from the whole of the UK

*In 1979 the Liberal Democrats did not exist, so figures are for the Liberal Party

Source: The Commons Library

Ed Davey, leader of the Liberal Democrats

The Liberal Democrats use a slightly different system that involves more of the grassroots party membership in the nomination process. In the leadership elections of 2019 and 2020, candidates had to be an MP and were required to have the support of 10% of Liberal Democrat MPs (which essentially meant one other MP in the two most recent contests) and the support of at least 200 members spread across at least 20 different local parties, to ensure widespread support across the party for any candidate. Once nominations close, the candidates campaign and are elected by all members of the party on a one member, one vote basis.

They use a system called the Alternative Vote, which should ensure a majority, although as in 2019 and 2020 the contests were between just two candidates (Jo Swinson and Ed Davey in 2019, Ed Davey and Layla Moran in 2020, after Jo Swinson lost her seat in the 2019 general election). In both contests the choice was between the two sides of the party, Ed Davey representing the more centrist, Orange Book group Liberal Democrats, while Swinson and Moran represented the more socially liberal and progressive wing of the party.

Synoptic links

In the UK, to become the head of government, the Prime Minister, you also need to be party leader. Therefore, the support of the party is crucial in maintaining the power of the Prime Minister. This idea is explored in more detail in Chapter 7.

Knowledge checks

Who is the current leader of each of these UK parties?
- The Scottish National Party
- The Green Party
- The Democratic Unionist Party (Northern Ireland)

What is unique about the Green Party leadership and how does the party justify this?
How many *female* UK party leaders are there currently?
How many black and minority ethnic (BAME) party leaders has the UK had?

Useful term

BAME General term covering Black, Asian and Minority Ethnic groups.

Contesting elections

At election time parties play a critical role. Apart from supplying approved candidates, the party organisations form part of the process of publicising election issues, persuading people to vote and informing them about the candidates. Without the huge efforts of thousands of party activists at election time, the already modest turnout at the polls would be even lower. Representatives of the parties are also present when the counting of votes takes place, so they play a part in ensuring that the contests in elections are fair and honest.

Synoptic links

The way in which parties contest and promote elections is explored in Chapter 4.

Political education

It is not only at election time that parties have an educative function. They are also continuously involved in the process of informing the people about the political issues of the day, explaining the main areas of conflict and outlining their own solutions to the problems that they have identified. Part of this process involves educating the public about how the political system itself operates. This can most clearly be seen in the way the Green Party raised awareness about environmental issues, while UKIP made the role and position of the EU a source for debate. Labour also raised awareness of the issues of low pay, zero hours contracts and funding the 'bedroom tax', all of which have introduced these ideas to members of the public who are not directly affected by them.

This function is becoming less important. To some extent the media and think tanks have taken over in supplying information to the public, but the growth of the internet and social media has also marginalised the parties. Pressure groups, too, play an increasing role in informing the public. Even so, parties do present the electorate with clear choices in a coherent way.

Reinforcing consent

Finally, parties also have a 'hidden' function, but a vital one nonetheless. This can be described as 'the mobilisation and reinforcement of consent'. All the main parties support the political system of the UK, that is parliamentary democracy. By operating and supporting this system, parties are part of the process that ensures the general population consents to the system. If parties were to challenge the nature of the political system in any fundamental way, this would create political conflict within society at large. Parties that challenge the basis of the political system are generally seen as extremists and only marginal elements in the system.

The funding of UK political parties

How parties are funded

The position on party funding is a complex one. This is because UK parties have multiple sources of finance. The main ways in which parties are funded are as follows:

- Collecting membership subscriptions from members
- Holding fundraising events such as fetes, festivals, conferences and dinners
- Receiving donations from supporters
- Raising loans from wealthy individuals or banks
- The self-financing of candidates for office
- Up to £2 million per party available in grants from the Electoral Commission (see below for details).
- Money granted to opposition parties in the Commons and Lords

It is immediately apparent that the larger parties have better access to funds than their smaller counterparts. While the Conservative Party attracts large donations from wealthy individuals and businesses (other parties do too, but on a much smaller scale), Labour receives contributions from trade unions. These amounted to nearly 60 per cent of the party's total income in 2014–15. This figure, though, has reduced as the rules for union donations are changing, essentially making it

> **Study tip**
>
> When answering questions on parties, do not confuse *features* with *functions*. Features are the main characteristics of parties, while functions concern their roles and objectives.

easier for individual union members to opt out of contributing to the party. In 2017 Labour saw its share of funds from trade unions drop to just over 11 per cent of its total income.

In November 2013, the former Labour MP for Rotherham, Denis MacShane, pleaded guilty at the Old Bailey to false accounting, after submitting false receipts for £12,900 and was sentenced to six months in prison

Smaller parties, by contrast, have no such regular sources of income. Add to this the fact that they have small memberships and we can see their disadvantage. It is understandable that donors are less likely to give money to parties whose prospects of ever being in power are remote. Those donors who do give to small parties are essentially acting out of idealism rather than any prospects of gaining influence.

The funding of parties was regulated in 2000 by the **Political Parties, Elections and Referendums Act**. Among other regulations, this made the following stipulations:

- People not on the UK electoral roll could no longer make donations (thus reducing foreign influence).
- Limits were placed on how much could be spent on parliamentary elections.
- Donations over £500 had to be declared.
- Donations over £7500 were to be placed on an electoral register.

This regulation stressed transparency rather than any serious limits on the amounts being donated. State funding was rejected as a solution at that time, and election spending controls were extremely generous.

These regulations were further developed in the wake of the MP's expenses scandal with the 2009 **Political Parties and Elections Act**, which gave the Electoral Commission the power to investigate and impose fines, restricted donations from non-UK residents and imposed tighter regulations in the run-up to elections.

Why party funding is controversial

Before looking at the issues surrounding party funding, we should consider how much parties actually receive. Table 2.2 shows the income of significant UK parties.

Table 2.2 Party funding in the third quarter of 2019 (covering the general election)

Party	Total reported	Donations accepted (excl. public funds)	Public funds accepted	Total accepted
Conservative and Unionist Party	£5,805,980	£5,338,696	£424,749	£5,763,445
Green Party	£115,839	£77,650	£28,189	£105,839
Labour Party	£5,512,406	£2,829,146	£2,646,940	£5,476,086
Liberal Democrats	£3,345,220	£2,916,505	£381,496	£3,298,001
Plaid Cymru — The Party of Wales	£141,531	£0	£141,531	£141,531
Scottish National Party (SNP)	£332,662	£130,738	£201,924	£332,662

Party	Total reported	Donations accepted (excl. public funds)	Public funds accepted	Total accepted
The Brexit Party	£3,390,000	£ 3,390,000	£0	£3,390,000
The Independent Group for Change	£8,332	£8,332	£0	£8,832
The Liberal Party	£28,000	£28,000	£0	£28,000
UK Independence Party (UKIP)	£168,501	£168,501	£0	£161,501
Women's Equality Party	£59,000	£59,000	£0	£59,000
Total	£19,388,844	£15,417,900	£3,827,870	£19,245,771

Source: www.electoralcommission.org.uk/media-centre/latest-quarterly-figures-published-political-party-donations-and-loans-great-britain-q3-2019

The figures in Table 2.2 illustrate clearly that funding favours the two biggest parties, putting small parties at a great disadvantage when it comes to fighting elections, thus creating political inequality. Beyond this, the question of party funding has a number of issues that are even more serious. The controversies include the following:

- Funding by large donors represents a hidden and unaccountable form of political influence. Parties are not allowed to change specific policies or propose legislation as a direct result of donations, but donors must expect some kind of political return for their investment. This might be true of trade unions and the Labour Party and of business interests and the Conservatives.
- Aspects of funding may well verge on being corrupt — morally, if not legally, at least. Some donors may expect to receive an honour from party leaders, such as a peerage or knighthood, in return for their generosity. This is sometimes known as 'cash for honours'. It cannot be proved that it exists, although between 2006 and 2007 the issue was investigated by police. While it was not taken further by the Crown Prosecution Service, suspicions remain.
- The steady decline of party memberships has meant that parties are even more reliant upon donors, which further opens up the possibility of corruption and the purchasing of political influence.

The Electoral Commission, which monitors the income of political parties in the UK, has reported examples of large donations to parties. Some interesting examples are these:

- Between 2015 and 2017 the Conservative Party received £11.3 million from prominent figures and companies in the financial sector.
- In the same period the Conservatives received £3.6 million from property companies.
- One individual (hedge fund proprietor Angus Fraser) donated £1,137,400 to the Conservative Party during this period.
- The Unite trade union gave £657,702 to the Labour Party early in 2017.
- At the same time UNISON, the public service union, donated £376,242 to Labour.

So, such individual donations are not only seen as undemocratic forms of influence, but often carry some other kind of controversy. Similarly, trade union donations to Labour have been criticised on the grounds that members of unions are not given a clear enough choice as to whether their subscriptions should be spent in that way, though with the percentage of income coming from the unions having now dropped to 11 per cent, this may be a less valid criticism than before 2017. It is also said that Labour is unduly influenced by union leaders because so much of its income comes from them.

Case study

Being held to account

In 2016 Labour was fined £20,000 by the Electoral Commission for breaching finance rules. The investigation was launched after £7614 was found to be missing from the party's election return for the costs of Ed Miliband's 'tombstone'. The investigation went on to identify 24 other undeclared election expenses, totalling £109,777. At the time, Bob Posner, the commission's director of party and election finance, said: 'The Labour party is a well-established, experienced party. Rules on reporting campaign spending have been in place for over 15 years and it is vital that the larger parties comply with these rules and report their finances accurately if voters are to have confidence in the system.' In a statement, the Commission said it was pushing the government for an increase in the maximum £20,000 penalty available to it for a single offence 'to an amount more in proportion with the spending and donations handled by large campaigners'.

In 2017, following some rule changes, the Conservative Party was fined £70,000 for breaches in its expenses reporting for the 2015 general election. The Commission found that the Conservatives had failed to correctly report £104,765 of campaign expenses and incorrect reporting of a further £118,124. Commission chairman Sir John Holmes said the Tories' failure to follow the rules 'undermined voters' confidence in our democratic processes' and said there was a risk that political parties were seeing such fines as 'a cost of doing business'.

Activity

Do some research into the 2019 general election. Were any accusations issued or fines imposed over party funding or spending?

Alternative funding structures and restrictions

There is a strong case for saying the way in which parties are funded in the UK is undemocratic and in need of reform, a conclusion reached by the Phillips Report of 2007 entitled 'Strengthening Democracy: Fair and Sustainable Funding of Political Parties'. The report suggested that state party funding based on vote share or membership size would make party politics in the UK fairer and more democratic.

The problem, however, is that there is no agreement about what to do. There are four basic solutions:

1. Impose restrictions on the size of individual donations to parties. To be effective, the cap would have to be relatively low.
2. Impose tight restrictions on how much parties are allowed to spend. This would make large-scale fundraising futile.
3. Restrict donations to individuals, i.e. outlaw donations from businesses, pressure groups and trade unions.
4. Replace all funding with state grants for parties, paid for out of general taxation.

As we have seen, there already is some state funding of parties in the UK. All main parties receive funds from the Electoral Commission. These are called Policy Development Grants (PDGs) and can be used to hire advisers on policy. Over

£2 million is available for this purpose. Table 2.3 gives the figures for allocation of Policy Development Grants for the 2018–19 session.

Table 2.3 Policy Development Grants 2018–19

Party	Value of grants
Conservative Party	£476,554.05
Democratic Unionist Party	£172,865.63
Labour Party	£476,554.05
Liberal Democrats	£476,554.06
Plaid Cymru	£175,137.86
Scottish National Party	£201,613.43
Total	**£1,979,279.08**

Source: The Electoral Commission

In addition there is **Short money**, which is distributed to all opposition parties to fund their parliamentary work in the House of Commons, and **Cranborne money**, which does the same for opposition peers in the House of Lords. The aim of this money is to help fund opposition parties in their role of scrutiny, as the government of the day has the support of the civil service to help prepare for parliamentary statements.

Short money heavily favours large parties because it depends upon how many seats parties have won at previous elections as well as how many votes they received. Thus, in 2019, Labour received more than £8 million, while the SNP only received £825,589.25 (see Table 2.4 for a fuller breakdown). Interestingly, UKIP refused over half a million pounds in Short money after winning one seat in 2015. The party's one MP (Douglas Carswell) suggested it was corrupt and designed to favour established parties. So, state funding of parties already exists. The real question, though, is whether state funding should *replace* private donations altogether.

Table 2.4 Short money allocations, 2019/20

Party	Total
DUP	£238,266.31
Green Party	£115,256.12
Labour Party*	£8,045,182.44
Liberal Democrats	£658,202.22
Plaid Cymru	£104,059.44
Scottish National Party	£825,589.25

* Labour's total includes £840,712.01 funding for the Office of the Leader of the Opposition

Source: House of Commons, HR and Finance

Much of the debate about party funding relates to state financing. However, although several political parties favour this, there is little public appetite for it. Taxpayers are naturally reluctant to see their taxes being used to finance parties at a time when attitudes to parties are at a low ebb. However, state funding remains the only solution that could create more equality in the system. As long as funding is determined by 'market forces', it is likely that the large parties will be placed at a significant advantage.

Useful terms

Short money Named after Ted Short, the politician who introduced it, Short money refers to funds given to opposition parties to facilitate their parliamentary work (research facilities, etc.). The amount is based on how many seats and votes each party won at the previous election.

Cranborne money Named after Lord Cranborne, it refers to funds paid to opposition parties in the House of Lords to help with the costs of research and administration to help them scrutinise the work of the government.

Synoptic links

One of the key functions of Parliament is to scrutinise the work of the government. Cranborne money and Short money help opposition parties to carry out this function more effectively, as covered in Chapter 6.

The other popular policy idea is to eliminate the abuses in the system. This involves full transparency, limits on how much business and union donors can give, and a breaking of any link between donations and the granting of honours. While there are merits to the idea of introducing full state party funding, it seems unlikely that this will arrive in the UK in the near future. Far more likely is the idea that individual donations should be limited. Greater transparency has largely been achieved, but the problem of 'cash for honours' or the suspicion that large organisations can gain a political advantage through donations persists. Action may well centre on a 'deal' between Labour and the Conservatives. Labour might sacrifice some of its trade union funding in return for caps on business donations. The Liberal Democrats, with their unwavering support for state funding, will have to remain on the sidelines for the time being.

Debate

Should UK parties receive full state funding?

Arguments for

- It would end the opportunities for the corrupt use of donations.
- It would end the possibilities of 'hidden' forms of influence through funding.
- It would reduce the huge financial advantage that large parties enjoy and give smaller parties the opportunity to make progress.
- It would improve democracy by ensuring wider participation from groups that have no ready source of funds.

Arguments against

- Taxpayers might object to funding what can be considered 'private' organisations or parties with views they find offensive.
- It would be difficult to know how to distribute funding. Should it be on the basis of past performance (in which case large parties would retain their advantage) or on the basis of future aspirations (which is vague)?
- Parties might lose some of their independence and would see themselves as organs of the state.
- State funding might lead to excessive state regulation of parties.

 Having considered the merits of the arguments, you need to decide which side is the most convincing. This may be because the single most convincing argument is on one side, or because of the overall weight of the arguments on one side.

Key terms

Left-wing A widely used term for those who desire change, reform and alteration to the way in which society operates. Often this involves radical criticisms of capitalism made by modern liberal and socialist parties.

Right-wing Reflects support for the status quo, little or no change, stressing the need for order, stability and hierarchy; generally relative to conservative parties.

Established political parties

Right-wing and left-wing politics

Before looking at specific established parties, it is worth considering the idea of 'left-wing' and 'right-wing' parties as these terms are often used to define and compare political parties. The terms '**left-wing**' and '**right-wing**' should be treated with some caution; although they are commonly used in everyday political discussion to describe an individual's or a group's political stance, they are not very precise expressions.

Communism   Socialism   Liberalism   Conservatism   Fascism

Figure 2.2 The political spectrum from left-wing to right-wing

It is because they are so vague that we need to be careful using these terms. In fact, it is usually best for the sake of clarity to avoid them. It should also be noted that left and right descriptions of politics (see Figure 2.2) will vary considerably from one country to another. Nevertheless, we can construct a scheme that gives a reasonable picture of the left–right divide in the context of UK politics. Many issues do not fall easily into a left–right spectrum such as environmental issues, but we can usefully consider economic issues and social issues to illustrate the distinctions. These are shown in Table 2.5. The spectrum could refer to any dividing issue in politics, but typically we tend to use it to refer to the role of the state, with those on the left wanting a larger role for the state and those on the right wanting a smaller role for the state.

Table 2.5 The left–right divide in UK politics

	Left	Centre-left	Centre	Centre-right	Right
Economics and trade	State economic planning and nationalisation of all major industries State regulation of large industries that exploit consumers or workers Relaxed approach to government borrowing; much state investment in infrastructure	Elements from both centre and left	Largely free-market economy with some state regulation Pragmatic approach to government borrowing to stimulate economic growth	Elements from both centre and right	Strong support for totally free markets No state intervention in the economy
Income and employment	Redistribution of income to create more economic equality Strong trade unions, and protected rights for workers Protectionism for domestic industries Anti-EU	Elements from both centre and left	Pro-free trade Mild redistribution of income, with some poverty relief Pro-EU and in favour of the so called 'soft Brexit' option	Elements from both centre and right	Very low levels of taxation Avoidance of excessive government borrowing to stimulate growth Protectionism for domestic industries Free labour markets, with weak protection for workers Anti-EU and in favour of the so called 'hard Brexit' option

	Left	Centre-left	Centre	Centre-right	Right
Social issues	Strong support for the welfare state Stress on equal rights and protection for minority groups Tolerance for alternative lifestyles such as same-sex marriage, surrogate motherhood Liberal attitude to crime and its remedies	Elements from both centre and left	Welfare state to concentrate on the most needy Support for multiculturalism Mixed attitudes to crime — typically a liberal attitude to minor crime but a hard line on serious crime The state should promote individualism	Elements from both centre and right	A more limited welfare state, with caps on the total amount of benefits available to families and tougher criteria for the claiming of benefits Anti-immigration — support for strict controls Opposed to multiculturalism Traditional attitude to moral and lifestyle issues Stress on patriotism and national interest

The origins and development of the Conservative Party

Conservatism in the UK has its origins in the conflict that raged during the seventeenth century over the role and authority of the monarchy. Those who supported royal authority (as opposed to Parliament) were known as royalists, but eventually came to be known as 'Tories'. During the seventeenth century, it became clear that the supporters of Parliament and democracy in general (mostly known as 'Whigs') were gaining the upper hand over royalists. However, a new conflict began to emerge as the Industrial Revolution gathered pace in the middle of the nineteenth century.

With industrialisation and the growth of international markets, the capitalist middle classes began to grow in size and influence. Their rise challenged the traditional authority of the aristocracy and the landed gentry, the owners of the great estates whose income was based on rents and the products of agriculture. The middle classes were represented largely by the Whigs and the landed gentry by the Tories, who were beginning to be described as 'conservatives'. They were described as conservatives because they resisted the new political structures that were growing up and wished to 'conserve' the dominant position of the upper classes whom they represented.

As the nineteenth century progressed, conservatism began to develop into something closer to the movement we can recognise today. Sir Robert Peel (prime minister 1834–35 and 1841–46) is generally acknowledged as the first Conservative Party prime minister. He and Benjamin Disraeli (prime minister 1868 and 1874–80) formed the party, basing it on traditional conservative ideas. The party's main objectives were to prevent the country falling too far into inequality, to preserve the unity of the kingdom and to preserve order in society. It was a pragmatic party, which adopted any policies it believed would benefit the whole nation.

The political background to the Conservative Party is best understood by considering two traditions. The first is often known as 'traditional conservatism' and it dates from the origins of the party in the nineteenth century, which would develop into '**one-nation** conservatism'. The other tradition emerged in the 1980s. It is usually given one of two names — '**New Right** conservatism' or 'Thatcherism', after its main protagonist, Margaret Thatcher (prime minister 1979–90).

Key terms

One nation A paternalistic approach adopted by conservatives under the leadership of Benjamin Disraeli in the nineteenth century and continued by David Cameron and Theresa May in the twenty-first century, that the rich have an obligation to help the poor.

New Right There are two elements — (i) the neo (or new) conservatives who want the state to take a more authoritarian approach to morality and law and order, and (ii) the neo-liberals who endorse the free-market approach and the rolling back of the state in people's lives and businesses.

Traditional conservatism leading to one nationism

Originating in the late part of the eighteenth century, traditional conservatism emerged as a reaction against the newly emerging liberal ideas that were the inspiration behind the revolutions in North America (1776) and France (1789). Conservative thinkers such as Edmund Burke (1729–97) became alarmed at the rise of ideas such as freedom of the individual, tolerance of different political and religious beliefs, representative government and a laissez-faire attitude towards economic activity (that is, the state avoiding significant interference in the way in which wealth is distributed in society). Conservatives believed that such a free society, with so little control by government, would lead to major social disorder.

Thereafter, conservatives continued to consistently oppose the rise of any new ideology, so, later in the nineteenth century, the rise of socialism was opposed. This anti-socialist position remained in place until the 1970s, when it reached its height under Margaret Thatcher.

However, conservatism is not merely a reaction to any dominant ideology; it acknowledges that society must evolve and conserve the best elements of the past. In this sense the Conservative Party looks to enact limited reforms to release the pressure building for a major upheaval or radical movement, as occurred during the French Revolution. The key principles of the one-nation conservatives are outlined in Figure 2.3.

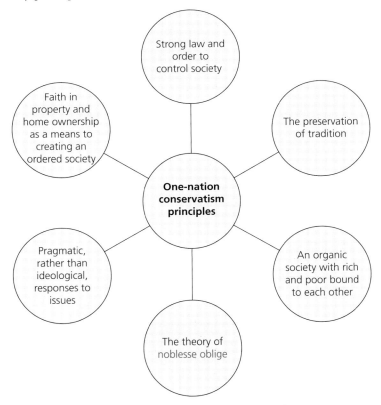

Figure 2.3 The key principles of one-nation conservatives

Useful term

Noblesse oblige A term meaning those of a higher social standing, i.e. the nobility, have a moral duty or 'obligation' to help those of a lower social standing who, through no fault of their own, have fallen on hard times or found themselves in a difficult situation.

New Right conservatism (Thatcherism)

The term 'New Right' was used to describe a set of political values and ideas, largely emerging in the USA in the 1970s and 1980s, which were adopted by many conservatives throughout the developed world. It was a reaction both against the socialist ideas gaining some ground in Europe, Asia and South America, and against traditional conservative values that were seen as too weak to deal with contemporary economic and social policies. It was associated in the USA with Ronald Reagan (president 1981–89) and in the UK with Margaret Thatcher (prime minister 1979–90).

The movement can be divided into two different aspects. These are neo-liberalism and neo-conservatism. Most, but not all, conservatives of the New Right subscribed to both sets of ideas, though often they leant towards one more than the other.

Principles of neo-liberalism

Neo-liberalism is associated with the economic and social philosophers Friedrich Hayek (Austrian, 1899–1992) and Milton Friedman (American, 1912–2006). The main beliefs of neo-liberalism are outlined in Figure 2.4.

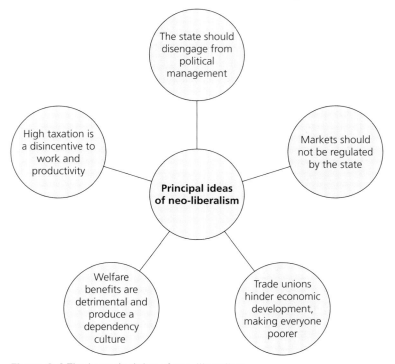

Figure 2.4 The key principles of neo-liberalism

In practical terms, neo-liberal politicians within the modern Conservative Party propose reducing direct taxes, privatising industries that have been taken over by the state (such as transport and energy), reducing welfare benefits so that they are only a 'safety net' for those who have no means of supporting themselves, and curbing the powers of trade unions. In addition, they propose allowing the economy to find its own natural level, even during recessions, rather than the state actively trying to control economic activity.

Principles of neo-conservatism

Ironically, while neo-liberalism proposes the *withdrawal* of the state from economic activity, neo-conservatism proposes a strong state, albeit a small one, yet both are considered part of the same New Right movement. There is a link between them in that neo-liberalism proposes a very free society and this opens up the possibility of disorder. Neo-conservatives, therefore, seek to maintain authority and discipline in society. The main beliefs of neo-conservatism are outlined in Figure 2.5.

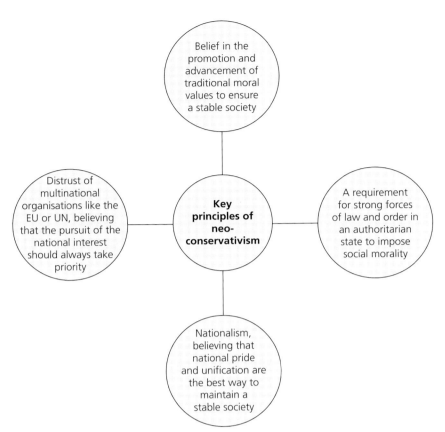

Figure 2.5 **The key principles of neo-conservatism**

We can see that neo-conservatism has much in common with traditional conservatism in that it promotes traditional 'national' values and sees order as a key value to be maintained by the state. However, while most conservatives accept that different lifestyles should be tolerated, neo-conservatives seek to impose a single national culture on society.

Conservative ideas and policies today

The economy

When the Conservative Party returned to power in 2010, it was faced with an economic crisis, the main aspect of which was a huge and growing budget deficit. Successive governments had been spending considerably more than their taxation receipts. The national debt was huge. In March 2010 it stood at £1.13 trillion. This led to the party adopting a rigorous approach to economic management. Above all, its economic policy was dominated by the aim of eliminating budget deficits (i.e. having a balanced budget) and reducing the national debt. The belief is that only a balanced budget can promote economic growth.

Under Theresa May, after 2016 the goal of a balanced budget was abandoned as a medium-term goal. It was seen as unattainable and as inhibiting economic growth. However, party policy remains pragmatic and cautious about economic policy. Public expenditure, the party stresses, must be kept under careful control. It was to save the economy that the Conservative Party in 2020 authorised unprecedented levels of national borrowing to fund the lockdown caused by the Covid-19 pandemic. It was a pragmatic response to maintain stability in society and ensure businesses and

Activity

Research the following people and outline what part they played in the development of the New Right:

- Sir Keith Joseph
- Enoch Powell
- Geoffrey Howe

employment could survive the shutdown without creating a high level of welfare recipients that might lead to a dependency culture.

The party retains a neo-liberal position in its attitude to markets, which was one of the reasons for many neo-liberal figures in the Conservative Party giving support to leaving the EU in order to end EU regulation of the British economy. The party believes government policy should always promote free markets and free trade. It is also to this end that it is determined to curb the power of trade unions to keep labour markets free.

The attitude of the Conservative Party to taxation is partly neo-liberal and partly one nation. On the one hand, personal and company taxation should never be excessively high as this will inhibit enterprise and wealth creation (a neo-liberal view). The party seeks to reduce corporate taxation as much as is feasible. On the other hand, the party has accepted that taxation on lower incomes is too high and risks creating higher levels of poverty and dividing the nation (a one-nation view). It therefore seeks to take many more people out of taxation altogether. The burden of tax has been shifted towards middle-income groups.

Law and order

The party retains the view that prison and stern punishments are the best deterrent against crime. It therefore believes that sentencing policy should be in the hands of elected government and not unelected judges. The party is opposed to 'liberal' ideas about crime and punishment, and opposes such proposals as the legalisation of drugs and the excessive use of 'community' sentences where offenders do not go to prison but instead make amends in their community.

Conservatives stress the need for security and see it as the first duty of government to protect its citizens. In the fight against terrorism, therefore, they accept that civil liberties (privacy, freedom of movement and expression) may have to be sacrificed in the interests of security. For this reason, in both the 2017 and 2019 Conservative Party manifestos, the party pledged to alter the Human Rights Act to ensure it had the ability to deal with national threats such as terrorism, prioritising social stability and safety above individual rights. In 2020 the Conservative government also felt the need to limit various freedoms to ensure the success of lockdown measures during the Covid-19 pandemic.

Welfare

Modern Conservative policy concentrates on the need to ensure that welfare benefits are not a disincentive to work. The government has introduced a stricter system of means testing to prevent unemployment being seen as a preferable option. Two other policies seek to restore the balance between work and benefits. One is the introduction of a living wage (or minimum wage) as a greater reward for work at lower levels of pay. The second is an overall cap on total welfare benefits for families, so that unemployment is less attractive.

Party policy is committed to maintaining the welfare state and safeguarding the NHS and the education system. However, the party believes that these two services should be subject to competition and market forces, and that private-sector enterprises should become involved in the provision of services. This, it believes, can increase efficiency so that services can improve without increasing expenditure on them.

Foreign affairs

Conservatives support the North Atlantic Treaty Organization (NATO) and the UK's close alliance with the USA. However, they also believe that the UK's best national interests lie in retaining an independent foreign policy. If it is in the UK's interest, they believe that the country should intervene in foreign conflicts. The party is committed to retaining the UK's independent nuclear deterrent in the form of Trident submarine-based weapons. After considerable internal conflict, the party has decided to reduce the UK's generous contributions to international aid.

External influences

The Conservative Party has had a long history of being the party of business and as such, groups that represent business interests often exert a powerful influence over Conservative Party policy making and direction. Most notable are the Confederation of British Industry (CBI) which will work with the party to promote British business interests, and the British Banking Association (BBA) which also pushes the Conservative Party to follow policies that will support the banking industry, notably persuading the previous Conservative Prime Minister David Cameron to not impose fines and increased controls on banks following the financial crash of 2008. That being said, both the CBI and BBA warned against Brexit and certainly against a hard-line Brexit, which shows the limits of external influences.

The Conservative Party also has a history of being influenced by major press barons, dating back to the pre-War figures of Lord Beaverbrook and Lord Rothermere. In the 1980s, Rupert Murdoch (owner of News International which owned the *Sun* and the *Times*) was a key supporter of Margaret Thatcher and helped to shape the anti-union stance of the Conservative party as he fought the printing unions. Meanwhile, in 2001, the then owner of the *Daily Telegraph*, Conrad Black, was made a Conservative Life Peer having been nominated by then leader William Hague, while Viscount Rothermere IV, owner of the *Daily Mail*, has continually been an active supporter of the Conservative Party. It is therefore not surprising that so much of the printed media in the UK supports the Conservative Party and the press barons expect the party to support many, if not all, of their positions.

Table 2.6 Current Conservative policies

Policy Area	Right tendency	Centre-right tendency
Economic management	Support for free trade and deregulation of business; lowering of taxation on all with trade agreements negotiated on a bi-lateral basis.	The economy should be managed to ensure society does not become divided, with support for furlough and lower rates of taxation for the poorest.
Social justice	Support for a traditional, Christian centred society based on individual value and clear hierarchies and social structures.	Ensuring the provision of a safety-net during the Covid-19 crisis and a desire to 'level-up' society and the idea of social mobility. Support for more progressive ideas, such as gay marriage.
Industry	Extensive privatisation of all areas. Decrease union protections. Limited government intervention.	Government investment in infrastructure, like HS2 and new airports to facilitate private industry. Financial support for struggling businesses to avoid economic hardship. Acceptance of some necessary state services that can work with the private sector.
Welfare	Seeks to reduce, if not end, welfare provisions, including proposals to privatise part of the NHS. More stringent tests of welfare recipients.	Support for the NHS as part of the national identity. The need for welfare provision, but with a purpose to encourage people into work for social stability.
Law and order	Support for strong law and order to ensure that society operates in a traditional, well-ordered way.	Support for law and order, but also for individual freedoms and liberties and not be imposed in a draconian way.

Policy Area	Right tendency	Centre-right tendency
Foreign policy	Focus on British nationalism, opposed to membership of supranational organisations like the EU. Support for a strong and independent military capacity.	Favours international co-operation to facilitate free trade and to intervene in foreign disputes that may not directly impact on British interests, such as international aid and peace keeping services as part of supranational organisations.
Environment	Opposition to environmental policies that may hinder business. At the extreme end, may be climate change deniers.	Support for environmental initiatives and a focus on reducing the UK's carbon footprint.
Constitutional reform	Opposition to progressive reforms of the New Labour era. Support for traditional systems and laws, including the monarchy, the House of Lords and the abolition of the Human Rights Act.	Seeks to 'fix' problems left over from previous reforms, rather than removing those reforms. Fixing what exists in small ways, such as EVEL in devolution (see Chapter 5), rather than major reforms.

Conservative Party factions

Although we tend to think of political parties as being unified and cohesive, they are often split into different **party factions**. This division will usually be over an idea, such as between Eurosceptics and pro-Europeans in the Conservative Party or between the left wing and right wing of the party. These factions sometimes coalesce into official groups that act like a party within a party, trying to direct the overall policies and attitude of the party as a whole.

In 2019, divisions over Brexit resulted in the Conservative Party dividing itself, with 21 Conservative MPs being suspended from the party and others leaving the party to join the Independent Group for Change or the Liberal Democrats, while many more chose not to stand for re-election. Factions therefore play an important role in shaping political parties and influencing how unified they are. The key Conservative Party factions are outlined in Table 2.7.

Since 2019, we can also identify some emerging factions within the Conservative Party, with a group of libertarian MPs who voted against lockdown measures and other measures that restricted individual choice and freedom, seeing it as part of an over-powerful state. Meanwhile, MPs from the 'red wall' constituencies have focused on pushing for government action in a 'leveling-up' agenda. This highlights how factional divisions evolve as issues and circumstances change and explain why parties are continually evolving.

Table 2.7 Conservative Party factions

Faction	Core ideas	Key members
Cornerstone	Traditional values: Christian, nationalist and focused on family values Reactionary, opposed to social reforms such as same-sex marriage and legal abortion	Edward Leigh Jacob Rees-Mogg
Conservative Way Forward	Thatcherite, neo-liberal Retention of free markets through low taxation and deregulation Opposed to trade union power and welfare provision	Gerald Howarth Liam Fox
Tory Reform Group	One-nation conservative, seeking national unity and believing that too much economic inequality is divisive	Ken Clarke (now in the Lords)

The origins and development of the Labour Party

Until the twentieth century the working classes (many of whom did not gain the right to vote until 1884) were largely represented by a collection of MPs and peers from both the Liberal and Conservative parties. When trade unions became legalised towards the end of the nineteenth century, however, the working class at last had organisations that could represent their interests. It was therefore logical that the unions should begin to put up candidates for election to the UK Parliament. But the unions were not a political party and did not seek power. A new party was needed. In fact, two parties of the left emerged.

Creation of the Labour Party

The main Labour Party was created in 1900 and was very much an offshoot of the trade union movement. It was funded by the unions and many of its members were union leaders and members. Before that, in 1893, a socialist party had already been founded, known as the Independent Labour Party (ILP). In 1906 the ILP formed an agreement with the Labour Party. They agreed not to put up parliamentary candidates against each other in the same constituencies. However, this agreement was short-lived and the two parties began to go their separate ways.

The ILP was a genuinely socialist party, committed to the overthrow of capitalism and its replacement by a workers' state, albeit by peaceful, democratic means. The Labour Party, by contrast, was a more moderate socialist party that did not propose a workers' state but simply wished to improve the conditions of the working class and to control the excesses of capitalism. The state, as envisaged by Labour, would seek to reconcile the conflicting interests of the working class with those of their employers. Both parties still contained extreme socialists, but the distinction was essentially that the ILP was purely socialist while Labour was a more moderate form of socialist party, generally known for democratic socialism, that is socialism that worked within a democratic framework. By 1970, though, the ILP had ceased to exist as a separate party and since then has acted more as a faction within the Labour Party.

Many of the characteristics of the development of the Labour Party can still be seen today. The party continues to be financed largely by trade unions, and union leaders play a major role in the party organisation. Although the ILP no longer exists, its traditions can still be found among a persistent group of left-wingers who form a faction within the party. Many of this faction were responsible for the election of Jeremy Corbyn, a prominent left-winger, as party leader in 2015. Some of them still promote the ideas that formed the basis of the ideology of the old ILP.

Labour since the Second World War

The 1945 general election was something of a turning point for the Labour Party as it achieved full majority control of the Commons for the first time. From then on Labour became the UK's second major party and regularly competed with the Conservative Party for power. However, in the 1980s the party suffered two huge defeats at the hands of Margaret Thatcher's Conservative Party. This ultimately resulted in a split in the party. Some left to form a new party, the Social Democratic Party (SDP); some, led by Michael Foot and Tony Benn, wished to return to Old Labour values and move further to the left; others, led by Neil Kinnock and John Smith, however, saw the future of the party lying in more moderate policies, towards the centre of politics. This branch of the party became known as 'New Labour', and its policies were characterised as 'Third Way'. After John Smith's sudden death in 1992, Tony Blair became leader, closely supported by Gordon Brown, Robin Cook and Peter Mandelson. Blair led the party to three election victories in 1997, 2001 and 2005.

Labour and the unions

As we saw above, the Labour Party was formed out of the trade union movement. Ever since then, there has been a strong relationship between Labour and the trade unions in the UK. Until the 1960s, the Labour Party and the trade union movement were seen as indistinguishable, or as two-sides of the same coin, the party acting as a mouthpiece or political arm for the unions. During this time, many Labour MPs had come to politics through union support and union politics and the trade unions, by requiring all members to enrol and pay a membership subscription to the Labour Party, exercised considerable financial influence and control over the party. During the Attlee years of 1945–51, the Labour government introduced many reforms that the unions demanded and would help to establish the power of the unions in British industry for years to come, including legislation around strikes and pay.

However, during the 1960s, the differences between the trade unions and Labour politicians began to grow. Until the 1960s both had supported full employment and protections for industrial workers, but during the 1960s, as the process of de-industrialisation occurred and the UK economy began moving away from heavy industry, full unemployment became much harder for politicians to achieve. Inflation, caused in part by pay increases and protections for British industry, also made it difficult for the Labour Party in power to support all the demands of the unions. Increasingly, the union leaders became political figures in their own right and would challenge the Labour Party leadership, the question, asked by the Conservatives throughout the 1970s, was who governs, the unions or the elected politicians?

Until the 1980s, Labour Party leaders were elected from among the MPs and only MPs could vote, but the Labour Party Constitution and any Labour Party policy had to be approved by the party as a whole. The union leaders had a block vote, meaning the leader would cast the votes for all their members, so the leaders of the largest trade unions dominated discussions and decisions on party policy. This made it increasingly difficult for more moderate, social democrat members to exert influence and was one reason for the Labour Party moving to the left during the 1980s and maintaining its commitment to nationalisation (covered in Clause IV of the Labour Constitution) and full employment at a time when the economy was changing, and these ideas were becoming less popular. In 1981, Labour also introduced an electoral college system for leadership elections, which would allocate a third of the votes to the unions, with their leaders exercising a block vote, a third to MPs and a third to party members. This meant some party members received multiple votes (as MPs, members of a union and/or members in their own right). As union members were also party members, this increased the control and power of the unions in determining the party leadership and direction of the party and was one reason for the social democrats leaving the Labour Party.

In the 1980s, as the power of the traditional unions declined, the reputation of the unions declined and the Conservative government introduced tighter controls and regulations on unions, so their membership and funds declined, reducing their overall influence. Following the humiliating defeat of 1983, the Labour Party began to take on the unions, with leader Neil Kinnock publicly chastising the worst excesses of union behaviour and his successor John Smith removing the block vote. Tony Blair would take it further, persuading the membership to remove Clause IV from the Labour constitution and seeking new forms of funding from business and other non-union sources. Although the unions remained closely associated with the Labour Party, their strength and influence during the New Labour era was substantially reduced.

Since 2010, despite Ed Miliband replacing the electoral college with a straightforward one member one vote system, the unions have seen an increase in their influence on the post-New Labour party. Although they do not control the votes of their members, an endorsement of a candidate for leadership or for particular policies can have a powerful influence over their membership and their support helped to secure victories for Ed Miliband and Jeremy Corbyn over more centrist candidates. The importance of union funds have also played an increasingly important part in Labour Party finances, while under Corbyn, union leaders, particularly of the powerful UNITE union, worked closely with the Labour leadership and influenced key appointments to the party hierarchy and policy decisions. Although things may change under Sir Keir Starmer, the unions continue to play an influential role in Labour Party politics.

Core values and ideas of the Labour Party

As we have seen, Labour's story can be divided into two parts. The first, the **Old Labour** period, runs from the early days until the 1990s. The second, the **New Labour** period, runs from the early 1990s until the present, when the party may well be splitting once again.

Old Labour

Critics loosely describe the traditions of the Labour Party as 'socialism'. This is an illusion; Labour was never a socialist party. It did not propose a workers' state and has never attempted to abolish capitalism. As mentioned above, it is better thought of as a democratic socialist party. The best way to understand the Old Labour tradition is to look at its general values and then at its actual policies, focusing on the period from 1945 to 1983.

Old Labour values

- An essential value is *equality*. Labour used to support redistribution of income to reduce the worst inequalities. A better characterisation of equality for Labour is 'social justice'. Labour has also always supported formal equality, i.e. equal treatment under the law.

- Old Labour supporters tend to see society in terms of *class conflict*, arguing that the interests of the two great classes (working and middle class) cannot be reconciled, so governments must favour the interests of the disadvantaged working class.
- Recognising that total equality was not feasible, Labour championed *equality of opportunity*, the idea that all should have equal life chances no matter what their family background.
- *Collectivism* is a general idea shared by socialists of all kinds. It is the concept that many of our goals are best achieved collectively rather than individually. It includes such practical applications as the welfare state, trade unionism and the cooperative movement.
- Old Labour saw common ownership mainly in terms of public ownership of major, strategic industries, run by the state on behalf of the people, with the idea of nationalisation of key industries.
- *Trade unionism* is also important. Old Labour recognised that workers were weak compared with employers. Support for powerful trade unions was, therefore, vital in restoring the balance of power between employers and workers.
- Old Labour believed that the central state could play a key role in controlling economic activity and in securing social goals. This may be described as *statism*. By placing such responsibilities in the hands of the central state it ensured equality of treatment for all.
- Finally, *welfarism* is important. This is the idea, associated with collectivism, that every member of society should be protected by a welfare system to which all should contribute.

Old Labour policies and actions

Old Labour had two main periods of power during which it could convert some of its values into practical reforms. These were 1945–51 and 1964–79. In those periods, at various times, Labour converted values into political action in the following main ways:

- The welfare state, including the National Health Service, was created in the 1940s.
- Trade unions were granted wide powers to take industrial action in the interests of their members.
- Major industries were brought into public ownership (nationalisation) and state control in the interests of the community and the workers in those industries. Among the industries were coal, steel, shipbuilding, rail and energy.
- Taxes on those with higher incomes were raised in order to pay for welfare and to redistribute income to the poor.
- Comprehensive education was introduced in the 1960s to improve equality of opportunity.
- Discrimination against women and ethnic minorities was outlawed in the 1960s and 1970s. Equal pay for women was introduced.

New Labour

Tony Blair and his cohort of leaders, from 1994 to 2010, supported by the economic philosopher Anthony Giddens, creator of the 'Third Way', developed a new set of moderate policies, often described as New Labour.

New Labour values

New Labour was opposed to the ideas of the 'hard left' in Labour and sought instead to find a middle way (the 'Third Way') between socialism and the free-market, neo-liberal ideas of the Conservative Party under Margaret Thatcher. Its main values were as follows:

- New Labour thinkers rejected the socialist idea of *class conflict*, arguing that all members of society have an equal right to assistance from the state.
- The party accepted that *capitalism* was the best way of creating wealth, so markets should remain largely free of state control.
- Nevertheless it was recognised that capitalism could operate against the interests of consumers, so it should be regulated, though not controlled. The state should be an *enabling state*. Allowing the economy to create wealth and giving it support where needed, but the state should not, on the whole, engage in production itself.
- New Labour de-emphasised collectivism, recognising that people prefer to achieve their goals individually. *Individualism* was seen as a fundamental aspect of human nature.
- *Equality of opportunity* was stressed. Education and welfare would create opportunities for people to better themselves.
- *Communitarianism* is the concept that although people are individuals with individual goals, we are also part of an organic community and have obligations and duties in return for our individual life chances. This is a weaker form of collectivism.
- The party recognised that the UK was deeply undemocratic and that rights were inadequately protected. It therefore was committed to *political and constitutional reform*.

New Labour policies and actions

New Labour was known as much for what it did *not* do as for what it *did* do. Despite calls from trade unions to have their powers — largely removed in the 1980s — restored, Labour governments refused. Similarly, pressure to bring privatised industries back under state control was resisted. Blair and his Chancellor, Gordon Brown, also resisted the temptation to restore high taxes on the wealthy and on successful businesses to pay for higher welfare, preferring to use public borrowing to facilitate their policies. At first, when the UK experienced an economic boom in the late 1990s and early 2000s, the extra spending could be sustained, but when the economy slowed down after 2007 the debts mounted up.

New Labour's political programme included the following policies:

- Huge increases in expenditure on the NHS
- Similarly large investment in education, especially early years education
- Reductions in corporate taxation to encourage enterprise
- An extensive programme of constitutional reform including the Human Rights Act, devolution, freedom of information and electoral reform in devolved administrations
- Through the tax and welfare system, various policies to reduce poverty, especially child and pensioner poverty
- Encouraging employment by introducing 'welfare to work' systems.

Labour ideas and policies today

Labour policy is still evolving following the dramatic split in the party in 2015–16. It is therefore advisable to view party policy from the point of view of the two wings of the party, the left and the centre-left. Table 2.8 explains the main tendencies in a number of key policy areas.

Knowledge check

Which policies of the current Labour Party may be considered as 'Old Labour' policies?

Table 2.8 Current Labour policies

Policy area	Centre-left tendency	Left tendency
Economic management	A pragmatic view including targets to reduce public-sector debt	Expansionist: high public expenditure should be used to promote investment, improve public services and create jobs
Social justice	Some adjustments to taxation to promote mild redistribution of income from high- to low-income groups	Radical tax reforms to promote significant redistribution of income from rich to poor
Industry	Industry to remain in private hands and be regulated by the state	Large infrastructure industries to be brought into public ownership (nationalised) Strong regulation of private-sector industries and finance
Welfare	Supports a strong welfare state and well-funded health service and education However, welfare benefits to be capped to ensure work pays and prevent abuse of the system	Strong support for the NHS and state education Abolition of university tuition fees More generous welfare benefits to help redistribute real income
Law and order	A mixture of authoritarian measures and 'social' remedies to crime	Emphasis on social remedies to crime
Foreign policy	Retention of a UK independent nuclear deterrent Strong support for NATO and the alliance with the USA	Largely 'isolationist', favouring non-intervention in world conflicts Abolition of the independent nuclear deterrent
Environment	Strong support for environmental protection and emissions control	The same as the centre-left
Constitutional reform	Some reforms are supported, including an elected second chamber and a proportional electoral system	More radical reforms, possibly including abolition of the second chamber and more independence for local government

Labour Party factions

As with the Conservative Party, the Labour Party is made up of various factions that compete for control of the party. They have been particularly important when it comes to leadership contests, as shown following Jeremy Corbyn's resignation in 2019. Under Jeremy Corbyn, the far left of the Labour Party came to dominate at most levels of the party structure, and its 2019 election manifesto was heavily influenced by this faction of the party. When it was rejected by the public in the 2019 general election, Corbyn chose to stand down as leader of the party.

In the ensuing leadership contest, the choice centred around the far-left candidate Rebecca Long-Bailey and the soft-left candidates Sir Keir Starmer and Lisa Nandy. Starmer won the contest in April 2020 with 56.4 per cent of the vote. As a soft-left Labour politician, Starmer falls somewhere between the far left and the Blairite factions of the party. Starmer is believed to be someone who can steer Labour more towards the centre ground of politics while retaining the support of the far left. Having been elected in the midst of the Covid-19 pandemic, at the time of writing in 2020 he had had little time to establish himself, but he had the challenging task of trying to unite a fractured party. His swift sacking in June 2020 of Rebecca Long-Bailey as shadow education secretary following her endorsement of an article that contained anti-Israeli sentiments shows he is willing to tackle the issue of antisemitism in Labour, even if it alienates the far left of the party. In October 2020, he went even further by suspending his immediate predecessor as leader of the Labour Party and figurehead of the left-wing, Jeremy Corbyn, for his reaction to the

Equalities and Human Rights Commission's report into antisemitism in the Labour Party during his leadership. Corbyn claimed in a statement that 'the scale of the problem (of antisemitism in the Labour Party) was also dramatically overstated for political reasons by our opponents inside and outside the party, as well as by much of the media.' The key Labour Party factions are outlined in Table 2.9.

Table 2.9 Labour Party factions

Faction	Core ideas	Key members
Momentum	Far left-wing, seeking wealth redistribution through taxation, public ownership of key industries, and the abandonment of the nuclear deterrent	Jeremy Corbyn John McDonnell Rebecca Long-Bailey
'Blairites'/Social Democrats	Centrist, key supporters of New Labour and the Third Way, as described above	Yvette Cooper Hilary Benn Stephen Kinnock
Blue Labour	Focused on working-class issues and employment Socially conservative, believing in traditional 'British values', anti-large-scale immigration, pro-free markets, but with protection for UK industry from foreign competition	Maurice Glasman Rowenna Davis Frank Field

The origins and development of the Liberal Democrats

The Liberal Democrats as a party is the product of an amalgamation of two parties in 1988. These were the Social Democratic Party, which had split off from the Labour Party and contained a group of moderate social democrats who felt that Labour had moved too far to the left, and the Liberal Party, which was a century old at that time.

- The Liberal Party had existed since 1877. It emerged as a coalition between Whigs and radicals. Its first leaders were Lord Palmerston and William Gladstone. The party was as important as the Conservatives until the 1920s, when it began to decline. By the end of the Second World War it had been eclipsed by the Labour Party. Until the 1990s it then played a very minor part in UK politics and returned only a handful of MPs to the UK Parliament. Despite its period in the political wilderness, the Liberal Party had become a home of radical political ideas under the leadership of Jo Grimond from 1956–67, many of which were ultimately adopted by the two main parties.
- The SDP was formed in 1981. It soon began talks with the Liberal Party. The problem for the two parties was that they were competing for the same voters. At the 1983 general election the Liberals and the SDP made an electoral pact whereby they would not put up candidates against each other. The pact was known as the Alliance. However, the plan failed and the two parties won fewer than 30 seats between them at that election and again in 1987. The decision was taken, therefore, to merge completely and the Liberal Democrats was born in 1988.

The Liberal Democrats reached the height of their electoral success in 2005 when the party won 62 seats. It was in 2010, however, that the real opportunity came. With no party winning an overall majority, the Liberal Democrats had a choice of whether to join with Labour (which had just been rejected by the voters), join the Conservatives (which sat on the opposite side of the political spectrum on many issues) in a coalition, or refuse to participate in the government, losing any opportunity to influence events first-hand and leaving a vacuum of power at a time of a major financial crisis. With no 'good' options, its leader, Nick Clegg, chose the Conservatives and they found themselves in government for the next 5 years.

Nick Clegg (right) and David Cameron in the rose garden of 10 Downing Street following the coalition agreement of May 2010

During the coalition years, the Liberal Democrats argue that they were able to act as a positive influence in government, injecting many of their own 'green' policies, introducing the pupil premium to support funding in schools and taking millions of low-income people out of paying tax altogether, as well as preventing the Conservative government from implementing, according to Liberal Democrat supporters, more extreme and less acceptable policies. Despite these arguments for success in the coalition era, what might have proved to be a resounding breakthrough turned in to a disaster for the party. The electorate decided to punish the Liberal Democrats for broken promises (mainly over a commitment not to raise university tuition fees, a commitment it dropped straight away) and working with the Conservative Party, which alienated many ex-Labour supporters who had voted Liberal Democrat to prevent a Conservative victory. As a result the party won only eight seats at the 2015 election. Nick Clegg resigned as leader and was replaced by Tim Farron, and the Liberal Democrats was once again a minor party, as it had been for 60 years between the 1930s and the 1990s.

Early in 2019 it looked as though the Liberal Democrats might be recovering and becoming a major force again in UK politics. Under the leadership of Vince Cable, in the European Parliament election held in June 2019, the Liberal Democrats secured 19.6% of the national vote and won 16 seats, second only to the Brexit Party and ahead of the other main parties, with Labour securing 13.6% of the vote and 10 seats, while the Conservatives achieved 8.8% of the vote and 4 seats. By October 2019 the Liberal Democrats had increased its number of MPs to 21 thanks to gaining four seats in the 2017 general election and then accepting a slew of former Conservative and Labour MPs who had defected from their original parties. Despite this increase and claims from its new leader, Jo Swinson, that the Liberal Democrats could be the next party of government, the party was again rejected when it came to the 2019 general election — it was reduced to 11 MPs, with Jo Swinson losing her own seat to the SNP.

Knowledge check

Why did the Liberal Democrats lose so many seats in 2015?

Core values of Liberal Democrats

The Liberal Democrats is not just a liberal party. As its title suggests, it also espouses social democratic values and ideas. Its values come, therefore, from a mixture of the two traditions. The social democratic values are largely those described above in the section on New Labour. Here we look at the liberal side of its position. The main liberal values adopted by the party include the following:

- *Liberty* is the core liberal value. Of course, complete freedom is not feasible in a modern society, so liberals confine themselves to believing that the state should interfere as little as possible in people's private lives. Privacy, freedom and individual rights must, they insist, be protected. The stress on liberty was a feature of nineteenth-century fundamentalist liberalism, often associated with **classical liberals**. In the latter part of the nineteenth century and the twentieth century, liberals expanded their ideas outside the protection of liberty and began to accept a wider role for the state in promoting welfare and social justice. These were known as new or **modern liberals**.
- Liberals also pursue *social justice*. This means three things. First it means the removal of unjustifiable inequalities in incomes in society, second it means equality of opportunity, and third it means the removal of all artificial privileges to which people might be born.
- *Welfare* is now a key liberal value. The liberal view is that people cannot be genuinely free if they are enslaved by poverty, unemployment or sickness, or the deprivations of old age. State welfare, therefore, sets people free.
- Liberals are highly suspicious of the power of government. They therefore believe that the power of government should be firmly controlled. The main way in which this can be achieved is by limiting the power of government via a strong constitution. This is known as *constitutionalism*.
- Liberal Democrats are *social reformers*. They strongly support the rights of women, the disabled, ethnic minorities and the LGBT+ community. They have also been strong supporters of same-sex marriage.
- The party has always been concerned with the causes of *human rights* and *democracy*, so it has supported constitutional reform. This aspiration is often described as *liberal democracy*.
- *Multiculturalism* is a key theme among liberal values. Different cultures and lifestyles should be welcomed and granted special rights. This links to the liberals' *pluralist* outlook on society.
- A modern value concerns the *environment*. Liberals believe that human life will be enriched by a healthy physical environment and by biodiversity.

It should be stressed that many of these so called liberal values are also held by many members of other political parties, notably those on the centre-left. Indeed,

Key terms

Classical liberals
Classical liberalism is a philosophy developed by early liberals who believed that individual freedom would best be achieved with the state playing a minimal role.

Modern liberals Modern liberalism emerged as a reaction against free-market capitalism, believing this had led to many individuals not being free. Freedom could no longer simply be defined as 'being left alone'.

many of them have become core British values. For example, the ideas of John Maynard Keynes in economic policy and William Beveridge in developing the NHS and welfare state, both of whom were prominent Liberals in the middle of the twentieth century, still exert a strong influence on many current politicians from all parties, as well as the modern Liberal Democrats. What distinguishes liberalism from other political traditions is that liberals place these values higher than all others. For example, the rights and liberties of individuals are so precious that they should be sacrificed only in exceptional circumstances.

Liberal Democrat ideas and policies today

The party is still trying to recover from its poor performances in the general elections of 2015, 2017 and 2019 and the loss of its leader to electoral defeat. With the Covid-19 pandemic, the Liberal Democrat leadership contest was delayed and the party was run on a temporary co-leadership basis, leaving it in an uncertain position with a lack of clear direction. By the end of 2020 Ed Davey had been elected as the leader and has been attempting to steer the party in a new direction.

The economy

Liberal Democrat economic policy is not especially distinctive. However, it does propose the rebalancing of the UK economy so that wealth and economic activity are spread more widely round the country. On the whole, Liberal Democrats are pragmatic about economic management. For example, government budget planning should not operate in such a way as to favour one section of society over another. Thus, in times of economic recession, the poor in society should be protected and the wealthy should bear the brunt of tighter economic policies. Taxation should always be fair, based on ability to pay, and should redistribute real income from rich to poor. To this end, in 2019, the party introduced a policy of a 1p-in-the-pound tax increase to help fund the NHS and a policy of free childcare for all two- to four-year-olds.

Law and order

Two principles characterise Liberal Democrat policy:

- Wherever possible, the law enforcement system, including prisons, should seek to rehabilitate offenders as much as punishing them. Liberal Democrats believe that most crime has social causes and these causes should be attacked.
- The system of law and order must not become so over-authoritarian that human rights are threatened. There must be a balance between civil liberties and the need for peace and security.

The European Union

The Liberal Democrats would have preferred the UK to stay inside the EU. Despite initially accepting the result of the 2016 referendum, although wanting the UK to remain as part of the single market, in 2019 the party reversed position and pledged to revoke Article 50 and stop Brexit.

Welfare

Education and health are key priorities for the Liberal Democrats. It believes spending on both should be protected and increased whenever the quality of services is threatened, leading to a 2019 pledge to increase the number of teachers in England by 20,000. The party believes the benefits system should be designed to encourage work and should be fair, favouring those who cannot support themselves. Poorer pensioners and single parents should be especially supported.

Foreign policy

Though Liberal Democrats support NATO and its aims, they are suspicious of excessive interference by the UK in conflicts abroad. They would abandon the renewal of the Trident nuclear submarine missile system. They strongly support the use of international aid. Wherever possible, international conflicts should be settled through the United Nations rather than through direct military intervention.

Liberal Democrat factions

It may seem odd to describe a party with only 11 MPs as having factions, but the Liberal Democrat Party still had a party membership of over 106,000 as of 2019, the largest in its history, and the factions that existed before 2015 still exist across the party and shape the leadership contests within Parliament. The main factions are outlined in Table 2.10.

Table 2.10 Liberal Democrat Party factions

Faction	Core ideas	Key members
Orange Book Liberals	Traditional liberal values of free markets and the withdrawal of the state from excessive interference Focus on individual liberties	Ed Davey
Social Liberals	Policies concerning social justice, with wealth redistribution from rich to poor through taxation and welfare provision	Tim Farron Jo Swinson Layla Moran

Discussion point

Look at Table 2.11. Consider the various policies of the three established parties and evaluate how similar they are and how different they are on key areas.

You may wish to focus on the following ideas:
1 The extent of agreement between the main parties.
2 The differences that exist as a result of ideological belief.
3 The differences that exist as a result of practical or political factors, such as targeting key groups.

Synoptic link

Manifestos are where party policies are set out and presented and they are usually created with a view to helping a party win an election by targeting key groups of voters, which will be covered in Chapter 4.

Activity

Look at Table 2.11 and the key policies from the 2019 general election of the established parties. Choose one of the mainstream parties and consider what these policies show about which factions of the party were dominant and which groups of supporters were being targeted.

Table 2.11 Key policies from the 2019 general election manifestos

Policy area	Conservative	Labour	Liberal Democrat
Economy	No income tax, VAT or national insurance rises	Raise minimum wage to £10 an hour Nationalise key industries Scrap charitable status for private schools	A penny income tax rise for the NHS Tax frequent flyers Freeze train fares Give zero-hour workers a 20% raise
Law and order	Hire an extra 20,000 police officers by 2022	Restore prison officer levels to 2010 levels End private prisons in the UK	Legalise cannabis
Welfare	Increase the number of nurses by 50,000 Pensions to rise by 2.5% a year Spend £6.3 billion on 2.2 million disadvantaged homes Continue the rollout of Universal Credit Create 250,000 extra childcare places	Increase the health budget by 4.3% Stop state pension age rises Introduce a national care service Scrap Universal Credit Introduce free bus travel for under-25s Build 100,000 council homes a year	Free childcare Recruit 20,000 more teachers Build 300,000 new homes a year
Foreign affairs	Leave the EU in January 2020 Introduce a points-based immigration system	Hold a second referendum on Brexit Give EU nationals the right to remain	Stop Brexit Resettle 10,000 refugees a year

Emerging and minor UK political parties

The growth of other parties in the UK

Although UK politics is usually dominated by two main parties, other parties play an important role in British democracy. None of these other parties has managed to break through like Labour did in the 1920s, but in their own ways they have all had some influence in the British political system. From the creation and administration of devolved institutions, to Brexit and the **co-option** of environmental ideas by the more established parties, alternative parties have shaped and developed the political debate in the UK. Perhaps most importantly, they exist to provide the electorate with more choice during elections.

The Scottish National Party (SNP)

As the fortunes of the UK's traditional third party, the Liberal Democrats, have declined since the 2010 general election, those of the Scottish National Party have blossomed. The SNP won enough seats in the 2007 Scottish parliamentary election to form a government. The party has formed the government of Scotland ever since. At Westminster, by contrast, it made little headway until 2015. Then, in the 2015 general election, the SNP won 56 of the 59 Westminster seats on offer in Scotland. It was an extraordinary result. Scottish voters were disillusioned with the main British parties, while many were interested in greater autonomy or even independence for Scotland. Perhaps the greatest achievement of the SNP was in persuading David Cameron to allow a referendum on Scottish independence in 2014 and acquiring greater levels of devolved responsibility in the Scotland Act of 2016. Although the party lost 21 of its Scottish seats in the 2017 general election, it recaptured most of them in 2019, returning to 48 seats and entrenching its position as the third major party in Westminster as well as the dominant party of Scotland.

The UK Independence Party (UKIP)

In politics, parties always need to be careful what they wish for. UKIP was launched as a party in 1993, in reaction to the development of the European Union (see Chapter 8 for more on this process). UKIP made its great electoral breakthrough in the 2015 general election. The party had already made progress in local elections and elections to the European Parliament, but this was the first time it had made a major effort in a general election. However, the outcome of its success was rather different from that of the SNP. UKIP won 12.6 per cent of the popular vote. However, because its support was so dispersed, this was converted into only one parliamentary seat. Thus, UKIP made an impact and took many votes away from the other main parties, but as a group it remained on the fringes of the political system.

The public support and threat to the major parties were enough, however, to convince David Cameron to pledge to an in/out referendum in 2016. UKIP not only got the referendum it wanted, it also got the result it had been campaigning for. Yet, having achieved its greatest triumph, the party then lost its purpose. In the 2017 general election the party's vote slumped, it lost its only parliamentary seat, and its leader, Paul Nuttall, resigned. The party seemed to be on the verge of extinction and in 2019 sank to just 0.1 per cent of the vote share. UKIP continues to exist as a party, but with Brexit happening, it no longer appears to have a role in British politics.

The Brexit Party

Launched in January 2019 by former UKIP leader Nigel Farage, the Brexit Party was established to put pressure on the Conservative government to achieve Brexit. The party was seen as a new version of UKIP and attracted many of UKIP's former supporters and backers. The Brexit Party served as a warning and a threat to the Conservative Party to get Brexit done or risk losing support, a move that put pressure on Theresa May and would eventually push the Conservative Party to the right under Boris Johnson. In May 2019 the Brexit Party achieved 36 per cent of the vote in the EU elections, a similar percentage to that which gave David Cameron a majority in the 2015 general election, though with half the turnout. This gave it 29 of the UK's 72 seats in the European Parliament. However, since the appointment of Boris Johnson as the Conservative leader, Brexit Party support has collapsed. It contested fewer than half the seats in the 2019 general election and achieved only 2 per cent of the vote. As with UKIP, as its central issue becomes resolved, it is likely that the Brexit Party will cease to have a relevant role in UK politics.

> ### Synoptic link
>
> UKIP and the Brexit Party are bound up with Euroscepticism. In order to understand their beliefs and their objections, you will need to have a good idea about how the institutions of the EU work and the nature of sovereignty in the UK, both of which are covered in Chapter 8.

The Green Party

The Green Party had a similar experience to UKIP, though on a smaller scale. The Greens' share of the vote rose from 1 per cent in 2010 to 3.8 per cent in 2015. The party won just one seat, Brighton Pavilion, where Caroline Lucas, co-leader of the party at the time, remains popular. Although the Greens have failed to make a major electoral breakthrough, Caroline Lucas has become a vocal and popular MP who has given her party a major platform in Parliament, while the adoption of more environmental policies by the major parties can, in part at least, be seen as a reaction to the growing support for the Green Party. Although Lucas is the most visible figure of the Green Party, she is not the current leader, as the Green Party tend to rotate their leadership often and practice a policy of shared leadership, with Jonathan Bartley and Sian Berry leading the party since 2018, neither of whom are currently MPs.

The Democratic Unionist Party (DUP)

Due to its unique situation and history, Northern Ireland has developed its own party system. The vote there is largely split between nationalist parties that want Northern Ireland to join with the Republic of Ireland and unionist parties that want Northern Ireland to remain part of the UK. The existence of different parties in Northern Ireland shows how different parties can allow different parts of the UK to follow their own paths and deal with their own problems. These issues are covered more in Chapter 5 on devolution. On its own, Northern Ireland does not have enough MPs to make a crucial difference in Westminster votes, however the fiercely unionist DUP, with 10 seats, was able to have tremendous influence in politics during Theresa May's time as prime minister as it made a supply-and-demand agreement that helped her government to survive, for a time at least, and ensured its voice had to be heard.

What does this reveal to us? While the UK remains dominated by the two main parties, several other parties are now making an impact, in terms of running parts of the UK through devolved institutions, parliamentary seats, votes or policy-making. It is especially true that in the national regions voters are offered a greater choice of party. In Scotland, for example, five parties have significance and offer a realistic choice. These are the SNP, Labour, Conservative, Liberal Democrats and Greens. In Wales four parties compete for significant influence, while in Northern Ireland the electoral system guarantees that at least five parties have a share in government.

Policies of the SNP and the Green Party

The SNP

The Scottish National Party is a centre-left party. Its main policies are as follows:

- The overall objective is complete independence as a sovereign state within the European Union.

- For as long as Scotland remains within the UK, the party supports constitutional reforms such as an elected second chamber, the introduction of proportional representation for general elections and votes for 16-year-olds.
- The party is social democratic and supports social justice. When Scotland has control over its own direct taxes it intends to redistribute real income from rich to poor. The party also supports the idea of the living wage.
- The party is opposed to the UK retaining independent nuclear weapons and favours the cancellation of Trident.
- The SNP has abolished university tuition fees paid by students within Scotland. It sees education at all levels as a key component of equality of opportunity. It has reintroduced the Educational Maintenance Allowance (EMA) for students above the age of 16. This has been abolished in England.
- Environmental protection is a key policy. SNP policies are almost as strong as those of the Green Party.
- The party supports the welfare state and would protect generous state provision of health, education and social security benefits.
- The SNP is also staunchly pro-EU and has been vocal in opposition to Brexit.

The Green Party

The Green Party obviously has environmental concerns at the centre of its policies. In other areas it has a left-wing stance. Among its radical policies are the following:

- Large numbers of new, low-cost, environmentally friendly homes should be financed or built by government to solve the housing crisis.
- There should be massive new investment in public transport.
- University tuition fees for students should be abolished.
- The party proposes an extensive programme of constitutional reform to make the UK more genuinely democratic.
- It proposes a wealth tax on the top 1 per cent of the income ladder, a living wage of £10 per hour and a special tax on large banks making excessive profits.
- In 2019 it adopted a policy to introduce a universal basic income of £89 per week.
- The party is opposed to the maintenance of Trident and the use of all nuclear power.
- It supports the legalisation of cannabis.

Study tip

Note that the impact of small parties is not just felt in terms of parliamentary seats won, but in terms of how many votes they may take away from more established parties. This forces those parties to react by modifying their policies to prevent their support leaking away.

Discussion point

Which UK party do you think is (a) the most left-wing and (b) the most right-wing?

You may wish to consider the following ideas:

1 Which factor most determines a position on the political spectrum.
2 How the policies stand as a combination.
3 What the ideology might be behind the policies.

UK political parties in context

Party systems

A **party system** describes the features of a political system in relation to the parties that operate within it. The term 'system' describes both how many parties there are and how many of those parties make a significant impact. The party system can help us understand how a political system works and it can also help us to explain change. This has been especially true in relation to the UK in recent times. Descriptions of different kinds of party system are provided below.

Table 2.12 illustrates where the various party systems can be found and some of their features.

Table 2.12 Examples of party systems

Party system	Countries	Features
One-party system	China, Cuba, North Korea	All three countries describe themselves as communist states. The Communist Party is the only legal party
Dominant-party system	Scotland (SNP)	The SNP holds nearly all the UK parliamentary seats and has governed Scotland since 2007, holding many more seats than any other party in the Scottish Parliament
Two-party system	USA	Democrats and Republicans hold virtually all elected positions at all levels of government in the USA
Two-and-a-half-party system	Canada	The Liberal Party of Canada and the Conservative Party of Canada form the two dominant parties, with the New Democratic Party emerging as a serious third party
Multi-party system	Italy, Germany	Italy has so many parties it is remarkably unstable. Governments regularly tend to fall Germany has a four-party system with the Christian Democrats and Social Democrats dominating, but they have to form coalitions with either the Greens or the Free Democrats. The afD (Alternative for Germany) party also has 94 seats

One-party system

This is where only one party is allowed to operate. This is normally associated with highly authoritarian regimes and we would not consider them to be democratic in the generally accepted sense of the word.

Dominant-party system

Here we are referring to democratic systems that do allow parties to operate freely, but where only one party has a realistic chance of taking governmental power. Such systems are highly stable, though there is a lack of accountability and competition.

Two-party system

Only two parties have a realistic chance of forming a government. It implies that two parties win the vast majority of the votes at elections and most of the seats in the representative assemblies of the state.

Two-and-a-half-party system

These are systems where there are two main parties that contest elections but also a sizeable third party. Usually, these third parties can be seen as holding the balance of power between the two main parties, much as the Liberal Democrats did in 2010 in the UK. Countries like Canada and Australia appear to fit this model. While the

UK could be seen as a two-and-a-half-party system, it is rare that the third party actually forms part of the government. This means it is not entirely accurate to consider the UK in this way, despite the presence of the Liberal Democrats and then the SNP as sizeable third parties.

Multi-party system

These are common in Europe and growing more so. As the name suggests, there are several parties competing for votes and power. There is no set number to define a multi-party system, but the key is that more than two parties have a realistic chance of being a part of the government and governments tend to be made up of coalitions. Although these systems can look a lot more fragile and unstable than dominant or two-party states, there is actually far less volatility as many of the same parties will regularly find themselves in government, time and again.

Synoptic links

Party systems should not be looked at in isolation from the electoral systems used in various countries, as there are strong links between electoral and party systems. This section should therefore be read in conjunction with the material in Chapter 3, which describes various electoral systems and their impacts. Generally, FPTP tends to create a two-party system (e.g. the USA) and PR tends to lead to a multi-party system, while majoritarian and hybrid systems tend to be more unpredictable in the party system they create. This means that while it can be helpful to consider a link between electoral systems and party systems, there is a lot more to it than a simple case of the electoral system causing the party system.

The party system in Westminster and beyond

The dominance of two parties in the UK has varied over the long term. Table 2.13 illustrates the dominance of two parties in terms of *seats* in the period 1979–2019. However, Figure 2.6 demonstrates that two-party dominance in terms of *votes* remained less pronounced until such dominance was restored in June 2017, and appears to have remained in 2019.

Table 2.13 Two-party dominance in the UK, 1979–2017

Election year	Conservative seats	Labour seats	Largest third-party seats	% of seats won by two main parties
1979	339	269	11	95.8
1983	397	209	23	93.3
1987	376	229	22	93.0
1992	336	271	20	93.2
1997	165	418	46	88.4
2001	166	413	52	87.8
2005	198	356	62	85.6
2010	307	258	57	86.9
2015	331	232	56	86.7
2017	318	262	35	89.2
2019	364	203	48	87.2

Source: The House of Commons Library

That other parties have been unable to convert their increasing proportion of votes won into significant numbers of seats is almost wholly due to the electoral system, which discriminates against them. It is therefore accurate to say that the UK remains a two-party system in terms of *seats* but is a multi-party system in terms of *votes*. As Figure 2.6 shows, however, two-party dominance showed signs of returning in 2017 and in 2019. The impact of the FPTP system is discussed in detail in Chapter 3.

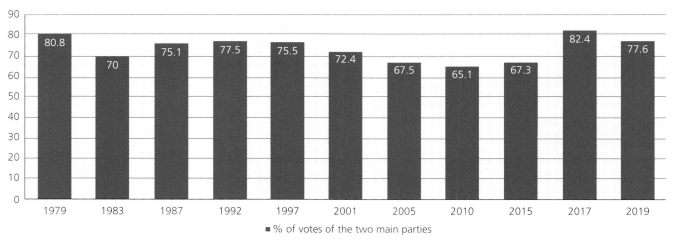

Figure 2.6 Vote share of the two main parties combined

Looking at the different regions of the UK, we see a similar model of one-party dominance or two-party competition. Since 2007 the SNP has dominated elections to the Scottish Parliament, making Scotland a dominant-party system. There are opposition parties, but neither the Conservatives nor Labour have had a realistic prospect of forming a government since then, despite the use of the additional member system (AMS) voting system. Since devolution was introduced to Wales, Labour has always been the main party, sometimes securing a majority, sometimes working in a coalition and sometimes as a minority government, again suggesting a dominant-party model. London mayoral elections have been dominated by the two-party system, with only Conservative and Labour candidates ever finishing in the top two, although elections to the Greater London Assembly are more mixed. Northern Ireland has its own party system, which appears at first to be multi-party with a number of parties being required to form a coalition to create a government, but the system is dominated by the DUP and Sinn Féin, suggesting a two-party dominance. Therefore, we may describe the UK as a whole as a multi-party system with many parties in different areas competing to gain power, but in terms of electoral results the UK elections reveal a dominant or two-party system at work.

Knowledge check

What kind of electoral system is used (a) in Northern Ireland and (b) in Wales and Scotland?

Factors that affect party success

Here we briefly examine three aspects of parties that go some way to determining why they may succeed or fail: leadership, unity and media exposure. We should first, however, consider why small parties have so much difficulty achieving a breakthrough. This is shown in Table 2.14.

Table 2.14 Why small parties find it difficult to make an impact and how they can nevertheless succeed

Why small parties fail	How small parties can succeed
They lack funding	They may find wealthy benefactors to support them, as occurred with UKIP after 2010
The electoral system may discriminate against them	In devolved regions, proportional representation helps small parties
	If a party can develop concentrated geographical support, it can break through in a region, such as the SNP
They lack media exposure	A strong, charismatic leader may help to gain public support, as occurred with Caroline Lucas for the Greens and Nigel Farage for UKIP and the Brexit Party
They lack organisation in communities	They may gain widespread popular support with populist ideas, as UKIP achieved
	They can build good community teams and can successfully compete in local government elections which then helps them in subsequent general elections
People consider voting for them to be a wasted vote	In proportional systems, fewer votes are wasted

Synoptic links

The factors that determine party success are closely tied to the factors that affect electoral success, and so the roles of the party leader and the media will be considered in greater depth in Chapter 4.

Leadership

This is crucial. Voters respond to the quality of the individual who leads a party and who therefore, in the case of the two main parties at least, is a potential prime minister. The qualities that voters prefer, disregarding for the moment their political beliefs, include:

- Experience
- Decisiveness
- Ability to lead
- Media image
- Intelligence
- Apparent honesty

In the past we have seen leaders who damaged the prospects of their party, such as Gordon Brown, Ed Miliband and Nick Clegg, and others who enhanced their party's fortunes, such as Margaret Thatcher and Tony Blair (at least until they both fell from grace). However, it was among smaller parties that leadership became important in 2015. Nicola Sturgeon, the SNP leader, made a hugely favourable impression in TV debates, as did Nigel Farage of UKIP and the popular and strong leader, Charles Kennedy, who took the Liberal Democrats to their most successful result in 2005. Farage was a master of the media, ensuring that his party was constantly in the news, while Sturgeon enjoyed very positive public approval ratings in the opinion polls.

Leaders do not win or lose elections, but from 2010 to 2019, and in elections to devolved assemblies, there is no doubt that party fortunes were influenced by the performance and image of their leaders.

Unity

It is often said by political commentators that a disunited party has no hope of being elected. The facts appear to bear this out. Some examples in both directions can illustrate this:

- In the 1980s, the Conservative Party united around the leadership of Margaret Thatcher while Labour was split between its left and right wings. In fact, the party did literally split in 1981. This resulted in two huge victories for the Conservatives at the 1983 and 1987 general elections.
- In 1997, Labour was an almost totally united party around the banner of New Labour under Blair. The Conservatives under John Major had been wracked by internal division, mainly over the UK's position in Europe. The result was a crushing victory for Labour.
- In 2015, the united Conservative Party dominated the disunited Labour Party. However, in the 2017 general election campaign, Labour succeeded in uniting around a radical manifesto, which resulted in a dramatic improvement in its fortunes.
- In 2019, having suspended moderate Conservatives from the party and required all candidates to sign a pledge to back getting Brexit done, the Conservative Party was able to unify around the central issue of Brexit while the Labour Party found itself divided over the issues of Brexit, antisemitism and the radical nature of its manifesto.

The evidence is therefore compelling that the commentators are right. United parties will always have a huge advantage over disunited parties.

The media

Whatever the true policies of a party are, the electorate will often be influenced by the image of the party as portrayed in the media, notably the printed press, such as newspapers. The newspapers tend to line up predictably at election time. Research suggests that newspapers only reinforce existing political affiliations and do not change minds, but there remains the probability that their campaigns may well persuade people to vote for the party their newspaper supports. Some wavering voters may also be swayed.

There is usually a correlation between the political views of the readership of a newspaper and the political stance of the paper itself. However, this may be because readers tend to buy newspapers with whose views they agree. There is scant evidence that newspapers significantly influence voting. Indeed, in the 2017 general election the vast majority of newspapers backed the Conservatives, mainly because their owners backed Conservative policies, but to little avail, as the party lost its Commons majority following a Labour resurgence.

Television and radio broadcasters, such as the BBC and ITV, have to be neutral and balanced by law. Nevertheless, TV in particular does give exposure to party leaders. TV debates have had an impact on the fortunes of the parties. In 2010, for example, Liberal Democrat leader Nick Clegg's performance in the TV debates was widely praised. Partly as a result, the Liberal Democrats did well enough to enter into a coalition with the Conservatives. In 2015, by contrast, Labour leader Ed Miliband performed poorly and this was a factor in Labour's failure to win the election. Equally, in 2017, Theresa May's refusal to participate in televised debates and Jeremy Corbyn's surprisingly good performance helped to shift the media perceptions of the two leaders.

A viewer watching Ed Miliband in the 2015 opposition leaders' debate

There has also been a growth in the role of social media, with Twitter and other social platforms becoming a key way for parties to promote themselves and critics to attack politicians they oppose. Social media allows politicians and political actors to speak directly to the public, bypassing many of the rules about media coverage and the influence of traditional media outlets. However, social media is difficult to control and its effects and importance are difficult to quantify. Perhaps more worryingly, social media can be used to target voters in intrusive and forceful ways, even in ways that are, at best, ethically dubious, as highlighted by the reports surrounding the collection of personal data from millions of Facebook users by Cambridge Analytica. Data was taken not only from those who had agreed to take their survey but also from their friends and contacts on Facebook to utilise and sell for political advantage.

The media is not decisive, nor is leadership and nor is a party's level of unity, but, put together, they *are* influential. However, it is still the performance of the government and the policies of the opposition parties that determine the outcome of elections.

Debate

How influential is the media in determining party success?

The case that it is influential

- The media is the prism through which public perceptions of the parties are created.
- The winning party usually has the support of most print newspapers.
- Since 2010, the leadership debates have become key moments in general election campaigns.
- Parties are increasingly developing resources to use social media to influence voters as well as utilising social media to collect data and target voters in increasingly sophisticated ways.
- Leaders spend time cultivating positive media images.

The case that it is not influential

- Influential media tends to reflect, rather than lead, attitudes to parties.
- Despite nearly all print newspapers opposing him, Jeremy Corbyn performed well in the 2017 general election.
- There is little evidence to suggest that leadership debates have affected public perception or changed minds.
- Social media tends to act as an echo chamber and rarely changes opinions or attitudes towards parties.
- Other factors, like leadership and policies, may be more influential.

It is worth considering the long- and short-term influence of the media, considering how far the media can actually change ideas in the short term, compared to long-term perceptions that the media can develop.

A summary of the role of parties in the UK

How well do parties enhance representative democracy?

Political parties play a number of key roles in the UK's representative form of democracy. These are, in particular, the following:

- They are vital in the selection of candidates for office. Without parties, candidates would campaign as individuals, which would make it difficult for voters to understand what collective policies might result from their decisions.
- They mobilise support for political *programmes*, not just individual policies. This is known as aggregation. Without such aggregation, politics would become incoherent.
- Parliament itself relies on party organisations to operate in an effective way. The parties organise debates and ensure that ministers are called to account. They also organise the staffing of parliamentary committees.

On the other hand, parties can also distort representation. The governing party is always elected without an overall majority of the national vote and yet it claims to have the mandate of the people. The 'winner takes all' nature of party politics may result in government that is too partisan and does not seek a consensus of support for policies. The coalition government between 2010 and 2015 was a rare example of parties cooperating with each other.

Parties also tend to reduce issues to 'binary' decision-making. That is, they tend to claim that one type of decision is either wholly wrong or wholly right. In reality this is rarely the case, but adversarial party politics tends to create these kinds of false choices.

A consideration of the role of positive and negative aspects of parties in representative democracy is given in Table 2.15.

Table 2.15 The role of parties in the UK's representative democracy

Positive aspects	Negative aspects
They provide open opportunities for people to become active in politics. They are inclusive and make few demands on members	Adversarial party politics is negative in that it denies the creation of consensus and reduces issues to false, simplistic choices
They make political issues coherent and help to make government accountable	Parties claim legitimacy through their electoral mandate even when they are elected to power with a minority of the popular vote
They help to make elections and the operation of Parliament effective and understandable to the public	Parties can become elitist so that small leadership groups dominate policy-making to the detriment of internal democracy
They identify, recruit and 'train' people for political office and leadership	They limit the pool of talent for political leadership to members only

Summary

Having read this chapter, you should have knowledge and understanding of the following:

→ What political parties are and do, and why they are so central to an understanding of how government and politics work in the UK

→ How parties are funded, the main issues concerning party funding and what proposals for reform have been offered

→ How political parties and their leaders fit into the left–right spectrum in UK politics

→ How the main political parties developed historically and what are the main ideological principles behind them

→ The nature and impact of smaller parties in various parts of the UK

→ The nature of the different party systems that exist within the UK, why they differ and the significance of those differences

→ In what ways and to what extent small parties make an impact on UK politics

→ The main reasons why some parties are successful and others are less successful

Key terms in this chapter

Classical liberals Classical liberalism is a philosophy developed by early liberals who believed that individual freedom would best be achieved with the state playing a minimal role.

Left-wing A widely used term for those who desire change, reform and alteration to the way in which society operates. Often this involves radical criticisms of capitalism made by modern liberal and socialist parties.

Modern liberals Modern liberalism emerged as a reaction against free-market capitalism, believing this had led to many individuals not being free. Freedom could no longer simply be defined as 'being left alone'.

New Labour (Third Way) A revision of the traditional Labour values and ideals represented by Old Labour. Influenced by Anthony Giddens, the 'Third Way' saw Labour shift in emphasis from a heavy focus on the working class to a wider base, and a less robust alliance with the trade unions.

New Right There are two elements — (i) the neo (or new) conservatives who want the state to take a more authoritarian approach to morality and law and order, and (ii) the neo-liberals who endorse the free-market approach and the rolling back of the state in people's lives and businesses.

Old Labour (social democracy) Key Labour principles embodying nationalisation, redistribution of wealth from rich to poor and the provision of continually improving welfare and state services, which largely rejected Thatcherite/free-market reforms or a Blairite approach.

One nation A paternalistic approach adopted by conservatives under the leadership of Benjamin Disraeli in the nineteenth century and continued by David Cameron and Theresa May in the twenty-first century, that the rich have an obligation to help the poor.

Party system The way or manner in which the political parties in a political system are grouped and structured. There are several variants that could apply to the UK; these include one-party dominant, two-party, two-and-a-half-party and multi-party systems.

Right-wing Reflects support for the status quo, little or no change, stressing the need for order, stability and hierarchy; generally relative to conservative parties.

Further reading

Websites

Information about all political parties can be found on their websites. This is also true of important party factions:

- Conservative Way Forward: **www.conwayfor.org**
- Cornerstone Group: **http://cornerstonegroup. wordpress.com**
- Tory Reform Group: **www.trg.org.uk**
- Momentum: **www.peoplesmomentum.com**

Information about party regulation and funding can be found on the Electoral Commission site: **www. electoralcommission.org.uk**

More information is on the UK Parliament site: **www. parliament.uk**

Books

Bale, T. (2016) *The Conservative Party: From Thatcher to Cameron*, Polity Press

Cole, M. and Deigham, H. (2012) *Political Parties in Britain*, Edinburgh University Press

Cook, A. (2012) *Political Parties in the UK*, Palgrave Macmillan

Davis, J. and Rentoul, J. (2019) *Heroes or Villains: The Blair government reconsidered*, Oxford University Press

Driver, S. (2011) *Understanding British Party Politics*, Polity

Kogan, D. (2018) *Protest and Power: The battle for the Labour party*, Bloomsbury Reader

Thorpe, A. (2016) *The History of the British Labour Party*, Palgrave — up-to-date but goes back to the origins of Labour

Practice questions

1

Source 1

A

It would appear that the era of two-party dominance in Westminster is over as the main parties tear themselves apart over the issue of Brexit. With neither party holding a majority in Westminster, smaller parties have become so much more important in influencing the course of events. The formation of the Independent Group for Change, made up of moderate Labour and Conservative MPs, the swelling numbers of Liberal Democrat MPs and the success of the Brexit Party in the EU elections show that the old loyalties to the two main parties are collapsing, to say nothing of the ongoing success of the SNP in Scotland and voice of the DUP in Northern Ireland. Who knows where we will go from here, but it appears that a future election could well throw up a truly multi-party system in Westminster, with anyone in power.

Source: A political commentator writing in September 2019

B

Once again, political reality trumps media hype and minor party optimism. The 2019 election saw the Liberal Democrats lose seats, the Change group get wiped out and the Brexit Party fade into irrelevance as the Conservatives and Labour secured more than three-quarters of the vote and over 87 per cent of the Westminster seats. Was this ever really more than a contest between those two parties? The SNP once again showed its domination of Scotland, while the DUP and Sinn Féin remain the dominant voice in Northern Ireland. It is hard to see the UK as anything other than a two-party system.

Source: A political commentator writing in December 2019

Using the sources, evaluate the view that the UK is now a multi-party system.

In your response you must:
- *compare and contrast different opinions in the sources*
- *examine and debate these views in a balanced way*
- *analyse and evaluate only the information presented in the sources.* (30)

2

> **Source 2**
>
> While it may be wrong to suggest that ideology and the beliefs of party members play no role in the success of party politics today, it would be fair to suggest that they have been marginalised in preference for strong party leadership. Recent elections have highlighted the difference a strong party leader, who can unify the party, grab media attention and instil confidence in an expectant nation, can make compared to a listless, dithering leader who fails to lead and make decisions. Of course it helps to be leader of one of the two main parties, but the success of the SNP under Sturgeon and UKIP and the Brexit Party under Farage show just what a smaller party can do with a charismatic leader. A strong leader can boost party membership, inspire activists to go knocking and campaigning on their behalf, and persuade donors to part with ready cash to fund them. Without a strong leader, I doubt very much that any of these other aspects of party success could be achieved. A strong leader will choose and focus on the right issues at the right time and deliver a message the people want to hear and deliver electoral success time and time again. It is no longer about the party, it is all about the leadership.
>
> Source: A political commentator

Using the source, evaluate the view that the abilities of the leader is the most important factor in explaining party success in the UK.

In your response you must:
- *compare and contrast different opinions in the source*
- *examine and debate these views in a balanced way*
- *analyse and evaluate only the information presented in the source.* (30)

3 Evaluate the view that the Labour Party remains committed to its traditional values and beliefs.
You must consider this view and the alternative to this view in a balanced way. (30)

4 Evaluate the recent divisions that exist between the Labour and Conservative parties over the economy, law and order and foreign affairs.
You must consider this view and the alternative to this view in a balanced way. (30)

5 Evaluate the case for introducing the complete state funding of parties.
You must consider this view and the alternative to this view in a balanced way. (30)

6 Evaluate the significance of two parties, other than the Conservative and Labour parties, in UK politics.
You must consider this view and the alternative to this view in a balanced way. (30)

3 Electoral systems

In the 2010 general election, David Cameron's Conservative Party won 36.1 per cent of the vote; enough to remove the incumbent Labour government of Gordon Brown but not enough to form a single-party majority in the House of Commons. In the 2015 election, the Conservatives achieved 36.9 per cent, which was enough to secure a slim majority in Parliament, but in 2017 Theresa May lost that majority despite winning 42.4 per cent of the vote. In 2019, Boris Johnson secured a landslide majority of 80 seats but achieved only 43.6 per cent of the vote.

Let's just think about those figures and what they mean. In 2015, a party that won only a third of the vote secured a majority of the representation. In 2017, that same party increased its vote by 5.5 per cent, yet the Prime Minister was considered to have 'lost' the election. In 2019, the new Prime Minister increased the vote by 1.2 per cent and had an emphatic victory. How can it be that the percentage of votes does not tally with the outcome and that small changes in votes can lead to major differences in outcome? The answer comes from the way in which votes are translated into seats in a representative body through an electoral system, in this case first-past-the-post, or FPTP.

In this chapter we consider various electoral systems used in the UK, how different electoral systems work and their importance in the political system, as well as the controversies over how they operate. Then we examine the impact of the growing use of referendums in the UK. Even if it is true, as some critics have claimed, that voting in elections changes little, referendums do change things, as the EU vote of 2016 proved so dramatically.

Boris Johnson as Prime Minister following his 2019 election victory

Objectives

In this chapter you will learn about the following:

→ How the different electoral systems used in the UK operate

→ How to compare FPTP with other electoral systems as they are used in the UK

→ The advantages and disadvantages of each individual electoral system

→ The debates on why different electoral systems are used in the UK

→ The relationship between electoral systems and party representation

→ What a referendum is and how they have been used in the UK

→ The impact of referendums on political life in the UK since 1997

→ The arguments for and against the use of referendums in a representative democracy

Different electoral systems

Functions of electoral systems

Before learning about the types of electoral systems used in the UK, it is worth considering what the functions of an electoral system are. Knowing its functions will help you to evaluate and compare the effectiveness of each electoral system by considering whether it is fit for purpose.

The main functions and importance of elections include the following:

- Elections are used to choose representatives. In a democracy, legislators and decision-makers have to be elected.
- Elections are the most important way in which citizens become involved in politics. For many it is their *only* form of political participation.
- Elections are a time when government and elected representatives can be called to account. During an election campaign the candidates must justify what they and their party have done.
- Democracy demands that the people have choice over who represents their ideas and interests. Elections should provide that choice.
- Elections have an educative function. During election campaigns the public can become better informed about the key political issues that face their locality, region or nation.
- Elections provide a *mandate*. The winners in an election are granted democratic legitimacy, the political authority to carry out the political programme that they are proposing.

In addition, it is worth considering how the process of holding an election in the UK, regardless of which electoral system is used, helps to promote democratic legitimacy and ensure the functions given above are achieved.

As Figure 3.1 shows, there are many positive features of the way UK elections are held, suggesting they are highly democratic. However, this process for holding elections only relates to the casting and counting of votes. How those votes are turned into representation depends on the system that is used. Each system has its advantages, but also has disadvantages that can make it appear less fair or less desirable than other systems. This is the key debate that surrounds electoral systems and which one is best for the UK to use in general elections. It is therefore important for you to decide if you think the positives of each electoral system outweigh the negatives, and how they relate to each other.

Study tip

When considering the effectiveness of an electoral system, break down the key functions and decide which you consider to be the most important and why. Prioritising the most important function for you will help you evaluate the strength of the arguments for each system and reach a well-reasoned judgement.

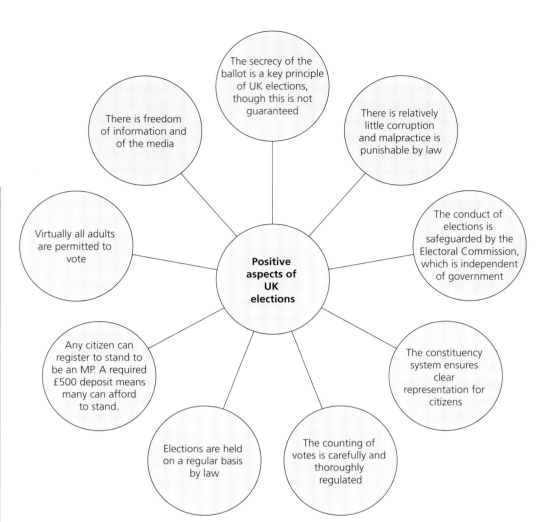

Figure 3.1 Democratic aspects of UK elections

Types of electoral systems

There are four main electoral systems that you will study: **first-past-the-post
(FPTP)**; **additional member system (AMS)**; **single transferable vote (STV)**;
and **supplementary vote (SV)**. Table 3.1 gives more information about each system
and where in the UK it is used. Each is an example of a type of electoral system
and understanding the different types of electoral system will help you evaluate the
reasons behind the different workings of each system.

Table 3.1 Types of electoral system

Type of system	Definition		Electoral system	Where used in the UK
Plurality	To win a seat, a candidate only requires one more vote than any other candidate, meaning they do not need to secure an absolute majority		FPTP	General elections Local council elections in England and Wales
Majoritarian	Used to elect a single candidate, these systems are designed to attempt to secure an absolute majority for the winning candidate		SV	London Mayor Other metro mayors By-elections for STV

Type of system	Definition	Electoral system	Where used in the UK
Proportional	A system that attempts to allocate seats in direct proportion to votes cast. As such they are multi-member constituencies	STV	Northern Ireland Parliament Scottish local government
Hybrid	A system that mixes two other types of system, such as plurality and proportional	AMS	Scottish Parliament Welsh Senedd Greater London Assembly

First-past-the-post

The main electoral system used for general elections in the UK is first-past-the-post. FPTP is a straightforward electoral system where voters simply pick a single candidate to represent their **constituency**. The votes are counted and whoever gains the most votes wins the seat, regardless of the total number of votes cast or the relative proportion. It is therefore a '**plurality**' system. The key features of FPTP are outlined in Figure 3.2.

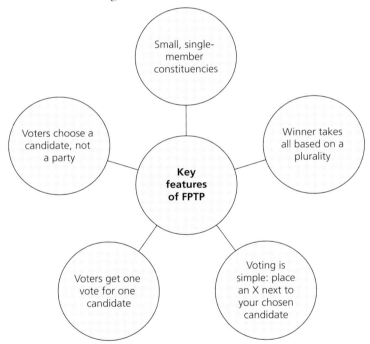

Figure 3.2 **Key features of FPTP**

For general elections in the UK, FPTP operates by dividing the country into 650 constituencies of roughly equal size. The average adult population of a constituency is 75,000, though there is some variation and, of course, the geographical size of constituencies varies considerably. Tightly populated London constituencies are much smaller in size than sparsely populated constituencies in rural areas. Ideally, constituency populations should be relatively homogeneous (uniform), to allow for effective representation by their single representative.

To illustrate this, Tables 3.2 and 3.3 show the results in two constituencies in 2019. Table 3.2 shows a candidate winning an **absolute majority** in the Arundel and South Downs constituency (57.9 per cent), while Table 3.3 shows a candidate winning merely a plurality (41.9 per cent) in the Lanark and Hamilton East constituency.

Table 3.2 Arundel and South Downs constituency 2019 election result

Candidate (party)	Votes won	Percentage of total votes
Griffith (Conservative)	35,566	57.9
Bennett (Lib Dem)	13,045	21.2
Sankey (Labour)	9722	15.8
Thurston (Green)	2519	4.1
Wheal (Independent)	566	0.9

Source: The House of Commons Library

Table 3.3 Lanark and Hamilton East constituency 2019 election result

Candidate (party)	Votes won	Percentage of total votes
Crawley (SNP)	22,243	41.9
Haslam (Conservative)	17,056	32.1
Hilland (Labour)	10,736	20.2
Pickard (Lib Dem)	3037	5.7

Source: The House of Commons Library

These results reveal two key issues with FPTP that impact on the effectiveness of the democratic representation offered: that some constituencies are safe seats, meaning there is little real choice for voters, while others are minority seats, meaning only a minority actually choose the winning candidate. In 2019, 421 of the 650 seats were won by an absolute majority, with 207 won by between 40 per cent and 50 per cent, and 22 won by less than 40 per cent. However, if we look back at the 2015 general election, we see a more worrying outcome: only 319 out of 650 MPs won an absolute majority of the votes in their constituency. Fifty MPs even secured a seat with less than 40 per cent of the popular vote in their constituency. So, most elected MPs had to admit in 2015 that more people voted *against* them than *for* them.

There is one final issue to consider about the level of support MPs secure, and that is the issue of turnout. The results shown in Tables 3.2 and 3.4 and indeed for all statistics around winning margins only deal with the percentage of votes actually cast. When turnout is considered and we broaden it to all potential voters, we have an even less representative set of results. For example, in Arundel and South Downs, with a 2019 turnout of 75.4 per cent, the winning candidate only secured 43.5 per cent of the electorate's support, someway short of an absolute majority, while in Lanark and Hamilton East, with a 2019 turnout of 68.34 per cent, the winning candidate only received support form 28.6 per cent of the total electorate. This underlines the importance and impact of turnout, covered in Chapter 1, on electoral outcomes and democratic support.

> ### Synoptic link
> The issue of turnout and how it impacts on the legitimacy of candidates and democratic representation is covered in Chapter 1.

Key issues relating to FPTP
As with all electoral systems, there are good and bad points to using FPTP, as shown in Table 3.4. However, there are three highly significant issues that relate to FPTP in particular and form the main basis for criticising the system. Understanding what these issues are, and their impact on election outcomes, will help you to evaluate the strength of the arguments for and against reform.

Table 3.4 Advantages and disadvantages of FPTP

Advantages of FPTP	Disadvantages of FPTP
It is a simple system and voters can understand exactly what they are voting for	It gives an advantage to parties that have concentrated support in certain regions
It helps to ensure representatives are closely bound to the needs and concerns of their constituency	It is disadvantageous to parties whose support is dispersed widely
It helps to ensure small extremist parties find it difficult to gain representation	It favours the large parties and prevents serious challenges from small parties
It tends to produce an outright winner, that is, a party that has an overall majority in the House of Commons and therefore produces a clear mandate	There is a 'winner's bonus', where the biggest party tends to win more than its proportionate share of the vote. In 2019 the Conservatives won 43% of the votes, which was converted into 56% of the seats
It is traditional, having been used as the main voting system throughout British electoral history, and is part of British political tradition	In some recent general elections (2010, 2015 and 2017), the system failed to produce a decisive government majority

The importance of concentrated support

The FPTP electoral system has the effect of favouring those parties that have their support concentrated in certain areas. This can be demonstrated if we consider the overall general election result in 2019. This is shown in Table 3.5.

Table 3.5 The result of the UK general election, November 2019

Party	% of vote won	% of seats won	Seats won	Notes
Conservative	43.6	56.2	365	Conservative support is concentrated in south and central England and across anti-EU areas of the north
Labour	32.2	31.2	203	Labour support is concentrated in industrial, urban northern England and Wales and across London
Scottish National Party	3.9	7.4	48	The SNP only contests the 59 seats in Scotland
Liberal Democrats	11.5	1.7	11	Lib Dem support is widely dispersed across the whole UK
Democratic Unionist Party (DUP)	0.8	1.2	8	The DUP only contests seats in Northern Ireland
Sinn Féin	0.6	1.1	7	Sinn Féin only contests seats in Northern Ireland
Plaid Cymru	0.5	0.8	4	Plaid Cymru only contests seats in Wales
Brexit Party	2.0	0	0	Brexit Party support was thinly spread across England and Wales
Green Party	2.7	0.2	1	Green Party support is widely dispersed
Others	2.2	0.2	1	Very dispersed

Source: The House of Commons Library

The reason why this occurs is that parties with dispersed support, such as the Green Party, the Liberal Democrats (as well as Labour and the Conservatives in Scotland), UKIP and the Brexit Party, find it very difficult to secure a high enough total to win in individual constituency contests and hence gain few seats in Parliament.

In England and Wales, Conservative support is heavily concentrated in the south-east and rural areas, while for Labour it is concentrated in major cities and urban regions, especially in the north of England, south Wales and London, so it tends to gain a lot of seats in these areas. However, it was notable that in 2019 the Conservatives

Knowledge check

Which parties have benefited from the lack of proportional representation in FPTP?

were able to make inroads into the so called 'red wall' of northern industrial seats in England due to those areas' support for Brexit. The SNP also does well under FPTP because its support is concentrated in one region. Therefore, the SNP holds 48 seats in Westminster compared with the 11 the Lib Dems has, despite getting only 3.9 per cent of the vote in 2019 compared with the Lib Dems' 11.5 per cent.

This discrepancy can be emphasised by considering how many votes, on average, it took for each party to secure the election of a candidate. This can be calculated by dividing the total number of votes won by each party nationally by the number of seats the party won, the 2019 calculations being shown in Table 3.6.

Table 3.6 Average votes needed to elect one member in the 2019 general election

Party	Number of votes per seat won
Green Party	866,400
Liberal Democrats	336,000
Labour	50,800
Conservative	38,300
Plaid Cymru	38,300
DUP	30,500
Sinn Féin	26,000
SNP	25,900

Source: The House of Commons Library

Table 3.6 shows great disparities between how efficiently the parties turned votes cast for them into seats won by them. The Northern Ireland parties (Democratic Unionist and Sinn Féin) have low averages largely because they are evenly matched and because of low turnouts. However, it is striking how big a disparity there is between the Green Party and the Liberal Democrats with extremely high averages, and the SNP with its very low average. We can also see that the Conservative Party had an advantage over Labour, which was decisive in the 2019 election.

The parties that lose out under FPTP in this system tend to favour reform, while the main 'winner' (the Conservatives) has the least interest in reforming the system to remove such discrepancies.

Discussion point

Evaluate the view that the advantages of FPTP outweigh the disadvantages.

In your discussion you may want to consider the following ideas:
1. Whether the high degree of local support it ensures outweighs its unrepresentativeness.
2. Whether preventing extremist parties from gaining representation outweighs the undemocratic exclusion of smaller parties.
3. Whether a system that prioritises local concerns is better than a system that would prioritise national ones.

Activity

The 2015 general election revealed some of the key flaws with the FPTP electoral system. Research the number of seats each party won and the percentage of the votes it won. How can this information be used to support the arguments against FPTP? Does 2015 provide more convincing evidence for or against using FPTP for general elections? Why?

Safe seats

A **safe seat** is a constituency where it is almost certain that the same party will win the seat at every general election. The Electoral Reform Society (**www.electoral-reform.org.uk**) estimated that 316 seats out of 650 were safe seats in 2019, representing nearly half the country.

The implications of there being so many safe seats include the following:

- Parties will pay little attention to safe seats in the election campaigns, so voters will receive less information.
- MPs sitting for such safe constituencies are less accountable for their actions because they have virtually no chance of losing their seat at the next election.
- Voters in safe seats may feel their votes are 'wasted' because they have no realistic chance of influencing the outcome. This may be the case whether they support the winning party or one of the losing parties.
- It means that votes are, effectively, not of equal value. Votes in safe seats are worth less than votes in seats that are keenly contested, where voters may have more of an impact.
- The Electoral Reform Society estimated that in the 2019 general election, 70.8 per cent of the voters, numbering 22.6 million, were effectively casting 'wasted' votes because they had no role in influencing the outcome in their constituencies.

Marginal seats

There is no precise definition of a **marginal seat**. However, in general, marginal seats are those where the outcome of an election is in doubt. Such seats are likely to change hands from one party to another quite frequently. It is, therefore, often said that elections are won and lost in these marginal seats. In the 2019 general election there were 141 marginal seats in the UK, defined as one where the last winning candidate led by 10 per cent or less from the nearest challenger (source: The House of Commons Library).

The implications of the existence of marginal seats include:

- Parties concentrate their efforts on marginal seats, so voters there receive much more attention and information.
- Votes in marginal seats are more valuable than votes in safe seats as the voters in marginal seats are more likely to influence the result.
- The individual candidates become more important in marginal seats. In safe seats the qualities of individual candidates matter little, but in marginals they can be crucial.
- Marginal seats may result in 'tactical voting'. A tactical vote is when a voter who supports a party that is unlikely to win a constituency switches allegiance to one of the other parties in the hope of influencing the outcome, usually by blocking the less favoured party.

The case for and against FPTP

As an electoral system, FPTP has its supporters and its detractors. Among the supporters are established members of the two main parties. This is not surprising as the Conservative and Labour parties are the main beneficiaries of the system, although the Labour Party has considered changing its policy position towards reform of the system. It is also not surprising that it is small parties that support a change to the system. Even the Scottish National Party, which now benefits so strongly from FPTP, supports reform. Pressure groups such as the Electoral

Key terms

Safe seat A seat in which the incumbent (the person who already holds the seat or position) has a considerable majority over their closest rival and is largely immune to swings in voting choice. The same political party retains the seat from election to election.

Marginal seat A seat held by the incumbent with a small majority/plurality of the vote.

Reform Society and Unlock Democracy are also prominent campaigners for change.

One of the most important debates in contemporary UK politics is whether FPTP is desirable and should be retained, or whether it is undemocratic and should be replaced. The arguments on both sides are summarised in the debate below.

Study tip

In addition to the positive and negative features of FPTP, you should consider the arguments in the light of the main alternative voting systems.

Debate

Should FPTP be retained?

The case for retention

- It is easy to understand and produces a clear result in each constituency.
- The result is usually known very quickly.
- It produces one single representative for each constituency and so creates a close constituency–MP bond.
- The accountability of the individual MP is clear to the electors.
- The system tends to produce a clear winner in a general election, i.e. a single party with a parliamentary majority. This helps to promote strong, stable, decisive government.
- It helps to prevent small parties breaking in to the system. This is useful if the small parties are undesirable 'extremists'.

- Arguably FPTP has stood the test of time. Abandoning the system would be a dangerous step into the unknown.
- A switch to a different system might have all sorts of unintended consequences.
- In 2011, a referendum decisively rejected a proposal for change.
- In elections with complex concerns — as occurred in 2017 when the Brexit issue was combined with other social and economic matters — FPTP gives voters the opportunity to choose a candidate based on their individual attitude to such issues, rather than merely according to their party allegiance.

The case against retention

- The overall outcome is not proportional or fair. Some parties win more seats than their support warrants, while others win fewer than they deserve.
- It means that many votes are effectively wasted because they can have no impact on the outcome in safe seats. Many seats become part of party 'heartlands', where there is no possibility of a realistic challenge from other parties. It also produces 'electoral deserts', where there is effectively no party competition.
- Votes are of unequal value in that votes in safe seats are less valuable than votes in marginal seats (see Table 3.6). UKIP votes were of hugely less value than Conservative votes in 2015.
- It encourages some voters to vote tactically and so abandon the party they really want to support.

- It prevents new parties breaking into the system and so produces political 'inertia'.
- It has, since 1945, always resulted in the winning party securing much less than half the popular vote. In 2015 the winning Conservative Party was elected with just 36.9 per cent of the popular vote, meaning 63.1 per cent of the voters had voted against the governing party. In 2005 Labour won the election with a majority of 66 from only 35.2 per cent of the popular vote. This calls into question the legitimacy of the government.
- FPTP always used to deliver governments with a majority of the seats in the House of Commons. However, in 2010 and 2017 the system failed to do this, returning governments without such an overall majority. If it is failing to achieve its main objective in modern times, this suggests it should be replaced with a fairer system.

Look over the points for both sides of the debate and consider how important each one is to you and why. Refer back to the functions at the start of this section and consider if FPTP is fit for purpose in relation to your most important function.

Other electoral systems in the UK

The additional member system

The additional member system is a hybrid system that combines FPTP with a proportional representation system, in this case one called '**closed party list**'. It is used in Scotland and Wales and for the Greater London Assembly. A proportion of the seats (which varies from region to region) is awarded through FPTP. The rest are awarded by a regional closed party list system. This means that every voter has two votes. One is for a constituency candidate in the normal way, the other is for a party. In Scotland, 73 seats are elected by the FPTP method while 56 are elected via the list system. In Wales, 40 seats are constituency-based and 20 decided by the list system.

So, some of the elected representatives have a constituency to look after, while others do not. The latter have been elected from the party lists and are free of constituency responsibilities. No real distinction is made between the two, though the senior party members tend to be elected from lists rather than in constituencies.

AMS is something of a compromise. It is designed to make a system *partly* proportional, but also preserves the idea of parliamentary constituencies with an MP to represent them. It helps smaller parties, but also favours the larger ones. It achieves two objectives at the same time, preserving the idea of constituencies and a constituency representative, but producing a much more proportional result than FPTP.

> ### Useful term
>
> **Closed party list** A proportional electoral system where voters vote for a list of candidates provided by a party. Based on the proportion of the votes a party receives it will be awarded a number of seats from across large multi-member constituencies. The order of the candidates is determined by the party, with the higher preferences being the ones most likely to secure seats.

> ### Study tip
>
> A common mistake is to believe that proportional representation is an electoral system. It is not. It is a *description* of *various* electoral systems that tend to produce a proportional outcome between the parties.

How AMS works in Scotland and Wales:

- A proportion of the seats are elected using FPTP, as for UK general elections.
- The remaining proportion of seats are elected on a proportional system based on several regions of the country. This is known as the regional closed party list part of the system.
- There is an important variation in the regional list part of the vote. The variable top-up system adjusts the proportions of votes cast on the list system based on the over- or under-representation parties have experienced in the FPTP seats. This is a complex calculation, but, in essence, what

An example of an AMS ballot paper for the Scottish Parliament happens is that the seats awarded from the list system are adjusted to give a more proportional result.

- Parties that do less well in the constituencies (typically Conservatives or Greens) have their proportion of list votes adjusted upwards. Those that do proportionally well under first-past-the-post (typically Labour in Wales and the SNP in Scotland) have their list votes adjusted downwards.
- The overall effect of variable top-up is to make the total result in seats close to proportional to the total votes cast in both systems.

Debate

Would AMS be a good system to use for Westminster elections?

Advantages

- It produces a broadly proportional outcome and so is fair to all parties.
- It gives voters two votes and so more choice.
- It combines preserving constituency representation with a proportional outcome.
- It helps small parties that cannot win constituency contests.

Drawbacks

- It produces two classes of representative — those with a constituency and those elected through the lists. The latter tend to be senior.
- It is more complex than first-past-the-post. Having two votes can confuse some voters.
- It can result in the election of extremist candidates.

Do the advantages of AMS outweigh the disadvantages? Consider what you would prioritise as the most important function of an electoral system. Would AMS achieve this function more or less effectively than FPTP? Based on that, and the earlier points about FPTP, which system would you consider the stronger for UK general elections?

When we look at the results of elections under AMS we can see that a party wins some of its seats through constituency contests and the rest from the regional list elections in which voters choose a party rather than an individual. Table 3.7 shows the results of the elections to the Scottish Parliament in 2016. It shows how well each party performed in both the constituency elections and the list elections. Bear in mind that the seats on the regional list system are manipulated to produce a more proportional result.

Table 3.7 Results of the elections to the Scottish Parliament, 2016

Party	Constituency seats won	Regional list system seats awarded	Total seats won	% seats won	% votes won in the regional lists
SNP	59	4	63	48.8	41.7
Conservative	7	24	31	24.0	22.9
Labour	3	21	24	18.6	19.1
Green Party	0	6	6	4.7	6.6
Liberal Democrats	4	1	5	3.9	5.2
Others	0	0	0	0	4.5

Source: The Electoral Commission

We can see that the proportion of seats won by each party is quite close to the proportion of votes each of them won in the party list contest, so the result is broadly proportional. We can also see that all the smaller parties won very few constituency seats. Conversely, the SNP won 59 out of the 73 constituency seats available. Had this election been conducted under the first-past-the-post system, the SNP would have dominated by winning 104 (81 per cent) out of 129 seats! Under AMS, the SNP won 48.8 per cent of the seats on 41.7 per cent of the popular vote — a much more proportional outcome.

Activity

Research the result of the elections to the postponed Welsh Parliament and the Scottish Parliament held in 2021. Which parties seemed to benefit most from the AMS system and which party or parties did not? Explain your answers with the use of statistical evidence from the results.

The single transferable vote

The single transferable vote is the system used in Northern Ireland for its Assembly elections and for local council elections. It is an example of a proportional system.

It is a complex system, especially when it comes to the counting and the establishment of the result. This is how it works. Some detail has been omitted for the sake of brevity, but the following features are what you need to know:

- There are typically six seats available in each constituency.
- Each party is permitted to put up as many candidates as there are seats, i.e. up to six. In practice, parties do not adopt six candidates as they have little chance of winning all six seats available. Four is the normal maximum number from each party.
- Voters place the candidates in their order of preference by placing a number 1, 2, 3, etc. beside their names. They can give a preference for all possible candidates, or only those candidates they support.
- Voters can vote for candidates from different parties or even all the parties, though few do.
- At the count, an *electoral quota* is calculated. This is established by taking the total number of votes cast and dividing it by the number of seats available plus 1. So, if 50,000 votes were cast and six seats are available, the quota is:

$$50,000 / (6 + 1 = 7)$$

This works out as 7143. One is then added, giving a final figure of 7144.
- Initially, all the first preferences are counted for each candidate. Any candidates who achieve the quota are elected automatically.
- After this stage the counting is complex. Essentially, the candidate that came last is eliminated and the second and subsequent preferences from their ballot papers are added to the other candidates. If this results in an individual achieving the quota, they are elected.
- This process continues until six candidates have achieved the quota and are elected.

Northern Ireland Assembly Election
Name of constituency

You can make as many or as few choices as you wish
Put the number **1** in the voting box next to your first choice
Put the number **2** in the voting box next to your second choice
Put the number **3** in the voting box next to your third choice **And so on**

FIRST, Candidate
Address
Party

SECOND, Candidate
Address
Party

THIRD, Candidate
Address
Party

FOURTH, Candidate
Address
Party

FIFTH, Candidate
Address
Party

SIXTH, Candidate
Address
Party

An example of an STV ballot paper from the Northern Ireland Assembly election

The complex counting system is designed to ensure that voters' preferences are aggregated to make sure that the six most popular candidates *overall* are elected. The overall outcome tends to be highly proportional, with each party achieving its fair share of the votes and seats.

An example of a single-constituency contest from the Northern Ireland Assembly election in 2016 illustrates how this works. The results shown in Table 3.8 are for Fermanagh and South Tyrone.

Table 3.8 Results of Fermanagh and South Tyrone constituency, 2016

Party	Number of candidates offered	Candidates elected
DUP	2	2
Sinn Féin	4	2
Ulster Unionists	2	1
SDLP	1	1
Others	6	0

Source: The Electoral Commission

The quota was 6740. Only one candidate (Arlene Foster of the DUP) achieved the quota on first preference votes.

Table 3.9 shows the overall results in Northern Ireland in 2017. Here we can clearly see how proportional the outcome is. Every one of the main parties won approximately the same proportion of seats as the proportion of first preference votes gained. It is also interesting to see how many parties won some representation.

Table 3.9 Results of the Northern Ireland Assembly election, 2017

Party	Seats won	% seats won	% first preference votes won
Democratic Unionists	28	31.1	28.1
Sinn Féin	27	30.0	27.9
Ulster Unionists	10	11.1	12.9
SDLP	12	13.3	11.9
Alliance	8	8.9	9.1
Others	5	5.6	10.1

Source: The Electoral Commission

Debate

How effective is STV as an electoral system?

Advantages

- It produces a broadly proportional outcome.
- It gives voters a very wide choice of candidates. The second and subsequent choices of the voters are taken into consideration in the counting.
- Voters can vote for candidates from different parties and show a preference between candidates of the same party.
- As there are six representatives per constituency, each voter has a choice of those to represent them and usually can be represented by someone from the party they support.
- It helps small parties and independent candidates to be elected.

Drawbacks

- It is quite a complex system that some voters do not understand.
- The vote counting is complicated and can take a long time.
- It can help candidates with extremist views to be elected.
- With six representatives per constituency, the lines of accountability are not clear.

Consider again your priorities for an electoral system. Based on this, how persuasive do you find the argument that STV should be used for Westminster elections? Your evaluation always depends on what you choose to prioritise.

The supplementary vote

The supplementary vote is a system used to elect a single candidate in a constituency. It is designed to produce a winner who can claim to be supported overall by a majority of the voters. In the UK its main use is to elect city mayors. It could be used to elect MPs but there is little support for this kind of reform. Most reformers prefer the idea of proportional representation rather than the supplementary vote.

Election of the Mayor

Vote once (X) in column one for your first choice
Vote once (X) in column two for your second choice

		column one **first choice**	column two **second choice**
1	FIRST, Candidate Party	☐	☐
2	SECOND, Candidate Party	☐	☐
3	THIRD, Candidate Party	☐	☐
4	FOURTH, Candidate Party	☐	☐
5	FIFTH, Candidate Party	☐	☐
6	SIXTH, Candidate Party	☐	☐

An example of an SV ballot paper

Voters have two choices, a first and second choice. If any candidate achieves an overall majority, i.e. 50%+, of the first choice or round, they are automatically elected. If this does not happen, the top two candidates go into a second round of counting. All the others drop out. The second-choice votes of the eliminated candidates are added to the first choices for the top two candidates. As there are only two candidates left, one of them is almost certain to achieve an absolute majority. So, the winner has an overall majority of a combination of first- and second-choice votes. Table 3.10 illustrates how this worked in the election of the London Mayor in 2016. Sadiq Khan, the eventual winner, achieved only 44.2 per cent of the first-choice votes, but 161,427 voters put him as their second choice, which was enough to give him an absolute majority. An assessment of SV is outlined in Table 3.11 while Table 3.12 shows a comparison of alternative electoral systems.

Table 3.10 Elections for Mayor of London, 2016

Candidate	Party	1st round votes	% votes	2nd round votes	% total
Sadiq Khan	Labour	1,148,716	44.2	1,310,143	56.8
Zac Goldsmith	Conservative	909,755	35.0	994,614	43.2
Sian Berry	Green	150,673	5.8	–	–
Caroline Pidgeon	Lib Dem	120,005	4.6	–	–
Peter Whittle	UKIP	94,373	3.6	–	–
Seven other candidates	Various	173,439	6.6	–	–

Source: The Electoral Commission

Table 3.11 An assessment of the supplementary vote system

Advantages	Disadvantages
The winning candidate can claim an overall majority	The winning candidate may be chosen as a second-choice candidate
It is relatively simple for voters to understand	It will probably entrench and promote the two-party system
Voters have great opportunity to express their support for more than one party	Third parties will be more excluded from winning seats than under FPTP

Discussion point

Evaluate the view that UK general elections would be improved by the use of SV.

To discuss this you may wish to consider the following issues:
1　Would using SV make a substantial difference to the current composition of the House of Commons?
2　What impact might SV have on the representation of the parties?
3　Is gaining an absolute majority worth candidates possibly being elected as a 'second choice'?

Now you have learnt about the four key electoral systems used in the UK, you are in a position to be able to compare them. Table 3.12 gives you an overview of the key comparisons to consider.

Table 3.12 A comparison of alternative electoral systems to FPTP

FPTP	STV	AMS	SV
A clear winning candidate	Weaker owing to multi-member constituencies	Mixed; still has a clear winning candidate, but also top-up candidates	Much the same, possibly stronger as it will secure a majority of final votes
Usually secures a clear winning party and mandate	Much less likely to achieve this, and more likely to rely on coalitions	Less likely to secure single-party government, but Wales and Scotland have consistently had single-party dominance and Scotland did have a majority following 2011	Possibly more likely to achieve this as it promotes the two main parties
It is easy to understand	Much more complicated and time-consuming Generally, it has worked where used	A mixture of two systems operating on different principles makes it slightly more complicated It has worked where used	Slightly more complicated, but not much It has worked where used
Stops extremist parties and other small parties winning seats	Much more likely that smaller or extremist parties will gain some form of representation	More likely that smaller parties will gain some form of representation, but limited	Would actually make it harder for smaller and more extremist parties to gain a seat
Lack of proportionality	Far more proportional representation of parties	The worst aspects of proportionality are tempered by the top-up seats, though not eliminated	Possibly even less proportional than FPTP
Strong constituency link	Much weaker constituency link	Strong for the FPTP part, but weaker for the top-up seats	Exactly the same

Electoral system analysis

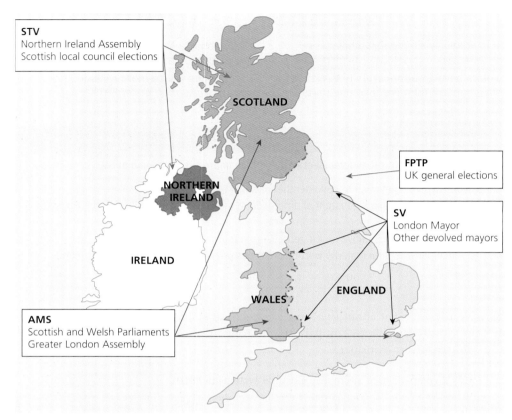

Figure 3.3 Where different electoral systems are used in the UK

Why are different electoral systems used in the UK?

Before 1997, the only electoral system used in the UK was FPTP, but with the election of Tony Blair's New Labour government, new electoral systems were introduced across the UK as shown in Figure 3.3. New Labour had a manifesto commitment to electoral reform because:

- after 18 years in opposition, it wanted to prevent another long tenure of Conservative domination
- the party had pledged to modernise British democracy and bring it more into line with other European countries
- before the election, there was a concern that the party might not gain an absolute majority and would need to form a coalition with the Liberal Democrats, who were committed to electoral reform.

However, having benefited from FPTP, New Labour abandoned plans for national electoral reform, settling instead for launching the **Jenkins Commission** to investigate possible reform. However, to honour its manifesto commitment, New Labour did seek to introduce alternative electoral systems in newly devolved regions as part of its constitutional reform programme.

Useful term

Jenkins Commission
An independent investigation into the best form of alternative voting, commissioned in September 1997 and run by Roy Jenkins. It reported in September 1998, proposing the use of AV+ instead of FPTP. Its proposals were not adopted.

Beyond the manifesto commitment to electoral reform, decision-makers in New Labour hoped to produce a party system that was most desirable in each of the different contexts. For example, STV was adopted in Northern Ireland to reflect the fact that it is a highly divided community and that all the different communities should be represented via a multi-party system. As a result, five different parties achieved significant representation in the Northern Ireland Assembly in 1998.

In Scotland, after devolution in 1997, there was a concern that the then-dominant Labour Party would dominate the country in elections if FPTP were retained. The change to AMS ensured that the main English parties could not dominate in the same way as they did in Westminster. The shift also helped to fulfil the Labour manifesto pledge to introduce electoral reform by experimenting in the devolved bodies. Despite Conservative opposition to electoral reform, it was the AMS system that allowed the Scottish Conservatives to begin rebuilding support in Scotland, while the SNP, rather than Labour, was able to become the dominant party, even securing a majority in 2011. In Wales, AMS has worked to a degree, preventing absolute Labour dominance, though it is usually the dominant party.

Despite the introduction of alternative systems, FPTP seems destined to remain as the system used for general elections for some time to come. This is largely because the political establishment (in both main parties) takes a broadly conservative view of the issue. Most senior politicians prefer the status quo and fear the unknown, as represented by proportional representation, as well as the fact that the two main parties are the ones that usually benefit from FPTP.

Study tip

The type of voting system can have an impact on the resulting party system. Make sure you understand how the different electoral systems used in the UK have affected the party system in the regions and nationally.

The role and importance of elections

Elections are not as simple as they first appear. In many ways they are the key feature of a democracy, allowing the people to choose and remove those in power and ensuring the people are engaged and represented in the political process. Different electoral systems achieve this in different ways, but beyond the theory, how far do elections, as they currently operate across the UK, enhance democracy? The key issues for and against elections as a tool for democracy are reviewed in the Debate box below.

Synoptic link

When considering the importance of elections you should also consider the wider principle of mandate and manifesto, covered in Chapter 4.

Do elections enhance democracy?

Positive arguments

- Elections allow the electorate to hold the outgoing government to account. There is a clear choice between the government and other parties.
- Elections create representative assemblies in an organised way and at regular intervals.
- There is widespread public confidence that elections in the UK are well regulated and that the outcomes are genuine expressions of the will of voters.
- Under FPTP, elections usually produce strong and stable governments, with majorities in the House of Commons.
- UK elections provide strong constituency representation so that voters are confident that there will be representation of their interests.

Negative arguments

- Voters may feel that a vote for a smaller party is wasted, so the choice is not as wide as may appear to be the case.
- Elections can cause social rifts. Partisan tensions during heated elections can lead to personal and vitriolic attack, as outlined in the 2017 and 2019 general elections.
- There is a danger that too many elections will lead to voter apathy and a decline in turnout, particularly with excessive numbers of second-order elections, like to devolved bodies and local councils.
- Under FPTP, elections produce majority governments that are, nevertheless, supported by a *minority* of the electorate.
- While elections to devolved assemblies are generally proportional, elections to the Westminster Parliament are not proportional, exaggerating the popularity of large parties and discriminating against small parties.

While elections can cause some issues in a democracy, overall it is impossible to consider an effective representative democracy without them, so your judgement is likely to be that, despite the flaws, elections do enhance democracy.

Caroline Lucas, among others, has suggested that the type of electoral system impacts voter turnout, thereby undermining democracy as people become disillusioned by the negatives of FPTP. However, turnout in second-order elections has not risen as a result of electoral reform. While some safe seats do see lower turnouts, it is not a consistent picture, suggesting that other factors are more important in determining turnout than the electoral system used.

The electoral system and government

We must not separate the nature of elections from the nature of government. The voters, although they are choosing from a selection of candidates, are actually deciding which party they would prefer to form the government. Indeed, that is what is probably uppermost in their minds when they enter the polling station. The problem is, as we have seen, that the majority of voters do not get the government they voted for.

On the other hand, elections under FPTP in the UK have proved to be effective in producing single-party governments with clear mandates and accountability. This, though, may be beginning to change. Three consecutive elections, 2010, 2015 and 2017, failed to produce a government with a decisive majority, and twice with no single-party majority at all. This may lead us to conclude that the traditional link between FPTP and single-party government has been weakened.

If the UK were to adopt proportional representation for national elections, the country would have to get used to the idea of multi-party government. We might also have to get used to the idea of unstable government. The experience of coalition government in the UK in 2010–15 was mixed. It was stable and lasted for 5 years with few major defeats in Parliament. However, there was also widespread concern that the junior coalition partner, the Liberal Democrats, did not have sufficient influence, so government was still dominated by one party. There is some evidence that voters were unhappy with the experience of coalition, feeling betrayed by the Liberal Democrats for going into power with the Conservatives, which helped to explain why they turned decisively against the Liberal Democrats in the 2015 general election by voting against all but eight of their candidates.

It is unlikely, after this, that many third parties will be keen to join a larger party in such a coalition in the future. Even so, it is difficult to judge the *real* level of support for small parties because so many voters do not support them as they fear it will be a wasted vote. This was revealed in the 2017 election when, faced with the major issue of Brexit, voters flooded back to the two main parties, with their combined vote share being 82.4 per cent, the highest since 1970. However, the weaknesses of the resulting **minority government** highlight how undesirable such a system might be at times of great national debate.

If FPTP were to be replaced by a proportional electoral system, it is likely to make it harder for any party to win an overall majority in Parliament. This can be seen as a desirable outcome as it would prevent governments being excessively powerful. To govern as a minority government or **coalition**, a government would have to seek a consensus on every issue and democracy would, possibly, be better served.

Critics, however, point to the instability this could produce, with governments frequently falling and having to be re-formed, as occurs in some European states. Without a parliamentary majority, governments would lose decisiveness and be unable to deliver their electoral mandate.

With its focus on two-party domination, SV is more likely to continue resulting in single-party government, though the enormous majorities that can be achieved under FPTP may become less common. This is not inevitable though, as in 2020 Copeland and Middlesborough both elected independent mayors using SV, so it is possible for independent candidates or small parties to win seats, even though how far this might occur if SV were used for general elections is unclear.

Electoral systems and party systems

The UK party system is dealt with in more detail in Chapter 2, but there is some debate over whether or not the type of electoral system determines the type of party system that we have.

> ### Discussion point
>
> Evaluate the view that the electoral system determines the party system.
>
> Areas to discuss may include:
> 1. If we believe the UK does have a two-party system, is that because of the nature of the FPTP electoral system, or because that is how people prefer it, a simple choice between two main parties?
> 2. Is it more accurate to suggest the UK now has a multi-party system in terms of voter support but *not* in terms of parliamentary seats?
> 3. If we changed the electoral system, would a multi-party system immediately take over?

Certainly there has been a general trend towards a decline in support for the two largest parties over the past few decades and a rise in support for smaller parties such as UKIP and the Brexit Party, the SNP and the Green Party, as can be seen in Figure 3.4. Another way of considering this is to ask whether the UK political system is now **pluralistic**, with voters seeking parties that are more focused on their particular concerns. If this is the case, the two-party system must be doomed, irrespective of the electoral system. On the other hand, voters may ultimately shrink from a multi-party system and return to a preference for a two-party choice, as appeared to be the case in 2017.

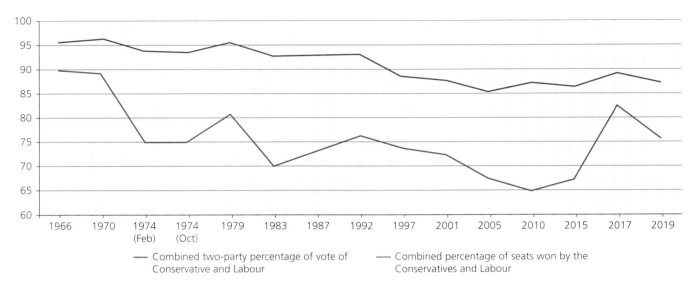

Figure 3.4 Percentage of votes won by the two main parties compared to the percentage of seats won by them

Nevertheless, introducing proportional representation (like AMS or STV) for UK general elections would be more likely to produce a multi-party system. Parties such as the Green Party, the Liberal Democrats and Plaid Cymru would probably win significant numbers of seats. The larger parties, Labour, the Conservatives and the SNP, would win fewer seats than currently. For some, this is a desirable outcome as it would provide a pluralist, more representative result and mean that voters are better represented. On the other hand, it might give an opening for extremist parties and possibly create a chaotic political system with too many competing parties.

The introduction of SV, as a majoritarian system, is more likely to result in securing the two-party dominance of the Conservative and Labour parties, and the SNP in Scotland, perhaps making it even stronger than the existing system. Alternative parties may gain more votes but would find it harder to gain seats by securing absolute majorities.

Electoral systems and voter choice

How much choice do voters really want? If the UK adopted a proportional representation system for general elections, would more voters opt for smaller parties? As things stand, voters tend to be *forced* into voting for either Labour or the Conservatives because any other vote would automatically be wasted. They are also sometimes forced to vote tactically, opting for a less preferred choice to influence the outcome.

Under a proportional system, nearly every vote counts and every vote is of equal value and the need to vote tactically would be reduced. With some systems, such as STV, voters are even able to discriminate between candidates of the *same* party. The lack of choice under FPTP is quite stark but we do need to ask whether voters want more choice, especially if greater choice leads to less stable government. Critics point out that proportional systems are more difficult to understand. They also say that the loss of the close relationship between MPs and constituencies would be a blow to democracy. As such, the public desire for such reforms may be limited.

Ultimately the supporters of proportional representation see the debate as democracy versus over-powerful government, equality versus discrimination. Those who favour the retention of FPTP see the issue in terms of order versus chaos, strong versus weak government. A majoritarian system like SV would give voters greater choice in their first preference and ensure more votes counted in the result, but it is likely that the overall choice would be reduced to two key candidates.

What do the voters want?

It is hard to know how voters feel about electoral reform; the major parties rarely refer to it and it appears to be low down the list of voters' priorities at election time. One possible piece of evidence that can be used to suggest the British public do not want electoral reform is the outcome of the 2011 referendum on AV, where the public rejected replacing FPTP with the **alternative vote** electoral system, a system used, for example, in Australia and the Labour Party leadership elections.

AV is not commonly used in the UK (being reserved to by-elections in STV systems) and is unlikely to be considered in the future. Nevertheless, the result of the 2011 referendum should not be taken to indicate that public opinion is opposed to reform. There were several reasons why the voters rejected AV that were unrelated to their desire for some kind of change:

- The proposal was promoted by the Liberal Democrats (in coalition government with the Conservatives) and the party was very unpopular at that time. It is therefore estimated that many voters used the referendum to show dissatisfaction with the Liberal Democrats rather than to reject AV.
- AV is a complex system, so many voters rejected it because they did not understand it.
- The pro-reform campaign was poorly run while the anti-reform campaign was well organised and funded.

> ### Useful term
>
> **Alternative vote** A majoritarian system that uses preferential voting, with the candidates with the fewest votes being eliminated until eventually one candidate secures an absolute majority.

> ### Activity
>
> Research the 2011 national referendum on the possible introduction of AV and consider the following questions:
> - What was the majority rejecting the proposal?
> - What was the overall turnout?
> - What does this suggest?

> ### Discussion point
>
> Evaluate the view that the UK should abandon the first-past-the-post electoral system for general elections and replace it with another system currently used in the UK: SV, AMS or STV.
>
> Three key areas to consider are:
> 1. What effect would change have on the party system, and would such change be desirable?
> 2. What effect would change have on government formation, and would such change be desirable?
> 3. What effect would change have on the UK's democracy and on the experience of voters, and would such change be desirable?

> ### Study tip
>
> It is common for exam questions relating to electoral systems to ask you to consider changing the current general election system of FPTP to an alternative. To evaluate this, you need to be able to consider if the change would be better or worse than the current FPTP system. This is your opinion and should relate to what you believe is the most important function of an election, which will allow you to reach a reasoned judgement.

Referendums

Referendums in the UK

Useful term

Referendum A vote, which may be national, regional or local, in which qualified voters are asked a single question about a proposal, where the answer is either 'yes' or 'no'.

Synoptic link

Referendums are the main form of direct democracy used in the UK and should therefore be considered in conjunction with Chapter 1. All referendums held in the UK relate to some form of constitutional reform, so also relate to reforms considered in Chapter 5.

Referendums are a form of direct democracy that allows the public to decide on an issue presented to them by the government. They can be seen as an alternative or superior form of democracy when compared to elections, but in the UK they may be seen as a means of enhancing representative democracy and allowing the public a say on a key issue that has constitutional significance. Before 1975 **referendums** were almost unknown in the UK political system. (An attempt to hold a referendum in Northern Ireland in 1973 had failed as half the community boycotted it.) In 1975, however, there was a national referendum on whether the UK should remain a member of the European Economic Community (the forerunner of the European Union), which the country had joined just 2 years earlier.

Although referendums were used again in 1979 to determine whether or not Scotland and Wales should become devolved (with neither gaining the required number of votes to pass on that occasion), the use of referendums did not become commonplace in the UK until the election of New Labour in 1997. After that, referendums started to be used to establish constitutional reforms, promote democracy and test public opinion. In 2011, the coalition government held the second national referendum, on whether to change the UK's general election system to the alternative vote method, as part of the compromise measures that both parties agreed in the Coalition Agreement of 2010.

Few of these referendums were controversial and their use was not a source of major concern. However, since 2011, two tremendously significant referendums have been held that have, in different ways, had a profound impact on the way politics and government operate in the UK. Those two referendums were the 2014 vote on whether Scotland should become an independent state and the 2016 referendum, the outcome of which meant the UK had to leave the EU.

How referendums operate

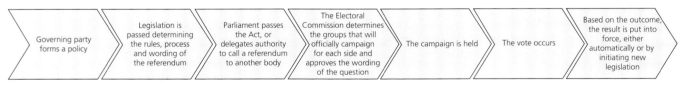

Figure 3.5 The process of holding a referendum

A referendum can be simply defined (see Figure 3.5). It is a vote, conducted at local, regional or national level, in response to a question that usually has a simple 'yes' or 'no' answer, although in the case of the 2016 referendum voters were asked to choose between 'leave' and 'remain'.

Unlike elections, referendums are *ad hoc*, occurring only when Parliament chooses to allow them to be held. Legally, referendums are only advisory and a means of testing public opinion. This is because of parliamentary sovereignty, which means only Parliament has the legal authority to enact constitutional reforms. As such, referendums have usually been used to test public approval of proposed legislation that has already been passed by Parliament, making it clear what is being voted for, as was the case with devolution and the AV referendum.

However, referendums do hold democratic, or popular, sovereignty, which means Parliament is very unlikely to ignore the result of a referendum, even though it has the legal right to do so. This was why, despite 80 per cent of MPs supporting remaining in the European Union, 77 per cent voted to trigger Article 50 to leave the EU as a result of the popular opinion expressed through the referendum. In this case, there was no clear plan, but the government felt the will of the people had to be respected, even though legally it did not have to respect their opinion.

The key reason why a referendum might be held is that, for some reason, it is felt preferable that the people themselves should resolve an issue rather than the elected representatives of the people. Why should a people's vote be preferable to a vote by an elected assembly? After all, it can be argued, in a representative democracy we elect people to make decisions on our behalf, to use their judgement and to mediate between competing demands. Why should we make the decisions ourselves? Examples of these reasons are considered in Table 3.13.

Table 3.13 Why referendums have been called in the UK

Reason for referendum	Example
To entrench a constitutional reform	Extension of devolution to Wales, 2011
To test public opinion	North East Assembly, 2004
To resolve a conflict within a political party	The EEC referendum of 1975 The EU Referendum of 2016
To resolve a conflict between parties sharing power	The AV referendum, 2011
To resolve a conflict within the wider community	The Good Friday Agreement, 1998
To achieve a political goal	The Scottish independence referendum, 2014

Table 3.14 details most of the important referendums that have been held in the UK since 1997. There have been several other regional and local referendums held over the period that are not shown in the table. These have involved such issues as the introduction of congestion charges in city centres and whether or not a city or metro area should have an elected mayor. These referendums had variable outcomes.

Table 3.14 Referendums in the UK

Year	Issue	Level	Why held	% yes	% no	% turnout
1997	Should additional powers be devolved to Scotland and a Scottish Parliament established?	Scotland	A fundamental change in the system of government needed popular consent	74.3	25.7	60.4
1997	Should additional powers be devolved to Wales and a Welsh Assembly established?	Wales	A fundamental change in the system of government needed popular consent	50.3	49.7	50.1
1998	Should the Belfast Agreement be implemented?	Northern Ireland	This required support across the whole divided community	71.1	28.9	81.0
2004	Should additional powers be devolved to northeast England and a regional assembly established?	Northeast England	A fundamental change in the system of government needed popular consent	22.1	77.9	47.7
2011	Should further devolved powers be given to the Welsh Assembly?	Wales	A fundamental change in government power needed popular consent	63.5	36.5	35.6

Year	Issue	Level	Why held	% yes	% no	% turnout
2011	Should the UK adopt the alternative vote system for general elections?	National	The coalition government was divided on the issue of electoral reform	32.1	67.9	42.2
2014	Should Scotland become a completely independent country?	Scotland	A fundamental question about who governs Scotland	44.7	55.3	84.6
2016	Should the UK remain a member of the EU?	National	A fundamental constitutional question. The governing Conservative Party was split on the issue. Also to meet the challenge of UKIP	48.1	51.9	72.2

Activity

Consider the referendums shown in Table 3.14.
- Why was the turnout so high in any two referendums and so low in two others?
- Do some research and outline two reasons why Scotland voted against independence.
- Do some research and outline two reasons why the UK voted against a change to AV.

A number of regulations govern the conduct of referendum campaigns. In national and regional referendums, there is official recognition of the bodies that campaign on each side of the question. Expenditure on referendum campaigns is regulated to ensure that each side spends approximately equal funds. This is done by the Electoral Commission. The Electoral Commission also works to ensure that the sides in the campaign do not issue false information and organises the counting of votes.

Referendums and elections

Referendums are quite different from elections. The key differences between the two devices are outlined in Table 3.15.

Table 3.15 Differences between elections and referendums

Elections	Referendums
Held regularly, by law	*Ad hoc*, if and when Parliament chooses
Concern multiple policies or issues	Usually single issue
Usually have multiple choices and potential outcomes	Usually a simple, binary choice and single outcome
Legally binding	Not legally binding
People vote to fill an office or choose a government	People vote to decide an issue

Knowledge check

Which public body organises and regulates the conduct of referendums?

Despite the differences there is one crucial similarity. Both referendums and elections grant legitimacy to decisions. In the case of an election, the winners claim a mandate for their policies; with referendums, the electorate is directly granting authority to government to implement a specific decision.

The impact of referendums

There is a maxim in politics that governments should never call for a referendum unless they are confident about what the answer will be. There are two reasons why this is a sensible principle. The first is that, normally, governments use referendums as a way of securing direct consent for major policies they want to introduce. A good example was devolution of power to Scotland, Wales and Northern Ireland in the late 1990s. Devolution was, effectively, a policy of the Labour government of the day, but it needed to be reinforced by confirmation in a referendum. The government was also confident that it would win the three votes. So it proved.

A group of protesters demanding a second referendum on the proposed Brexit deal

The second reason is that, if a government supports one side of a referendum debate, it will be placed in a difficult position if it loses. It is a severe blow to its authority. This occurred in 2016 when the UK voted to leave the EU. The Prime Minister, David Cameron, felt his position was untenable and so resigned. The wider result was a complete change in the government's stance on Europe and many ministers lost their positions or resigned.

It goes without saying that the 2014 vote on Scottish independence was what the government of the UK hoped for. However, it did have a major impact on the politics of devolution. The closeness of the outcome was a huge boost to the Scottish National Party. Before that, during the campaign, as the outcome was thrown increasingly into doubt, all three main English parties were forced to promise Scotland greater powers for its Parliament and government. The government won the vote, but it was too close for comfort. The Scottish referendum did not result in independence, but it did result in a major shift in power towards Edinburgh. Then, 2 years later, when the UK voted to leave the EU, a fresh Scottish crisis ensued. The problem was that 62 per cent of Scottish voters voted to *stay* in the EU. This meant that the Scots were being dragged out of the EU against their expressed will. The result has seen renewed calls for a second referendum on Scottish independence so that people in Scotland can choose to stay in the EU.

So, referendums can change things whatever the outcome. They can promote political change and they can also remove policies from the immediate political agenda, as occurred when electoral reform was soundly rejected in 2011. Whether such impacts are desirable is a matter of some debate.

The case for and against the use of referendums

Until recently, the tide of public opinion seemed to be turning in favour of the use of referendums, especially after the vote on Scottish independence in 2014. That referendum was deemed a success in that it involved the vast majority of the people of Scotland and its result was emphatic enough to settle the issue for some time to come. Then the referendum on the UK's membership of the EU changed attitudes again. The result, which shocked the political establishment and was totally unexpected in the light of the opinion polls, demonstrated how divided a society the UK had become.

In this sense it settled the issue — the UK had to leave the EU — but it also led to fears that the substantial minority that had voted to remain were being completely ignored and marginalised. Furthermore, many commentators suspected that many of those who voted to leave were not voting on the issue of the EU itself but on their broader concern that their voices were not being heard by the political system based in London. The EU poll revealed many of the concerns that people have expressed about referendums. Of course, the winning side had a different perspective. For them it was a hugely successful exercise in popular democracy. Conventional politics had been defeated by the will of the majority. For this one exciting moment the people had made history for themselves. Thus, we can see two contrasting opinions about the use of referendums. Whether they are a positive or negative method of settling major issues is a matter of debate, but following the controversies of the 2016 referendum and the depth of division that referendum has created throughout the country, that continues to be felt long after the result, future governments are likely to be very wary before sanctioning a future referendum.

Knowledge check

What is meant by the term 'tyranny of the minority' and how does it relate to referendums?

Debate

Should referendums be used to settle political issues?

Arguments for

- Referendums are the purest form of democracy, uncorrupted by the filter of representative democracy. They demonstrate the pure will of the people, as occurred in the EU vote.
- Referendums can mend rifts in society, as occurred with the decisive result of the 1998 vote on the Belfast Agreement.
- Referendums can solve conflicts *within* the political system and so stave off a crisis. This was especially the case with the EU referendums in both 1975 and 2016.
- Referendums are particularly useful when the *expressed* (as opposed to *implied*) consent of the people is important, so that the decision will be respected. This was very true of the votes on devolution in 1997.
- The people are arguably much more informed than they ever were in the past. The internet and social media in particular have facilitated this. This makes them more capable of making decisions for themselves rather than relying on elected representatives.

Arguments against

- The people may not be able to understand the complexities of an issue such as the consequences of leaving the EU or adopting a new electoral system.
- Referendums can cause social rifts. This arguably occurred in both 2014 in Scotland and 2016 in the EU referendum.
- There is a danger that the excessive use of referendums may undermine the authority of representative democracy. This has been a particular danger in some states in the USA.
- A referendum can represent the 'tyranny of the majority'. This means that the majority that wins the vote can use their victory to force the minority to accept a change that is against their interests. The Scots, who voted strongly to stay in the EU in 2016, claimed they were being tyrannised by the English majority.
- Voters may be swayed by emotional rather than rational appeals. It may also be that they are influenced by false information.
- Some questions should not be reduced to a simple yes/no answer; they are more complicated. The 2011 question on electoral reform is an example of this. Perhaps several *different* options should have been considered, not just one.

Both sides have strong cases, but you need to decide which case is the strongest, even if you qualify it. A clear judgement based on the strongest point or the total weight of the points offered needs to be explained in relation to the other side.

Study tip

When answering questions on referendums and direct democracy, it is essential to use the context and impact of at least four relevant referendums to illustrate your discussion. The case study below provides four examples and how to use them to cover a range of examples.

Case study

Are referendums a better form of democracy?

Extension of devolution to Wales, 2011

To combat the issue of unequal devolution across the UK, all major parties in 2010 promised to commit to further devolution to Wales. The coalition government developed a set of proposals that would give the Welsh Assembly primary legislative powers and further administrative powers. These proposals were then put to the people of Wales to decide in a referendum in 2011.

This referendum gave people a choice that had not existed in 2010, whereby they could vote against further devolution, and removed the issue of devolution from other pressing issues, such as economics following the financial crash of 2008, thus providing a better form of democracy.

The low turnout, particularly when compared with the turnout in the 2010 and 2015 general elections, suggested a lack of voter engagement. The result mirrored the outcome of the manifestos of the major parties, suggesting it was not necessary and not a popular form of democracy.

AV referendum, 2011

As part of the Coalition Agreement, a compromise was reached between the desire of the Liberal Democrats for electoral reform to a more proportional system and the desire of the Conservative Party to avoid any such reform. As such, they agreed to hold a referendum on whether or not to replace FPTP with AV for general elections.

Holding the referendum allowed the public to voice their opinion on electoral reform that had not been supported to any great degree by the two main parties and had not been prominent in any electoral campaign, and gave the opportunity to test public opinion on a major constitutional issue.

The low turnout and decisive vote for FPTP suggested that electoral reform was not an issue that concerned the electorate and should have been left to elected representatives to decide. A report on the outcome of the AV referendum by the University of Essex suggested that: 'If people are not politically engaged and have little understanding of a proposed change, but nonetheless feel that they have a duty to vote, an easy solution to their choice problem is to support the status quo.' This would suggest that a majority who lacked political engagement and education voted to keep FPTP because they did not understand the proposed change, suggesting that referendums should not rely on the opinions of a public that is not concerned with the issues involved.

Scottish independence referendum, 2014

In 2011 the SNP won an outright majority in the Scottish parliamentary elections, giving it a mandate to carry out its manifesto pledge to hold an independence referendum. Westminster passed an Act allowing the Scottish Parliament to hold an independence referendum within a year.

The turnout was higher than in any elections, showing an increase in public participation and engagement. The issue was widely and hotly debated, increasing political education and engagement in Scotland, so much so that increases in turnout have been seen in subsequent elections. The people of Scotland were able to determine their own course on a clear issue, in a way that the relatively small number of Scottish MPs in Westminster could not.

The Scottish government chose the timing of the referendum (shortly after Glasgow held the Commonwealth Games, a week after the anniversary of the Battle of Stirling Bridge, and in the year of the 700th anniversary of the Scottish defeat of the English at the Battle of Bannockburn) to influence Scottish voters. It agreed to allow 16- and 17-year-olds to vote in the referendum, partly because they were far more likely to vote for independence than older voters. Such tactics suggest that referendums are subject to political manipulation.

EU referendum, 2016

As the result of UKIP gaining support in opinion polls and increasing tensions within the Conservative Party, David Cameron pledged to hold an in/out referendum on membership of the EU.

The turnout was higher for the referendum than in any election held in the UK since 1992. The fact that a majority of those who voted chose to leave, while the majority of MPs representing them wished to remain, suggests that referendums are useful for keeping representatives in tune with public opinion. That the only serious party advocating leaving the EU in 2015 (UKIP) received 12.6 per cent of the vote shows that elections are compromises and that clearer, more popular choices can be made through referendums.

The fact that there was no prepared plan has led to political upheaval and tensions since the referendum. A range of complex issues has emerged that suggest a referendum is

→

too simplistic a tool to determine major constitutional issues. The police service reported a significant increase in the number of hate crimes against minority groups perpetrated in the two months following the referendum, suggesting a tyranny of the majority was taking place. Accusations were made that the public was misled by false campaigning and manipulation.

Taken as a set, these four referendums provide key arguments for and against the greater use of referendums, suggesting that in some ways referendums are better than elections but in other ways they are worse. However, it is better to consider referendums as a part of the democratic process and as a means of enhancing or undermining the system of representative democracy currently used, rather than as an alternative to it.

Referendums and representative democracy

When we consider the use of referendums as a democratic device, we should be comparing them with the way in which decisions are reached by elected representatives. The advantages of representative democracy when it comes to making key decisions are outlined below.

- Representatives are more likely to adopt a rational approach and resist emotional reactions to questions. For example, many voters were concerned about immigration in the EU referendum and were responding to appeals to their patriotism and the perceived dangers to 'British values' posed by migrants entering communities. Elected representatives, on the other hand, could weigh up the benefits as well as the problems of high numbers of migrants.
- Elected politicians have experts to help them make decisions. They can ensure that the information on which they base their judgements is accurate. Most people rely on the media for their information, which is at best conflicting and at worst dubious.
- Elected representatives have to concern themselves with the competing interests of both the majority and minorities. Voters, on the other hand, usually only have to think of their own interests. In the Scottish independence referendum, the people did not need to consider the implications for the UK as a whole. In the same way, voters in the EU referendum did not need to consider the implications for Northern Ireland of a leave vote.
- There is an expectation that, as professional politicians, MPs are in a better position to make a reasoned judgement than ordinary people with less knowledge and understanding of the complex issues. Judgement and good sense are qualities we consider when we elect them.

Set against these considerations, we can more clearly see the disadvantages of referendums. All the advantages listed above are lost when a referendum is used. Those who opposed the use of a referendum in 2016 on whether the UK should leave the UK were adamant that such a key decision should have been made by Parliament and not by the people, who were poorly placed to exercise a rational judgement.

Summary

Having read this chapter, you should have knowledge and understanding of the following:
- ➜ How the various electoral systems used in the UK work
- ➜ How to debate the advantages and disadvantages of each system used in the UK
- ➜ How to make comparisons between FPTP and the other electoral systems
- ➜ How referendums work and the way in which they have been used in the UK
- ➜ The positive and negative aspects of referendums
- ➜ An understanding of why different electoral systems are used in the UK
- ➜ How different electoral systems impact on party representation

Key terms in this chapter

Additional member system (AMS) A hybrid electoral system that has two components or elements. The voter makes two choices. Firstly, the voter selects a representative on a simple plurality (FPTP) system, then a second vote is apportioned to a party list for a second or 'additional' representative.

Coalition government A government that is formed of more than one political party. It is normally accompanied by an agreement over policy options and offices of state.

First-past-the-post (FPTP) An electoral system where the person with the highest number of votes is elected. Victory is achieved by having one more vote than other contenders — it is also called a plurality system.

Marginal seat A seat held by the incumbent with a small majority/plurality of the vote.

Minority government A government that enters office but does not have an absolute majority (50%+1) of the seats in the legislature (Parliament).

Safe seat A seat in which the incumbent (the person who already holds the seat or position) has a considerable majority over their closest rival and is largely immune to swings in voting choice. The same political party retains the seat from election to election.

Single transferable vote (STV) A system that allows voters to rank their voting preferences in numerical order rather than simply having one voting choice. In order to obtain a seat, a candidate must obtain a quota. After the votes are cast, those with the fewest votes are eliminated and their votes transferred. Those candidates with excess votes above the quota also have their votes transferred.

Supplementary vote (SV) A majoritarian system where the voter makes two choices (hence the term 'supplementary'). If one candidate obtains over 50 per cent on the first vote then the contest is complete; if no candidate attains this level, all but the top two candidates are removed. Then the supplementary choices are redistributed and whoever from the remaining two gets most votes wins the seat.

Further reading

Websites

The Electoral Commission regulates elections and referendums. Its site is objective:
www.electoralcommission.org.uk
The Electoral Reform Society is a campaign group. Its website may not be objective, but it contains a great deal of accurate factual information: **www.electoral-reform.org.uk**
Another campaign group that discusses electoral systems is Unlock Democracy:
www.unlockdemocracy.org.uk
The official parliamentary website also has information: **www.parliament.uk**

Books

Up-to-date and accessible books on electoral systems and electoral reform are thin on the ground. A few are:
Cowley, P. and Ford, R. (2014) *Sex, Lies and the Ballot Box*, Biteback and (2016) *More Sex, Lies and the Ballot Box*, Biteback — these offer a series of articles that relate to voting systems and their impact
Farrell, D. (2011) *Electoral Systems*, Palgrave Macmillan — this is excellent, though it contains information about systems used around the world, so the UK experience is only one part

Geddes, A. and Tonge, I. (eds) (2015) *Britain Votes*, Hansard Society — this deals with the performance of the parties and voting behaviour, and also analyses the effects of the first-past-the-post electoral system

Renwick, A. (2011) *A Citizen's Guide to Electoral Reform*, Biteback — this deals well with all the arguments about various systems

The Electoral Reform Society's Annual Reports — these contain analyses of recent elections and how the various UK electoral systems worked

Practice questions

1

Source 1

Proportional representation systems are a danger to British democracy and values. Such systems would allow extremist parties to enter the mainstream of British politics, while also making it more likely that such parties may have the balance of power in forming coalitions as mainstream parties fail to secure working majorities. How could a government based on a coalition have a clear mandate to govern and be held to account if there is no winning party? Such a system is unwanted by the British public and would be time-consuming and overly complex. Yes, some people might point to fairer representation, a more diverse range of parties and views and greater democracy, but how can it be an improvement with such limited accountability? They may suggest people would engage more if they felt their votes would count more, but where is the proof that turnout has increased in those regions that use alternative voting systems? If anything, the complex nature of such systems is likely to alienate voters or confuse them, undermining the very principles they claim to champion. Proportional representation will do no more than tear up centuries of British traditions.

Source: A political commentator writing in 2020

Using the source, evaluate the view that proportional representation would undermine elections to the House of Commons.

In your response you must:
- *compare and contrast different opinions in the source*
- *examine and debate these views in a balanced way*
- *analyse and evaluate only the information presented in the source.* (30)

2

Source 2

A

Since its introduction into the UK in 1998, AMS has become firmly embedded in British regions and worked effectively. While retaining the constituency representation that many voters value, it has ensured more proportional outcomes, with minor parties gaining in representation where they may otherwise have been excluded. Single-party government is still possible, but is less likely, with a need for consensus and cooperation between various parties ensuring a more representative form of politics that better reflects the desires of voters. Considering the fact that it is only used for second-order elections, the impressive turnout figures for Scotland suggest it would work very well at Westminster.

Source: A supporter of AMS

B

Since 1998 Labour has been in every government in Wales, while the SNP has dominated politics in Scotland since 2007. Far from improving democracy, AMS appears to have entrenched single-party dominance in these regions, with a lack of accountability resulting. How can representation be effective when different representatives answer to different constituencies? How can it be right that a handful of Green MSPs can hold such power in Scotland when so few voted for them? AMS is a reform that is only wanted by the few and will serve no great benefit.

Source: An opponent of AMS

Using the sources, evaluate the argument that adopting AMS would enhance UK general elections.

In your response you must:
- *compare and contrast different opinions in the sources*
- *examine and debate these views in a balanced way*
- *analyse and evaluate only the information presented in the sources.*   (30)

3 Evaluate the view that electoral reform in the UK since 1997 has been ineffective.
   You must consider this view and the alternative to this view in a balanced way.   (30)

4 Evaluate the view that STV should be used for elections to the House of Commons.
   You must consider this view and the alternative to this view in a balanced way.   (30)

5 Evaluate the view that referendums are a worse form of democracy than elections.
   You must consider this view and the alternative to this view in a balanced way.   (30)

6 Evaluate the view that referendums have had little impact on UK politics since 1997.
   You must consider this view and the alternative to this view in a balanced way.   (30)

4 Voting behaviour and the media

If we were to ask what the typical Labour voter looks like, what their characteristics are, we might draw this picture:

- They are slightly more likely to be a woman.
- She will be under 40 years old.
- She works in the public sector, perhaps in the NHS or as a teacher or social worker.
- Her income is average, perhaps just above average or much lower than average.
- She is from an ethnic minority.
- Her parents probably also voted Labour.
- She lives in the north of England.

The typical Conservative voter has the opposite profile.

If explaining the results of elections were that simple, we could predict the outcome by just counting the number of people who fit into such typical profiles. Unfortunately, this is too simplistic. There may be discernible *tendencies* among various sections of the population, but there are so many other factors at play that such social factors need to be treated with great caution. They do, however, explain voting patterns, and members of the parties who campaign during elections are well aware of these tendencies and therefore which groups or individuals they need to appeal to in order to win votes.

Different press reactions to the exit poll for the 2019 general election

This chapter will examine the many factors that determine the outcomes of elections. These include the image of the parties, the past performance of parties, the changing perceptions of parties and their leaders, as well, of course, as how the public judge the performance of the incumbent government. The media also has an impact, as we shall see, although this is difficult to judge. All these factors help to determine what is known as voting behaviour in UK elections.

Voting behaviour

Social factors

When it comes to explaining how people vote, we tend to group people by different social factors, such as class, gender, age, ethnicity and geographic region. These are not easy groupings to determine and, of course, not everyone in a group thinks or votes in the same way, but we, like politicians, are trying to identify general trends and patterns across groups to help explain electoral outcomes as well as patterns of change in UK politics. Politicians will try to target and 'win' key groups to help them win an election. Therefore, we must begin by considering the differing social groupings and how and why their role and importance have changed.

Social class

Before we look at the influence of **social class** on voting, we need to explain the normal classifications used by researchers in this field. Table 4.1 shows the meaning of four different classifications. The table also shows what proportion of the total population is in each class, according to the last national census in 2011.

Useful term
Social class The way in which social researchers classify people on the basis of their occupations and, to some extent, their income. Class can influence various forms of typical behaviour, including political attitudes and voting trends. It should be noted that the term 'class' has a contested definition and can be used to include many different aspects. Here we consider social class as economic, but it can also include cultural and social capital which relate to different concepts.

Table 4.1 How people are classified in the UK

Classification	Description	Typical occupations	% of population in each classification
AB	Higher and intermediate managerial, administrative, professional occupations	• Banker • Doctor • Company director • Senior executive	22.17
C1	Supervisory, clerical and junior managerial, administrative, professional occupations	• Teacher • Office manager • IT manager • Social worker	30.84

Classification	Description	Typical occupations	% of population in each classification
C2	Skilled manual occupations	• Plumber • Hairdresser • Mechanic • Train driver	20.94
DE	Semi-skilled and unskilled manual occupations, unemployed and lowest-grade occupations	• Labourer • Bar staff • Call centre staff • Unemployed	26.05

Source: 2011 census

Of course these are simply standard academic definitions of class and many people will identify as being from a class that is not in line with these objective definitions. We must remember, therefore, that we are basing class on an objective set of criteria that may differ from peoples' personal definitions of class or how they identify themselves. If we had asked the question 'How much does a person's class influence the way they vote?' back in the 1960s and before, the answer would have been something like, 'We can predict with great accuracy how a person will vote if we know to which social class they belong.'

Possibly as many as 80 per cent of people voted the way their social class indicated. Even in 1979, class voting was probably the most important factor in determining how people voted. Most AB voters favoured the Conservatives while the DE classes mostly voted for Labour. The C1 class was typically, though not overwhelmingly, Conservative, and C2 was mostly Labour. This meant, of course, that the political battle was largely fought among two types of voter: those whose class identity was not clear and those who did not vote the way their class indicated that they might. This latter group were known as '**floating voters**', i.e. it was difficult to predict how they would vote. It also meant that the C1 and C2 classes were the key groups in determining electoral outcomes, so small swings by the floating voters in these two classes could have a major impact on the electoral result. Consequently, the main parties would tailor their election **manifestos** towards these groups of voters, based on class identity.

The reasons why class used to be strongly associated with voting trends are fairly straightforward. Three links stand out:

- Voting was a part of a person's class identity. To be middle or upper class was to be conservative; to be working class meant you would support the party of the working class.
- Both major parties developed strong, deep roots within communities, so there was a culture of voting for one party or another. The wealthy commuter belt around London, for example, was steeped in Conservative attitudes, while the poorer east of London had a strong sense of being a Labour-led community. Such roots were strengthened by Labour's associations with strong trade unions.
- There was a selfish reason. The Conservative Party was perceived to govern more in the interests of the middle class and the better off, while Labour developed policies to help the working class and the poor. It was therefore rational to choose the party associated with your class.

In recent years, class voting has declined noticeably. This is not to say it has disappeared, but it is certainly less pronounced. Figure 4.1 shows patterns of voting for Labour by those in class DE, while Figure 4.2 shows the link between class AB and Conservative voting. A close link between class and party support is often described as 'voting attachment'.

Useful term

Floating (or swing) voter
A voter who tends to vote unpredictably in different elections and who is liable to change the way they vote fairly often and who does not identify closely with any party.

Key term

Manifesto In its manifesto, a political party will spell out in detail what actions and programmes it would like to put in place if it is successful in the next election — a set of promises for future action.

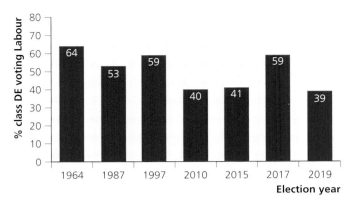

Figure 4.1 Class DE voting for Labour

Source: Ipsos MORI/Earlham Sociology/Ashcroft polling

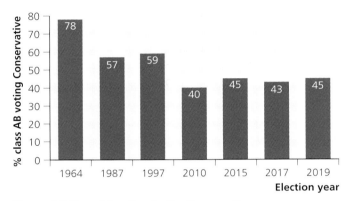

Figure 4.2 Class AB voting for the Conservatives

Source: Ipsos MORI/Earlham Sociology/Ashcroft polling

We can see from Figures 4.1 and 4.2 that voting on the basis of class has declined, despite a slight resurgence for Labour among DE voters in 2017 which fell back in 2019. The main features shown by these statistics include the following:

- In 1964, 64 per cent of class DE voted for Labour, as we would expect.
- There was still a tendency for some, up to a third, of the old working class to vote Conservative, but the correlation between class and voting remained strong.
- Conservative support among the working class was understood to be the result of a factor known as 'deference'. This was a tendency for some members of this class to 'defer' to or respect those whom they considered to be their superiors, i.e. members of the upper and middle classes, who were perceived to be Conservatives.
- Some lower-middle- and working-class voters *aspired* to be middle class and so voted Conservative as evidence of their aspiration.
- The correlation between class AB and Conservative voting has always been stronger. There have been fewer swing voters in this class.

Nevertheless, the decline in voting on class basis has been marked, with the proportion of AB voters who voted Conservative falling from 78 per cent in 1964 to only 40 per cent in 2010, with a small recovery since. To some extent this was a reflection of 'New' Labour's achievement in attracting middle-class support away from the Conservatives, but the decline in class-based voting habits has deeper roots. Among the causes of the decline are these:

- A trend known as **class dealignment** has been important. This is a tendency for fewer people to define themselves in terms of class. In other words, social class has declined in importance within UK culture.
- The main parties, including the Liberal Democrats, have tended, especially since the 1980s, to adopt policies that are 'centrist' and consensual, and can therefore appeal to a wider class base, largely in the centre of society.
- There has been a rise in the influence of other factors. This has tended to replace social class as a key factor in voting behaviour. This helps to explain why the Conservative Party was able to win in each class group in 2019, as a result of a commitment to Brexit and alienation of traditional Labour, and explains the absence of class distinctions in Table 4.2. Indeed, as this table shows, in the 2019 general election class seemed to be almost irrelevant in determining how people cast their votes, with the Conservatives winning across all classes, with similar percentages in the AB, C1 and C2 classes voting Conservative and Labour, with only a small narrowing of the gap among the DE class.

Table 4.2 Class-based voting in the 2019 general election

Class	% voting Conservative	% voting Labour	% turnout
AB	45	30	68
C1	45	32	64
C2	47	32	59
DE	41	39	53

Source: Ipsos MORI

The general election of 2019 also reveals another electoral issue related to class: turnout. The pattern in 2019 continued the long-term trend of turnout increasing with social class. There is no single reason for this; it could relate to levels of education, or to perceptions about how much their vote is valued, to a class-based feeling of disconnection from the political system at the lower levels. This creates a major problem for a party that depends on votes from the C2 and DE classes; not only have these classes been shrinking as a proportion of the population, but their potential voters are also much less likely to vote in elections. This is a difficulty for the traditional Labour Party and explains New Labour's success as it prioritised C1 and AB voters, while a return to more traditional Labour values under Jeremy Corbyn did not bring decisive electoral success. The lower levels of turnout also mean parties, when campaigning, will focus more on issues important to the higher classes as they are more likely to vote, creating a self-perpetuating cycle of DE voters not voting because they are ignored by politicians, who in turn ignore DE voters as they do not vote.

One final note on class-based voting. We can see that relatively small percentages of the social groups vote for third parties across the UK, but they do reflect some class identity with key issues. Table 4.3 demonstrates the influence of class on voting for smaller parties. It uses the 2015 election, when UKIP made its breakthrough. This seems to suggest that the Liberal Democrats were punished less for joining with the Conservatives by the AB class than they were by the C2 and DE classes, while the issue of leaving the EU was far more important in the C2 and DE classes than it was for the AB and C1 classes. This perhaps explains why former UKIP voters have turned to the Conservative Party and why the Conservatives are now able to win across all class groups. Meanwhile, Green support seems to be consistent across class lines.

Table 4.3 Class and voting for other parties, 2015 general election

Party	Class AB (%)	Class C1 (%)	Class C2 (%)	Class DE (%)
Liberal Democrats	12	8	6	5
UKIP	8	11	19	17
Green	4	4	4	3

Source: Ipsos MORI

Debate

Does social class still matter in UK elections?

Yes

- Issues over levels of taxation and welfare payments, closely linked to class, still distinguish the main parties.
- A number of voters still identify with a party based on their social class.
- Regional voting tends to reflect class-based issues such as wealth and poverty.
- Social mobility (or lack of it) and inequality remain major concerns for many voters.

No

- Major issues, such as Brexit and immigration, cross class lines.
- The size and importance of the working class have declined, making it less of an electoral force.
- Increasing levels of home ownership and better educational opportunities make it harder to determine class affiliation.
- To be successful, modern parties must appeal to a variety of social classes, not just one or two of them.

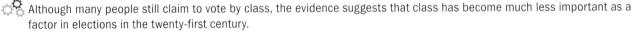 Although many people still claim to vote by class, the evidence suggests that class has become much less important as a factor in elections in the twenty-first century.

Party voting

Traditionally, most voters in the UK would closely align with a particular political party based on their class, and would remain loyal to that party, regardless of any other factor. This was tightly bound up with social voting (detailed above), but even when people moved from one class to another as a result of social mobility, they would remain loyal to the party of their family and background. That meant parties could rely on their **core voters** to support them and the two main parties would be guaranteed a certain degree of support thanks to party loyalty.

However, with changes in the economic and class basis in the UK since the 1970s, we have seen a rise in **partisan dealignment**, which has meant that the core vote for the major parties has been shrinking. It also means that voters are less likely to support the party that we would expect them to vote for. So, increasing numbers of DE and AB voters are not voting Labour or Conservative, respectively, as we would have predicted.

This was seen most notably in the 2017 and 2019 elections, where high numbers of voters who in the past had voted for one party (usually Labour, but sometimes Conservative) switched allegiances over issues like Brexit. As such, it is difficult to argue that people cast their votes today primarily out of loyalty to a particular party.

Gender

When it comes to election campaigns and party strategies, the role of gender is clearly one that is believed to be important. Of course, by gender, what parties often mean is 'women', as there is little evidence of any specific tactic to reach out to male voters. Over the years, parties have tailored strategies to target female voters, such as Labour's 2017 pledge to conduct a gender-impact assessment on all policies and legislation, Labour's 2015 Woman to Woman pink minibus which visited 75 constituencies to encourage more women to vote (though this was itself criticised for being a sexist campaign). Also, David Cameron's and Tony Blair's commitment to increase the number of female MPs and the policy of using all-women short lists for local council elections and party positions helped promote the visibility of women in politics at all levels. While politicians clearly believe that such strategies are important in winning elections, it is not at all clear that there is an identifiable 'women's vote', as women, like men, tend to vote based on issues other than their gender. It is perhaps unsurprising then that the gender gap in party voting is usually quite small, with men and women voting in similar percentages for political parties, with swings between parties over elections seeming to follow a similar trend, as shown in Table 4.4.

Table 4.4 Gender and voting in general elections

Year	% voting Conservative		% voting Labour		% voting Liberal/Alliance/ Liberal Democrats	
	Men	Women	Men	Women	Men	Women
1979	43	47	40	35	13	15
1983	42	46	30	26	25	27
1987	43	43	32	32	23	23
1992	41	44	37	34	18	18
1997	31	32	45	44	17	18
2001	32	33	42	42	18	19
2005	34	32	34	38	22	23
2010	38	36	28	31	22	24
2015	38	37	30	33	8	8
2017	43	40	35	42	10	9
2019	46	43	31	34	12	12

Source: Ipsos MORI

However, given the relative size of the population, a few percentage points with one group could make a significant difference and we have seen a clear shift in the pattern of female voting since the 1970s, compared with male voting. In the 1970s and early 1980s, women were more likely to vote Conservative than Labour, with 47 per cent of women voting Conservative in 1979 and 46 per cent in 1983, compared with just 35 per cent and 26 per cent voting Labour in those respective elections. This led to gaps of 12 points and 20 points in gender-based voting between the main parties. In contrast, male voters in these elections had gaps of 3 points and 12 points between Conservative and Labour. However, since 1997, women have become more likely than men to vote Labour, although the gaps are small.

Why has this change happened? Mostly it seems to relate to the changing role of women within society and changes that have occurred within the parties. In the 1970s and early 1980s, the Conservative Party was seen as the party of 'housewives', which sought to keep prices low in order to allow mothers to run an effective home. This strategy won Ted Heath the 1970 election and was an issue that Thatcher would emphasise throughout her time as Prime Minister. Labour, in contrast, was dominated by traditional trade unions and workers from heavy industries that were largely male and focused on the rights and benefits of working men, rather than issues relating to the home. Since the 1980s, more women have entered the workplace and become involved in more issues and areas than the role of a traditional 'housewife'. This also saw women begin to make up higher percentages of the staff in some traditional Labour-voting occupations, such as call centre staff, shop workers and the caring professions. Perhaps more importantly, the collapse of Britain's industrial base forced the Labour Party of the 1990s to consider a greater range of issues, many of which related better to women voters than the old industrial struggles.

While on most issues there is little to distinguish between male and female voters, there are some issues that are prioritised slightly more by female voters than male voters. These include health and education, while men are slightly more likely to favour foreign intervention, nuclear power and nuclear weapons. These differences relate to gendered roles in society that have, in many ways, created the gender gap in voters. With caring professions, such as nursing, social work and teaching traditionally being seen as 'female' roles, it is perhaps not surprising that gendered attitudes in society impact the issues which are prioritised by women. In a similar way, the gendering military and defense roles have made these appear more important to men. While the difference is small, the different attitudes of the two main parties to these issues may explain the swing by women from the Conservatives to Labour over the past 40 years.

One final thing to note is that simply looking at gender as an issue hides other vast social differences, most notably age. In 2019, women in all age categories supported Labour slightly more than men did, but differences became much less stark as groups got older. In the 18–24 age range, 65 per cent of women voted Labour, compared with 46 per cent of male voters.

Gender, then, is perhaps too large a social grouping to make effective explanations for electoral results, though small changes in gender voting can have large implications on the national scene.

Education

Education has become a major dividing line in UK politics, though one that has so far received less consideration for its impact. In the past there was a relatively small percentage of people sitting A Levels, let alone going to university, which made the impact of education too difficult to quantify in any meaningful way. Until the 1960s, those with more academic qualifications were likely to be more middle class and so, as class-based voting dominated, were more likely to vote Conservative. Those in heavy industry and more working-class professions did not require higher levels of education, and so, if there was an effect, those with more qualifications were more likely to vote Conservative. However, since the growth of university courses and access to courses for more people (in 2017–18, 50% of all school leavers went on to some form of higher education for the first time), as well as the social factors that saw an end to many of the old industrial jobs, the impact of education on politics began to change.

There is no established explanation, but evidence does seem to suggest that education does appear to have a 'liberalising' affect, with voters more likely to vote Labour or Liberal Democrat the higher their level of qualification. For example, in 2017, 47 per cent of people who held a degree were likely to vote Labour or Liberal Democrat, while 36 per cent would vote Conservative and 4 per cent UKIP. Of those with no formal qualifications, only 23 per cent were prepared to vote Labour or Liberal Democrat, while 53 per cent would vote Conservative and 17 per cent UKIP. For those with GCSEs it was 31 per cent who would vote for the left-wing parties and 62 per cent for the right-wing parties, while those with A Levels saw 39 per cent voting left-wing and 49 per cent voting right-wing. This would indicate a clear education shift and has, in many ways, changed the dynamic of the political parties. It explains why the Conservatives have gained support from working class areas with lower levels of educational qualifications, who may traditionally have voted Labour, while Labour has gained support amongst more of the AB, C1 classes with higher educational qualifications. This perhaps explains why Labour has become more focused on socially liberal policies which traditional working-class voters would not have supported.

However, before we assume that education is the key factor, it is worth remembering that the Conservative Party was the single most popular party across all educational groups in 2019. Yes, their support decreased with each new level of educational achievement, but they were still 8 per cent ahead of Labour with voters who held a degree, so education is a factor, but not, alone, the crucial division in British politics.

Age

While the issue of gender is unclear, age reflects a much clearer dividing line in British politics, as Table 4.5 shows.

Table 4.5 Age and voting in five general elections

	1979			1997			2015			2017			2019		
Age range	% Con	% Lab	% All*	% Con	% Lab	% LD*	% Con	% Lab	% LD*	% Con	% Lab	% LD*	% Con	% Lab	% LD*
18–24	42	41	12	27	49	16	27	43	5	18	67	7	21	56	11
25–34	43	38	15	28	49	16	33	36	7	22	58	9	23	54	12
35–44**	46	35	16	28	48	17	35	35	10	30	50	9	30	46	14
45–54				31	41	20	36	33	8	40	39	9	41	35	13
55–64***	47	38	13	36	39	17	37	31	9	47	33	9	49	28	12
65+				36	41	17	47	23	8	59	23	10	57	22	11
Total, all ages	45	38	14	31	43	17	37	30	8	42	40	7	67	14	11

* In 1979 there was no clear third party, which would emerge with the SDP/Liberal Alliance in the 1980s before becoming the Liberal Democrats in 1988

** In 1979 the figures are for age 35–54

*** In 1979 the figures are for age 55+

Source: Ipsos MORI/Ashcroft polling

Knowledge check

Study Table 4.5. What has been the trend in voting behaviour by the over-65 age group?

When considering age voting, we have a pattern that almost directly mirrors the trend in class voting. Although in 1979 Labour tended to have slightly more support among younger voters than older ones and the Conservatives increased in support up the ages, the difference was not large. Indeed, in the youngest age group of 18–24, there was no difference in the proportion voting Labour and Conservative. Age was therefore not a major factor in explaining voter choice, while class was, though people did still become more likely to be pro-Conservative as they got older. However, since 1997, age appears to have become the main dividing line in British politics, replacing class as the dominant statistic. Indeed, Figure 4.3 highlights the age at which people become more likely to vote Conservative than Labour.

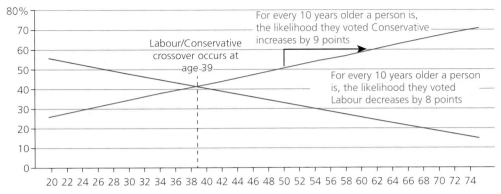

Figure 4.3 Age as a key indicator for how people voted in 2019

Source: YouGov

Why has age become the main dividing line in voting behaviour? Once again, it reflects the shifting economic position of Britain. In the 1980s and 1990s, Britain moved from an industrial economy to a service-based economy, meaning fewer jobs in traditional 'class-based' sectors such as factories and mining and much more focus on white collar, office-based jobs, where position, which comes with age, counts for more. Also, as property ownership has risen dramatically, more people are able to own their own home and seek more middle-class protections. However, this has made it much harder for younger people to buy a home of their own and to become less reliant on support from the state. The rise of the gig economy too, with more flexible but less well-protected jobs, is focused on younger workers, meaning they are less interested in more traditional policies offered by the Conservative Party. This may explain the dramatic shift from class to age as the key element of voting behaviour, but there are other possible explanations for why younger voters are more likely to vote Labour than Conservative and, in Scotland, why they are more likely to vote SNP.

- Younger people tend to be more progressive and less 'conservative' in the more general sense of the word.
- As people age and acquire more assets, there is a tendency to focus on more rational, self-interested issues, rather than wider social issues.
- Perhaps it is more compelling to suggest that younger people have fewer responsibilities, and can therefore indulge in more outward-looking ideas, whereas later in life the responsibilities of a career, a family and property ownership may lead to more cautious views.
- Voting by the young for what may be described as more progressive parties is more understandable. The young tend to adopt more progressive ideas based on greater levels of reform and change, for example about environmental protection, social justice and democratic reform.
- To reinforce the link between progressivism and the young, it has been noted that a large proportion of new members of the Labour Party in 2015–16, most of whom joined to support Jeremy Corbyn, were young voters.

One final thing to consider is the role of turnout. Younger voters are less likely to vote than older voters. Even in 2017 with a reported 'youthquake', the proportion of younger voters turning out to vote was only 54 per cent, 15 per cent below the average turnout and far below the turnout of the over-65s. As a result, parties tend to prioritise the concerns of older voters over those of younger voters, a lesson learnt by Ed Miliband in 2015 when he pledged to scrap university tuition fees and limit pension increases, and lost the election quite heavily as a result. By contrast,

Activity

Look at Tables 4.4 and 4.5 and Figure 4.3. Identify four statistics that most effectively illustrate the impact of gender and age on voting behaviour.

in pledging to maintain the 'triple-lock on pensions', the Conservatives appealed to older voters and won a majority in the same election.

Ethnicity

As with any social issue, there is no automatic reason why any political party should be favoured by people of any given ethnicity as in an equal society, race should not play a major role in politics. This idea would relate to the stated beliefs of all registered parties to support a racially tolerant and mixed society, with members of all ethnic groups having the same rights in the UK and the fact that all official parties claim to oppose discrimination against black and minority ethnic (**BAME**) people. While that may be the race neutral theory, the history of the established parties has shaped BAME attitudes, with the Conservative Party having a history of rhetoric and supporters opposing immigration and equal rights while Labour passed major anti-discrimination legislation when in power in the past. Equally, in recent times, there is a perception that while some parties, including the Conservatives, UKIP and the Brexit Party, may claim to be racially tolerant, their actions and policies could be seen as being discriminatory, as evidenced by the Conservative actions in the Windrush scandal of 2019. The evidence from elections seems to bear this out and show that BAME voters on the whole, and some groups esepcially, favour the Labour Party, often by quite large margins. Table 4.6 shows BAME voting in recent elections'.

> **Useful term**
>
> **BAME** Short for 'Black, Asian and Minority Ethnic'. It is a policy term that is used in the UK but is now increasingly opposed by people of ethnic minority background, mainly for 'lumping' all minority groups together and obscuring differences between minority groups. Many ethnic minorities now prefer to be referred to as people of colour.

Table 4.6 Ethnicity and voting

Election	% BAME voting Conservative	% BAME voting Labour	% BAME voting Liberal Democrats
1997	18	70	9
2010	16	60	20
2015	23	65	4
2017	21	65	6
2019	20	64	12

Source: Ipsos MORI/Ashcroft polling

The bias towards Labour is both clear and consistent. There was a reduction in the gap in 2015–19 and this has been confirmed by British Future, a think tank that studies attitudes towards migration and ethnicity. British Future suggests that the ethnic bias among some groups against the Conservatives may be waning. Of course, such judgements are hard to qualify as people tend not to vote on the basis of their ethnicity. Trying to consider BAME voters as a single group suggests there is something different about BAME voters compared with white voters, which is not true, though race does create a significant lens through which other issues are viewed (including for white voters) when it comes to issues like housing, employment and immigration. As such, we must consider factors beyond ethnicity to explain why a majority of BAME voters tend to vote Labour.

One possible reason why BAME voters tend to favour Labour in general is related to economic and historic factors. Historically, immigrants from BAME backgrounds came to the UK and were employed in the major cities working in industrial roles, which made them natural allies of the Labour Party and encouraged the Labour movement to adopt more policies to protect its supporters. The concentration of BAME voters in large cities, which tend to be more socially liberal, also explains their tendency to favour Labour over the more rural Conservatives. Equally, the Conservative Party had a reputation in the 1960s and 1970s for being the anti-immigration party, with Enoch Powell's 'Rivers of blood' speech and support for traditional, white-centred nationalism, which the Conservative Party has struggled to shake.

Within the UK population, the proportion of BAME people in social classes C2 and DE is greater than the proportion of white people in those social classes. This therefore suggests that on the basis of class voting, BAME people are more likely to hold left-wing preferences. There are a few reasons for this, mostly relating to systemic prejudices and biases inherent in British society, but it means that racial issues are closely related to class issues. Increasing social mobility and the movement of more BAME voters into the AB and C1 classes perhaps account for the recent increase in Conservative support among BAME voters. However, for as long as wider social inequality sustains a link between ethnicity and social class, it should not be surprising that BAME voters will favour the Labour Party with its history of being the party of the working class.

As with other factors, turnout is also impacted by ethnicity, with BAME voters being much less likely than white voters to turn out to vote. That, and their concentration in urban centres that are traditional Labour strongholds, means that their concerns and issues are often overlooked in election contests.

Discussion point

In 2015, all the BAME MPs for Labour were elected from constituencies that had majority BAME voters, while the BAME Conservative MPs came from the safest Conservative seats with the fewest minority voters. Discuss why you think this might be and what it reveals about ethnicity and party politics in elections.

To discuss this you may wish to consider the following issues:
1   What this reveals about race-based voting in the UK
2   What it suggests about partisan voting in the UK
3   How these points might relate to party politics

Synoptic link

All these social factors relate directly to the policies and reputations of the political parties. Make sure you have a strong understanding of how parties try to target key groups of voters and how that impacts on their policies, as outlined in Chapter 2.

Region

Here we are looking at voting bias in various regions of the UK. However, we need to be cautious. Wealth, income and prosperity are not evenly distributed in the UK. We know that the southeast corner of England is much wealthier than the rest of England. We also know that there are regions of great deprivation, including the far southwest and northeast of England and South Wales as well as several decayed city centres and areas where traditional industries have declined. It would be surprising, therefore, if such areas did not favour left-wing policies, as proposed by Labour. The same is true of Scotland, though there it is not Labour that is the beneficiary of voting in depressed areas but the Scottish National Party. In other words, regional variations may in fact be class variations rather than geographical ones.

Table 4.7 demonstrates the variability of regional voting in the UK.

Table 4.7 Voting by region, the 2015 and 2017 general elections

Region	% Conservative		% Labour		% Liberal Democrats		% Green		% SNP or Plaid Cymru		% UKIP	
	2015	2017	2015	2017	2015	2017	2015	2017	2015	2017	2015	2017
North of England	31	37	43	53	7	5	3	1	n/a	n/a	15	3
South of England	46	54	26	29	10	11	5	3	n/a	n/a	13	2
Midlands	43	50	32	42	6	4	3	2	n/a	n/a	16	2
London	35	33	44	55	8	9	5	2	n/a	n/a	8	1
Scotland	15	29	24	27	8	7	1	0	50	37	2	0
Wales	27	34	37	49	7	5	3	0	12	10	14	2

Source: The House of Commons Library

Knowledge check

Study Table 4.7. Answer these questions:
1 Which region is most dominated by the Conservatives?
2 In which two regions does Labour do best?
3 Where is Conservative support weakest?
4 Where do the Liberal Democrats do best?

We can draw a number of conclusions about regional voting in recent general elections, based on the statistics in Table 4.7:

- The south of England (outside London) is very solidly Conservative.
- Labour dominates in London, holding the largest percentage of Labour votes in the country.
- The Conservatives are also dominant in the English Midlands, though slightly less so than in the south.
- Labour leads in the north of England but this is not a decisive lead.
- Scotland has moved from being a Labour stronghold before 2010 to being dominated by the SNP.
- Labour is the dominant party in Wales, though that is based largely on the industrial south of Wales.
- Liberal Democrats have little support outside London and the south of England.

There is a sense that Labour does have deep roots and strong local party organisations in the north of England and in Wales, so it is inevitable that the party will poll well in those regions. Similarly, voting for the Conservative Party in the south of England is understandable since in rural and suburban areas, the Conservatives have long dominated the political culture. These regional variations are, therefore, real factors. However, as we saw above, much of the variation can be traced to economic rather than regional influences. There are many more depressed and declining areas in the north of England, Wales and Scotland.

Northern Ireland has its own, unique political culture and the major English parties do not compete there. We shall not, therefore, consider voting behaviour in Northern Ireland as this is a highly specialised field.

The election in 2019 saw an interesting shift in regional voting as the Conservatives gained seats from the Labour Party in its traditional northern areas. This was mostly down to the issue of Brexit, leading Boris Johnson to declare that people in those regions had 'lent him their votes'. This shows that single issues can often trump social factors in elections.

Synoptic link

Regional voting is so important in the UK because of the FPTP voting system, which benefits those that have concentrated support in key areas, as explored in Chapter 3.

A summary of social and demographic factors

At the beginning of this section we drew an idealised view of a 'typical' Labour voter and their opposite number as a Conservative supporter. Of course, this was a huge generalisation as we have seen above. This is because everyone is an individual. Many voters make up their own mind about which party to support, whatever their class, age or ethnic background and no matter where they live. They are influenced by many other factors than their personal circumstances. It is important that you can consider and analyse the relative importance of each of these social factors and compare them across three elections, using the case studies at the end of this chapter. You need to ensure you can show and explain changes over time and clearly identify which social factor you would consider the most important and why that would be. It is also worth considering whether social factors are more or less important than other factors, which we shall consider next. Table 4.8 summarises these factors and offers an estimation of how significant they are in predicting voting tendencies.

Table 4.8 The influence of social and demographic factors

Factor	Influence
Gender	There is little difference in voting habits between men and women, though there is a slight tendency for women to favour Labour
Age	This is a key factor. Older voters significantly favour the Conservatives. Younger voters have a Labour bias and also tend to support the Green Party, a trend that has been growing since 1997
Ethnicity	Ethnicity is a significant factor, though reduced by low turnout and heavy concentration in safe seats
Class	Class used to be the most important determinant of voting behaviour but is becoming much less influential
Region	There are wide regional variations in voting patterns. Scotland is the most remarkable, with the SNP currently in complete dominance. The south of England is solidly Conservative, leaving Labour with a mountain to climb in that region. Thanks in part to Brexit, the Conservatives have made inroads into Labour's former dominance of northern England

The influence of social and demographic factors is complicated by the effects of turnout. Turnout, the proportion of those qualified to vote who do actually cast their vote, can have a further impact, but that impact is variable, as we have seen. This is because turnout varies among different groups. The next section examines the importance of such differential turnout rates.

Synoptic link

The issue of voter turnout is part of a wider issue of political engagement and participation. These are examined fully in Chapter 1.

Activity

If you were a policy-maker, what policies might you adopt to attract the votes of the following demographics?
- The 18–24 age group
- Women
- Members of first-generation ethnic minorities
- Migrants from the EU
- The 45–64 age group
- The over-65 age group

Activity

Access the YouGov website and search for the article on 'How Britain voted in the 2019 general election'. What trends in party support can you find using the following demographic factors?
- Home ownership versus renting
- Education level
- Working in the private or public sector
- Income level

Turnout

Before considering the impact of turnout in detail, we should consider the turnout statistics over a considerable period of time. Figure 4.4 shows national turnout figures since 1945.

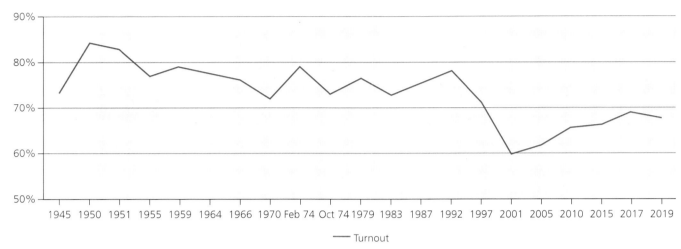

Figure 4.4 Turnout figures in UK general elections since 1945

We can see from Figure 4.4 that there was a sudden dip in turnout at the turn of the century. Before that the statistics were relatively healthy and in line with historical levels of turnout. Since then low turnouts have been a growing concern. Apart from the fact that low numbers turning up at the polls suggests a worrying level of disillusionment with politics, when small numbers actually vote it can erode the legitimacy of the elected government.

Two questions have to be asked about turnout at elections, especially general elections:

1 Why does turnout vary from one election to another?
2 Is there a long-term trend in turnout figures?

Variable turnout

The first question may be answered in a number of ways, but possibly the most important consideration is how close the election appears to be. There is a clear correlation to be seen in Figure 4.4. Turnout was relatively high (79 per cent) in February 1974 (the election was so close that it resulted in a hung parliament), in 1992 (which was a narrow win for the Conservatives), and in 2015 and 2017. The turnout figures for those last two elections were still historically low but represented a significant recovery from previous elections. Both elections were very close results. By contrast, 2001 and 2005 were foregone conclusions; Labour was going to win against a disunited Conservative Party, so turnout slumped. Even the landmark election of 1997, which swept Labour into power after 18 years in the wilderness, saw a fall-off in turnout. Again, the outcome was predicted months before the actual poll. So, we can say that the main factors in variable turnouts are how important the election might be and how close the outcome is forecast to be. This is supported by the fact that turnout is often far higher in marginal seats where the result is uncertain than in safe seats.

Age, class and turnout

Age

The impact of age on voting patterns is distorted by the fact that there is a great difference between turnout among different age groups. In general, younger voters are more reluctant to turn out and vote than older generations. This has the effect of exaggerating the impact of age on the outcome of elections and referendums.

Table 4.9 shows the turnout figures for five recent elections, broken down by age categories.

Table 4.9 Percentage voting turnout in UK general elections by age range

Age range	2005	2010	2015	2017	2019
18–24	37	44	43	54	47
25–34	49	55	54	55	55
35–44	61	66	64	56	54
45–54	65	69	72	66	63
55–64	71	73	77	71	66
65+	75	76	78	71	74
Overall	61	65	66	69	67

Source: Ipsos MORI

Between 1992 and 2015, voting among young people fell by approximately 19 per cent. By contrast, turnout among other groups, especially those aged 55+, held up well. The statistics shown in Table 4.9 demonstrate this recent trend. In 2005, a year of a very low national turnout, only 37 per cent of the youngest group of voters (18–24) actually turned up to vote. This compared with over 70 per cent in each of the two oldest age categories. In 2015, the turnout figure of the 18–24 age group, at 43 per cent, was well below the very high figures of 77 per cent and 78 per cent in the two oldest groups. All the statistics in the table show that the older a person is, the more likely they are to vote. Interestingly, in 2017 there was a much reported 'youthquake' where the press reported a large surge in young voters, which helped Labour to a much better than expected performance, perhaps costing the Conservatives a majority. As Table 4.9 shows, there certainly was a significant increase in turnout for the 18–24 age group, increasing by 11 per cent. Despite this, the 18–24 age group remained the age group with the lowest voter turnout and fell back in 2019 to 47 per cent, suggesting, perhaps, that the 'youthquake' had been more of a blip than a change.

If we combine these findings with the observation that the older age groups are much more likely to vote Conservative than the younger age groups (see Table 4.5), we can see that the age-demographic advantage that the Conservative Party enjoys is magnified by turnout trends.

There are a number of possible reasons for the decline in turnout among younger voters:

- There is widespread **disillusion** with conventional politics among the young. This may be caused by the fact that politicians have introduced policies that discriminate against this age group. However, it is also ascribed to general **apathy**, to a belief that politics has nothing to do with the things that concern the young and that voting will not make a difference.
- The young are increasingly finding alternative ways of participating in political activities, such as e-petitions, direct action and social media campaigns, thus moving away from conventional political activities.
- Younger people tend to be interested more in single issues than in broad political ideologies. This is reflected in low election voting figures and less interest in political parties, but increased participation in pressure group activity and online campaigning.
- Many young people feel the need to abstain. **Abstention** is when someone does not vote because they do not feel that any of the parties is worthy of their support; no party adequately represents their own views and aspirations.

However, voting turnout among the young at two recent referendums was relatively high. In the Scottish independence referendum, polling organisation ICM estimates that 75 per cent of 16- to 17-year-olds, 54 per cent of the 18–24 group and 72 per cent of those aged 25–34 turned up to vote. In the EU referendum, polling by Opinium suggests that 64 per cent of those aged 18–34 voted. Then, in the 2017 election, there was an increase in voting by younger age groups, most of whom supported the Labour Party. As a result, Labour's share of the national vote rose by about 10 per cent compared with 2015. This was enough to prevent the Conservative Party winning an overall majority in parliament, though it is worth remembering that the percentage of the young who turned out in 2017 was still far below that of older groups.

Class

Turning to turnout figures broken down by class, we can see a further significant effect. Table 4.10 reveals that members of class AB are much more likely to vote than members of class DE. So, combining voting preferences and turnout figures, we can see a further disadvantage suffered by the Labour Party. The class that favours Labour turns up to vote in much smaller numbers than their Conservative counterparts in class AB. When we add the class-turnout effect to the age effect on turnout, it is apparent that the Labour Party has a bigger problem trying to persuade its natural supporters to vote than it does to persuade people to support the party in the first place!

Table 4.10 Percentage voting turnout in UK general elections by social class

Social class	2005	2010	2015	2017	2019
AB	71	76	75	69	68
C1	62	66	68	68	64
C2	58	58	62	60	59
DE	54	57	57	53	53
Overall	61	65	66	69	67

Source: Ipsos MORI

Why those in social classes C2 and DE are less likely to vote than those in classes AB and C1 is a complex issue. However, if we accept that the main reason why people do not vote in general is that they do not feel the outcome will make any difference, we can find some clues. The other factor is that people will vote if they understand the issues and if those issues relate to them in some direct way.

Table 4.11 summarises the important points that arise from Tables 4.9 and 4.10.

Table 4.11 The links between turnout and demographics

Feature	Significance
In 2005 and 2015, the 18–44 age group in particular had especially low turnout figures	This poses a problem for Labour, which receives a higher proportion of support from younger voters
The 65+ age group shows high turnout figures in all elections	This gives a large advantage to the Conservative Party and UKIP (in 2015), which are most supported by this group
In all three elections, turnout among the AB class was much higher than among the DE class	This gave a large advantage to the Conservatives, who are more supported by classes AB, and discriminates against Labour, whose support is concentrated in classes DE
Overall turnout in the elections between 2005 and 2017 showed an upward trend after the drop from 1992 to 2001	This suggests that the 'participation crisis' may be gradually coming to an end, though this may be due to the fact that all these elections were relatively 'close'. The election in 2019 rebuts that, but it is too soon to see if that was an anomaly or not

Activity

Study the results of the 2015, 2017 and 2019 general elections in the constituency where you live. What were the turnout figures? Were they higher or lower than the national figure for each year? If they were particularly high or low, can you offer a reason for that?

We cannot leave the issue of voter turnout without noting that turnout at local elections, elections to devolved assemblies and for city mayors remains much lower than the numbers at general elections. In searching for a reason, we should recall the basic reason for voter apathy. This is that people will not vote if they do not believe their vote will make a difference. At local and regional government levels, there is a general perception that power is concentrated at the centre so that the representatives they are being asked to vote for have relatively little power of their own. This is further borne out by the fact that voting behaviour at local level is largely determined by *national* issues rather than local ones. This may change as devolved administrations in Scotland, Wales and Northern Ireland gain even more powers and as city mayors become more established, but for the time being, voting turnout at below central government level remains low in what are considered 'second order' elections.

Individual voting theories

The factors considered above suggest that voting behaviour can be understood based on group dynamics, but this rests on the theory that people in a certain group can be expected to vote in a certain way. While this may be true, it ignores the individuality of voters and the idea that people cast their votes based on who they are, not which group they belong to. We are now going to consider an alternative set of voting theories, which suggest that voting behaviour is best explained by individual decisions.

There are, broadly, three theories that help us to explain why individuals cast their votes as they do: valence, rational choice and issue-based voting.

Valence

A valence issue is one where voters make their decision based on the party or candidate they think is most likely to run the country effectively, especially in relation to the economy. Most political commentators today argue that valence is the most important predictor of voting behaviour, especially with the emergence of partisan dealignment, as described above. Leading political analyst Peter Kellner summed up valence like this:

> [M]illions of swing voters don't [take a strong view on individual issues], they take a valence view of politics. They judge parties and politicians not on their manifestos but on their character. Are they competent? Honest? Strong in a crisis? Likely to keep their promises?

> Source: YouGov, 29 July 2012

> ### Key term
>
> **Governing competency**
> The perceived ability of the governing party in office to manage the affairs of the state well and effectively. It can also be a potential view of opposition parties and their perceived governing competence if they were to secure office.

As well as general competence, voters pay special attention to economic competence. This includes how well they believe a party will manage the UK economy and how well they believe it has done this when in power in the past. This may be described as judgements about **governing competency**. Who will be most responsible with the taxpayers' money? Who will do most to spread wealth or promote growth? This is also sometimes described as economic voting. It is a powerful influence on voting behaviour, as one might expect.

Voters will look at the performance of the UK economy and decide which of the parties has done most to improve it, and which party has damaged the economy in the past. A prime example was the problem the Labour Party had after the economic crisis of 2008. Labour was blamed by many voters for contributing to the crisis and for allowing government debt to rise by an alarming amount. Labour defeats in the 2010 and 2015 elections were based partly on such economic voting. In 2016 the new Chancellor of the Exchequer, Philip Hammond, stated he would manage the economy on a 'pragmatic' basis so people could feel confident about his competence.

Leadership (discussed in more detail below) is a key valence issue. Voters like strong leaders with desirable personal characteristics. They will look at the leaders' past record as politicians, just as if they were applying for any job. Weak leaders are rarely supported. Labour's Ed Miliband suffered from being perceived as weak in the 2015 general election. Gordon Brown, who was defeated in the 2010 general election, was punished by the electorate for being seen as indecisive. The media tends to play on such perceptions and amplify them to feed a narrative and can turn a general attitude into a damaging attack on the competence of a leader.

When the next election comes around, there are likely to be two major issues that will affect valence voting: how the government responded to the Covid-19 pandemic and resulting financial impact, and the effect of Brexit as that unfolds.

There are other valence issues to consider, but the ones described above are most commonly cited by those who seek to analyse election results.

Valence issues can be summed up as follows:

- How generally competent the previous government was and how competent voters think other parties would be if in government.

- How economically competent the government was and the other parties are likely to be.
- How strong and 'prime ministerial' the leader of the party is.

Rational choice

Rational choice voting is a theory that suggests people vote based on what they consider to be in their own best interests. Typically, this might be focused on the economy, but it will relate specifically to what they will gain from the new government. In this sense, voters look at the manifestos and all the various policies and ideas and decide which party, overall, will govern most effectively in a way that suits them, as individuals. This decision is usually based on several issues and may be considered in relation to governing competency; people may feel certain policies by a party would make them worse off, but that party may be the rational choice if it will manage the economy more effectively. This ties in closely with the idea of liberal democracy and assumes that people are rational and will do what is in their own best interest. This reduces the voting decision to one based on logic, removing emotion from the process.

Issue voting

Issue voting suggests that voters will decide whom to vote for based on a single issue that means a great deal to them. This could be rational, such as a pensioner deciding to vote for the party that promises the highest rate of pensions, but usually it is seen as irrational, in the sense that someone may vote for a party based on this one issue, even though it might lead to them being worse off in many other ways. The issue could be environmental, or perhaps the issue of Brexit, where people will vote for a policy that may make them financially worse off but that they believe in. Trying to determine which single issues will secure the most votes for a party is difficult to do in advance, but parties that can successfully adopt popular issues have a better chance of winning over these voters. This can be seen in the main parties adopting more environmental policies to counteract the rise in support for the Greens, or the Conservative Party adopting an increasingly Eurosceptic attitude to head off the challenge of UKIP.

Sometimes events have a habit of shaping the single issue that dominates an election. The Manchester Arena bombing and London Bridge attacks in 2017 made the issue of terrorism and policing a major issue for voters in that election, while in 1983 the issue of the Falklands War seems to have persuaded some voters to back the Conservatives. Often the central issue will be more obvious, such as Brexit in 2019.

Factors affecting individual voting

When it comes to casting a vote, there are several factors at play that help a voter make their decision. These factors, outlined in Figure 4.5, relate closely to the individual voting theories. All of them play a role in determining how voters vote, but you should consider which one you feel is the most important in explaining the outcomes of the elections in your chosen case studies. It will also be important to know how the key factors listed below influenced the general elections you study.

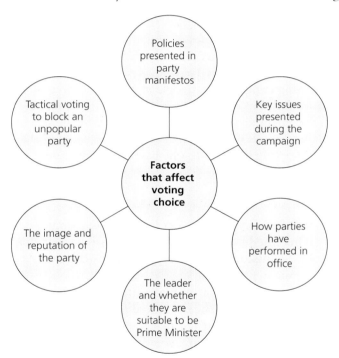

Figure 4.5 Factors affecting individual voting

Manifestos

Manifestos are lists of policies a party sets out to try to appeal to voters and persuade them to vote for it. Since 1945 manifestos have become increasingly specific and detailed. They still contain the broad beliefs of the party, but there are also many very particular intentions. In 2019, for example, the Labour Party manifesto was extremely specific. Its commitments included the following:

- A 4.3 per cent per year increase in the health budget.
- To renegotiate the Brexit deal with a new UK–EU customs union and close EU single-market alignment, which would then be put to a 'final say' referendum by the British people.

- To abolish tuition fees for university students and to restore maintenance grants.
- To increase the minimum wage to £10 an hour for everyone over 16 within a year.
- To stop pension age increases and review recent increases for stressful jobs.
- To introduce a National Care Service, including free day-to-day care for the elderly.
- To nationalise key industries, including the main energy providers, the water industry, Royal Mail and broadband provision.
- To end Universal Credit.
- To abolish the charitable status of private schools.

Other parties also make extremely specific commitments in their manifestos, such as the Conservative Party pledge to increase the number of nurses by 50,000, a guarantee that pensions will increase by at least 2.5 per cent each year and that that they would spend at least £6.3 billion on 2.2 million disadvantaged homes. The Liberal Democrats, meanwhile, promised to stop Brexit, increase income tax by one penny in the pound to fund the NHS, and to recruit 20,000 more teachers. As we can see, the general trends are similar, but the specifics vary between the main parties.

The development of such precise manifesto pledges is intended to give the electorate a clear set of issues and policies that the government will be committed to pursuing. In this sense, manifestos are an attempt to convince voters that it is in their rational interest or that the issue they value most will be delivered. By delivering on manifesto commitments, a party can claim to have delivered what they had promised to and therefore appeal to valence voting.

Manifestos also help to establish the doctrine of the **mandate**, which is a major feature of the UK political system. It is central to the relationship between the electorate, parties and government. The term 'mandate' can be described as consent, in that a mandate implies the consent of one person, allowing another to do what they feel is necessary for their welfare. In politics, therefore, a mandate represents the consent of the people, allowing a political party to do what it feels is necessary in the national interest if it succeeds in being elected to government.

The doctrine of the political mandate operates like this: when a party wins an election and forms the government of the UK or of one of the devolved administrations in Scotland, Wales and Northern Ireland, it has a mandate to carry out all the policy commitments contained in its election manifesto.

> **Key term**
>
> **Mandate** The successful party following an election claims it has the authority (mandate) to implement its manifesto promises and also a general permission to govern as new issues arise.

> **Synoptic link**
>
> The concept of the mandate is fundamental to democracy as it is how permission is granted from the people to the government. The mandate also gives the winning party the opportunity to set out its legislative agenda in Parliament, based on the manifesto promises it made, an issue explored further in Chapter 6.

It follows from this that the existence of party manifestos is vital to the operation of the mandate. The significance of the mandate doctrine is as follows:

- Electors can feel confident that they understand which policies they are consenting to when they cast their vote. This does, of course, assume that they have read and understood the party manifesto. If they have not, they are giving their consent in ignorance. Even so, the doctrine does assume that electors have full knowledge of the manifestos and so can make a rational judgement.

- The mandate strengthens government, in that the winning party gains legitimacy for its policies.
- The mandate means that Parliament (or devolved assemblies) can call government to account on the basis of the governing party's manifesto. If a government strays from its electoral mandate, Parliament and assemblies can feel justified in challenging the government.
- The mandate also gives electors the opportunity to judge the performance of government when election time comes round. They can ask the question, 'How successful was the government in delivering its mandate?'
- All the MPs from the winning party who are elected are 'bound in' by the mandate, as most voters vote for a party manifesto rather than an individual. Party leaders can therefore maintain discipline among members by emphasising to them that they were all elected on the same mandate.

The doctrine of the mandate does have some problems. These include:

- It depends upon a single party winning an election outright. When there is a coalition, as occurred in the UK election of 2010, two or more parties are involved and the actual content of the mandate is unclear. Similarly, if no party wins an overall majority, as occurred in June 2017, and a minority government is formed, the government cannot legitimately claim a mandate.
- Voters who have opted for one party do not necessarily agree with *all* its manifesto commitments. However, the mandate doctrine assumes the electorate has given its consent to the *whole* manifesto.
- Circumstances may change after a party takes power. This means it will have to amend its policies or abandon some or develop new policies as the Liberal Democrats had to do in 2010 when they abandoned their manifesto commitment to abolish university tuition fees to form the coalition.
- Some manifesto commitments may be rather vague and open to interpretation. This makes calling the government to account based on its manifesto difficult and open to dispute.

Essentially, voters must make a judgement on the collection of policies offered in a manifesto to help determine whom to award with an electoral mandate.

Party leaders

Here we are examining the impact of party leaders, and particularly *public perceptions* of the leaders, on voting behaviour. It has to be remembered that, at a general election, the voters are choosing a future prime minister as well as a ruling party and a local MP. The image and qualities of party leaders are crucial. Are they trustworthy? Are they decisive? Do they promote a strong image of the UK abroad? Can they keep the government united? Certainly in the 2015 general election, David Cameron enjoyed a much more positive image than Ed Miliband, his Labour opponent. Tony Blair began in 1997 with an extremely positive image, and won two more elections, but by 2007 his image was so tarnished that the Labour Party replaced him with Gordon Brown. Brown himself suffered a poor image, based on negative media portrayals and his reputation for indecisiveness. This contributed to Labour's election defeat in 2010.

The kind of qualities that the public normally cite as important in a leader include:

- Record in office (if they have been in office)
- Compassion
- Decisiveness

- Apparent honesty and sincerity
- Strong leadership
- Clear vision
- Communication skills

The press and broadcasters would certainly have us believe that the character and image of the party leaders are vital factors in the outcome of elections. The evidence, however, is less clear-cut. For example, in the 1979 general election, Labour Prime Minister James Callaghan led his Conservative opponent, Margaret Thatcher, by 20 per cent in popularity polls, but lost the election. Similarly, in 2010, Liberal Democrat leader Nick Clegg was the most popular of the party leaders, following impressive showings in televised leadership debates, but his party's share of the vote fell by 1 per cent and the party lost five of its parliamentary seats.

Yet there appears to be some evidence that party leaders and their popularity (usually described as 'satisfaction' and 'dissatisfaction' ratings by opinion polls) increasingly *do* swing elections. Table 4.12 seems to demonstrate this.

Table 4.12 Satisfaction ratings of party leaders, November 2019

Leader	% satisfied	% dissatisfied	Net satisfaction
Boris Johnson (Con)	39	40	−1
Jeremy Corbyn (Lab)	20	58	−38
Jo Swinson (Lib Dem)	23	39	−16
Nigel Farage (Brexit Party)	26	42	−16
Nicola Sturgeon (SNP)*	29	54	−25

* Rating among Scots only

Source: Ipsos MORI/TNS Scotland

At first sight Table 4.12 seems to confirm that party leadership ratings do make a difference. Boris Johnson had the best poll rating of all the leaders shown (though his net rating was still negative) and his Conservative Party did win the election with a large majority. Jeremy Corbyn suffered from the worst satisfaction ratings of any opposition leader since polling began and Labour did go on to suffer its worst defeat since 1983.

However, a look at the other leaders raises some questions about the role of party leaders. While Jo Swinson was personally not highly regarded, the Liberal Democrat vote share increased from 7.4 per cent in 2017 to 11.5 per cent in 2019. Nicola Sturgeon, meanwhile, suffered a negative polling image just before the election, yet the SNP regained a number of seats lost in 2017, suggesting in both these cases that the party, rather than the leader, was what mattered to voters. Nigel Farage is a slight anomaly. The Brexit Party was very much his party and its fortunes seem to have been tied directly to his image, which remains incredibly high among his supporters, but incredibly low among his opponents. Overall, though, it was the issue of Brexit that mattered more than the leader, and many Brexit voters turned to the Conservatives rather than the Brexit Party, suggesting the issue, rather than the leader, was more important to those voters.

The role of the party leader and how a changing reputation can impact on a party's fortunes can be seen by comparing the performances of Jeremy Corbyn in 2017 and 2019. During the 2017 general election a remarkable phenomenon occurred. Starting the campaign as an underdog, reviled by much of the press and opposed by many MPs in his own party, Jeremy Corbyn created a bandwagon effect, mostly among the young. There is little doubt that his resurgence was a major influence on the outcome of that election and the Labour revival. In this, he was perhaps helped by

Activity

Research the current satisfaction rating for the leaders of the Conservative, Labour, Liberal Democrat, Green and Scottish National parties. Which of them have positive and which have negative images, and why?

Theresa May performing poorly in her campaign, creating a negative media image and failing to take part in television debates. While 2019 was more dominated by the issue of Brexit, the contrast of Boris Johnson with Jeremy Corbyn highlighted the old flaws and, after 2 years and a lack of any concrete policies on Brexit on Corbyn's part, people seemed to believe that Johnson was the better choice to be leader of the nation. In both elections, Corbyn as Labour leader did make a difference to the votes.

Tactical voting

Synoptic link

The issue of tactical voting, in response to the idea of wasted votes, is a particular problem FPTP and the safe seats it creates, as covered in Chapter 3.

Tactical voting, as discussed in Chapter 3, is undertaken in special circumstances and in specific constituencies. In elections many votes are considered 'wasted' votes. This is because they will have no influence on the outcome. A common form of wasted vote is a vote for a different party, such as the Liberal Democrats or the Green Party. Those who support such parties may be frustrated by their inability to affect the result. Some, therefore, may abandon their first-party preference and vote instead for one of the parties that *does* have a chance of winning in a constituency. Such a tactical voter may vote for one of the main parties as their second choice so as to influence the final outcome, usually as a means of blocking the less favoured of the two parties.

A 2015 Conservative election poster designed to encourage some English voters to vote tactically for the Conservatives

Typical examples of tactical voting have been as follows:

- Labour supporters voting Conservative to keep out a UKIP candidate in a close UKIP–Conservative contest.
- Green Party supporters voting Labour to keep out a Conservative candidate in a close Labour–Conservative contest.
- Labour supporters in Scotland voting Conservative to keep out an SNP candidate in a close SNP–Conservative contest.

- Conservative Party supporters in Scotland doing likewise to keep out an SNP candidate in a close SNP–Labour contest.
- Plaid Cymru supporters in Wales voting Liberal Democrat to keep out a Conservative candidate in a close Conservative–Liberal Democrat contest.

It is difficult to estimate how much tactical voting occurs and even harder to establish whether it has any effect on electoral outcomes. John Curtice, a leading election expert at Strathclyde University, estimated that tactical voting *could have* affected the result in as many as 77 constituencies in the 2015 general election, but this does not mean that many actually *were* affected in the event. Ipsos MORI's research into voting at the 2010 general election suggested that as many as 10 per cent of voters chose their second preference and the figure was especially high among Liberal Democrat supporters, at 16 per cent. But this still does not answer the question of how much influence tactical voting has on the final results. Since 2015, activists have been using social media to try to target and persuade voters to vote tactically, normally from a left-wing point of view, trying to 'keep out the Conservatives'. Websites have also been created to help voters decide how to tactically cast their votes, though their impact is not clear as yet.

Circumstantially, the evidence from Scotland in 2015 suggests that tactical voting had no effect. Most attempts to persuade people to vote tactically involved keeping out SNP candidates, but the SNP won handsomely, suggesting there was no great impact. On the other hand, the slump in Liberal Democrat voting in the 2015 general election might in some way be explained by tactical voting, especially as Conservative media played on fears of an alliance between Labour and the SNP to persuade Liberal Democrat voters to vote Conservative to keep Miliband out. None of the parties has ever issued formal instructions for its supporters to vote tactically, and until this happens it may remain impossible to estimate its impact.

The influence of the media

Before looking at the media's influence on politics, we should consider the fact that the nature of the media is changing. Increasingly, people are accessing news online. Furthermore, social media is increasingly being used to disseminate information, opinion and even propaganda. The main issue concerning this change is that it has become more difficult to distinguish between fact and fiction. Some have called this the 'post-truth' era. The spreading of unsubstantiated facts and sometimes deception is thought to have been an influence in both the EU referendum campaign in the UK and the 2016 US presidential election. There has been an assumption that the truth will ultimately emerge, but there is no guarantee that this will happen.

Broadcasting

The term 'broadcast media' covers all television stations, such as the BBC, ITV, Channel 4 and Sky, and radio networks like Capital, LBC and regional radio stations. All broadcasters in the UK are bound by law to remain neutral and to offer balanced reporting of election and referendum campaigns. This means they have no intentional influence on voting behaviour. Although the BBC has sometimes been accused of a liberal or left-wing bias, nothing has been proved or substantiated.

It is just as well that there is a legal neutrality requirement as research indicates that television and radio remain the main sources of information for voters in UK election campaigns.

The leaders' debate between Johnson and Corbyn on ITV in 2019

We should not leave this subject, however, without referring to the televised debates that have now become a common feature of elections. The BBC, ITV, Sky and Channel 4 have all held leadership debates in recent elections and do so under scrupulous conditions, overseen by the Electoral Commission. However, it is not at all clear that leadership debates have any significant impact on the voters. We have already seen that Nick Clegg's spectacularly good performance in the 2010 debates still led to a decline in his party's share of the popular vote. Similarly, after a BBC Challengers' debate in April 2015, held just before that year's election and in which David Cameron did not take part, the opinion polls suggested that Ed Miliband narrowly won, even over the enormously respected Nicola Sturgeon, but Miliband's poor standing in leadership polling did not change and his party lost an election it was expected to win.

All we can really assert about broadcasting is that the parties use television and radio as an important way of getting their messages across, but they do not expect to gain any special advantage from it.

The press

When it comes to the press we are mainly referring to newspapers, whether in physical format or the more common online versions. The press can also refer to magazines and periodicals, but these are seen more as reviews rather than setting the debate and agenda in the way the daily newspapers do. Following the 1992 general election, when the Conservatives won a surprise victory after most predicted a Labour win, the *Sun* newspaper famously proclaimed, 'It's the Sun Wot Won It'. Certainly the *Sun* had run a relentless campaign against the Labour Party and especially its leader, Neil Kinnock, and the opinion polls, at first predicting a comfortable Labour victory, turned round near the election date and John Major's Conservatives secured a majority. Whether or not it was indeed the press that had changed voters' minds, however, is open to contention.

Unlike television and radio, there is no press regulation in terms of political bias and UK newspapers are highly politicised, influenced in part by the ownership of the individual papers but mostly by the beliefs of their readership. Today, more of the newspapers support the Conservative Party than other parties, and the two largest circulation tabloids, the *Sun* and the *Daily Mail*, both support the Conservatives, reflecting the attitudes of the owners of those papers. At first sight, therefore, we may conclude that newspapers *do* influence the way people vote. However, this may be an illusion.

While large sections of the public do believe that the press influences them, research suggests it does not. Instead the newspapers tend to *reflect* the typical political views of their readers, rather than leading them. This is likely to be something of a two-way process as the papers may also reinforce *existing* political attitudes, but there is no strong evidence that they can change them. Indeed, giving evidence to the Leveson Inquiry into press behaviour in 2012, the *Sun*'s owner, Rupert Murdoch, admitted that newspapers do not swing votes, they merely reflect readers' opinions. Perhaps the greatest evidence against the importance of the press is that in 2017 Jeremy Corbyn's Labour Party received 40 per cent of the national vote despite only having one paper (the *Mirror*) supporting it.

Social media

Although parties and government increasingly use social media as a way of communicating with the public and of 'listening in' to public opinion, it is too early to assess its influence. Certainly, as it is an open medium, unlike the press and broadcast media, it is more difficult for any one party or political group to gain any special advantage, although there does seem to be a case that those with access to large resources can use online ads and tools like Twitter bots to influence the outcome.

Synoptic link

There is a fuller discussion of the role of digital democracy in Chapter 1.

Unlike the broadcasters, the web is unregulated so there are opportunities for any group to gain some political traction. It is especially useful to small parties such as the Greens and UKIP, which do not have the resources, in terms of membership and national organisation, to be able to compete with the large parties in conventional campaigning, though neither has been able to use it to make a major breakthrough in terms of electoral success.

Social media's real impact lies in the way that it is reported by more mainstream or traditional outlets. Increasingly, broadcast media and press media report on debates and issues that have been raised on social media, which are subject to less scrutiny and accountability. As traditional media reports on opinions and issues popularised on social media, so those ideas become established in the mainstream political debate and help to influence political opinions and choices. This is a relatively new phenomenon, but the publication of the report into Russian interference during the

EU referendum and subsequent general elections in the UK does seem to indicate that social media provides a platform that groups with vested interests can use to influence the political debate in the UK, though not necessarily with the best of intentions. These issues have only become more apparent as time has developed, with companies like Cambridge Analytica using peoples' personal data to target and influence them into voting a particular way on behalf of clients. The development of 'fake news', both actually false news and conspiracy theories, as well as the assumption that any news that a person does not agree with or support must be fake, have come much more clearly into the mainstream of political discussion, stifling political discussion and, in the worst cases, causing actual harm. People have attacked 5G phone masts and actively encouraged others to avoid health measures and vaccines. This only highlights the dangers for democracy and politics that social media can present.

Does the media influence election outcomes?

As we have considered above, the media certainly plays a role during election campaigns, but is it an influential one? Some examples of the role the media has played during election campaigns are outlined in Table 4.13.

Table 4.13 Has the media influenced election outcomes?

Election	Media moment	How it may have influenced the outcome	How it may *not* have influenced the outcome
1979	The *Sun* issued a headline saying 'Crisis? What crisis?' as Jim Callaghan's reaction to the Winter of Discontent	Jim Callaghan never actually said this, but it caught the public imagination, suggesting that he was out of touch with public opinion, thereby swinging opinion against a formerly popular PM	Polls showed that Callaghan remained personally popular throughout the election and was actually far ahead of Thatcher. He was actually Labour's biggest asset
1997	The *Sun* switched support from the Conservatives to Labour	The *Sun* (and most of the press) publicly switched support to Blair and New Labour, persuading many former Conservative voters to vote Labour instead	Polls suggested Labour was on course for a large victory anyway. The *Sun* and the press were simply reacting to the existing situation
2010	The first televised debate was won by Nick Clegg	Leader of the Liberal Democrats, Nick Clegg, was reported as having won the first leaders' debate, raising his profile at the expense of Cameron, perhaps costing the Conservatives the votes needed for an outright majority	The Liberal Democrats only increased its share of the vote by 1% nationally and actually lost seats, suggesting any impact from the debate was, at best, limited
2015	The televised leaders' debate	Ed Miliband fell off the stage and gave an over-excited 'Hell yes, I'm tough enough' response to a question, making him appear less prime ministerial than David Cameron	Opinion polls suggested the debate made no real difference to voting intentions, merely confirming existing intentions
2019	Facebook advertising	At the start of December 2019, the Facebook Ad Library showed the Conservatives had 2500 live paid-for adverts, while Labour only had 250, and the Conservatives went on to seal a huge victory over Labour	At the same time, the Liberal Democrats had 3000 paid ads on Facebook, more than the Conservative Party, yet it lost seats

The role of the media between elections

Beyond elections the traditional role of the media was to report on events as they happened and provide a commentary to help explain events to the public while allowing them to make their own, informed decisions. The media could also act as a forum for public debate and discussion, challenging public ideas and acting as a bridge between the electorate and their representatives. Democratically, this not only helped to maintain an informed and well-educated public, but also to act as a check on representatives and democratic institutions by scrutinising their work through investigations. This role is still important, as we have seen by the way in which the *Daily Telegraph* investigation of the MPs expenses scandal in 2008 helped to hold many MPs to account, while the media reporting of the Windrush scandal in 2018 saw the resignation of the then Home Secretary and a public investigation into the matter.

However, since the 1980s the role of the media has been changing and it now plays a quite different role in modern politics, which has opened it to criticism and attacks and even accusations that it may be undermining democracy. The key changes relate to the following ideas:

- The tabloid press has always been partisan, but it has become more so with the rise of social media platforms.
- The tabloid press has become more focused on scandal and mocking politicians it opposes, rather than providing information and informed debate.
- The media, in the way it tends to prioritise negative political stories over positive ones, has contributed to a national attitude of cynicism towards politics and politicians.
- By focusing ever more on leaders and personalities in politics, the media has turned politicians into celebrities instead of focusing on their abilities and public service.
- Linked to the previous point, the media has helped make politics a form of mass entertainment.
- The development of 24-hour news since the 1990s has caused media outlets to create stories and issues to fill the slot and seen relatively minor issues become far more prominent than they really should be.
- The rise of online media platforms has led to increasingly partisan and uninformed debate proliferating, with opinions often being stated as fact. This in turn has seen the more traditional media outlets lose control of the agenda and follow such online trends.
- The rapid growth of social media has helped spread political education and awareness, but often at a more superficial and less engaged level than traditional forms of media. This can be especially problematic as social media can create echo-chambers where people only seek out opinions that match and confirm their existing beliefs.

These changes have largely resulted from the commercial development of the media. Newspapers have always been about competition, but in the past they could rely on a high level of loyalty from their readership. Since the 1980s, competition between papers has become much fiercer, while the BBC and ITV have been joined by commercial news outlets in television and broadcast media. The rise of social media has seen an even greater increase in competition between news agencies and forced the traditional media to respond. The result is that the media has become far more commercially minded and now seeks to engage the public by entertaining them as much as by informing them. This has led to greater populism in the press and a focus on 'catastrophes', 'scandals' and 'enemies' as these are far more likely to attract consumers than news stories. As an example, think of the way in which Prime Minister's Questions each Wednesday is portrayed as a 'gladiatorial battle' between the Prime Minister and the Leader of the Opposition, with a weekly winner and loser. This is hardly the informed scrutiny of the government the media traditionally reported on.

Then again, the media is doing this to attract viewers, readers and subscribers; basically, the general public. More often than not, the media tends to respond to public opinions and moods because that is what the public wants and expects to hear. The press may add weight and exacerbate the points, but it tends to reflect public attitudes rather than shape them. Would the press focus so much on celebrities if people did not want to read about them? The fact that the PMQs is by far the most viewed element of Parliament's weekly agenda suggests there is little public desire for detailed reporting on the real, rather dull, work in committee rooms. Of course, this is likely to be something of a two-way process, with the media helping to shape some attitudes that it then caters to, such as examples of some sections of the media portraying immigrants and members of the LGBT+ community in a negative light, which become the attitudes their consumers want reported upon.

Some examples of the way in which the press has influenced politics between elections include the following:

- Sleaze: Between 1992 and 1997 the press investigated and reported on a number of scandals relating to Conservative MPs, such as 'cash for questions' and numerous affairs from MPs leading a campaign for the public to go 'back to basics' in terms of morality. It led to the Conservative Party being associated with 'sleaze' and damaging its reputation with voters long before the 1997 election.
- The war in Iraq: The case for war in Iraq in 2003 was already controversial, but when the BBC reported that the case for war had been hyped up by using a 'dodgy dossier' to 'sex up' the issues involved it became a full-blown scandal that resulted in the death of weapons inspector David Kelly. Although the Hutton Inquiry later cleared the government, the accusations undermined trust and confidence in Blair, leading him to be seen as a liar and puppet of the USA.
- The expenses scandal: In 2009 the *Daily Telegraph* published details of expenses claimed by MPs and peers, many being wrongfully made or for inappropriate items, such as a duck house. This subjected all MPs and peers to a high level of scrutiny and forced many to resign, while others were subjected to police investigations, arguably as a result of the media scrutiny.
- 'Enemies of the people': In November 2016 the *Daily Mail* published a headline claiming three high court judges were enemies of the people for ruling that Parliament, not the PM, should trigger Article 50. The attack was widely condemned (though not by the government) because it politicised and endangered the judiciary. It also inflamed tensions and political divisions over the issue of Brexit.
- Political participation: Since 2016, reports have indicated that many of those in the UK who participate in politics principally do so through social media platforms rather than more traditional methods, such as delivering leaflets and organising petitions.

Public opinion polls

Public opinion polls have been a feature of British political life since the 1940s, when the first poll was carried out by the Gallup organisation. Gallup (an American company that had started polling in the USA before the Second World War) predicted that Labour would win the 1945 general election, much to the surprise of most commentators of the day. Gallup was right and from then on polls became increasingly used to gauge political opinion. Since their early days, however, opinion polls have become both more influential and increasingly controversial.

Two questions need to be asked about opinion polls. The first is: Do they have any effect on voting behaviour? The second is: If they do, how much does it matter

Useful term

Public opinion poll Poll carried out by research organisations using a sample of typical voters. They are mainly used to establish voting intentions, but can also be used to gauge leaders' popularity and the importance of specific issues in voters' minds.

that they have often proved to be inaccurate, meaning the electorate are being influenced by false information?

There is some evidence that polling figures can affect voting. The most striking piece of evidence concerns the 2015 general election. In that election, most opinion polls were predicting close to a dead heat between the two main parties, resulting in a second hung parliament. In that event, it was widely suggested, the SNP, which was heading for a huge victory in Scotland, would hold the 'balance of power'. In other words, there would probably be a Labour–SNP coalition. Indeed, the Conservative Party began to campaign on that basis, hoping to gain votes and win an outright victory.

In 2015, the polls overestimated the Labour vote and underestimated support for the Conservatives. The question is: Was there a late surge for the Conservatives, fuelled by voters wishing to avoid a hung parliament with the SNP in control? Certainly, the Conservative Party campaigned on that basis. Similarly, the polls were showing the Liberal Democrats doing poorly in the campaign. Did this lead to further defections from the party as increasing numbers of voters decided to vote tactically and so depressed the Liberal Democrat vote even further?

The opinion polls were also inaccurate in the June 2017 general election. Most showed a Conservative lead varying between 5 per cent and 12 per cent, outcomes that would have won the party a comfortable parliamentary majority. In the event, however, the Conservatives came in barely 2 per cent ahead of Labour and there was a hung parliament. Survation and YouGov were the only polls among many that predicted such a result.

That opinion polling can be inaccurate is in little doubt. The polls on the whole were significantly wrong in the 2014 Scottish independence referendum, the 2015 and 2017 general elections and the 2016 EU referendum, and though right, underplayed the Conservative victory in 2019. They may also have affected voting behaviour, though we cannot be sure as there is scant research data available. In any event, there has been so much concern about their impact that the British Polling Council investigated their performance. It concluded that inaccurate sampling and statistical methods were to blame. The polls, said the report, will inevitably overestimate Labour support as they are currently conducted. The report did, however, stop short of recommending banning the publication of polls in the run-up to elections in case they influence voting.

Activity

Research how accurate or inaccurate the opinion polls were during the 2019 general election.

Debate

Should the publication of opinion polls be banned in the run-up to elections?

For banning

- Opinion polls may influence the way people vote.
- Opinion polls have proved to be inaccurate so they mislead the public.
- Arguably politicians should not be captive to changing public opinion as expressed in the polls.

Against banning

- It would infringe the principle of freedom of expression.
- If publication of opinion polls is banned, they will become available privately for organisations that can afford to pay for them.
- Opinion polls give valuable information about people's attitudes that can usefully guide politicians.
- The polls would still be published abroad and people could access them through the internet.

The arguments for banning opinion polls are more related to concerns with how the polls are conducted and used, so it is probably worth suggesting that they should not be banned, but there is a case for ensuring greater accuracy and consistency.

A summary of factors affecting voting behaviour

The study of electoral behaviour is a complex one. Many different factors are at play in any election campaign. However, we can identify two key sets of influences. The first determines the size of each party's core voter base and explains why this may be increasing or reducing. The second concerns those voters who are not part of this core, so that their votes are 'up for grabs' — these are known as swing or floating voters.

Tables 4.14 and 4.15 summarise these factors and offer an evaluation of how important they are in explaining voting behaviour.

Table 4.14 Factors that determine the size of the parties' core vote

Factor	Explanation	Estimate of impact
Social class	Classes AB mostly support the Conservative Party, while DE voters are usually Labour supporters	Strong but in decline as a factor
Age	The older a voter is, the more likely they are to vote Conservative or UKIP. Younger voters favour Labour and the Greens	Strong and consistent
Region	The south of England is solidly Conservative, the Midlands is mixed and the north is mostly but not exclusively Labour country	Strong, though Brexit has had a role
Ethnicity	Black and Muslim voters strongly prefer Labour, but other ethnic groups are more mixed in their support	Significant but variable
Gender	Slight Labour bias among women	Limited
Class dealignment	Progressively fewer people are attached to a particular social class	This reduces the impact of class on party choice
Partisan dealignment	Progressively fewer people identify strongly with the aims of a particular party and more are becoming floating voters	The size of the main parties' core vote is shrinking gradually but not dramatically

Table 4.15 Factors that affect how floating voters will vote

Factor	Explanation	Estimate of impact
Valence	The image of the parties and their leaders and general beliefs about the competence of the parties	Very strong
Economic voting	Choice on the basis of the performance of the economy and how voters believe parties will manage the economy	Strong
Rational choice	Rational voters deciding which party will most benefit the community as a whole or themselves	Moderate
Issue voting	Voters deciding which issues are most important and which parties have the best policies relating to those issues	Moderate
Tactical voting	Voters opting for their second-choice party if they think their first choice is a wasted vote	Uncertain, probably moderate
Party leaders	People deciding who would make the best prime minister	Some significance but weak
The press	People being influenced by newspaper campaigns and preferences	Probably weak
Broadcast media	The way in which broadcast media reports on events and the trends it shows and what issues it draws attention to	Uncertain, but probably moderate due to impartiality
Social media	Depends on the person's use. If they are an active user it may have a strong bearing, although usually social media confirms existing intentions and attitudes. If they are not an active user, there may be some influence via other mediums	Varying, but potentially strong
Opinion polls	Voters may decide to change their mind because they do not want the outcome that is predicted in the polls	Probably weak

Case studies of key general elections

Over the following pages you will be given specific information about four general elections that you may wish to use as part of your three required for the exam. Whichever elections you choose to study, the key is to be able to explain the outcome for each election and its longer-term trends. Essentially, these case studies are your examples to explain and reach a judgement about the relative significance of factors affecting voting behaviour.

Study tip

When considering election results, it is important to look at changes since the previous election. These can be more revealing than the current statistics.

Case study

The 1979 general election

The Labour government, which had come to power with a small majority in 1974, had suffered the loss of a number of MPs and repeated crises. By 1979, public opinion had turned against Labour, allowing the Conservatives, under Margaret Thatcher, to make large gains and win a clear majority in the 1979 election (Table 4.16).

Table 4.16 Results of the 1979 general election

Party	Seats won	Change since 1974	% votes won	Change since 1974
Conservative	339	+63	43.9	+8.1
Labour	269	–50	36.9	–2.3
Liberal	11	+1	13.8	–4.5
Others	16	+5	5.4	–1.4

Turnout: 76%

Source: The House of Commons Library

Demographic issues

A key factor that emerged at around the time of this election was the sharp decline in the number of people describing themselves as 'working class'. This may have eroded Labour's vote. Conversely, the size of the middle class was growing, helping the Conservatives.

Valence and other issues

Perhaps the key issue in this election was the power of trade unions. In the winter of 1978–79 there had been a wave of strikes by public-sector workers, leading to bins being left unemptied, shortages of power and disruption of public transport. It became known as the 'Winter of Discontent'. Therefore, a key issue was: Which party was best placed to control union power? The answer for many floating voters was the Conservative Party.

The UK economy was also not in a good state. There was high inflation, growing unemployment and falling growth. The Conservative response was to plan a return to free markets and to curb union power. Prices would come down, they argued, if unemployment continued to rise so that wages would fall. The middle classes were especially attracted to such policies. Much of the change in voting preference was economic in nature, choosing the alternative Conservative manifesto promises over the more traditional Labour points.

Party leaders

Ironically, James Callaghan, Labour Prime Minister before the election, had a more favourable image than his opponent, Margaret Thatcher. He was seen as reliable and likeable, whereas she was seen as distant and too 'posh'. Thatcher had also been an unpopular education secretary in the early 1970s. Some members of the electorate may have been reluctant to vote for a female prime minister. Nevertheless, the Conservatives won, perhaps in spite of, rather than because of, their leader.

Turnout and its significance

Turnout was a little down on past trends, largely due to abstentions by the working class and union members, who were Labour's core voters. However, it held up well and was not a significant factor in the result. Opinion polls may have played a role in boosting Conservative turnout as they showed an increasingly close race.

The campaign

The 1979 election was the first campaign where the parties focused on using the media as a part of their campaigns, rather than the media passively reporting on what had happened, without trying to shape opinions. As such, the media played a more prominent role in this campaign than in any previous one and was a key focus in party strategies. Press conferences were timed to provide stories for the midday news, afternoon walkabouts by leaders were designed to coincide with the early evening news, and major speeches were timed to make the evening news.

→

As a result of opinion polls, which showed the public did not like adversarial campaigns, the two main parties avoided making attacks on each other, especially after criticism was levelled at the Conservatives following an insult-laden first attack ad. Both parties also ensured their more radical elements were kept silent.

While Callaghan certainly 'won' the campaign, coming across as more popular as a leader and closing the polling gap between Labour and the Conservatives, it was not enough to stop the Conservative Party winning the election.

Policies

The most important policies were Conservative pledges to curb the power of the trade unions and proposals for economic reforms. Other policies, like the right-to-buy scheme, were popular, but were of secondary importance to the policies that tapped into the mood for change following the Winter of Discontent.

How people voted

Table 4.17 shows how key groups voted in the 1979 general election.

Table 4.17 How key groups voted in 1979

Regional	All areas swung towards the Conservatives, but the swing was much more pronounced in southern England
Class	The Conservatives remained dominant with the AB and C1 voters. Labour won the C2 and DE vote, but the Conservatives gained swings of 11% and 9% respectively across these groups
Gender	Men were evenly split between the two main parties. Women showed a slight preference for the Conservatives
Age	Labour won among the 18–24 age group, but the Conservatives won across all other age groups. Labour support declined most in the 35–54 age group
Ethnicity	There is a lack of data for ethnicity in 1979 as BAME voters made up less than 5% of the population and parties and polling organisations did not regard them as significant enough to consider

Outcome

The Conservatives were to be in power for the next 18 years.

Case study

The 1997 general election

In 1992, the Conservatives had won a fourth election in a row with a majority of 23. However, since then their reputation as a party of government had suffered and the Labour Party had reformed itself into 'New Labour' under Tony Blair. The 1997 election resulted in the Conservative Party's worst electoral defeat since 1906 and Labour's largest-ever electoral victory (see Table 4.18).

Table 4.18 Results of the 1997 general election

Party	Seats won	Change since 1992	% votes won	Change since 1992
Labour	418	+145	43.2	+8.8
Conservative	165	−165	30.7	−11.2
Lib Dem	46	+30	16.8	−1.0
Others	30	+6	9.3	+3.4

Turnout: 71%

Source: The House of Commons Library

Demographic issues

Tony Blair and the Labour leadership recognised that the traditional working class, Labour's natural core vote, was diminishing in size and that the party could not rely on it to get into power; it simply did not have enough votes. They therefore decided to woo the middle classes, part of the Conservatives' core vote, by adopting centrist ('Third Way') policies. This was achieved to great effect. The young were also persuaded to vote Labour as it represented a break from traditional, out-of-date politics.

Valence and other issues

The image of the Conservative Party, which had been in power for 18 years, was a tired one, but above all the party was disunited, largely over Europe. It had also presided over a deep economic recession in the early 1990s, so competence was an issue. By contrast, Labour had no economic record to defend and appeared to be a younger, fresher party, united around a definable set of policies – the so called Third Way. The main problem for the Conservatives was that the electorate remembered the recession of the late 1980s and early 1990s and blamed the party for it. There was also a general sense that the Conservatives had mismanaged the economy.

Party leaders

There could hardly have been a greater contrast between the two main party leaders. Conservative Prime Minister John Major appeared to be grey, unexciting and weak, whereas Tony Blair was clearly in command, and was young and attractive, with a clear vision. The Liberal Democrat leader, Paddy Ashdown, also enjoyed a positive image, reflected in a good election for his party.

Turnout and its significance

In this election we saw the first signs of a long-term decline in turnout. The figure of 71 per cent seems healthy by modern standards, but was much lower than typical levels in the past. There does not seem to have been any impact on the result, but this was a watershed in political participation.

The political issues

There were two main salient issues — the NHS and the state of education. Both services had been in decline. Labour promised to make huge investments in both to raise standards. Chancellor Gordon Brown promised to be financially responsible, which was a strong message, especially as Labour had a bad reputation as a 'tax and spend' party. Labour was fortunate that an economic recovery was under way when it took office, so it could pay for the improvements in public services.

In addition, in the years 1992–94 John Major had presided over a Conservative Party hopelessly divided over the Maastricht Treaty, which transferred large amounts of power to the European Union. The hangover from this was still present in 1997, as was the reputation for 'sleaze'.

The campaign

The election campaign lasted six weeks, far longer than the 31-day average going back to 1959. Major hoped this would expose Labour divisions and put pressure on Blair. Both campaigns focused on the leaders, touring marginal seats and using campaign buses and planes. Labour's campaign was strict and well organised, run from the Millbank Media Centre. The use of media soundbites, like Blair's three priorities being 'education, education, education', helped galvanise the public. The Conservatives focused on fears of Labour restoring union power and Blair being untrustworthy. The Conservative Party, though, had to spend much of its time dealing with the issue of sleaze and financial corruption, which distracted from its wider campaign. The Conservatives tried to focus on a negative anti-Labour campaign, while Labour focused on a positive 'pro-Blair' campaign. In a sense, the campaigns did little to change the results, but the election did see an enormous step towards disciplined, media-focused electioneering modelled on American politics.

Policies

The Conservatives tried to focus policies on economic recovery, but the issue of Europe dominated and divided the party, especially with the presence of the Referendum Party. Labour focused on five specific pledges: to cut class sizes in schools, to introduce fast-track punishment for young offenders, to cut NHS waiting lists, to get 250,000 unemployed under-25s into work, and to cut VAT on heating and not raise income tax, all focused on presenting the New Labour Third Way.

Opinion polls

The opinion polls, having been spectacularly wrong in 1992, made a more concerted effort to survey a wider cross-section of society. They were much more accurate in 1997, but the legacy of distrust from 1992 meant that New Labour did not believe them and kept campaigning as if it might lose. There is also a case for suggesting that by showing New Labour as clear favourite to win a large majority, turnout may have been suppressed slightly.

How people voted

Table 4.19 shows how key groups voted in the 1997 general election.

Table 4.19 How people voted in 1997

Regional	Labour gained votes across all regions, bucking the trend towards a Conservative south and a northern Labour. The Conservatives were wiped out in Scotland and Wales and reduced to only 11 MPs in London, becoming a party of the English suburbs and shires
Class	Labour gained across all groups, most notably with C1 (+19%) and C2 (+15%), although the Conservatives still 'won' the AB vote and the C1 group was tied
Gender	Labour closed the gender gap, with men and women equally likely to support Labour
Age	The Conservatives remained dominant with the over-65s, but Labour won decisively among all other age groups
Ethnicity	Labour beat the Conservatives among white voters, with 43% of the vote, and BAME voters, with 70% of the vote. The Conservatives gained only 32% of the white vote and a mere 18% of the BAME vote

Study tip

In comparing elections, it is useful to see 1997 as a turning point where changes happened and to consider things as they were before and after 1997. For example, before 1997 women were slightly more likely to vote Conservative, but after 1997 women became slightly more likely to vote Labour. This will help you to establish and demonstrate change over time and make comparisons. You also need to be able to explain the reasons for these changes.

Case study

The 2015 general election

In 2010 the UK had seen the creation of its first coalition government since the Second World War. However, the Liberal Democrats paid a price and were almost wiped out as a parliamentary party, losing seats across England, mostly to the Conservative Party. Meanwhile, in Scotland, the SNP broke Labour's traditional dominance of the region, allowing it to replace the Liberal Democrats as the third party in Westminster, while the Conservatives won their first majority since 1992 (see Table 4.20).

Table 4.20 Results of the 2015 general election

Party	Seats won	Change since 2010	% votes won	Change since 2010
Conservative	330	+28	36.9	+0.8
Labour	232	−24	30.4	+1.5
SNP	56	+50	4.7	+3.1
Lib Dem	8	−48	7.9	−15.1
UKIP	1	+1	12.6	+9.5
Others	23	−5	7.5	+1.0

Turnout: 66%

Source: The House of Commons Library

Demographic issues

A significant feature of this election was that a good deal of Labour's core working-class vote was captured by UKIP, especially in depressed areas of the country. It was also notable that the Liberal Democrats lost all 15 of its seats in southwest England. This dispelled the idea of the Liberal Democrat core vote in that region. The Conservatives, by contrast, dominated the south of England as a whole. In Scotland the SNP nearly swept the board. Again, Labour's regional heartland was taken away.

Valence and other issues

With the exception of the Scottish National Party, all the parties suffered from a negative image. The Conservatives were still divided over Europe and immigration. The Liberal Democrats were blamed for many of the shortcomings of the coalition government of 2010–15, while Labour was not trusted with the economy. Labour's competence was constantly called into question. The media may have had an impact by warning about the dangers of a likely Labour–SNP coalition, as indicated by opinion polls. Some voters turned away from this prospect by switching to the Conservatives. It seems that UKIP took more votes from Labour than from the Conservatives. The Conservatives also benefited most from the collapse of voting for the Liberal Democrats. In Scotland, the SNP won 56 out of 59 seats and so decimated the other parties there. This ensured that Labour could not win an overall majority or even lead a coalition government.

Party leaders

All of the party leaders, apart from SNP leader Nicola Sturgeon, suffered a negative image with the public. David Cameron's image was, however, the least negative. Nick Clegg of the Liberal Democrats and Ed Miliband of Labour were both short of public respect. It seems probable that images of party leaders played little part in the outcome.

Turnout and its significance

Turnout recovered a little in this election, though it was still well down on historic levels. Perhaps as many as 2 million former non-voters turned up at the polls, many supporting UKIP. If this was the case, the turnout of supporters for the other main parties must have been very low.

→

The political issues

A key issue was economic management. The Conservatives were still more trusted with the economy than Labour. The election was dominated, however, by the prospect of a referendum on UK membership of the EU. The Conservatives were rewarded (as was UKIP) for their support of such a referendum.

The Scottish referendum of 2014 was a key event. The campaign in that referendum put the SNP in a powerful position in Scotland. As we have seen above, fear of a possible Labour–SNP coalition may have brought many votes to the Conservatives.

Campaigns

This was the second campaign in which televised debates featured, but David Cameron refused to debate with Ed Miliband head-to-head, agreeing only to a seven-person leadership debate, which proved bland and made it difficult for Miliband to stand out. Although Labour stressed the NHS and many celebrity endorsements in its party-political ads, the Conservative focus on working families and asking about 'Labour's magic money tree' seemed more effective. As the campaign progressed, both sides started to stress the dangers of a possible coalition, one between Labour and the SNP and the other between the Conservatives and UKIP. The Conservatives focused heavily on campaigning against the Liberal Democrats in England as a means of preventing a Labour/SNP government, while in Scotland the SNP targeted both Labour and the Liberal Democrats for failing to consider Scottish issues.

Policies

Labour focused on five key pledges: to balance the books and cut the deficit; to freeze energy bills until 2017; to recruit 20,000 more nurses and 8000 more GPs; to prevent immigrants claiming benefits for 2 years; and to reduce university tuition fees to £6000 per year. The Conservatives pledged to make welfare cuts, but refused to say where these cuts would be until after the election. The Conservatives also pledged to hold an in/out EU referendum, tax-free income for those on minimum wage, and up to 20 hours a week of free childcare.

Opinion polls

The opinion polls, increasingly reliant on digital methods, got it wrong by underestimating the number of elderly voters who would turn out and be more likely to vote Conservative. This had consequences for the campaign, as Labour chose to fight a campaign on Labour policies rather than the dangers of a second Conservative term, while the widely reported polls suggesting there could be a Labour and SNP coalition may have influenced tactical voting and helped to increase Conservative turnout.

How people voted

Table 4.21 shows how key groups voted in the 2015 general election.

Table 4.21 How people voted in 2015

Regions	England, outside major cities and industrial centres, overwhelmingly voted Conservative, at the expense of the Liberal Democrats. Labour and the Liberal Democrats were nearly wiped out in Scotland by the SNP, which took 56/59 seats
Class	The AB, C1 and C2 categories all voted Conservative, though Labour did win with the DE class. In fact, Conservative votes among the AB and C1 groups actually increased from 2010
Gender	The Conservatives won a similar share of the vote among men and women, with 37% and 38% respectively. Labour did have a clear gender gap, with more women than men supporting Labour
Age	The age divide continued. Labour won with the 18–24 age group and tied the Conservatives with the 40–49 age group, but the Conservatives won with all other groups, achieving a 20% advantage over Labour with the over-65s
Ethnicity	Labour won 67% of the Afro-Caribbean vote compared with the Conservatives' 21%, and also among people of Indian subcontinent origins, though by closer margins. However, the Conservatives did win decisively with people of Asian origins outside of India

The outcome

The shock of defeat completely destabilised the Labour Party. There was a huge left-wing backlash against the failed policies of the leadership, so the party elected extreme left-winger Jeremy Corbyn in place of Ed Miliband. This precipitated divisions within Labour and a split in the parliamentary party. The outcome also saw the establishment of the SNP as the third party of the UK and the Liberal Democrats reduced in power.

The 2019 general election

Following the close election and minority government of 2017, both main parties had seen divisions grow. It was not clear how the 2019 election would play out, but Labour's vote collapsed across the UK, allowing the Conservatives to win their largest majority since the 1987 election (Table 4.22). In Scotland, the SNP made significant gains, having fallen back in 2017.

Table 4.22 Results of the 2019 general election

Party	Seats won	Change since 2017	% votes won	Change since 2017
Conservative	365	+47	43.6	+1.2
Labour	203	−59	32.2	−7.8
SNP	48	+13	3.9	+0.8
Lib Dem	11	−1	11.5	+4.2
Greens	1	−	2.7	+1.1
Others	22	−1	6.1	+1.3

Turnout: 67%

Source: The House of Commons Library

Demographic issues

A significant feature of this election was the shift in working-class support across northern industrial areas as former Labour and UKIP voters, tired of the Brexit deadlock, voted for the Conservatives over the Labour Party, reducing the 'red wall' of northern support. Labour continued to focus on younger and working-class voters with a more socialist manifesto, while the Conservatives focused on appealing to older and working-class voters angry over the failure to deliver Brexit.

Valence and other issues

The main issue to dominate was the handling of Brexit by the main parties. Parliament had been in chaos since the last election and the public had been frustrated by all sides' inability to move forward on the issue. As such, the clear message being offered by the new Prime Minister, Boris Johnson, on the issue became the main issue of the campaign, compared with the vague and complicated messages of the Labour Party.

Beyond Brexit, the issue of antisemitism dominated the Labour Party in the run-up to the election, as did the formation of new parties, including Nigel Farage's Brexit Party and the creation of the Independent Group for Change. The decision of the Brexit Party not to contest seats won by the Conservatives in 2010 and then not to contest

seats with strong Conservative Brexit supporters meant the 'right' predominantly only had one party to choose on the issue. The Liberal Democrats had seen their total of MPs increase to 21 with defections from both Labour and the Conservatives, but this only seemed to alienate supporters who did not want to be associated with members of one of the main parties. The issue of Scotland and Scottish independence also saw the SNP gain more seats.

Party leaders

Boris Johnson benefited from having been Prime Minister and leader for a limited amount of time and stuck to a strategy of not appearing too often in the media, instead relying on his existing reputation and media profile from his time as London Mayor. Unlike Theresa May in 2017, he appeared focused and more determined and clear in his policy-making than Jeremy Corbyn. Nicola Sturgeon continued to enjoy a positive image in Scotland, but Liberal Democrat leader Jo Swinson alienated the public by appearing to lack sincerity and for pledging to repeal Article 50.

Turnout and its significance

Turnout saw a slight drop in 2019 from the 69 per cent of 2017, perhaps because this was a less close election than previously but also perhaps because a December election discouraged some voters. However, it was not enough to make a major difference to the outcome.

The campaign

The campaign was much more subdued than previous ones, partly because of the winter campaign and partly because Boris Johnson seemed to be absent for much of the campaign, appearing in public relatively rarely for a serving leader. Although he did participate in a head-to-head debate with Corbyn, the main objective seemed to be to let the issue of Brexit persuade voters. In this sense, Johnson's main role was not to create headlines. Corbyn, meanwhile, managed a solid campaign with lots of public speeches but was unable to whip up the enthusiasm he had in 2017. Social media attack ads were more prevalent in this campaign than in any other and the Electoral Commission seemed unable to do much about it.

Political issues

The election cycle was dominated by the extension of the deadline for leaving the EU from October 2019 to January 2020

→

and then the inability of Parliament to find a solution to Brexit. Former big names from both main parties urged members not to vote for their party, including Michael Heseltine and John Major on the Conservative side and Alastair Campbell for Labour. During the campaign, floods hit parts of England and Johnson was criticised for not declaring them an emergency, while journalists from the Labour-supporting *Mirror* newspaper were banned from the Conservative campaign bus. Shortly before the election, a NATO summit was held in Watford, allowing Johnson to appear with other world leaders, including Trump.

Opinion polls

Opinion polls were again under scrutiny with the memory of 2015 still fresh. The polls seemed to indicate a Conservative win throughout, which perhaps shaped the campaigns as one for the Conservatives to lose rather than for Labour to win. However, the polls did underestimate the scale of the Conservative victory.

How people voted

Table 4.23 shows how key groups voted in the 2019 general election.

The outcome

The Conservatives won a clear majority of 80 seats, giving them a strong mandate and ensuring the continued domination of the Conservative Party since 2010. It was the first sizeable majority since the 2005 election and saw Jeremy Corbyn resign as Labour leader.

Table 4.23 How people voted in 2019

Regions	The Conservatives dominated rural and southern England (outside London) and made gains in industrial northern areas and in Wales. The SNP gained seats in Scotland to continue domination there, while the Conservatives lost seven seats, but remained the second-biggest party with six seats. There was a slight rise in nationalist parties in Northern Ireland
Class	The Conservatives won across all social classes and the percentage split between Labour and the Conservatives remained pretty consistent across all groups. The Conservatives won a higher proportion of votes among the C2 and DE classes than they did with the AB and C1 groups
Gender	There was a slight gender gap between the Conservatives and Labour, but this was mainly the result of a large gap in the 18–24 age group, with 65% of women in this group voting Labour, compared with 46% of men. Elsewhere, the gender gap was far less pronounced, with only a slight female bias towards Labour
Age	Age was once again the main dividing line in the British electorate, with younger voters, especially women, far more likely to vote Labour while the older a person became the more likely they were to vote Conservative, with 39 being the tipping point
Ethnicity	BAME voters continued to favour Labour over the Conservatives, leading 64–20%, while the Conservatives won the white vote over Labour 48–29%. However, Labour found itself 9% and 10% down on 2017 with both groups

Summary

Having read this chapter, you should have knowledge and understanding of the following:
→ The meaning of party core votes and the factors that affect the size of the core vote for each party
→ Changes in people's attachment to social class and to party
→ What factors affect the way floating voters behave
→ The influence of party leadership on election outcomes
→ The role and importance of mandates and manifestos in election campaigns
→ The influence of the media — broadcast, print and social
→ The part played by opinion polls in elections and their significance
→ Summaries of the outcomes of at least three key elections
→ How to explain the factors that caused the outcomes of three general elections

Synoptic link

Prime ministers will often use media events like NATO summits to appear presidential and help build public confidence in their credentials as a leader, which is explored in Chapter 7.

Key terms in this chapter

Class dealignment The process where individuals no longer identify themselves as belonging to a certain class and for political purposes fail to make a class connection with their voting pattern.

Disillusion and **apathy** A process of disengagement with politics and political activity. Having no confidence in politics and politicians as being able to solve issues and make a difference. Manifested in low turnout at elections and poor awareness of contemporary events.

Governing competency The perceived ability of the governing party in office to manage the affairs of the state well and effectively. It can also be a potential view of opposition parties and their perceived governing competence if they were to secure office.

Mandate The successful party following an election claims it has the authority (mandate) to implement its manifesto promises and also a general permission to govern as new issues arise.

Manifesto In its manifesto, a political party will spell out in detail what actions and programmes it would like to put in place if it is successful in the next election — a set of promises for future action.

Partisan dealignment The process where individuals no longer identify themselves on a long-term basis by being associated with a certain political party.

Further reading

Websites

The best information about voting behaviour can be found on the websites of opinion poll organisations. These include:

- YouGov: **yougov.co.uk**
- Ipsos MORI: **www.ipsos-mori.com**
- Polling Digest: **@Pollingdigest on www.twitter.com**

Information about voting and elections can also be found at the Electoral Commission site: **www.electoralcommission.org.uk**

Books

Cowley, P. and Ford, R. (2015) *Sex, Lies and the Ballot Box* and *More Sex, Lies and the Ballot Box* — a series of articles on the role of polling, voting behaviour and voting systems

Norris, P. (2010) *Electoral Engineering: Voting rules and political behaviour*, Cambridge University Press — an authoritative work

Macmillan/Palgrave publishes a book about every general election by various authors. Each one is packed with facts, figures and analysis. These are titled *The British General Election of 1979*, *The British General Election of 1983*, etc.

1

Source 1

When Margaret Thatcher claimed there was no such thing as society, she was right, at least as far as elections are concerned. Where once membership of a social class would have been the most important factor in a person choosing how to cast their ballot, class no longer plays any meaningful role for voters. Nor can we claim that people decide their vote on the basis of ethnicity, age, gender, or even geography. People have become much more individual, choosing how to vote on the basis of what they want and their own individual priorities. Their attitude to a party leader, their view on a key issue, how the government has behaved, even what the opinion polls say, play a much more important role than any kind of social identity. Recent elections show that people, not groups, decide what they want and vote accordingly. Politicians may still try to target groups, but to assume that women, the young and BAME voters cast their votes on the basis of who they are is profoundly patronising and may explain the alienation of large sections of society from politics.

Using the source, evaluate the view that social factors remain the most important factor in explaining election outcomes.

In your response you must:
- *make reference to at least three general elections, one from 1945–92, 1997 and one from 2001–19*
- *compare and contrast different opinions in the source*
- *examine and debate these views in a balanced way*
- *analyse and evaluate only the information presented in the source.* (30)

2

Source 2

Opinion polls are a vital component of election campaigns and remain important to politicians and the people. Politicians use them to keep in touch with the people and adapt their policies to the will of the people, ensuring democracy and representation are happening. Polls allow the media to justify its support or give strong warnings for what may be about to come, and allow all reporters and the public to see how the campaign is going, reflect on how other people are thinking of voting and decide how best to cast their vote on that basis. Polls also help political strategists plan the campaign more effectively, choosing how to marshal limited resources, identify target seats and key groups of voters, play to their strengths and learn from their mistakes. Without opinion polls, would there really be any point in having a campaign? Yes, mistakes happen and sometimes the polls get it wrong, but is it the polls that are inaccurate or are voters hiding their intention? Do polls not help to drive up turnout by showing a close election? If people choose to stay at home based on an opinion poll, is that the fault of the poll or is it a lack of social responsibility among the electorate?

Using the source, evaluate the view that opinion polls do not play a meaningful role in election campaigns.

In your response you must:
- *compare and contrast different opinions in the source*
- *examine and debate these views in a balanced way*
- *analyse and evaluate only the information presented in the source.* (30)

3 Evaluate the view that class is the most important social factor in
   explaining voting behaviour.
   You must consider this view and the alternative to this view in a balanced way.
   In your answer you must make reference to at least three general elections,
   one from 1945–92, 1997 and one from after 1997.   (30)
4 Evaluate the view that elections are divisive and undermine national unity.
   You must consider this view and the alternative to this view in a
   balanced way.   (30)
5 Evaluate the view that media image is more important than policies or
   strategy for influencing voting behaviour in the UK.
   You must consider this view and the alternative to this view in a
   balanced way.   (30)
6 Evaluate the view that elections have become increasingly
   unpredictable and uncertain.
   You must consider this view and the alternative to this view in a
   balanced way.   (30)

UK GOVERNMENT

To complete your study for Component 2 you will need to study the Non-Core Political Ideas in addition to the chapters in this book.

On most days in Washington DC, a queue of Americans can be seen waiting to enter the National Archives Building. What they are anxious to see is the original version of the Constitution of the USA. It is revered in America with almost religious fervour, so it is not surprising that so many want to visit. Written in the hot summer of 1787 in Philadelphia, the US Constitution appears on a single document, signed by the men who created it and who are known in the USA as the 'Founding Fathers'.

In the UK, monuments to the UK Constitution are small-scale affairs with few, if any, queues of patriotic Brits waiting to see the important documents, such as the memorial at Runnymede to commemorate the signing of Magna Carta in 1215, or the opportunity to see an original version of Magna Carta housed in Salisbury Cathedral. But the Magna Carta is only one small and marginal part of the UK Constitution. In truth, in the UK, there is no single document that people can queue to look at that marks the 'UK Constitution'. Why? The UK Constitution is not to be found in one document, it was never created at one moment in time and there are no laws that are supreme, as there are in the USA. Instead, what is supreme in the UK is the parliament of the day, and constitutional rules are contained in a variety of forms that have emerged over centuries of constitutional development.

The memorial at Runnymede to commemorate the signing of Magna Carta in 1215

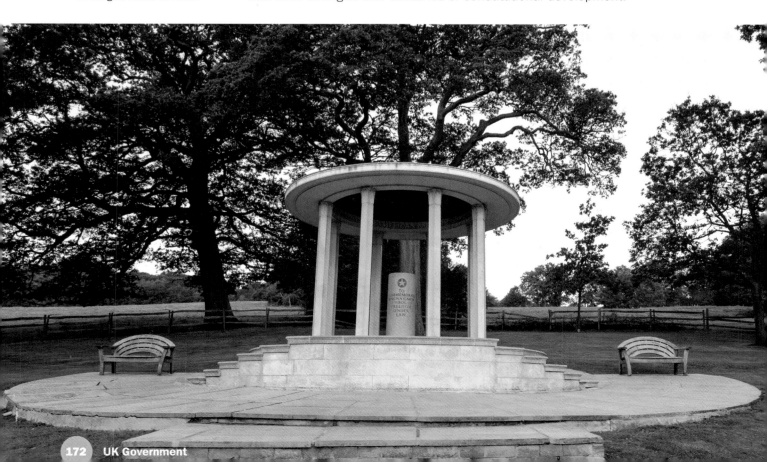

What is a constitution?

Before we examine the nature of the UK's Constitution, we need to establish what a **constitution** actually *is*. In short, a constitution is defined as the rules of the political game, which outline who has what powers and responsibilities, and how government and the people are expected to behave and interact with each other. It is, if you will, a contract between the government and the people.

Virtually every country in the world operates its political system within the constraints of a constitution. In most cases, the constitution of the state is written in a single, authoritative document that has been agreed on some particular occasion. Such constitutions are described as **codified**.

Three democratic countries — the UK, Israel and New Zealand — operate without such a specifically codified constitution. Instead the constitutions of these countries have evolved and are made up of a number of different sources. Such constitutions are described as **uncodifed**.

The functions of a constitution

Constitutions, whether codified or not, are a vital aspect of most stable political systems. All constitutions, no matter where they exist, perform the same set of functions. These are as follows:

- They determine how political power should be distributed within the state. This includes **federal** states where power is divided between the central government and regional institutions, as in the USA, and **unitary** states where ultimate power lies firmly in one place, as with the UK's parliamentary system.
- Constitutions determine the balance of power between the institutions of government, such as the Prime Minister, Cabinet, Parliament, the monarchy, the courts and the people.
- Constitutions also establish the political processes that make the system work. This includes the relationships between institutions and the rules that govern how they operate.
- A democratic constitution normally states what the limits of governmental power should be, in other words what the competence is of government. The UK Constitution is unusual in this sense as it places no limits at all on the competence of Parliament.

Key terms

Constitution A set of rules determining where sovereignty lies in a political system, and establishing the relationship between the government and the governed.

Uncodified/codified An uncodified constitution is not contained in a single written document, unlike a codified constitution which is written in a single authoritative document.

Unitary/federal A unitary political system is one where all legal sovereignty is contained in a single place, unlike a federal system where legal sovereignty is shared between a national government and regional governments.

With a system of **parliamentary sovereignty**, Parliament can, theoretically, do what it likes. By contrast, other democratic governments like the USA, France, Germany, South Africa and Japan are circumscribed by their constitutions.

- Just as constitutions limit governmental power, so they assert the rights of the citizens in relation to the state. Most countries that at least claim to be democratic have some kind of 'bill of rights', a statement that prevents the government from trampling on the civil liberties of its citizens.
- Constitutions establish the rules by which nationality is established, in other words who is entitled to be a citizen and how outsiders may become citizens. This also implies that a constitution defines the territory that makes up the state.
- Finally, we must remember that constitutions must be amended from time to time. It is therefore essential that a constitution contains within itself the rules for its own amendment. The UK is, once again, unusual in this respect as its Constitution can be changed in a number of ways. One is through a simple parliamentary statute; another is by the slow evolution of unwritten rules, known as **conventions**, or the publication of new official guidance or legal decisions. Normally, states have special arrangements for amending their constitution. In France and Ireland, for example, a referendum is needed to approve a change. In the USA it is necessary to secure a two-thirds majority of both houses of Congress and the approval of three-quarters of the 50 states that make up the Union. The UK has no such 'higher' methods of amendment; its Constitution has evolved naturally over the course of history and can be amended quite simply.

So, we have established the main functions of a constitution. Issues that concern any of these matters are therefore described as 'constitutional' in nature. Now we will examine the UK Constitution and the country's constitutional arrangements.

The nature and sources of the UK Constitution

Stages in the development of the UK Constitution

The UK Constitution has developed gradually over time. To some extent this is an unseen process, so slow and subtle that we hardly notice it. Constitutional change is something that concerns lawyers and politicians but few members of the public. From time to time, however, an event takes place that everyone notices. These events form the main landmarks in the development of the UK Constitution, and include the following:

- **Magna Carta, 1215** Little in Magna Carta has survived, save for a few **common law** traditions and some principles that have been turned into statute law. However, it was a key moment in history. It established that the **rule of law** should apply and the monarch should operate within the framework of law. It was to be centuries before this principle became normal practice, but Magna Carta was an important staging post in the development of constitutional rule.
- **The Bill of Rights, 1689** This Act of Parliament resulted from the replacement of King James II by the joint monarchy of William III and Mary. Parliament was anxious that the new monarchs would not exceed their powers, so the Bill of Rights effectively stated that Parliament was sovereign and would have the final word on legislation and the government's finances.

- **The Act of Settlement 1701** The Act established the legal rules governing the succession to the throne. It also stated that the monarch should be a member of the Church of England. However, its main significance was that it established the monarch's position as the ruler of the whole of the United Kingdom of England, Scotland, Wales and Ireland (Northern Ireland after 1921).
- **The Acts of Union 1707** These two Acts abolished the separate Scottish Parliament and so established the modern nation of Great Britain. Following a similar Act of Union in 1801, the United Kingdom of Great Britain and Ireland was established, altered to the United Kingdom of Great Britain and Northern Ireland following the independence of the Irish Free State in 1921. Of course the devolution of power to Scotland in 1998 brought back the Scottish Parliament, although it is still not the sovereign body in that country.
- **The Parliament Acts 1911 and 1949** These two Acts settled the relationship between the House of Commons and the House of Lords. Before 1911 the two houses were, in theory at least, of equal status. In 1911, however, the House of Lords lost its powers to regulate the public finances and could only delay legislation for 2 years; it could no longer veto proposed legislation for good. The 1949 Act reduced the delaying period to 1 year. As a result, the House of Commons is very much the senior house.
- **The European Communities Act 1972** This was the Act that brought the UK into the European Community, which later became the European Union. The UK joined in 1973. This Act is now consigned to history, as the UK voted to leave the EU in 2016. It was, however, for nearly 50 years, a key feature of the UK Constitution.
- **The European (Notification of Withdrawal) Act 2017** This gave parliamentary consent to the UK's exit from the European Union.

There have been many other key moments in constitutional development and many of them are described in this chapter. Those described above have been perhaps the most prominent in the history books. They trace the changing role of the monarchy and the growth in the authority of Parliament and the improved protection of citizens' rights.

The nature of the UK Constitution

Now we will consider some key principles concerning the UK Constitution.

It is uncodified

The UK Constitution is not codified. It is not contained in a single document. This is not the same as saying that it is unwritten; in fact much of the UK Constitution *is* now written. For example, the European Convention on Human Rights is a well-known document. Laws passed by Parliament, which make up about 80 per cent of the Constitution, are also written. To be codified, a constitution has to have three features:

- It must be contained in a single document.
- It must have a single source and therefore have been created at one moment in history, even if it has since been amended.
- The constitutional laws contained in it must be clearly distinguished from other, non-constitutional laws.

Virtually all modern democratic countries have a codified constitution. Uncodified constitutions will have multiple sources (discussed below) and by their very nature are more flexible and easier to change, as well as being slightly more confusing to understand.

Activity

Research the 1689 Bill of Rights. In what ways did it establish parliamentary sovereignty? How many of the rights remain applicable today?

Knowledge check

Which milestones in the development of the UK Constitution occurred in the twentieth century?

Study tip

It is important to know the historical development and landmark Acts of the British Constitution, but in most cases they are historical and would not be relevant to an answer addressing modern constitutional reforms.

Key term

Unentrenched/ entrenched An unentrenched constitution has no special procedure for amendment, unlike an entrenched one which requires separate rules and procedures for amendment.

There are, of course, advantages to being uncodified; the greater degree of flexibility allows the political system to be more responsive to changing attitudes and ideas and the constitution can be updated much more easily than a codified one. As such, uncodified systems tend to evolve and release the build-up of public pressure without the need for a major uprising or revolutionary action. On the downside, uncodified constitutions are less effective at protecting rights and this can lead to major upheavals based on short-term populist ideas. For example, following the Dunblane massacre in 1997, the UK government was able to very quickly introduce sweeping gun regulations across the UK following a public outcry; compare this to the USA and the inability to impose gun reforms despite public outcries over mass shootings, and you might feel that an uncodified constitution is the way to go. However, change 'gun regulation' to 'speech regulation' and we perhaps see the weaknesses of an uncodified constitution. In this situation, after, say, a major terrorist crisis, the UK government could easily suspend the right to free speech, in a way that the US government simply cannot, no matter how much they might want to stop protests or the free exchange of ideas.

It is unentrenched

It is worth remembering that all constitutions are a mixture of written elements and unwritten elements, but uncodified constitutions have a high proportion of unwritten elements. What really separates codified and uncodified constitutions is the fact that the former are **entrenched** and the latter are **unentrenched**.

Entrenchment means constitutional rules are very well protected and difficult to change, usually requiring some sort of super-majority of the public. This protects a constitution from short-term amendment, which is important because constitutional changes make a fundamental and important difference to the political system of a country. In democracies with entrenched constitutions, constitutional reforms are removed from the hands of a *temporary* government and requirements are put in place to ensure:

- that there is widespread popular support for a reform
- that it is in the long-term interests of the country.

But the situation in the UK is unusual as it is not possible to entrench constitutional principles and laws. This is because it is the UK Parliament, rather than the Constitution, that is sovereign. The sovereignty of Parliament asserts that each individual Parliament cannot be bound by its predecessors, nor can it bind its successors. This means, in effect, that every new Parliament is able to amend the Constitution as it wishes. All Parliament has to do to reform the Constitution is to pass a new parliamentary statute, using the same procedure as for any other statute; there is no 'higher standard' for constitutional laws.

As the government in the UK is normally able to dominate Parliament using its majority in the House of Commons, this means it can effectively control the Constitution. An example of this executive power was demonstrated when the UK Parliament passed the Human Rights Act in 1998. This incorporated the European Convention on Human Rights into UK law. It became binding on all political bodies other than Parliament itself. No special procedures were needed. A fundamental change to the British Constitution was made through a simple Act of Parliament. It could occur in this way because the UK Constitution is not entrenched.

The problem of the UK's failure to adopt any system of entrenchment was illustrated by the curious case of the **Fixed Term Parliaments Act 2011**, described in the case study.

Case study

How difficult is it to entrench constitutional reforms in the UK?

Fixed-term parliaments

After 2010 the new coalition government wished to change the Constitution to introduce fixed-term parliaments. It proposed a law stating that each new Parliament should sit for a fixed term of 5 years before the next general election. The Fixed Term Parliaments Act 2011 would take away the unwritten convention that the Prime Minister could name the date of the next general election. It was intended that this law should be permanent, in other words every new Parliament would have a life of 5 years. However, it is not possible to entrench laws in the UK. In practice, this means that a future Parliament could simply repeal or amend the Fixed Term Parliaments Act and either shorten or lengthen its life, or perhaps give back to the Prime Minister the right to choose the date of the next general election.

This means we cannot say that the idea of fixed-term Parliaments is a permanent feature of the UK Constitution. It *may*, however, become a convention (an unwritten long-term arrangement that is binding) that each new Parliament will set the date of the next general

election at 5 years into the future. It could, in other words, become part of the Constitution without entrenchment. Of course, the fact that we had elections in 2017 and 2019, around 2 years or so apart, suggests that the Act has not had much of a lasting impact! This highlights the way in which any attempt at entrenchment by one Parliament can easily be overcome by the next Parliament.

Use of referendums

Despite the inability to entrench constitutional laws in the UK, it is becoming common practice to hold a referendum when constitutional change is proposed. This was done for the devolution of power to Scotland and Wales in 1997, the introduction of elected mayors in London and a number of other locations, to approve the Good Friday Agreement in Northern Ireland in 1998, and on UK membership of the EU in 2016. Where a referendum has produced a 'no' result, as was the case with the vote on a change to the electoral system in 2011 or on Scottish independence in 2014, a constitutional change cannot realistically take place. The effect of such referendums is to entrench constitutional developments. It is inconceivable that the changes would be reversed without another referendum to approve such a reversal. Thus, the UK is moving gradually towards a system of entrenchment through referendum.

It is unitary

In a unitary constitution, political power is centred in one single place and all other regions and political bodies are inferior to that body. In countries like France, all power is centred around the national political institutions based in the capital city and all laws apply equally across all areas of the country. A federal constitution, by contrast, such as exists in the USA or Germany, divides legal power between a central body and regional bodies, meaning there are different areas of power controlled by the national body from the capital and regional bodies that govern a local area.

Traditionally, the UK was a unitary system and, technically, this remains true as the Westminster Parliament retains ultimate sovereignty and power. However, devolution has created a sort of **quasi-federalism** in the UK. Nonetheless, legally, Parliament could choose to return sovereignty to Westminster and end devolution. Politically, this would be highly unpopular, but it demonstrates why the UK remains a unitary system, rather than a federal system. The issue of devolution will be discussed later in this chapter.

> **Useful term**
>
> **Quasi-federalism** A system of devolution where it is so unlikely or difficult for power to be returned to central government that it is, to all intents and purposes, a federal system even though it is not in strict constitutional terms.

Discussion point

Evaluate the view that the UK should become a fully federal system.

Three key points to consider are:
1 The ways in which the UK already appears to be a federal system and why full federalism would be more or less desirable.
2 How such a federal system might work, considering pros and cons.
3 The problems that might arise from introducing such a system and whether these are better or worse than the current arrangements.

Parliament is sovereign

Writing in the nineteenth century, A.V. Dicey declared that there were 'twin pillars' that upheld the British political system; one of these pillars was parliamentary sovereignty, the other was the rule of law. The concept of sovereignty is vital to an understanding of how constitutions work. This is because a large part of any constitution is concerned with the distribution of sovereignty. A working definition of legal sovereignty would be: *Ultimate power and the source of all political power, as enforced by the legal system and the state.* This means that a body that is granted legal sovereignty within a constitution holds power that cannot be overruled by any other body. In many ways, too, an entrenched, codified constitution is also legally sovereign. If any individual or body challenges or abuses constitutional power, they must expect to be limited and sanctioned by the legal system. In the UK, Parliament is sovereign, meaning it has the ultimate legal authority to do things and cannot be overruled by anything else, though it may choose to 'lend' some of its sovereignty to other bodies, like the EU or devolved bodies. This means there are, in effect, few ways of limiting the power of Parliament.

Power should not be confused with legal sovereignty. Power is a more flexible concept and it can be added to or reduced or removed by bodies that hold sovereignty. The distribution of power within a political system is constantly changing, while the distribution of sovereignty is usually fixed by the constitution. Power is more to do with real-world politics; Parliament is sovereign because it has the legal right to end devolution in Scotland, but politically this would be very damaging to its reputation so it cannot exercise this right. Similarly, with Brexit, many MPs wished to remain in the EU and Parliament had the legal right to reject triggering Article 50 but, politically, it had to bow to the will of the people, sometimes known as 'popular sovereignty'. The different types of sovereignty are outlined in Table 5.1.

Synoptic link

The issue of sovereignty and its location is dealt with in more detail in Chapter 8.

Table 5.1 Types of sovereignty

Type of sovereignty	Meaning	Exemplar
Legal	Where ultimate and supreme legal authority lies: in the UK this is Parliament	Parliament can suspend or modify part of the Human Rights Act through derogation, as it did in 2005 with the use of control orders that otherwise would have breached the compatibility of the Counter Terrorism Act 2005 with Article 5 of the HRA
Political	The ability to exercise legal power, dealing with the reality of matters rather than pure principles	Parliament may wish to abolish the Human Rights Act but doing so would be met by opposition in the UK and internationally, making it difficult to do
Popular	The will of the people must be listened to and acted upon in order to maintain peace and ideas of British democracy	Despite a large majority of MPs at the time wishing to remain in the EU, Parliament voted to trigger Article 50 because it was the will of the people expressed through a referendum
Pooled	Parliament may choose to share its legal authority with other bodies, such as the EU, the UN or NATO	When the UK was a member of the EU, Parliament shared power over certain issues with the EU, allowing EU institutions to make decisions that Parliament would then follow

A word of caution is needed here. In the UK we often refer to the monarch as 'the sovereign'. This is potentially confusing as it implies that ultimate political power lies with the monarch. This may have been true before the seventeenth century, but it no longer represents anything like reality. It is, in other words, a historical curiosity, a throwback to a former age. Although Head of State, the sovereign is not *actually* sovereign.

The rule of law is paramount

The other 'twin pillar' of the British political system is the rule of law. This ensures that the UK Constitution is based on the law and that everyone is bound by those same laws. This leads to a rational and well-ordered society where, in theory, everyone is treated equally and receives the same consequences if they break such laws. This is fundamental to the concept of democracy and rights in the UK.

> **Synoptic link**
>
> The rule of law is integral to democracy in the UK, as seen in Chapter 1, and is overseen by the judiciary, discussed in Chapter 8.

The sources of the UK Constitution

With the UK Constitution being uncodified, it is made up of a range of different sources, which are described below.

Statute laws

Statute laws are Acts of Parliament that have the effect of establishing constitutional principles. They are the highest legal authority in the UK, taking priority over any other sources of the UK Constitution. The **Human Rights Act 1998** is an example, but we can add the **Parliament Act 1949**, which established limitations to the powers of the House of Lords, and the **Scotland and Wales Acts of 1998**, which devolved power to those countries. Further examples of constitutional statutes are shown in Table 5.2.

One of the distinctive features of the UK's constitutional arrangements is that a constitutional statute looks no different from any other statute. As Parliament is sovereign and can amend or repeal any statute, all statutes look alike and have the same status. The wording of a constitutional statute does not contain the words 'This is a Constitutional Statute'. In most other countries a constitutional statute is clearly differentiated from other laws and is superior to them.

Constitutional conventions

A convention is an unwritten rule that is considered binding on all members of the political community. Such conventions could be challenged in law but have so much moral force that they are rarely, if ever, disputed.

Many of the powers of the Prime Minister are governed by such conventions. It is, for example, merely a convention that the Prime Minister exercises the Queen's power to appoint and dismiss ministers, to conduct foreign policy and to grant various honours, such as peerages and knighthoods, to individuals. It is also a convention (known as the Salisbury Convention) that the House of Lords should not block any legislation that appeared in the governing party's most recent election manifesto.

> **Key term**
>
> **Statute law** Law passed by Parliament.

> **Study tip**
>
> To decide whether an Act of Parliament is a constitutional law or a regular law, think about what it affects. If it impacts on some part of the political process, like devolution, elections or how Parliament operates, it is likely a constitutional law. If it is to do with regular things, like taxes, roads, schools or hospitals, it is not likely to be a constitutional law.

Authoritative works

Much of the UK Constitution is based on key historical principles that have become effectively binding because they have been established over a long period of time. In order to clarify what these principles are and what they mean, legal scholars and experts over time have written works as guidebooks and explanations of them. In establishing what these principles mean and how they operate, these **authoritative works** have become a part of the UK's Constitution. The most important of these constitutional theorists include Blackstone (parliamentary sovereignty) and A.V. Dicey (rule of law), while the rules on how to form a coalition are now an authoritative constitutional work, having been drawn up by the then Cabinet Secretary Gus O'Donnell, in 2010.

The most important of the principles outlined in such authoritative works is the sovereignty of Parliament. We could add a similar concept, which is *parliamentary government*, the principle that the authority of the government is drawn from Parliament and not directly from the people. The *rule of law* is a more recent development, originating in the second part of the nineteenth century. The rule of law establishes, among other things, the principles of equal rights for citizens and that government is itself bound by legal limitations.

Common law

'Common law' is a largely Anglo-Saxon principle. It refers to the development of laws through historical usage and tradition. Judges, who occasionally must declare and enforce common law, treat it as any rule of conduct that is both well established and generally acknowledged by most people.

The most important application of common law has concerned the protection of basic rights and freedoms from encroachment by government and/or Parliament. The rights of people to free movement and to gather for public demonstrations, for example, are ancient freedoms, jealously guarded by the courts. So, too, was the principle that the Crown could not detain citizens without trial.

For the most part, common law principles have been replaced by statutes and by the European Convention on Human Rights, which became UK law in 2000 as the Human Rights Act. But from time to time, when there is no relevant statute, the common law is invoked in courts by citizens with a grievance against government.

Treaties

The UK Constitution is also comprised of international **treaties** and agreements that become binding on UK politics. Perhaps the most famous of these is the European Convention on Human Rights, which was signed by the British government in 1956. This meant the British government agreed to be bound by the terms of the European Convention and subject to the European Court of Human Rights. This subjects the British government to international law and agreements that it is compelled to honour, although it was not brought into UK law until the passage of the Human Rights Act in 1998.

Key terms

Authoritative work A work written by an expert describing how a political system is run; it is not legally binding but is taken as a significant guide.

Treaties Formal agreements with other countries, usually ratified by Parliament.

Study tip

The European Convention on Human Rights and the European Court of Human Rights are both parts of the Council of Europe which completely separate from the European Union.

Most of the relationship between the UK and the EU was defined and developed through the signing of treaties, such as the Maastricht Treaty in 1992 and the Lisbon Treaty of 2009. In this situation, Parliament allowed some of its sovereignty to be passed to the EU, and in the new treaties negotiated post-Brexit, the relationship between the UK and the EU will be redefined.

In the same way, the UK has agreed to be bound by international agreements relating to climate change (in the Paris Climate Agreement), the UN, the World Bank, the World Health Organization and NATO.

Activity

Research the Lisbon Treaty of 2009. In what ways did it alter the UK Constitution? Why might some people see it as a loss of sovereignty? What issues for democracy would be raised by the fact it was simply agreed to by the Prime Minister, not by Parliament?

Knowledge check

Identify three statute laws, three conventions and three treaties that have changed the UK Constitution.

Study tip

When answering questions where you are evaluating the effectiveness of constitutional reforms, you should compare the outcomes with the original intentions, as described below.

Table 5.2 The main sources of the UK Constitution

Type of source	Examples
Statute law	Equal Franchise Act 1928 — established full and equal voting rights for women
	Life Peerages Act 1958 — introduced the appointment of life peers to add to the hereditary peerage
	Human Rights Act 1998 — incorporated the codified European Convention on Human Rights into UK law
	Scotland Act 1998 — established a Scottish Parliament with legislative powers
	House of Lords Act 1999 — abolished all but 92 of the hereditary peers in the House of Lords
	Freedom of Information Act 2000 — introduced the right of citizens to see all official documents not excluded on grounds of national security
	Fixed Term Parliaments Act 2011 — replaced the Prime Minister's power to call an election at any time with the rule that elections should take place every 5 years, unless Parliament passes a vote of no confidence in the government
Constitutional conventions	The Salisbury Convention — states that the House of Lords should not block any legislation that appeared in the governing party's most recent election manifesto
	Collective responsibility — means that all members of the government must support official policy in public or resign or face dismissal (this is occasionally suspended for national debates such as the referendum on UK membership of the EU in 2016)
	Government formation — based on the convention that, following an election, the Queen must invite the leader of the largest party in the Commons to form a government
Authoritative works	The sovereignty of Parliament — establishes the supremacy of Parliament in legislation
	The rule of law — states that all, including government itself, are equal under the law
	Constitutional monarchy — a principle that the monarch is limited in their role and can play no active role in politics
	The 'O'Donnell Rules' of 2010 — establish how a coalition government may be formed
Common law	Most common law concerns principles of rights and justice. These have mostly been replaced by the European Convention on Human Rights. However, some of Parliament's powers and procedures are contained in common law. Interestingly, the definition of homicide still resides in common law
	The prerogative powers of the Prime Minister, exercised on behalf of the monarch, are essentially common law powers, which have never been codified
Treaties	The Treaty of Lisbon, 2007 — granted further powers and agreed to an effective EU constitution, including a process for leaving the EU, contained in Article 50
	The Paris Climate Change Accords, 2015 — saw the UK commit itself to an obligation to cut carbon emissions

Constitutional reform in the UK since 1997

The principles of constitutional reform

Before examining constitutional reform in detail, we should consider *why* such reforms have taken place. Constitutional reform has not occurred simply because politicians have believed it would be a good idea or that it would be popular with the public. Indeed, reforming the Constitution is not usually a great vote winner! So why do they do it? The answer lies in deep beliefs they hold that the UK political system can be improved; constitutional reform is normally designed to achieve improvement and address problems that have arisen in preceding years. In this sense constitutional reform in the UK is about 'pressure release' and allows the system to adapt and evolve rather than face a revolutionary crisis.

There are several ways in which reforms since 1997 have hoped to achieve such improvements. For New Labour, which came to power in 1997 after 18 years in opposition, the main motivations were set out clearly in its manifesto, while similar reasons have motivated subsequent demands for reform as well, particularly for the Liberal Democrats in the coalition years of 2010–15. Much of the reform since 2015 has been centred on resolving issues created by previous reforms or in dealing with the consequences of key political events, notably the EU referendum result of 2016.

The four main aims of the New Labour government for reform of the Constitution were:

- Democratisation, by removing undemocratic institutions and processes.
- Decentralisation, by dispersing power more widely and creating devolved regions to run their own affairs.
- Stronger protection of rights, by bringing the European Convention on Human Rights into UK law and other measures of rights protection.
- Modernisation, by removing outdated processes and bringing the UK into line with other, more modern democracies.

Since New Labour introduced its first wave of reforms in 1997, reform has either been about responding to the consequences of those initial actions or building on the original aims, as we shall see below. As an unintended reform, Brexit was not, initially, an aim for any of the major parties.

Before you begin looking at the specific reforms of the New Labour era (1997–2010), the coalition era (2010–15) and the post-2015 era, Table 5.3 provides a summary of many of the key constitutional reforms to help give you an overview.

Synoptic link

The strengthening of rights in the UK is a key way in which democracy has been reformed and developed in the UK and has given much greater power to the Supreme Court to act as a check on both Parliament and the government, covered in Chapter 8.

Table 5.3 A general summary of constitutional reform in the UK since 1997

Area of reform	Nature of the reform
House of Lords	Abolition of voting rights of most hereditary peers
House of Commons	Limited changes to the select committee system Fixed-term parliaments (though not entrenched) New control for backbenchers over the Commons agenda
Human Rights Act	Inclusion of the European Convention on Human Rights into British law, effective from 2000
Electoral reform	The introduction of new electoral systems for the Scottish Parliament, Welsh and Northern Ireland assemblies, elections to the European Parliament, for the Greater London Assembly and for elected mayors
Freedom of information	Introduction of freedom of information, effective from 2005 Public right to see official documents
City government	Introduction of an elected mayor and assembly for Greater London and devolution of powers over health and social care to the Manchester mayor and government
Local government	Introduction of a cabinet system in local government and the opportunity for local people to introduce elected mayors by referendum
Devolution	Transfer of large amounts of power from Westminster and Whitehall to elected bodies and governments in Scotland, Wales and Northern Ireland
Party registration and the Electoral Commission	A new Electoral Commission was set up to regulate elections and referendums, including the funding of parties This reform also required the first ever registration of political parties
Reform of the judiciary	The 'political' office of Lord Chancellor was abolished. The holder was no longer head of the judiciary or Speaker of the House of Lords. This function was replaced by an office known as Justice Minister The House of Lords Appeal Court was replaced by a separate Supreme Court in 2009 Senior judicial appointments are controlled by an independent committee of senior judges
Fixed-term parliaments	The 2011 Act removed the Prime Minister's power to determine the date of general elections. Each parliament should, except under exceptional circumstances, last for 5 years
English votes for English laws (EVEL)	A change in parliamentary procedure that means that MPs from Scottish constituencies will not be allowed to vote to impose policies on England or England and Wales
Recall of MPs	If an MP is imprisoned or suspended from the House of Commons for misbehaviour, a petition signed by 10% of the voters in a constituency can trigger a by-election
The UK and the EU	In 2016 a referendum determined that the UK would leave the EU in the following years. This will have wide-ranging constitutional impacts

Reform, 1997–2010

When the Labour Party was elected to power in 1997 it had grand plans for reform of the Constitution. Furthermore, it was committed to completing most of the changes within 5 years. This was indeed ambitious. One of the reasons for the rush was that the party had a huge 179-seat majority in the House of Commons and so felt it could push through the reforms with minimal opposition. It was almost inevitable that such a large project could not be completed, but the government had implemented a high proportion of its proposals by the time it lost office in 2010.

Devolution

Devolution was a key element of Labour's post-1997 reform programme and perhaps is most significant in terms of the consequences for the UK political system. It is covered in greater detail later in this chapter.

> **Key term**
>
> **Devolution** The dispersal of power, but not sovereignty, within a political system.

Parliamentary reform

House of Lords

The government of 1997 wanted to reform the House of Lords quite radically but had to move in two stages in order to persuade the House of Lords to vote for its own reform!

1 The first stage was the plan to remove **hereditary peers**. In other words, there would be an all-appointed chamber of life peers and Church of England bishops. There was, however, some obstruction to this and the government had to compromise by allowing 92 of the 753 hereditary peers to retain their seats, known as the Cranborne Compromise.

2 Stage two was to replace the remaining House of Lords with an elected, or partly elected, chamber. However, this ran into more obstruction and a lack of political consensus. The measure was, therefore, taken off the agenda.

Although the **House of Lords Act 1999** reduced the number of hereditary peers to 92, the House of Lords threatened to use its powers to obstruct and delay reform. The government left these 92 hereditary peers in place in return for the Lords' compliance.

It should be emphasised that although the 1999 Act was a limited reform, it did have the effect of making the Lords a largely appointed chamber as well as reducing the total number of eligible members, from 1330 to 669 in March 2000. The much higher proportion of peers who held their position *on merit* rather than by birth meant that the Lords became a more professional and efficient body.

Synoptic link

Reforms to Parliament will be discussed in greater detail in Chapter 6 as they have had a direct impact on the workings and procedures of Parliament.

House of Commons

The main reform under New Labour concerned the departmental select committees of the House of Commons. These committees of backbench MPs scrutinise the work of government departments. They are becoming more important and have enjoyed some enhancement in status. In 2004, the chairs of the committees were awarded additional salaries to raise their status. In 2010, one of the last acts of the outgoing Labour government was to introduce a system for electing members of the select committees by the whole chamber of the House of Commons. Before the reform they had been largely selected by party leaders. The other chairs of these committees are then allocated to parties based on their proportion of MPs in the House of Commons and then elected by the MPs within the party, in order to reduce the influence of party whips. These reforms have increased their independence.

Also, in 2010, a Backbench Business Committee was established. This gave MPs control of over twenty parliamentary days to debate issues of their choosing. This represented a small increase in backbench influence and control.

Useful term

Hereditary peers
Members of the aristocracy who owe their title to their birth, in other words they inherit their title, usually from their father. Some titles go back deep into history. Ninety-two such peers have a right to sit in the House of Lords.

Rights reform

Human rights reform

In 1998, the UK Parliament passed the Human Rights Act, possibly the most significant development in the protection of human rights in the UK since the Bill of Rights of 1689. Its provisions came into force in 2000. The Act incorporated the European Convention on Human Rights into UK law. The convention was made binding on all public bodies, including the government. UK courts were given the power to enforce the convention whenever it became relevant in any case coming before them.

In order to preserve the principle of parliamentary sovereignty, the convention is not strictly binding on the UK Parliament, though any laws that contravene the convention can be passed only if the government declares an overwhelming reason why it is necessary to do so. In practice, therefore, the terms of the ECHR are now binding in all parts of the UK.

It is worth remembering that the UK Parliament did not lose sovereignty by passing the Human Rights Act. Parliament can repeal it at any time and so the HRA would no longer be enforced in British courts. Indeed, the Conservative government that came to power in 2015 was committed to replacing the Act with a new *British* Bill of Rights.

The Freedom of Information Act 2000

Before 2000, much of the work of the government was done in secret, behind closed doors, and so there was a lack of transparency. The **Freedom of Information Act 2000** was introduced as a watered-down version of similar measures in operation elsewhere in Europe. The security services were exempt and the Act gave the government the right to conceal information if it felt it might prejudice the activities of government. In other words, the onus is on the 'outsider' to prove that a document or other information should be released.

When the Freedom of Information Act was passed, human rights campaigners thought it was too weak. Experience tells us otherwise, however. One major development illustrates its power. In 2008 a request was made to the Information Tribunal to release details of expenses claims made by MPs. Parliament attempted to block the request through the High Court, but failed. The information was released to the *Daily Telegraph*. When the revelations were released by the newspaper, it became clear that there had been widespread abuse of the generous expenses system. As a result, hundreds of MPs were accused of 'milking' the system for their own benefit. The results of the revelations were far-reaching. Many MPs decided to give up their seats, and some MPs and peers faced criminal prosecutions, with six MPs and two peers being found guilty and sentenced to prison terms or, in one case, a two-year supervision-and-treatment order on the grounds of mental health. Parliament was subjected to widespread ridicule and public condemnation, and the expenses system had to be radically reformed.

Synoptic link

Further discussion on the nature and status of rights in the UK, as well as the ways in which they are protected, can be found in Chapter 1, and the power they have given the judiciary to check the other branches of government is discussed in Chapter 8.

Activity

Research the 'MPs' expenses scandal' of 2009–10. Describe the issue and explain why the Freedom of Information Act was important in it.

The *Daily Telegraph* reports about MPs' expenses resulted from a Freedom of Information request and had huge repercussions for politics in the UK

Electoral reform

In 1997, Labour was worried that it might not gain enough seats to form an outright majority and might need the Liberal Democrats to help form a coalition. One of the main priorities for the Liberal Democrats was the issue of electoral reform, so New Labour introduced a commitment to electoral reform in its manifesto. In the end, Labour did not need the Liberal Democrats, but having committed itself to electoral reform it introduced a commission to investigate the matter, led by Lord Jenkins. When the Jenkins Commission finally reported its preference for an electoral system called AV+, the enthusiasm for electoral reform had waned, perhaps as a result of New Labour benefiting from the old FPTP system in the general election.

Labour carried out electoral reform when devolution was being considered. In developing devolution, it was decided at an early stage that the system used should reflect the party systems in those countries and should avoid domination by one or two parties, as was the case in England. It was therefore agreed that forms of proportional representation should be used. If FPTP had been adopted, Wales and Scotland would have been dominated by Labour and the Unionists would have controlled Northern Ireland. In the event, the additional member system was adopted for Scotland and Wales, while the single transferable vote was used in Northern Ireland.

Establishment of the Supreme Court

At the start of the twenty-first century, there was growing concern that the judiciary, that is, the senior levels of the court system, was in need of reform. This was largely as a consequence of the close relationship with and control over the judiciary that the government was showing through the role of Lord Faulkner in investigations into government actions surrounding the Iraq War. The **Constitutional Reform Act 2005** was passed to address these issues.

> ### Discussion point
>
> Evaluate the view that the judiciary remains in need of major constitutional reform.
>
> Three key points to consider include:
> 1. The role and purpose of the judiciary (found in more detail in Chapter 8).
> 2. What reforms have occurred and how effective the changes have been.
> 3. What reforms would be needed and their possible consequences.

The Constitutional Reform Act had three main effects.

1. **Separation of judiciary and government** Most importantly, it was seen as crucial that there be a clearer separation between the senior members of the judiciary and the government. In the past, the position of the Lord Chancellor had been ambiguous. He (or she) was a Cabinet minister and senior member of the governing party. At the same time, the holder of the office was head of the judiciary and presided over the proceedings of the House of Lords. This placed the Lord Chancellor in all three branches of government. Although the occupant of the post might protest that they understood the difference between their neutral *judicial* role and their *political* role as a Cabinet minister, suspicions persisted that one role would interfere with another. This perception of lack of **independence of the judiciary** had to be addressed in a modern system. The judicial role of the Lord Chancellor was therefore largely removed. The post was

Synoptic link

There is further discussion and detail about elections and electoral systems in Chapter 3. It is important to remember that the positives and negatives of different electoral systems feed directly into these constitutional arguments.

Useful term

Independence of the judiciary Under the separation of powers, the three branches of government — legislature, executive and judiciary — are independent; they have separate powers and can check each other's power.

combined with that of Justice Secretary, a Cabinet post, but the holder ceased to have a judicial role; they would be in charge of justice *policy*, but not *practice*.

2 **Supreme Court** The highest court of appeal had been the House of Lords when it met as a court rather than as part of Parliament. The senior 'Law Lords' would hear important appeal cases, often with great political consequences. There was a growing belief that it was not appropriate that members of the legislature (i.e. the Law Lords) should also be the highest level of the judiciary. In other words, it was seen as vital that law and politics should be completely separated to safeguard the rule of law. It was therefore decided to take the senior judges out of the Lords and to create a separate Supreme Court.

The Supreme Court was opened in the autumn of 2009 and began work immediately to establish its new independence. In reality, the change to a Supreme Court is partly a cosmetic exercise. The new court has the same powers as the old House of Lords. However, it is symbolically important and, so far, the rulings of the Court indicate that it is acting as a genuinely independent body.

3 **Appointment of senior judges** Finally, there was opposition to the continued practice of senior appointments to the judiciary being in the hands of politicians — mainly the Lord Chancellor and the Prime Minister. There was a constant danger that such appointments might be made on the basis of the political views of prospective judges rather than on their legal qualifications. A Judicial Appointments Commission (JAC) was therefore set up to ensure that all candidates should be suitable, using purely legal considerations. The government has the final say on who shall become a senior judge, but this must be after approval by the JAC. The most senior appointments — to the Supreme Court — have been placed in the hands of a non-political committee of senior judicial figures.

Synoptic link

The independence and neutrality of the judiciary are key aspects of democracy and are discussed in greater detail in Chapter 8.

Synoptic link

The relationship between different branches and how it has changed as a result of constitutional reforms will be dealt with more fully in Chapter 8.

Constitutional reform under the coalition, 2010–15

The election of 2010 saw something rare in British politics: the creation of a coalition government made up of two parties. It was such an unusual event that the process

The coalition government was headed by Conservative Prime Minister David Cameron and Lib Dem Deputy Prime Minister Nick Clegg

of establishing the coalition agreement was in itself a constitutional reform, setting out the process for forming future coalitions. Beyond this, it saw two parties sharing power with very different views on constitutional reform: a Conservative Party that only aimed to redress some of the issues left over from the New Labour reform years, and the Liberal Democrats, a party committed to widespread reforms that would radically alter the British Constitution.

Fixed-term parliaments

When the coalition government had to be formed in 2010, there was an overwhelming fear that it would be unstable and possibly not last very long. If the two parties came into conflict with each other, it would be easy for Parliament to defeat the divided government and force a general election. In addition, because the power to dissolve Parliament and call an election lay in the hands of the Prime Minister (by constitutional convention), this power could be used to 'bully' the Liberal Democrats into agreement. To avoid this undesirable possibility, the **Fixed Term Parliaments Act** was passed in 2011. It meant that there could be disagreement within government without the danger of it falling apart.

The Act was also designed to take away the Prime Minister's power to call a general election whenever they wished, thus giving their own party an advantage by surprising the opposition with a 'snap election'. As a result of the Act, an early election would occur only if a vote of no confidence in the government were to be passed and no new government could be formed within 14 days, or if a two-thirds majority in the House of Commons passed a motion for an early election.

Although the Act worked for the duration of the coalition, its limitations were exposed in April 2017 when Theresa May announced such a snap election, to take place on 8 June, and avoided the restrictions of the Act by introducing such a motion and achieving a two-thirds majority in the Commons. The opposition could not oppose the motion because it would have appeared weak in avoiding an election. A similar thing happened in 2019 when an election was triggered as part of the government's Early Parliamentary Act, 2019. This does suggest the Act has been of limited significance, though the very fact that the Prime Minister must seek Parliament's permission before calling an election shows a shift in the balance of power between the Prime Minister and Parliament.

Recall of MPs

The **Recall of MPs Act 2015** provided for constituencies to 'recall' an MP who had been involved in some kind of misbehaviour. It requires a petition supported by at least 10 per cent of the MPs' constituents to set the process in motion. It is a limited measure in that MPs cannot be recalled on the basis of their voting record or their policy statements. If an MP is imprisoned or suspended from the House of Commons, they may be subjected to a by-election, which they would be likely to lose and so lose their seat.

Although initially seen as so weak as to be ineffective by MPs including Douglas Carswell and Zac Goldsmith, the Act did lead to its first MP having to face a petition in 2018, as Ian Paisley Jr was charged with asking questions on behalf of a foreign government. In 2019 it saw two MPs lose their seats as direct results of recall elections: Fiona Onasanya in Peterborough for a custodial sentence of less than a year, and Christopher Davies in Brecon and Radnorshire for submitting a false or misleading expenses claim.

Further devolution to Wales

The **2006 Government of Wales Act** included a provision for further powers for 20 key areas to be transferred from Westminster to Wales, if the people of Wales requested it through a referendum. In order to redress the issue of **asymmetrical devolution** the coalition chose to hold such a referendum in 2011, in which 64 per cent of those who voted confirmed that further devolution was the preferred constitutional option for Wales (as opposed to the status quo or ending devolution).

The referendum result in 2011 meant that the Welsh Government automatically gained the power to make and pass primary legislation in the 20 key areas, including healthcare, education, transport and the environment, and would no longer need its decisions to be approved by Westminster.

The positive referendum result also led to the creation of the Silk Commission, which was tasked with investigating ways in which financial powers could be devolved and then which political and administrative powers could be further devolved to better meet the needs of the people of Wales. In 2012, the Commission proposed the transference of some financial powers, notably tax-raising measures, to the Welsh Assembly (as it was) to eventually give it control in raising 25 per cent of its annual budget. This was adopted in the Wales Act 2014 and began that year with the devolution of some small taxes, such as control over stamp duty and landfill tax. Other taxes would be devolved over time to the point where, in 2020, it would be given the power to vary income tax by up to 10 per cent.

The second part of the Silk Commission, relating to further devolution of administrative and political areas, was not reported until March 2014. The proposals made were much more widespread and proved more controversial than the recommendations already implemented. These issues would be discussed, but not acted upon until the Wales Act of 2017.

The House of Lords Reform Acts, 2014 and 2015

The 1998 House of Lords Reform Act had reduced the size of the House of Lords from 1330 to 669 by March 2000, but thanks to increasing numbers of appointments and longer life expectancies, the number of peers had increased to over 900 by 2014. In order to redress the problem that the House of Lords had grown to be the second-largest legislative body in the world (after the Chinese Communist Party Convention), the House of Lords Reform Act 2014 was passed to allow members of the House of Lords to resign or retire, which they had not been able to do previously, or to be removed for prolonged non-attendance or expelled for a serious criminal conviction. As of spring 2020, 105 peers had used this Act to resign or retire, while six had been expelled for non-attendance.

In 2015 a further Act was passed, allowing the Lords to remove a peer who had breached its code of conduct.

Constitutional reform since 2015

Extension of Scottish and Welsh devolution

In 2016 the new Conservative government was forced to grant further powers to the Scottish government. This was in response to the surge in nationalist feeling following a close result in the 2014 Scottish independence referendum. This gave a great deal of financial autonomy to Scotland and was known as 'devo-max'.

Useful term

Asymmetric devolution
A type of devolution where the various regions have been granted unequal amounts of power.

Activity

Do some research into the Silk Commission. What were its aims? What were its main proposals? How many of these proposals have been introduced?

Knowledge check

Identify three additional powers devolved to Wales as a result of the Wales Act 2014.

In Wales, some of the proposals from the second part of the Silk Commission were passed in the Wales Act of 2017, granting it similar levels of legislative and fiscal powers and responsibility to those of Scotland. In May 2020, the Welsh Assembly formally became the Welsh Parliament, or Senedd Cymru.

English votes for English laws

The success of devolution in Scotland and Wales had raised the issue of **the West Lothian Question**, with Scottish and Welsh MPs able to vote on matters that did not affect their own constituents. As an example, it was only as a result of Scottish MPs voting to increase student tuition fees to £3000 per year in 2002 that the measure passed, despite the fact that it would not affect Scottish students.

In order to solve this issue in Parliament, the Conservatives introduced English votes for English laws, which was an additional stage in the passage of legislation. If the Speaker determined that a bill only concerned England, or only England and Wales, then that bill could be vetoed by a vote of MPs representing only those regions. It did not, however, ensure that measures desired by a majority of English, or English and Welsh, MPs would be passed, as happened with extensions to Sunday trading in 2016. This passed the EVEL stage but was then defeated by a vote of the whole chamber, with SNP MPs voting against it, despite the fact Scotland already had longer Sunday trading hours and would not be directly affected by the bill.

Therefore EVEL acts as a tool for preventing measures that a majority of English, or English and Welsh, MPs oppose being forced upon them, but it does not allow them to pass measures they wish to see introduced, unless they can persuade Scottish MPs to support them.

City devolution outside London

During the coalition era of 2010–2015, attempts were made to introduce directly elected mayors to major cities across England, modelled on London. Parliament granted permission for the 12 largest cities in England to hold referendums on whether or not to have a directly elected mayor, while Doncaster would vote on whether or not to retain their elected mayor. Liverpool and Leicester councils chose to have directly elected mayors and Doncaster did vote to retain its elected mayor, but only one of the other ten major English cities (Bristol) actually voted to have directly elected mayors. Birmingham, Manchester, Sheffield, Newcastle and Leeds were among the nine major cities to reject the government's policy of spreading city-based devolution.

At the same time as these proposals, combined authorities were being introduced across England. Combined authorities, or metro-areas, are areas where previously independent local councils within a metropolitan area, such as the local councils that comprise Greater Manchester, could combine resources and share services across a wider area to make them more efficient and effective. These proved more popular as a means of cutting costs and provided an economic incentive.

The Conservative government that took office in 2015, and in particular its Chancellor, George Osborne, was committed to granting more autonomous powers to large cities as a means of tackling the longstanding issues of underperformance in England's major cities, including housing, business, and transport issues. In October 2015, Osborne announced that combined authorities would be allowed to keep all the revenue from business rates (local taxes levied on commercial businesses)

Useful term

The West Lothian Question An issue raised in 1977 by the Labour MP for West Lothian, Tam Dalyell, which concerned the problem of MPs sitting in the House of Commons and representing devolved areas, such as Scotland, Wales and Northern Ireland, being able to vote on matters that would only affect England, while English MPs could not vote on those issues that have been passed to devolved bodies, such as education and healthcare.

rather than giving it to the central exchequer, but only if they agreed to having a directly elected mayor who would be accountable and would chair the authority. This was enacted in the 'Cities and Local Government Devolution Act, 2016'. This represented a major step towards more devolved local government in England and, crucially, would be decided by the existing authorities, not by referendum, with a sizeable economic incentive to adopt them.

Since their inception as city-based devolved bodies, these institutions have developed into large areas of regional devolution far beyond individual cities. As of 2021, there are now eight of these metro-mayors where combined authorities have accepted a devolution deal and have elected mayors (West Yorkshire is a combined authority but has not yet agreed a devolution deal, though it will hold elections for an elected mayor in 2021). Following the original six metro-mayors first elected in 2017 (Cambridgeshire and Peterborough, Greater Manchester, Liverpool City Region, the Tees Valley, the West of England and the West Midlands) the Sheffield City Region elected its first mayor, Dan Jarvis, in 2018 while in 2019 Jamie Driscoll was elected as the first metro-mayor of the North of Tyne region.

These regions have grown far beyond city-based devolution, with the eight current metro-regions covering nearly 12 million people, or more than 21 per cent of the English population. When London is factored in, it means more than 20 million people, or nearly 36 per cent of the population of England are governed by some form of devolution. Table 5.4 shows the various powers and responsibilities they hold.

Table 5.4 Powers of city-based devolved bodies

Policy area	Cambrdigeshire and Peterborough	Greater Manchester	North of Tyne	Liverpool City Region	Tees Valley
30-year investment fund	£600 million	£900 million	£600 million	£900 million	£450 million
Industrial strategy	'Leading'	'Trailblazer'	'Rural productivity'	'Leading'	'Leading'
Education	Apprenticeship grants for employers Advisory panels Adult skills budget Post-16 education Oversight of skills	Apprenticeship grants for employers Advisory panels Adult skills budget Post-16 education Oversight of skills	Advisory panels Adult skills budget Oversight of skills	Apprenticeship grants for employers Advisory panels Adult skills budget Post-16 education Oversight of skills	Advisory panels Adult skills budget Post-16 education Oversight of skills
Housing	£170 million affordable housing fund	£30 million per year housing investment fund	Creation of a Housing and Land Board Compulosry purchase powers	Compulsory purchase powers	£6 million housing delivery team
Transport	Consolidated transport budget Local roads network Bus franchising £74 million transforming cities fund	Consolidated transport budget Local roads network Bus franchising £243 million transforming cities fund	n/a	Consolidated transport budget Local roads network Bus franchising £134 million transforming cities fund	Consolidated transport budget Bus franchising £59 million transforming cities fund
Health	Planning for helath and social care integration	Control of £6 billion health and social care budget	n/a	Planning for helath and social care integration	n/a

Policy area	Sheffield City Region	West Midlands	West of England	West Yorkshire
30-year investment fund	£900 million	£1.1 billion	£900 million	£1.1 billion
Industrial strategy	'Leading'	'Trailblazer'	'Leading'	n/a
Education	Apprenticeship grants for employers Adult skills budget Post-16 education	Advisory panels Adult skills budget Post-16 education Oversight of skills	Apprenticeship grants for employers Advisory panels Adult skills budget Post-16 education Oversight of skills	Adult education budget
Housing	Compulosry purchase powers	Compulosry purchase powers £6 million housing delivery team	Compulsory purchase powers	Access to a Brownfield remeditation fund of £400 million Compulsory purchase powers
Transport	Consolidated transport budget Local roads network Bus franchising	Consolidated transport budget Local roads network Bus franchising £250 million transforming cities fund	Consolidated transport budget Local roads network Bus franchising £80 million transforming cities fund	Key roads network Bus franchising £317 million transforming cities fund Smart ticketing
Health	n/a	n/a	n/a	n/a

Synoptic link

Referendums are increasingly used to entrench constitutional reforms and are dealt with more fully in Chapter 3, while the impact they have on sovereignty is considered in Chapter 8.

Leaving the EU

Perhaps the most significant constitutional reform since 2015 is an accidental one: leaving the EU. The full details of this will be addressed in Chapter 8, but the decision to leave the EU was not one that the government intended when it held a referendum on the issue in 2016. However, a majority of the people who voted, voted to leave and Parliament has pledged to honour this result. In undoing 47 years' worth of EEC and EU legislation and changing Britain's political, economic and international structures, Brexit is perhaps the biggest single constitutional change in the country's history. Although Britain formally left the EU in January 2020, the consequences and developments that result from this change will continue to impact in major ways on British politics for years, if not decades.

Activity

Investigate the requirements other nations have for introducing constitutional reforms. Most require 'higher' levels of support than the 52 per cent that voted to leave the EU. Why do you think that might be? Does this suggest that the British Constitution is better because it is able to respond easily to public demands or worse because major changes can be introduced by a small majority?

Devolved bodies in the UK and the impact of devolution on the UK

The nature of devolution

Devolution is an important constitutional development in the context of the UK. It is vital to understand precisely what the term means. A good definition, applied to the UK, looks like this:

> Devolution is a process of delegating power, but not sovereignty, from the UK Parliament to specific regions of the country. This is power that can be returned to Parliament through a constitutional statute. Therefore, it is a transfer of power without eroding the sovereignty of Parliament.

The road to devolution

The first calls for devolution emerged in the 1970s. The Labour government of 1974–79 considered the measure, largely under the influence of the SNP, which then had 11 MPs in Westminster. Labour had only a small parliamentary majority, so it relied on Liberal and SNP support for much of its time in government, to avoid losing votes of no confidence introduced by the then opposition leader, Margaret Thatcher. One of the prices of that support was the idea of devolution. The Liberal Party and the SNP believed that devolution would enhance democracy in the UK and bring government closer to the people. It was also in response to some early signs of nationalist sentiment in Scotland.

Labour was unenthusiastic but went ahead with referendums in Scotland and Wales. The proposal, however, was doomed from the start. The Conservatives, supported by many Labour MPs, insisted on a safeguard: that for devolution to take place, it would be necessary not only for the majority of Scots or Welsh to vote for it, but also that at least 40 per cent of the adult population approved, taking into account non-voters. Wales rejected the proposal anyway, while a majority of Scots who turned out voted in favour, but the number fell short of the 40 per cent threshold.

Devolution was forgotten for nearly two decades, as Margaret Thatcher and the Conservatives believed granting devolution to Scotland would be the first step on a road that would inevitably lead to Scottish independence. In the late nineties, however, circumstances changed. First there were renewed signs of growing nationalism in Scotland and Wales. Second, and coincidentally, Labour was elected to power with a huge mandate to reform the UK Constitution. Devolution was a key aspect of those reforms. Meanwhile, in Northern Ireland, a peace settlement had been reached between the rival Republican and Loyalist communities. To cement the peace, a devolution settlement with Northern Ireland was also negotiated. This would allow power sharing to be introduced so that a political settlement could underpin the military peace. Thus, by 1997, the stage was set for a full set of devolution proposals and referendums to approve them.

A summary of devolution throughout the UK is shown in Table 5.5.

Activity

Research the taxation powers of the governments of Scotland, Wales and Northern Ireland. In theory, what variations in taxation could be introduced across the UK? What variations have been introduced? Why?

Table 5.5 Devolution in the UK, 1979–2020

Year	Scotland	Wales	Northern Ireland
1979	Referendum on Scottish devolution. It fails to reach the threshold of 40% of the electorate approving	Referendum on devolution held in Wales. Only 20% vote in favour	Not proposed
1997	A referendum on devolution passes with a comfortable majority	A referendum on devolution passes with a narrow majority	The Belfast Agreement is successfully negotiated and agreed by leaders of the two communities
1998	The Scotland Act establishes further devolution, a Scottish Parliament and a Scottish Executive	The Government of Wales Act grants considerable administrative devolution to Wales. It also promises a referendum on income tax-raising powers	The Belfast Agreement is approved in a referendum The Northern Ireland Act is passed, granting devolution First elections to the Northern Ireland Assembly
1999	The first Scottish Parliament since 1707 meets and the Scottish government starts to operate	The Welsh Assembly is elected and holds its first meetings A Welsh Executive is also formed	Northern Ireland power-sharing government takes power
2002			The Northern Ireland Assembly is suspended. Central government takes over in the province
2007			Northern Ireland devolved government is restored
2014	A referendum on full independence is defeated by 55% to 45%	Wales Act gives the Welsh government powers over several taxes and promises a referendum on income tax-raising powers	
2015		The UK government announces the government of Wales may be given future powers over a proportion of income tax raised in the country without the approval of a referendum, despite this requirement existing in previous legislation	
2016	The Scotland Act grants wide-ranging financial powers to the Scottish government, including control over income tax rates		
2017		The Government of Wales Act passes, which devolved greater political, administrative and fiscal powers to the Welsh Assembly	In January, government in Northern Ireland collapses as power sharing breaks down
2020		In May the Welsh Assembly is officially renamed the Welsh Parliament	The government of Northern Ireland is restored in January following agreement between Sinn Féin and the DUP

Devolution in England

An English Parliament

With the success of devolution in Scotland, Wales and Northern Ireland, it may seem odd that there has not yet been a serious attempt to introduce devolution to England. Part of the issue is lack of demand; English nationalism has not taken hold or gained sufficient support to lead to a general demand for devolution in England. The other reason is more practical; England is too large to work as a devolved body within the UK. With 84 per cent of the UK's population and 95 per cent of the

UK's GDP, England would continue to dominate any national Parliament, meaning England effectively would control both the UK Parliament and its own English Parliament. In fully federal systems, even the largest regions, like California and Texas in the USA, have no chance of being able to dominate the federal government without the cooperation of many other states.

There is also the fact that many traditionalists, particularly in the Conservative Party, feel that Westminster already is an English Parliament and any further devolution to England would precipitate the break-up of the United Kingdom. As such, the only way devolution in England could occur is through regional or city-based devolution. This debate will be considered more fully below.

Regional devolution

At the time when devolution to the UK's three national regions was being implemented, the Deputy Prime Minister, John Prescott, was also floating the idea of devolution to the English regions. His plan was to devolve a similar amount of power to the regions as that being transferred to Wales, in other words administrative but not legislative or financial devolution. To test public opinion, a referendum was held in the North East region in 2004. The voters rejected the idea by a majority of 78 per cent to 22 per cent. This was so clear a verdict that the idea was promptly abandoned.

Under the coalition, cities, towns and districts were given the opportunity to elect mayors following a local referendum. However, few held referendums and fewer still voted in favour of an elected mayor (originally only 11 cities voted for an elected mayor).

Similarly, local authorities were given the option of changing to a 'cabinet' system of government. This involved the creation of a central cabinet of leading councillors from the dominant party or from a coalition, who might take over central control of the council's work, making key decisions and setting general policy. This would replace the former system where the work of the council was divided between several functional committees. But, as with elected mayors, the take-up for the new cabinet system has been patchy. Moreover, it is generally acknowledged that this kind of internal change does not tackle the real problems of local government. These are seen as threefold:

1 Lack of autonomy from central government
2 Lack of accountability to local electorates
3 Largely as a result of the first two problems, low levels of public interest in local government and politics

In addition to mayors, policing in England and Wales was to be devolved to local figures, through the creation of newly elected police crime commissioners (PCCs). It was hoped that PCCs would improve accountability for policing in local areas, but this has not been the case. Turnout at commissioner elections has been low and few people are even aware of who their local police commissioner is.

City government in London

In 1985 the Greater London Council, a powerful local government body with wide powers and responsibilities, was abolished by the Conservative government. Prime Minister Thatcher was determined to remove from power what she saw as a socialist enclave in the centre of Conservative Britain.

Labour, when it returned to power in 1997, was determined to restore government to London by creating a devolved Greater London Assembly. In addition, an innovation was to be introduced. This was the election of a mayor with a considerable degree of executive power. Elected mayors were unheard of in British history. In the past, mayors had been holders of ceremonial offices, appointed by councils and with no executive power at all.

Sadiq Khan, London Mayor since 2016

In 2000, following a decisive referendum in which the people of London approved the introduction of an elected mayor and assembly, elections were held for the two new institutions. However, the legislation seemed to ensure that neither would enjoy a hugely significant amount of political power.

The mayor controls the allocation of funds for different uses in London, funds that are distributed and administered by the elected assembly of 25 members. But at the same time, the assembly has the power to veto the mayor's budgetary and other proposals, provided there is a two-thirds majority for such a veto. Similarly, while the mayor has powers of patronage, controlling a variety of appointments, the assembly again has rights of veto. This was a classic example of the introduction of a system of 'checks and balances'. Furthermore, the electoral system used for the assembly, AMS (additional member system), as in Scotland and Wales, meant that there was no real possibility of a single party enjoying an overall majority. This ensured that the mayor would always face obstruction for controversial measures.

The office of London mayor was granted relatively limited power under the legislation. The holder of this office cannot be said to enjoy a similar position to powerful mayors in New York and Paris, for instance. However, within the limitations, it could be said that Ken Livingstone and his successors Boris Johnson and Sadiq Khan have been involved in several significant developments in London. Accepting that the mayor possesses influence rather than power, the three mayors so far have been wholly or partially instrumental in the following initiatives:

- Improved community policing
- The growth in the arts scene in the capital
- Vastly improved public transport

Knowledge check

Identify the two new electoral systems introduced for devolved assemblies in 1998.

Should cities be given more independent powers?

Arguments for

- Local democracy is closer to the people and will therefore more accurately reflect their demands.
- Local needs vary a great deal, so the 'one size fits all' suggested by central government control is not realistic.
- The UK as a whole is too 'London-centred', so autonomous local government may boost local economies and more evenly spread wealth and economic development.
- Demonstrating that local councils and mayors have significant powers will give a boost to local democracy.

Arguments against

- Central control means that all parts of the UK should receive the same range and quality of services.
- Central control of finance will prevent irresponsible local government overspending.
- Turnouts at local council and mayoral elections tend to be very low, so local government is not accountable enough.
- There is a danger that the traditional unity of the UK might be jeopardised.

 Think about the word 'should' in the question. Additional independent powers might be desirable, but *should* it happen? Why should it happen, or why not? Your opinion on this issue will allow you to evaluate and consider the stronger of the two arguments to reach an evaluative judgement, but make sure it is also based on evidence.

Scottish Parliament and government

Scotland was a special case. There had been administrative devolution in Scotland since the nineteenth century. What this meant was that a non-elected Scottish Executive administered various services in Scotland on behalf of the UK government. Matters such as education, health, local authority services and policing were managed separately in Scotland. The country also had its own laws and legal system. However, there had been no Scottish Parliament to pass these laws since 1707. Rather strangely, it was the UK Parliament in Westminster that made the laws for Scotland. It should be noted that nationalist sentiment was much stronger in Scotland than it was in Wales.

Scotland Act 1998

In 1997 a referendum was held in Scotland to gauge support for devolution, asking if they wanted a Scottish Parliament with law-making powers and if they wanted it to have tax-varying powers. The Scots voted overwhelmingly in favour to both questions, by 74 per cent to 26 per cent on a 60 per cent turnout. The following year the **Scotland Act** was passed, granting devolution. It was implemented in 1999 and the first Scottish Parliament was elected. The main powers that were devolved to this parliament, and the executive that was drawn from it, were as follows:

- Power over the health service
- Power over education
- Power over roads and public transport
- Power to make criminal and civil law
- Power over policing
- Power over local authority services
- Power to vary the rate of income tax up or down by 3 per cent

At the same time, a new electoral system was introduced for the Scottish Parliament. This was the additional member system. The government of Scotland would be formed by the largest party in the Parliament or by a coalition. The first minister, leader of the largest party, would head the executive government.

Scotland Act 2016

After the start of the twenty-first century, Scottish nationalism continued to grow in popularity, so much so that a referendum on full independence was held in 2014. Though the Scots voted 55.3 per cent to 44.7 per cent against independence, it was clear there was an appetite for greater devolution and, to head off the threat of the independence side winning, the Conservatives, Labour and Liberal Democrats all pledged to increase devolved responsibility and power to Scotland. This led to a second devolution stage, the **Scotland Act 2016**. This Act included the following measures:

- Widening of the areas in which the Scottish Parliament may pass laws
- Power over the regulation of the energy industry transferred to Scotland
- Control over a range of welfare services including housing and disability
- Control of half the receipts from VAT collected in Scotland
- Control over income tax rates and control over all receipts from income tax
- Control over air passenger duty and control over its revenue in Scottish airports
- Control over some business taxes

The Act represented a large transfer of powers and freedom of action. It means that the Scottish government now has an enormous amount of administrative, legislative and financial autonomy. It still stops well short of independence, but it does go a long way to making Scotland feel like a separate country, in charge of its own future.

The UK's decision to leave the EU in 2016, however, has once again destabilised the Scottish situation. The Scots voted by a large majority to remain inside the EU; 62 per cent to 38 per cent in favour of remaining. For many Scots, the only way to stay in the EU is to once again demand full independence.

Study tip

When discussing devolution, pay special attention to the differences between the powers granted to the Scottish, Welsh and Northern Irish governments.

Welsh Parliament and government

The referendum on Welsh devolution in 1997 was a close-run thing. The majority was only 50.5–49.5 on a low turnout of around 50 per cent, so only a quarter of the Welsh electorate actually voted in favour of devolution. It was therefore no surprise that considerably fewer powers were devolved to Wales than to Scotland.

Government of Wales Act 1998

The **Government of Wales Act 1998** set up an elected Welsh National Assembly, and a Welsh Executive to be drawn from the largest party in the Assembly and headed by a first minister. The Assembly had no powers to make or pass primary laws and the country was given no financial control. In other words, devolution to Wales in 1998 was purely administrative. As a result, the Welsh government could run a number of services but could not pass laws relating to those services. It did, however, have the power to decide how to allocate the funds it received from central government between the various services it oversaw.

The main areas of government devolved to Wales included:

- Health
- Education
- Local authority services
- Public transport
- Agriculture

Without its own means of raising finance, the Welsh government relied on an annual grant from the UK government.

Government of Wales Act 2014

Nationalist sentiment did not grow in Wales after the first stage of devolution. Nevertheless, demands for further devolution did begin to increase after 2010. The fact that the Liberal Democrat Party was part of the coalition government after 2010 helped the process as the Lib Dems supported further decentralisation of power. There were also fears that if considerable new powers were devolved to Scotland, the difference between the powers of the Welsh and Scottish governments would be too wide. There had been a small increase in devolved powers through the **Government of Wales Act 2006**, which also offered a future referendum to allow the people of Wales to approve further devolution, if they so wished. This led to the 2011 referendum, in which the Welsh voted 63.5 per cent to 36.5 per cent to approve support for further devolution. This resulted in the Silk Commission and the **Government of Wales Act 2014**, which included these provisions:

- There would be a referendum in Wales to decide whether the government of Wales should have partial control over income tax.
- The Welsh government was granted control over various taxes including business taxes, stamp duty charged on property sales, and landfill tax.
- The government of Wales would have limited powers to borrow money on open markets to enable it to invest in major projects and housing.

Government of Wales Act 2017

The Government of Wales Act 2017 gave greater powers to the Welsh Assembly and more autonomy in what was considered 'governing competency'. The key provisions were:

- Removal of the provision of the 2014 Wales Act to require devolution of taxation to be decided by a referendum
- The confirmation of fiscal measures passed to the Welsh Assembly, including the ability to vary income tax by 10p in the pound
- Greater freedom in borrowing
- The transfer of administrative and legislative responsibility for more areas, including energy efficiency and onshore oil and electricity production
- The creation of the Welsh Revenue Authority to collect Welsh-based taxes

The Welsh Parliament debating chamber

The Act also allowed the Assembly to be renamed the Welsh Parliament, which came into force in May 2020.

Northern Ireland Assembly and Executive

Northern Ireland is very different from Wales and Scotland and the devolution process reflected this. This is because the devolution settlement was part of the wider resolution of 30 years of conflict between the Republican (largely Catholic) and Loyalist (largely Protestant) communities. There had been devolved government in the province between 1921 and 1972, with a Northern Ireland Parliament (often known as Stormont, where it met) and government in control of such issues as education, welfare, health, policing, much criminal and civil law, housing and local government. With increasing sectarian violence breaking out in the 1970s, the Parliament was dissolved in 1972.

The Belfast Agreement, 1998

The Belfast Agreement, also known as the Good Friday Agreement, of 1998 restored the province's devolved powers. In place of the Parliament, an assembly was to be elected using the single transferable vote electoral system instead of first-past-the-post. STV was introduced to ensure that all sections of a divided society would be represented. Meanwhile the Northern Ireland Executive was based on power sharing, meaning there is no possibility that any single party could gain an overall majority and take control of everything. This meant that all major parties were *guaranteed* ministerial places. This was part of a device to try to reduce any possibility of future armed conflict. In addition, the government of Northern Ireland could only function if it was comprised of a coalition of the largest Republican and Loyalist parties.

Powers devolved to Northern Ireland include the following:

- The passage of laws not reserved to Westminster
- Education administration
- Healthcare
- Transport
- Policing
- Agriculture
- Sponsorship of the arts

In 2002, in the face of increased tension between the two communities and the failure of ministers from the two communities to cooperate with each other, the UK government dissolved the Northern Ireland Assembly. The suspension lasted until 2007. The ability of Westminster to dissolve a devolved body like this highlights the fact that devolution is not the same as federalism and that Parliament ultimately remains sovereign in the UK.

Devolution in Northern Ireland remains fragile, although the nationalists in the government continue to campaign for more powers to be devolved. However, problems emerged once again in 2017, following the Renewable Heat Incentive scandal, which saw Sinn Féin leave the power-sharing government, triggering two elections that failed to resolve the dispute. It was not until January 2020 that the two sides could agree to cooperate again and restore the devolved government. In such a volatile situation, it seems unlikely that further powers will be devolved soon.

Activity

Look at the varying levels of devolution granted to Scotland, Wales and Northern Ireland.
Identify the following:
- A power enjoyed by all three
- A power enjoyed by Scotland but not the other two
- A power enjoyed by Northern Ireland but not Wales

The impact of devolution on the UK

Devolution is pointless unless it reflects some kind of difference in how the countries of the United Kingdom wish to be governed. In the early days there did not seem to be a great deal of impact. This may have been because Wales and Scotland were both effectively governed by the Labour Party, the same party that was governing the whole of the UK. It was hardly surprising, therefore, that differences were not immediately apparent. Later, however, as the Scottish National Party grew in strength in the Scottish Parliament and Plaid Cymru gained a foothold in power in Wales, differences did begin to emerge. Meanwhile, in Northern Ireland, devolution was hugely significant, not least because the minority Republican community now enjoyed a share in political power.

A selection of the key impacts of devolution is shown in Table 5.6.

Table 5.6 Differences made by devolution: examples of how the countries differ from England

Country	Differences
Scotland	The dominant party is the Scottish National Party
	Personal care for the elderly is free
	Prescriptions are free (this is under threat)
	There are no university tuition fees for Scottish students
	There are greater restrictions on fox-hunting
Wales	No school league tables are published
	There are free prescriptions for everyone under 25
	There is free school milk for under-7s
	Greater help is provided for the homeless
	More free home care is provided for the elderly
	University tuition fees are capped at £6000 per year
Northern Ireland	The Republicans and Loyalists have to cooperate in government under permanent power sharing
	Gay marriage was not recognised until 2020
	There are greater restrictions on abortion
	Prescriptions are free (likely to change)

Perhaps the greatest example so far of the impact devolution has had on the UK is the way the different bodies responded to the Covid-19 pandemic. Differences in lockdown measures, travel arrangements, school closures and a host of other areas meant different parts of the UK responded in distinct and different ways. The full consequences of these different responses are not yet clear at the time of writing, but such varied responses would not have happened without devolution. These differences in approach have caused some opposition and raised the question of whether it is better to have such regional differences or uniformity in dealing with a crisis like Covid-19. Even more significant distinctions will come once financial powers are devolved and acted upon. With the ability to levy their own taxes and the freedom to spend the revenue as they wish, the devolved authorities will have the power to introduce more fundamentally distinct policies.

Has devolution been successful?

Assessment of devolution is possibly too early to be conclusive. We can attempt some sort of assessment and we can rehearse the arguments in favour of and against devolution, but it will take at least a generation before we can make a definitive judgement. One question has, in a sense, been answered, however. It was assumed that devolution would reduce nationalist sentiment and prevent the break-up of the United Kingdom. In Scotland this has certainly not happened. In fact, nationalism has grown, not declined, in Scotland since devolution was introduced. In Wales there has never been great enthusiasm for independence, and this remains true. Whether devolution is responsible for the continued indifference to nationalism is difficult to assess.

Debates on further constitutional reform

The UK leaving the European Union

Before we consider the future of constitutional reform, it must be emphasised that the decision to leave the European Union was by far the most significant, far-reaching reform of all as it will fundamentally change the way in which the UK is governed. It has also affected the politics of devolution. Although the UK did formally leave the EU in January 2020, the debates over the return of powers, the nature of the relationship with the EU and the constitutional impact on the devolved institutions will have consequences for many years to come.

Should reforms carried out since 1997 be taken further?

Reform of the judiciary

Possibly the greatest success has been the reform of the judiciary, created by the **Constitutional Reform Act 2005** and the implementation of the **Human Rights Act 1998**. The Supreme Court has been established and the senior judiciary is now seen as genuinely independent of government. This is regarded as a key element in the improved protection of human rights and brings the UK closer to modern conceptions of democracy. The Human Rights Act has also given the judiciary an important tool in protecting the rights of citizens over powerful government and allowed the courts to keep a check on the other branches of government, suggesting it has gone far enough.

However, while it has become more independent and willing to challenge the government, the power of the Court is limited as a result of parliamentary sovereignty and the lack of a codified constitution. Resolving this would require introducing a codified constitution, which would fundamentally alter the workings of UK politics, or at least entrenching the provisions of the Human Rights Act, which would give the Court greater powers.

> **Synoptic link**
>
> The reform of the judiciary and the impact on the relationship between the judiciary and Parliament is considered in greater detail in Chapter 8.

Devolution

Devolution has proved popular, especially in Scotland. Support for greater autonomy has also grown in Wales where the original referendum on whether to introduce devolution was only passed with a very narrow majority on a low turnout in 1997. There have been problems with the government of Northern Ireland, but at the very least devolution has helped to retain the fragile peace since the 1990s after 30 years of armed conflict.

However, this has created different rights and experiences for people living in different parts of the UK and created a sense of unfairness in the political system with the West Lothian Question. This has not really been solved by EVEL and the different levels of autonomy across the regions have only been partially addressed.

> **Study tip**
>
> Although you should be familiar with the main constitutional developments in the history of the UK, it is especially important to know the details of at least four constitutional reforms (other than exit from the EU) that have occurred within each specified timeframe (1997–2010, 2010–15, 2015 onwards).

As such, there is a strong case to be made for further devolution to England and the extension of devolved powers to Wales and Northern Ireland.

The Freedom of Information Act

Although this Act has been disliked by successive governments, it has proved invaluable in extending the media's ability to investigate effectively the work of government and other public bodies. It has also allowed citizens to prevent injustices by accessing the formerly secret information held about them, though the ability to prevent information being made public if it may harm national security gives governments the ability to circumvent it. In this sense, there may be a case for extending the Freedom of Information Act by removing the national security provisions to make it fully accessible.

The House of Lords

There has been some benefit in terms of making the Lords more professional and effective in checking government power and improving legislation. The removal of most of the hereditary peers did give greater legitimacy to the chamber and make it more willing to assert itself against governments with large majorities in the Commons.

However, the fact that 92 hereditary peerages remain, in what was only meant to be a temporary compromise, shows that reform of the Lords has not gone far enough. In many ways, the Lords could only be truly reformed by creating a fully elected chamber, but there seems neither the will nor the support for this move. At the very least, the removal of the remaining 92 hereditary peers would complete the first phase of New Labour's original House of Lords reform.

The House of Commons

Despite the attempts to redress the balance of power from the government towards backbench MPs, with the creation of elected select committee chairs and the Liaison Committee, the Commons remains largely dominated by the party in government, as long as it has an obedient and united majority. The only way to resolve this imbalance would be to ensure no single party dominated the chamber, by electoral reform, or by creating a much more robust constitutional structure that would clearly define the limits of the government in relation to Parliament. It is very unlikely that any sitting government will be keen to follow such a course of action, but a case could be made that it is necessary to introduce further reforms in order to limit the power of government, in relation to Parliament, a debate considered more fully in Chapter 8.

Electoral reform

Although electoral reform has been introduced in regional and second-order elections and is seen to be operating effectively in those areas, governments have failed to develop further electoral reforms for Westminster elections. Despite its manifesto commitment, New Labour failed to deliver this and since then parties that have benefited from it have been resistant to reform. There is a case for saying that the AV referendum of 2011 shows the public do not want electoral reform, but they were only given the option of one alternative system, and that was not one that anyone, even reformers, really wanted. As a result, there is a case for saying further reform is needed, perhaps by introducing reform to more local elections in England to familiarise voters with the alternatives.

> **Study tip**
>
> These are just the basic outlines of the debates around further constitutional reforms. For more detailed and fuller explanations, you will need to learn about them in the relevant chapters.

To what extent has constitutional reform since 1997 improved the state of UK democracy?

Democratic improvements

- The judiciary can now be said to be genuinely independent.
- Through regional and city devolution, power has been decentralised.
- Proportional representation for elections to devolved regional assemblies has improved representation.
- Elected mayors improve local democracy.
- Citizens' rights are now better protected.
- Freedom of information has been established.
- The increased use of referendums has extended popular democracy.
- The introduction of fixed-term parliaments has weakened executive power.

Democratic failures

- The electoral system of FPTP for general elections remains grossly unrepresentative in its outcomes.
- The House of Lords remains an unelected, undemocratic part of the legislature.
- The prerogative powers of the Prime Minister remain indistinct and largely unconfined.
- The largely unreformed House of Commons remains weak and unrepresentative.
- The UK Constitution remains uncodified, creating uncertainty and lack of public understanding, and retaining the danger of excessively powerful government.

While there are merits to both sides of the debate, consider which one you feel is the stronger or more convincing side, either based on one point that is more important than all others, or because overall the weight of the points makes that side the more convincing. This is your opinion and should be your clear judgement throughout any discussion.

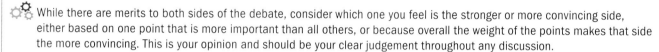

Discussion point

What further constitutional reform is needed to make the UK more democratic?

You should consider the following points:

1. What type of reform could be introduced and how would this impact democracy?
2. Is the reform really needed, or only desirable?
3. What consequences might the reform have and could they be worse than the current situation?

The debate over further devolution in England

There remains a question of whether devolution should be extended to English regions. It appears that regional devolution will be confined to only a few definable areas. Most, though not all of these, are based on cities. The debate about granting devolved powers to the English regions is summarised in Table 5.7.

Table 5.7 Should devolution be extended to the English regions?

Arguments for	Arguments against
It would extend democracy and improve democratic accountability by bringing government closer to communities	It would create a new layer of government that would be expensive
Devolved government could better reflect the problems specific to regions	It would create a need for too many elections, promoting voter apathy
It would help to prevent excessive differences between living standards and the quality of life in different parts of the UK	There are few signs of any great demand for such devolution

Of course, one such answer would be the creation of an English Parliament, but as mentioned earlier, England is far too dominant within the United Kingdom to make this a practical possibility.

The codification debate

Should the UK introduce a codified constitution? In the main, it is the Conservatives who oppose such a proposal. However, Labour governments have also avoided the issue. Labour, on the whole, has preferred to make *incremental* changes to the Constitution, gradually introducing new legislation such as the Human Rights Act, the Freedom of Information Act, the Devolution Acts and House of Lords reform to create a clearer set of arrangements. But, like the Conservatives, Labour does not believe the time is right for a fully codified Constitution.

Ranged against these conservative forces are the Liberal Democrats and small progressive parties and pressure groups such as Unlock Democracy and Liberty. They present a number of powerful arguments in favour of coming into line with nearly all other modern democracies by writing a new UK Constitution. In particular, they are concerned with the need for robust human rights and controls on the power of government.

Knowledge check

In order to be classified as 'codified', a constitution must have at least three features. What are these features?

The arguments for retaining an uncodified constitution

Flexibility

Supporters of the current arrangements say that the flexibility of the Constitution is a positive quality. The Constitution can, they say, adapt to a changing world without major upheavals. It is said that the UK's Constitution is 'organic'. This means that it is rooted in society, not separate from society. Thus, when society and its needs and values change, the Constitution can change automatically and without undue delay or confusion. Parliament can pass a new Act relatively quickly and new unwritten conventions can develop to take account of social and political change.

Some examples of such 'organic' and natural development can help to illustrate this quality:

- After the 9/11 attacks on New York and Washington in 2001, the threat of international terrorism became more acute. Had the UK had an entrenched and codified constitution, it would have been extremely difficult for Parliament to pass a wide range of anti-terrorist measures because there would have been too many constitutional constraints. The lack of a codified constitution meant that Parliament could do as it wished. Because of their codified constitutions, the USA and many European countries have had greater problems in introducing anti-terror legislation than the UK has.
- When the 2010 general election failed to produce an outright parliamentary majority for any one party, there was some confusion about what should be done in the absence of any codified rules. Such an event had not occurred for over 70 years. Nevertheless, the system was flexible enough to adapt. A new set of principles was quickly drawn up and a coalition government was formed relatively smoothly.

Executive power

When constitutional safeguards are weak or absent, government can be more powerful. This can be viewed positively or negatively. Supporters of the current uncodified Constitution argue that, on balance, it is better to have a government that can deal with problems or crises without too much inhibition. In the UK, the relationship between government and Parliament is flexible; in countries with codified constitutions it tends to be fixed, which can inhibit effective governance.

Conservative pragmatism

The typical conservative attitude to the UK Constitution suggests that it has served the country well for centuries. There have been no violent revolutions and no major political unrest. Change has occurred naturally and when it has been necessary rather than when reformers have campaigned for it. Furthermore, say conservatives, codifying the Constitution would be an extremely difficult exercise and the meagre benefits would not be worth the problems incurred.

The dangers of politicising the judiciary

A codified constitution would involve the courts, the Supreme Court in particular, in disputes over its precise meaning and application, making the courts even more political. For example, there would be conflicts over the exact powers of government, the nature of rights, or relations between England, Scotland, Wales and Northern Ireland. Bringing judges into political conflict puts the independence of judges into jeopardy, it is argued. Of course, disputes like these can arise with an uncodified constitution, but they could become more common following codification. In other words, the Constitution would become *judiciable*. The problem with such a development arises because judges are not elected and therefore not accountable. Critics point out that such political issues should not be resolved by judges; it is for elected representatives, they say, to make final decisions on constitutional meanings.

The arguments for introducing a codified constitution

Human rights

Perhaps at the top of the reformers' shopping list is the need for stronger safeguards for individual and minority rights. The UK has adopted the European Convention on Human Rights (by passing the Human Rights Act in 1998), but this remains weak in that it can be overridden by Parliament. Parliament remains sovereign and no constitutional legislation can remove that sovereignty. With a codified constitution, Parliament could not pass any legislation that offended human rights protection, offering far greater protection for the people.

Executive power

We have seen that conservatives and others have wished to retain the powerful position of government in the UK. Liberals and other reformers, however, argue that executive, governmental power is excessive in the UK. They say over-powerful governmental power threatens individual rights, the position of minorities and the influence of public opinion. A clear, codified constitution would, they assert, inhibit the apparently irreversible drift towards greater executive power. In particular, supporters of a codified constitution suggest that there are no real 'checks and balances'. It is argued that Parliament needs to have more codified powers to enable it to control government on behalf of the people.

Clarity

Most citizens of the UK do not understand the concept of a constitution. This is hardly surprising as there is no such thing as the 'UK Constitution' in any concrete form. There is, therefore, an argument for creating a single physical constitution so that public awareness and support can grow. If people know their rights and understand better how government works, it is suggested, this might cure the problem of political ignorance and apathy that prevails.

Modernity

As we have seen, the UK is unusual in not having a codified constitution. Many people regard this as an indication that the UK is backward in a political sense and has not entered the modern world. This became more pressing when the UK joined the European Community and has been reignited over the process of leaving the EU.

Rationality

As things stand, constitutional changes occur in an unplanned, haphazard, arbitrary way. If the Constitution were codified and entrenched, amendments and developments would be made in a measured, rational manner, with considerable democratic debate.

Debate

Should the UK introduce a codified constitution?

Arguments for

- It would clarify the nature of the political system to citizens, especially after changes such as devolution and House of Lords reform.
- The UK would have a two-tier legal system and so constitutional laws would be more clearly identified.
- The process of judicial review would be more precise and transparent.
- Liberals argue that it would have the effect of better safeguarding citizens' rights.
- It might prevent the further drift towards excessive executive power.
- The UK needs to clarify its relationship with the European Union.
- It would bring the UK into line with most other modern democracies.

Arguments against

- The uncodified Constitution is flexible and can easily adapt to changing circumstances, such as referendum use and the changing role of the House of Lords. If it was codified, constitutional changes would be difficult and time-consuming. It can also respond quickly to a changing political climate.
- Conservatives argue that it is simply not necessary – the UK has enjoyed a stable political system without a codified constitution.
- As the UK operates under a large number of unwritten conventions, especially in relation to the monarchy and prerogative powers, it would be difficult to transfer them into written form.
- The lack of constitutional constraints allows executive government to be strong and decisive.
- A codified constitution would bring unelected judges into the political arena.

 Which of these two sides do you feel has the stronger argument? Why do you think it is stronger? What makes that side stronger than the other side?

Activity

Research the Conservative proposals for a British Bill of Rights from the 2017 manifesto. How would these rights be different from the Human Rights Act? Would this make the British Constitution stronger or weaker?

Overview

Arguably, the UK is in the process, particularly since 1997, of creating a semi-codified Constitution, bit by bit. Increasingly large parts of the Constitution are now both written and effectively entrenched. The aspects of the Constitution that conform to this analysis are as follows:

- The European Convention on Human Rights, brought into law by the **Human Rights Act 1998**, is effectively a bill of rights. Although it could be fully or partly repealed in the future, this seems politically unlikely.
- The **Devolution Acts 1998** codify the powers enjoyed by the Scottish Parliament and the Welsh, Northern Ireland and Greater London assemblies.
- The public's right to see public information is codified in the **Freedom of Information Act 2000**.
- The status and conduct of political parties are now codified in the **Political Parties, Elections and Referendums Act 2000**.
- Important constitutional changes are effectively entrenched by the fact that they have been approved by referendum and can therefore be repealed only by referendum.
- The Electoral Commission has created a codified set of rules for the conduct of elections and referendums.

So the old boast, or perhaps criticism, that the UK Constitution simply evolves and remains in the control of the Parliament of the day is out of date. Since 1997, when the New Labour government started the process of constitutional reform, the UK Constitution has gradually become increasingly written and formal. Yet it remains fundamentally an uncodified constitution in the strict sense of the words.

Constitutional reform will have to continue in the UK. Apart from the effects of leaving the EU, the UK must deal with the continuing pressure for the decentralisation of power to the regions and away from London. If the increasing support for alternative parties persists, there will be continuing pressure for electoral reform. It is also difficult to imagine how the bloated House of Lords (standing at 800 members as of September 2020), dominated by people who have received the reward of a peerage for political loyalty, can continue in its present form, especially after Boris Johnson's controversial appointment of 36 new peers in July 2020 met with extreme media criticism.

Summary

Having read this chapter, you should have knowledge and understanding of the following:
- → What political constitutions are, what their functions are and what forms they typically take, with examples to illustrate this knowledge
- → Why the UK Constitution is different from most other constitutions in the modern world
- → How the UK Constitution came to evolve into its current form
- → How the unique nature of the UK Constitution affects the nature of the political system
- → Why considerable reform of the UK Constitution began in 1997
- → What reforms have been successfully accomplished, what reforms remain incomplete and what reforms still need to be considered
- → The nature of devolution and the prospects for its further development
- → The main arguments that suggest the UK Constitution is unsatisfactory, contrasted against what are perceived to be its strengths
- → What codification of a constitution is and how codification might affect government and politics
- → The nature of the debate as to whether it is now time to codify the UK Constitution

Authoritative work A work written by an expert describing how a political system is run; it is not legally binding but is taken as a significant guide.

Common law Laws made by judges in cases where the law does not cover the issue or is unclear.

Constitution A set of rules determining where sovereignty lies in a political system, and establishing the relationship between the government and the governed.

Conventions Traditions not contained in law but influential in the operation of a political system.

Devolution The dispersal of power, but not sovereignty, within a political system.

Parliamentary sovereignty The principle that Parliament can make, amend or unmake any law, and cannot bind its successor nor be bound by its predecessors.

Rule of law The principle that all people and bodies, including government, must follow the law and can be held to account if they do not.

Statute law Law passed by Parliament.

Treaties Formal agreements with other countries, usually ratified by Parliament.

Uncodified/codified An uncodified constitution is not contained in a single written document, unlike a codified constitution which is written in a single authoritative document.

Unentrenched/entrenched An unentrenched constitution has no special procedure for amendment, unlike an entrenched one which requires separate rules and procedures for amendment.

Unitary/federal A unitary political system is one where all legal sovereignty is contained in a single place, unlike a federal system where legal sovereignty is shared between a national government and regional governments.

Further reading

Websites

The best website to use for further information on constitutional issues is the Constitution Unit of University College London: **www.ucl.ac.uk/constitution-unit**
This excellent article published by the British Library traces the historical development of the uncodified constitution: **www.bl.uk/magna-carta/articles/britains-unwritten-constitution**
The government's own site looks at constitutional reform proposals: **www.gov.uk/government/policies/constitutional-reform**
For information on reforming the Constitution, look at the site for Unlock Democracy: **www.unlockdemocracy.org**

Books

Brazier, R. (2008) *Constitutional Reform: Reshaping the British political system*, Oxford University Press — a review of the early stages of constitutional reform
Bromley, C. *et al.* (2006) *Has Devolution Delivered?*, Edinburgh University Press — a good book on devolution
King, A. (2010) *The British Constitution*, Oxford University Press — perhaps the best book on this subject, written in 2010
Mitchell, J. (2011) *Devolution in the UK*, Manchester University Press — another good book on devolution

Practice questions

1

Source 1

Constitutional reform since 1997 has had four major objectives. These have been:
- to improve democratic legitimacy and accountability
- to decentralise power away from central government
- to provide better protection for human rights
- to bring constitutional arrangements up to date.

The record of these reforms has been mixed. In some cases, power has been successfully decentralised, with devolution being the key example. Human rights are undoubtedly better protected since the passage of the Human Rights Act and the Freedom of Information Act. The Constitutional Reform Act of 2005 has also guaranteed the independence of the judiciary.

On the other hand, the picture on legitimacy and accountability is less clear. The failure to significantly reform the House of Lords and the electoral system is a serious omission. Furthermore, the nationalists in parts of the UK complain that devolution has not gone far enough.

Using the source, evaluate the view that constitutional reform since 1997 has been successful.

In your response you must:
- *compare and contrast different opinions in the source*
- *examine and debate these views in a balanced way*
- *analyse and evaluate only the information presented in the source.* (30)

2

Source 2

The old unity of the UK has gone, to be replaced by a federal system that means the experiences of different citizens can be vastly different. A Scottish person may shop for eight hours on a Sunday, while those in England and Wales get a mere six! The sick must continue to pay for prescriptions in England, while in Wales and Scotland they receive them for free. We see in laws and administration that the different regions of England are heading in different directions, with ministers warring with the national government, which also happens to be the government of England, over the issue of Brexit and the Covid-19 pandemic. The major cities are now run as separate entities from the rest of England. Brexiteers may claim that sovereignty has returned to Parliament from the EU, but it seems more likely that it has lost sovereignty in turn to the regional bodies. In such a system it is hard to see the UK as anything other than a federal system.

Using the source, evaluate the extent to which the UK is now effectively a federal system.

In your response you must:
- *compare and contrast different opinions in the source*
- *examine and debate these views in a balanced way*
- *analyse and evaluate only the information presented in the source.* (30)

3 Evaluate the extent to which the UK is in need of a codified constitution.
   In your answer you should draw on relevant knowledge and understanding of the study of Component 1: UK politics. You must consider this view and the alternative to this view in a balanced way.   (30)

4 Evaluate the extent to which the constitutional reforms of 1997 to 2010 may be considered more important than the constitutional reforms that occurred after 2010.
   In your answer you should draw on relevant knowledge and understanding of the study of Component 1: UK politics. You must consider this view and the alternative to this view in a balanced way.   (30)

5 Evaluate the extent to which the next logical step for devolution is the creation of an English Parliament.
   In your answer you should draw on relevant knowledge and understanding of the study of Component 1: UK politics. You must consider this view and the alternative to this view in a balanced way.   (30)

6 Evaluate the view that the UK's constitutional arrangements undermine democratic principles.
   In your answer you should draw on relevant knowledge and understanding of the study of Component 1: UK politics. You must consider this view and the alternative to this view in a balanced way.   (30)

Study tip

Coverage of Core and Non-Core Political Ideas are available in other textbooks to complete your study for Components 1 and 2.

6 Parliament

About a mile south of Charing Cross, the geographical centre of London, on the north bank of the River Thames, stands the Palace of Westminster, commonly known as the Houses of Parliament. It is an iconic place. Parliament stands at the very centre of the UK political system. The UK Parliament is important both because of its history and because it is the sovereign body of the United Kingdom. All national legislation must pass through Parliament and obtain its approval. All power stems from Parliament; it can grant powers to an individual or a body, and it can take them away. All members of the government of the UK must also be Members of Parliament, in either the Commons or the Lords. Between general elections, the government must make itself accountable to the UK Parliament. (Note at this stage that we are calling Westminster the *UK* Parliament to distinguish it from the Scottish and Welsh parliaments. The term 'Westminster' is another way of distinguishing the UK Parliament from any other.)

Government in the UK sits in Parliament and ministers must attend regularly to lead debate, justify policies and accept criticism where it is due. Yet government is also separate from Parliament. The two great institutions have different roles. Government formulates policies whereas Parliament debates those policies and passes its opinion on them; government drafts legislation whereas Parliament scrutinises that legislation, suggests changes and occasionally may veto it; government ministers and their departments run the day-to-day affairs of the country while Parliament seeks to ensure that they do this efficiently, give good value for taxpayers' money and govern fairly. The relationship between government and Parliament is crucial to an understanding of how the political system works.

The Palace of Westminster on the bank of the River Thames, the home of the UK Parliament

Objectives

In this chapter you will learn about the following:

→ The structure and membership of the House of Commons and the House of Lords
→ The respective roles played by the House of Commons and House of Lords and the key differences between them
→ How legislation is passed
→ The work of MPs and peers, including their role in select committees
→ The relationship between Parliament and the executive
→ The changes that impact the role and importance of Parliament
→ The importance of parliamentary privilege

The House of Commons and House of Lords

The structure of the UK Parliament

The UK **Parliament** is divided into two houses, the **House of Commons** and the **House of Lords**, of which the House of Commons has become the more significant due to the fact that it is elected.

A parliament with two houses, or chambers, is described as **bicameral**. Most democratic systems in the world are bicameral. The usual reason for this, and this applies to the UK, is that it creates some kind of balance in the political system. Very often the second chamber has a different kind of membership to ensure better representation and to prevent the first chamber having too much power. The UK system evolved naturally, however, so the position of the second chamber, the House of Lords, is somewhat unclear. In systems with codified constitutions it is much more obvious why the second chamber exists.

The structure of the House of Commons

The main features of the House of Commons structure are as follows:

- There are 650 Members of Parliament (MPs), each elected from a constituency. Constituencies are of roughly equal size, normally containing between 60,000 and 80,000 voters. Most constituencies are in England (533). There are 59 constituencies in Scotland, 40 in Wales and 18 in Northern Ireland.
- Nearly all MPs in the UK represent a political party. Independent (non-party) MPs have occasionally been elected, but this is rare.
- MPs are divided into frontbench MPs and backbench MPs.
- Frontbench MPs are more senior. In the governing party, they are ministers and party officials appointed by the Prime Minister. Normally there are about 90 frontbench MPs on the governing side. The leading members (spokespersons and shadow ministers) of the main **opposition** parties are also described as frontbench MPs. There would normally be about 50 of these. The total number of frontbench MPs is therefore 140–50.
- **Backbench** MPs are very much the majority. They can be more independent than frontbench MPs but are still expected to show party loyalty. Members of smaller parties are also considered to be backbenchers.

Key terms

Parliament The British legislature made up of the House of Commons, the House of Lords and the monarch.

House of Commons The primary chamber of the UK legislature, directly elected by voters.

House of Lords The second chamber of the UK legislature, not directly elected by voters.

Opposition MPs and Lords who are not members of the governing party or parties.

Backbenchers MPs or Lords who do not hold any government office.

Useful term

Bicameral A legislative body made up of two (bi) chambers (camera).

Study tip

Do not confuse ministers and MPs. While it is true that frontbench MPs who are members of the governing party are all ministers, this does not mean that MPs in general are ministers. Backbench MPs are *not* ministers, whichever party they belong to.

- MPs do much of their work in committees. The main types of committee are select committees and legislative committees. The nature and work of these committees are described below.
- All main parties appoint **party whips** who work under a chief whip. The whips are mainly concerned with ensuring that MPs in their parties are informed about parliamentary business. They also try to ensure party loyalty and to persuade reluctant MPs to support their party's line. Whips may inform their party leadership how MPs are feeling about an issue and may warn of possible rebellions and dissidence.
- The proceedings of the House of Commons are presided over by the Speaker. This must be an MP who is elected by all other MPs. Though the Speaker comes from one of the parties, they are expected to put aside their party allegiance when chairing the Commons. The Speaker (there are also deputies who sit temporarily) is expected to organise the business of Parliament along with the party leaderships, to maintain order and discipline in debates, to decide who gets to speak in debates or at question times, and to settle disputes about Parliament's work. In the Parliament elected in 2019, Sir Lindsay Hoyle was the Speaker.
- Figure 6.1 shows the make-up of the House of Commons according to party in 2019.

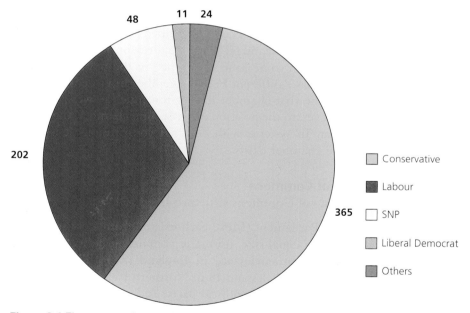

Figure 6.1 The party make-up of the House of Commons after the 2019 election

How to become an MP

It is highly challenging to be elected to the UK Parliament unless a person is adopted by a political party as its official 'prospective parliamentary candidate' (PPC). Therefore, to enter the House of Commons they must first be selected as a candidate by a party. Selection is carried out by local party constituency committees, which draw up a shortlist of proposed candidates and interview them, much as if they were applying for any job. Figure 6.2 shows the process of becoming an MP through a party.

Figure 6.2 **The process for becoming an MP (via a party)**

All proposed candidates will be party members, and many will have 'cut their teeth' in politics as a local government councillor. However, sometimes prominent national figures may be chosen because of their high profile. Labour leader Sir Keir Starmer is an example; he was Director of Public Prosecutions. It helps if the person lives in the constituency they are applying for, but this is not essential. The central party leadership will often try to influence local committees in their choice, but this does not always work. Having been selected as the party candidate, the person must then wait for the election and hope that they win their seat.

The structure of the House of Lords

The House of Lords is a curious body, largely because it has evolved gradually over history and because repeated attempts to reform it, to make it more democratic and rational, have failed. Its size is not regularised by law (currently comprising 800 members following the introduction of 36 new peers announced in August 2020), and the method of appointing its members remains dubious. Its structure looks like this:

- Ninety-two members are hereditary peers, people (nearly all men) who have inherited a title that entitles them to sit in the Lords. The number was determined in the **House of Lords Act 1999**. When a hereditary peer dies, their successor must be elected by all the remaining hereditary peers based on party affiliation. Although they are not professional politicians, hereditary peers in the Lords are expected to take their position seriously, attend and vote regularly, and take part in committee work.
- Twenty-six members are archbishops and bishops of the Church of England, known collectively as the 'Lords Spiritual'. This reflects the fact that Anglican Christianity is the established religion of the UK. Recently, however, leaders of other religions that flourish in the UK have also been appointed, though as **life peers** rather than as a permanent acknowledgement of the diverse religious beliefs held in the modern United Kingdom.
- The other members of the Lords, commonly known as life peers, are appointed. Technically, life peers are appointed by the reigning monarch, but this is one of the prerogative powers that has passed to the Prime Minister. Unlike hereditary peers, they cannot pass their title on to their children; it dies with them. Most life peers are nominated by the Prime Minister and the leaders of the other main

Useful term

Life peer A prominent member of society who is granted a peerage (becoming known as Lord or Lady, or Baron or Baroness). This entitles the holder to attend the House of Lords, take part in debates, and scrutinise and vote on legislation. The entitlement is for life unless they are convicted of serious misconduct.

parties. In other words, they are *political appointments* and this means they are expected to follow their party's line on most issues. There are also non-political peers appointed on the recommendation of non-government organisations and even by members of the public. There is a House of Lords Appointments Commission that decides which people shall be appointed (see Figure 6.4), and it can also veto unsuitable candidates nominated by party leaders.

- There is no firm constitutional principle concerning the balance of party members in the Lords. In general, there is a convention that parties can make nominations roughly in proportion to their strength in the House of Commons. Thus, since 2010, the Conservative Party has made more nominations than other parties. Before 2010 the Labour Party made more nominations than the others. But, as life peers are appointed for life, it can take many years to change the balance of party strengths in the House of Lords. Figure 6.3 shows the make-up of the House of Lords as it was at the beginning of 2021.

- The *political* make-up of the House of Lords is different from that of the Commons. In particular, it is always the case that the governing party does *not* have an overall majority of members. There are so many non-party members (as **crossbenchers** and non-affiliated) that there cannot be a government majority. In Figure 6.3, for example, we see that the Conservative Party had only 257 out of a total of 794 members in early 2021.

- There are frontbench spokespersons in the House of Lords just as there are in the Commons. The government must have representatives in the Lords as virtually all its business goes through both houses. Like their counterparts in the Commons, frontbench peers are expected to be especially loyal to their party leadership.

- The equivalent of the House of Commons Speaker is the Lord Speaker, a position held by Lord Fowler since September 2016.

- As in the Commons, much of the work of peers takes place in committees. There are legislative committees (in which all peers are allowed to participate) to consider proposed legislation and select committees. However, select committees in the House of Lords are much less significant than those in the Commons. The work and nature of legislative committees in the House of Lords are discussed below.

Useful term

Crossbencher A name given to members of the House of Lords who are not formal members of any political party and so are independent-minded. They organise as a group and are different from the non-affiliated members of the Lords.

Knowledge check

How many different types of Lords are there and how are the different types selected?

Synoptic link

The size and nature of the House of Lords is considered in relation to democracy in Chapter 1 and attempts at constitutional reform in Chapter 5.

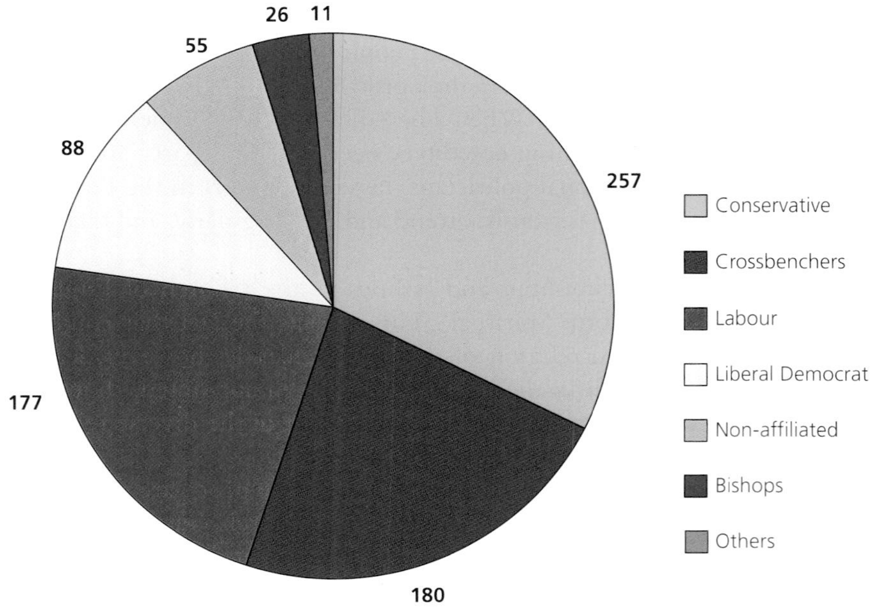

Figure 6.3 The party composition of the House of Lords at the beginning of 2021

Conservative — 257
Crossbenchers — 180
Labour — 177
Liberal Democrat — 88
Non-affiliated — 55
Bishops — 26
Others — 11

Nominated, either by current party leader, the House of Lords Appointments Commission or the public	Nominees are considered and vetted by the independent House of Lords Appointments Commission and the names are then passed to the Prime Minister	The Prime Minister considers the list and passes it to the monarch, who issues official Letters Patent	The newly created life peer presents their Letters Patent and is sworn in as a life peer in a short ceremony

Figure 6.4 The appointment process for becoming a life peer

Case study

How might the House of Lords be reformed?

While there is a strong consensus that there needs to be reform of the Lords, there is little agreement about what that reform should be. The debate almost wholly centres on the *composition* of a reformed chamber. There has been relatively little appetite for any significant change in the *powers* of the second chamber. This is for two reasons:

1   An increase in its powers would, it is feared, lead to an American-style situation where legislating could become too difficult. In the USA the two legislative chambers have similar status and powers. This makes legislation complex, long-winded and very often too difficult. Furthermore, a more powerful second chamber would simply duplicate the work of the current House of Commons, for no advantage.

2   If the second chamber had fewer powers than it currently exercises, the question would be asked, what is the point of it? So, the present powers of a second chamber are broadly supported.

A reformed second chamber would therefore continue to act as a safeguard against a government abusing its authority and becoming an elective dictatorship without sufficient check by the House of Commons. It would also remain as a revising and delaying chamber. Three main proposals have been supported:

1   An all-appointed second chamber
2   An all-elected second chamber
3   A combination of the two

There is a potential fourth idea, which is complete abolition of the second chamber. Although not common, New Zealand is a modern democracy that operates effectively with just one chamber.

The problems associated with the current House of Lords have been well recognised. They include the following:

● An unelected legislature is not democratic.
● Without elections members are not accountable for what they do.
● Occasionally the House of Lords thwarts the will of the government and House of Commons without democratic legitimacy.
● The appointment of life peers is open to abuse by party leaders, leading to charges of 'cronyism'. Party leaders use appointment to the Lords as a way of securing loyalty and

Synoptic link

Attempts to make the House of Lords more democratic are also elements of constitutional reform and proposals for future reforms link closely to attempts to improve democracy in the UK, discussed in Chapter 1.

→

rewarding their allies ('cronies'), which was a criticism levelled at Boris Johnson following his nomination of 36 new peers in 2020, including many Brexit supporters and his brother.

- Too many members of the current House of Lords are not active or are only semi-active.
- The chamber is too large.

Table 6.1 summarises the arguments for each of the three reform proposals.

Table 6.1 Arguments for a reformed second chamber

All-appointed	All-elected	Part-elected, part-appointed
People with special experience and expertise could be recruited into the legislative process	An elected second chamber would be wholly democratic	More democratic but with the advantages of appointed peers
The political make-up of an appointed body could be manipulated to act as a counterbalance to the government's House of Commons majority	If elected by some kind of proportional representation, it would prevent a government having too much power	PR elections would improve democratic representation of parties, while appointments could make Parliament more representative of the UK as a whole
Without the need to seek re-election members would be more independent-minded	Under PR, smaller parties and independent members would gain representation that they cannot win through FPTP in the House of Commons	It would create a chamber with independent peers and more accountability

There does remain the option of abolishing the Lords altogether. The supporters of this idea see the second chamber as a means of perpetuating privilege and spreading excessive patronage. They also point to its expense and to the fact that it has extremely limited democratic functions. But, despite the apparent logic of these reasons, abolition is unlikely to happen. Most democratic states are bicameral, with good reasons. There are always occasions when a special safeguard is needed against an alliance of government and a democratically elected legislature abusing their powers.

Discussion point

How should the House of Lords be reformed?

Three things to consider are:
1. The issues or problems with the current House of Lords.
2. The impacts the various alternatives would have on the House of Lords.
3. Whether the reforms would be better or worse than the current arrangements.

The functions and importance of the House of Commons and House of Lords

Parliament is made up of both the House of Commons and the House of Lords and both carry out the key functions of legislation, scrutiny and representation, but the ways in which they do this and their relative importance can vary. Here we consider the functions and how they relate to each chamber to help you draw a comparison between the two chambers.

Legislating: The Commons

This is the function of making, amending and passing laws. The process of passing laws is a formal set of procedures designed to ensure that legislation is acceptable to both houses and gives them an opportunity to suggest amendments. It is described in full later in this chapter.

Legislating is the UK Parliament's, and therefore the House of Commons', most important *constitutional* function. It involves the process of passing legislation and approving public finances. Most laws that are passed are created by the government of the day through the House of Commons and then, possibly, amended, before being passed and then progressing to the Lords. The process of considering proposed laws and then passing them ensures proposals from the government have democratic legitimacy and the support of the people, as expressed through the House of Commons. The people cannot be continually assembled to approve legislation or hold a referendum every time a new law is proposed, so the Commons does it for them. In this sense, therefore, the Commons is supporting government by granting it legitimacy for what it does. It strengthens government, rather than weakens it. While, officially, the House of Lords also has this function, by the **Salisbury Convention** the House of Lords will not block or oppose any terms contained within a winning party's manifesto, so the process of legitimising the government's laws is held mostly by the Commons.

> ### Synoptic link
>
> Legitimation is the key way in which representative democracy operates in the UK and is integral to understanding the role of Parliament as the key means of providing democratic legitimacy to the actions of government, as covered in Chapter 1.

An additional aspect of the legislation function of the Commons concerns public finances. Historically, monarchs usually sought the approval of the Commons when contemplating levying new taxes. This was essential to gain the **consent** of those who were to pay the taxes. Without it, tax collection would at best be difficult and at worst might end in armed revolt. The modern equivalent is that the House of Commons (this has not been a function of the House of Lords since the 1911 Parliament Act) must approve taxation and expenditure by the government every time a change is proposed. This process occurs every spring and summer after the Chancellor of the Exchequer has announced the annual Budget. It is extremely rare in modern times for the Commons to obstruct such proposals, but formal approval is always required.

While most of the legislative proposals are made by the government, there are occasions when backbench MPs develop their own legislation. This is called 'private members' legislation'. These private members bills can be introduced in one of three ways. Either through a **ballot**, where names of backbench MPs are drawn randomly on the second Thursday of a parliamentary session and allocated specific time to introduce and have the proposed bill discussed, or through the ten-minute rule. This is where MPs make a **ten minute** speech in support of a proposed bill (this is far less successful than the ballot method as time is so limited and is usually more about making a political point than a serious attempt to introduce a bill). Finally, they can be introduced through **presentation**, where the MP notifies the House of their intention to propose a new bill and then simply presents the title of the proposed bill to the House, but does not discuss it. An MP

Key term

Salisbury Convention The convention whereby the House of Lords does not delay or block legislation that was included in a government's manifesto.

Useful terms

Consent The idea that a proposed law or decision by the government is formally consented to by the people. This is vital in a democracy. However, in a parliamentary democracy, the elected Parliament can grant consent on behalf of the people.

Ballot In legislation, a means by which an MP may introduce a private members bill with the opportunity for full discussion and debate with allocated time.

Ten minute rule bills A means of introducing private members legislation by making a ten minute speech introducing the issue for consideration.

Presentation In legislation, a means of introducing private members bills by notifying the House and then formally 'presenting' the title of the proposed bill with no discussion or comment.

can present a bill to Parliament but their chances of seeing it through to law are small since the government has many opportunities to thwart such a procedure if it wants to. There are occasions when the government supports a Private Members' Bill, in which case it might pass. During the 2017–19 Parliament, a total of 15 Private Members' Bills achieved Royal Assent and became statute laws (14 via the Commons, one from the Lords), as outlined in Table 6.2.

Table 6.2 Successful Private Members' Bills, 2017–19

Final Act	Proposer
Assaults on Emergency Workers (Offences) Act 2018	Chris Bryant MP, Labour
Mental Health Units (Use of Force) Act 2018	Steve Reed MP, Labour
Homes (Fitness for Human Habitation) Act 2018	Karen Buck MP, Labour
Civil Partnerships, Marriages and Deaths (Registration etc.) Act 2019	Tim Loughton MP, Conservative
Organ Donation (Deemed Consent) Act 2019	Geoffrey Robinson MP, Labour
Parental Bereavement (Leave and Pay) Act 2018	Kevin Hollinrake MP, Labour
Parking (Code of Practice) Act 2019	Sir Greg Knight MP, Conservative
Prisons (Interference with Wireless Telegraphy) Act 2018	Maria Caulfield MP, Conservative
Stalking Protection Act 2019	Dr Sarah Wollaston MP, Conservative
European Union (Withdrawal) Act 2019	Yvette Cooper MP, Labour
European Union (Withdrawal) (No. 2) Act 2019	Hilary Benn MP, Labour
Animal Welfare (Service Animals) Act 2019	Sir Oliver Heald QC MP, Conservative
Health and Social Care (National Data Guardian) Act 2018	Peter Bone MP, Conservative
Lords Children Act 1989 (Amendment) (Female Genital Mutilation) Act 2019	Lord Berkeley of Knighton, Crossbench
Holocaust (Return of Cultural Objects) (Amendment) Act 2019	Theresa Villiers MP, Conservative

Activity

Research any three of the successful Private Members' Bills listed in Table 6.2. Why were they introduced? Why were they able to succeed? Are there any common trends across these Private Members' Bills?

Case study

The Voyeurism (Offences) Act, 2019

The Sexual Offences Act, 2003, had been passed before the development of the widespread use of camera phones and sharing of images online. Although the Act had criminalised some forms of voyeurism, it did not cover the new crime of 'upskirting' which became a major problem and intrusion in the 2010s. As such, it was difficult to prosecute the crime except under the common law provision of 'outraging public decency'.

In 2017, Gina Martin (a victim of upskirting) began a campaign to have the practice criminalized, gaining 58,000 signatures on an online petition, the support of the Labour Party and the then chair of the Women and Equalities Select Committee Conservative MP Maria Miller, the End Violence Against Women Coalition and others.

In response to this campaign, in March 2018, Liberal Democrat MP Wera Hobhouse introduced a Private Member's Bill entitled 'The Voyeurism (Offences) Bill' which would have banned the

→

practice of upskirting for the purposes of 'humiliating, alarming or distressing' the victim. The first reading was passed without objection on 6 March and Justice Secretary David Gauke and Prime Minister Theresa May indicated that the government would support the bill, giving it a higher chance of passage.

At the second reading, on 15 June (during a backbench business day) Conservative MP Sir Christopher Chope, called out 'oppose' which automatically sent the proposed bill for further debate, which would not occur until the next backbench business day on 6 July, when a similar objection by one MP would delay it further. Chope was widely criticised by members of all parties, though he claimed to be doing it on the principle that a government-backed bill should not be given time during backbench business days.

Following the public outcry from Chope's opposition, the government re-introduced it to the House of Commons on 21 June 2018 as a government bill (which cannot be delayed by one MP shouting oppose) and was accordingly given government time in the House of Commons. Being a narrow and uncontroversial proposal with cross-party support, the government bill passed through all stages in the Commons and Lords very quickly, gaining Royal Assent to become the Voyeurism (Offences) Act in February 2019, coming into force in April 2019.

In this way you can see the particular difficulties of Private Member's Bills becoming law, but how they can also serve to influence the government into taking action.

Legislating: The Lords

The Lords does not really legitimise legislation. While it is true that any legislative bill must be passed by the Lords before becoming law, this does not mean the Lords is granting consent as the Commons does. It does, however, give the Lords the opportunity to scrutinise proposed legislation, to give its opinion, possibly to ask the government and Commons to think again, and possibly to amend the proposals to improve them.

The Lords does have the power to delay a piece of legislation for up to one year. In effect, when it does this, the Lords is saying to government, 'Think again. We know we do not have the power to stop it but we have serious reservations about your proposal and want you to reconsider.' If the Lords does insist on trying to delay and prevent legislation that the government wishes to be passed, the Commons can vote to bypass the Lords and pass legislation without approval from the Lords after one year's delay. It rarely reaches this stage, but it has happened on four key occasions:

- **The War Powers Act 1991** allowing the UK government to prosecute war criminals even if the offences were committed outside the UK
- **European Parliamentary Elections Act 1999** establishing a new closed–list system for elections to the European Parliament
- **Sexual Offences Amendment Act 2000** lowering the age of consent for gay men to 16
- **The Hunting Act 2004** banning fox hunting with packs of hounds

In each case the Lords performed its function of asking the government and House of Commons to reconsider, but in each case it was the Commons and the government that prevailed.

Useful terms

Secondary legislation
Laws, regulations and orders mostly made by government ministers. They require parliamentary approval but do *not* have to pass through the full procedure. Most secondary legislation is not discussed in Parliament, but passes through automatically or 'on the nod'. Only the occasional controversial piece of secondary legislation provokes debate and a vote in Parliament. The House of Lords pays more attention to secondary legislation than the Commons does.

Delegated legislation
Laws and regulations made by ministers and other public bodies under powers granted by Parliament. On the whole they do not require parliamentary approval, though the UK Parliament reserves the right to review them.

Scrutinising secondary legislation

A great deal of (indeed, most) legislation emerging from government is actually **secondary legislation**. Secondary legislation, sometimes called **delegated legislation**, refers to any law-making or change to the law being made by any member of government that does not need to pass through normal parliamentary procedures. These are detailed aspects of law and regulations that ministers can make because they have been previously granted the power to do so according to a parliamentary statute. In other words, Parliament has *delegated* powers to ministers.

Another way of looking at this is to say that primary legislation means important laws that need parliamentary approval, while secondary legislation refers to more minor or specialised pieces of law that do not need to follow full parliamentary procedure. For example, primary legislation grants power to the secretary for transport to set speed limits on the roads, while secondary legislation concerns the minister actually setting those limits. Most such legislation takes the form of *statutory instruments*, sometimes referred to as Henry VIII clauses. These raised tremendous concerns in the proposed Brexit measures by Theresa May in 2018 and 2019 because her proposed legislation included a number of statutory instruments (more than 600) that would have allowed the government to make many decisions about Brexit, the return of powers and the rights of people without having to consult Parliament or be scrutinised. In March 2019, the government was trying to force through 500 such measures without scrutiny and proper checks, even threatening to remove the power of the Lords to vote against statutory instruments, thereby removing an important check on government power. Had the government been successful, it would, effectively, have given the executive the power to do anything it wanted in terms of dealing with the consequences of Brexit, which was one reason for the opposition to such proposals.

Statutory instruments are increasingly being used to make law. This is a matter of concern as most statutory instruments are not considered by Parliament, so there is little or no scrutiny and few checks on their use or quality. The House of Lords, however, does have more time and expertise to consider such legislation. The House of Lords Secondary Legislation Scrutiny Committee considers all secondary legislation and decides what proposals might cause concern. Where concern is expressed, the matter is brought to the attention of the whole house and from there referred to the Commons. Such referrals are rare, but they do provide an important discipline on government. Members of the Lords also share in the work of the Joint Committee on Statutory Instruments, which checks secondary legislation for errors in wording and meaning. This role is one where the Lords can claim to be more important than the Commons.

A key example of the House of Lords performing this function occurred in October 2015, when the Lords voted against a piece of secondary legislation that would have reduced the level of tax credits paid to low-income families. This action forced the government to amend the legislation until it was acceptable to peers.

Scrutiny of government: The Commons

Scrutiny is probably the most important *political* function of the UK Parliament, especially the House of Commons. As with consent, the government cannot be continuously accountable to the people. That only occurs at general elections. Instead, it is the Commons that calls government to account on a regular basis. This role has a number of aspects:

- It can take the form of criticising the government. This can occur on any parliamentary occasion, but usually happens during the sessions devoted to questions to ministers (every Monday and Thursday for an hour, based on a rota for 'Oral Questions') or Prime Minister's Question Time (PMQT) every Wednesday for half an hour.
- It can simply refer to the idea of forcing the government to justify its policies and decisions. If a minister knows they must face the Commons, they will be careful to prepare a good case for what they propose to do or what they have just done.
- Largely through the departmental select committees and the Public Accounts Committee (see below), members of the Commons have opportunities to investigate the *quality* of government, in other words how well we are governed, whether taxpayers' money is being well spent, whether government is efficient and rational, and whether policies have been well investigated. These committees are often critical of government and sometimes recommend alternative courses of action.
- Although the Commons is generally considered to be dominated by government, it can refuse to pass a piece of legislation. This occurred, for example, in April 2016, when the Commons voted against a new law extending legal opening hours for large stores on Sundays, much to the consternation of ministers. Voting against government legislation rarely occurs, but the mere threat that it *might* happen is usually enough to force government to think again. This occurred in March 2016 when the government withdrew a proposal to reduce entitlement to disability benefits in the face of widespread opposition from MPs. When this happens repeatedly, the government is weakened and made to appear ineffective, which can force a prime minister from power, as happened in 2019 with the pressure on Theresa May after repeated failures to pass proposals for Brexit through Parliament. To put it into context, between 2017 and 2019 May suffered 33 defeats, compared with only seven for Cameron in the 5 years of coalition and three for the period with a slim majority between 2015 and 2017.
- On extremely rare occasions, the Commons can remove a government by passing a vote of no confidence (see below). This dramatic action last occurred successfully in 1979 when the Labour government of the day was ousted, though Theresa May did survive a vote of no confidence in 2019.

In addition, the Commons must scrutinise the legislation proposed by the government (so this is both part of the legislative and the scrutiny functions of Parliament). All backbench MPs are required to serve on legislative committees (also known as Public Bill Committees). These committees examine proposed legislation (i.e. bills), often examining every line, to see whether it can be improved and whether additions or amendments can be made to protect the interests of minorities. It does not mean that Commons committees have the power to reject proposed legislation altogether. Only the Commons as a whole can do this.

The legislative committees are a weak aspect of the work of scrutiny in the House of Commons. This is largely because they are dominated by the government and its whips. Legislative committees rarely amend a piece of legislation without the approval of government. This is not to say it is an illusory function. There are occasions when proposals by groups of MPs are accepted by government.

Scrutiny of government: The Lords

The role of the Lords in carrying out scrutiny of the government is limited because nearly all senior ministers, especially Cabinet ministers, sit in the House of Commons. Government ministers do sit in the Lords to represent the government's position, but the Lords lacks the means and methods for really scrutinising the actions, beyond asking questions of this small number of junior ministers.

When it comes to scrutinising legislation though, the Lords does have some advantages. There are many members of the House of Lords who are experts in their field and who represent important interests and causes in society. When scrutinising legislative proposals, therefore, they have a great deal to offer.

Apart from general debates, the main way in which the Lords carries out scrutiny is through what is known as the 'committee stage' of a bill. At this stage, any peers may take part in debating the details of proposed legislation and may table or propose amendments. This is possibly the key role of the House of Lords. The committee stage often improves legislation, adds clauses that protect vulnerable minorities, clarifies meaning and removes sections that will not operate effectively. Occasionally the Lords may amend a bill so severely that the government is forced to drop it altogether.

Representation

Representation is the final key function of Parliament, but it can take very different forms, and debates often surround which type of representation is being considered; whether it is representation of constituency, groups, the national interest or social representation. It is worth remembering that all are valid forms of representation and an MP, or peer, must make a decision about which type of representation they will choose to be ruled by.

Constituency representation: The Commons

It is widely acknowledged as a great strength of the UK political system that every MP represents the interests of their constituency. This is a neutral, non-partisan role in that an MP is expected to take care of the interests of *all* constituents, no matter for whom they voted. It may involve lobbying a minister whose department is proposing something that is unpopular in the constituency, or it might involve raising the matter on the floor of the House of Commons where it will receive considerable publicity. It might even involve joining a local campaign of some kind.

Sometimes the interests of a constituency may run counter to government policy. This presents a dilemma for MPs from the governing party. What are they to do if a government policy may cause strong dissent in their constituency? This has occurred, for example, with the fracking debate. The Conservative government supports fracking but many constituencies with a Conservative MP representing them feel threatened by fracking. Usually MPs abandon their party loyalty on such occasions and lobby for their constituency. The party whips do not like this but usually allow an MP to put constituency before party allegiance. Similar problems have arisen for Conservative MPs in the Thames Valley over the proposed expansion of Heathrow Airport and across rural England in dealing with the development of HS2.

Knowledge check

Identify three ways in which Parliament carries out its role of scrutiny.

It also happens that individual constituents approach their MP for help if they are in dispute with a public body, such as HMRC over tax or the Department for Work and Pensions over welfare payments. Indeed, constituency work of this kind takes up much of the average MP's time. Most MPs hold regular 'surgeries' when constituents can bring their problems to the MP's attention. If MPs feel their constituents have a good case, they will try to put things right on their behalf. This function is often described as the redress of grievances.

Constituency representation: The Lords

The House of Lords, meanwhile, in theory only represents the interests of its own members; as such, it has no formal constituency to represent.

The representation of groups: The Commons

MPs do not only represent the narrow concerns of their constituents; they often also pursue the interests of a section of society or a particular cause. This is often the result of their background before they became MPs. For example, members of trade unions will tend to support their former fellow workers, while former business leaders will support their former industry. All pressure groups try to recruit MPs to their cause as it gives them exposure in Parliament. Organisations such as the Countryside Alliance, Friends of the Earth and Age UK enjoy the support of groups of MPs. Furthermore, increasingly, campaign groups encourage supporters to write to MPs in large numbers to try to further their cause. Modern examples of this concern opposition to such issues as fracking, HS2 (the building of a high-speed rail line from London to the Midlands and North), Heathrow expansion and banking regulation.

MPs have also formed themselves into groups to pursue a particular interest or cause by creating cross-party groups. Among these have been all-party parliamentary groups on these subjects:

- Ageing and older people
- Betting and gambling
- Counter-extremism
- Islamophobia
- Motor neurone disease
- Race and community
- Sex equality

These groups transcend party allegiance and seek to exert collective pressure on government over key issues. They have varying degrees of success and while they may choose to represent a group, they are not necessarily elected by these groups.

Of course, the most important groups represented in the Commons are the political parties. Due to the FPTP electoral system and the two-party nature of UK politics, the Commons does not accurately represent the proportion of votes the various parties win, as illustrated by the distortion of party representation in the 2019 general election:

- The Conservatives won 43.6 per cent of the votes but 56.1 per cent of the seats.
- The Liberal Democrats won 11.5 per cent of the votes, an increase of 4.5 per cent from 2017, yet went from 12 to 11 seats (immediately after the election).
- The SNP won 3.9 per cent of the total number of votes yet secured 48 seats.

Nevertheless, the Parliament elected in November 2019 was much more representative of party support among the voters than the 2015 general election, and Conservative gains had more to do with the collapse in Labour support from 2017 to 2019.

Synoptic link

When discussing the nature and role of the House of Lords, reconsider the discussion of representative democracy in Chapter 1.

Activity

Find out who your local MP is. Look at their website. Identify three local issues they have been concerned with.

The representation of groups: The Lords

In some ways, this form of representation is better carried out by the Lords than the Commons. Since the removal of most of the (mainly) Conservative hereditary peers in 2000, no single party has an overall majority, allowing for a greater range of opinions and views to be represented. As the Lords do not need to concern themselves with the needs of a constituency or getting re-elected, they can focus on issues that affect a group from across the whole of society. In addition, since many of those sitting in the House of Lords come from non-political backgrounds, they represent a wider range of experience than the House of Commons.

Furthermore, the Lords allows a much larger range of political opinion to be represented than the Commons, partly because the power of the whips is much weaker, so party control is much weaker, but also because peers can represent small parties that find it difficult to win seats in the Commons. UKIP actually had five peers in the Lords until 2019, when they chose to sit as independents. In addition, a number of non-affiliated and crossbench peers means much greater freedom to speak about more marginal interests and to represent a wider range of groups.

National debate: The Commons

From time to time a great national issue arises that stands above party politics. Often it is an issue that concerns foreign policy and the use of the armed forces, but it has also involved the signing of foreign treaties. On major issues like an armed conflict or a time of national crisis such as the Covid-19 pandemic, it is Parliament (both houses) that is called on to debate the issue and to express the *national* will. Here, Parliament is often seen at its best, when party allegiances are set aside, when powerful speeches are heard, and when the representatives of the people can be heard above the noise of party conflict.

Yvette Cooper, former Labour frontbencher and current Chair of the Home Affairs Select Committee, asking a question during a Commons debate on Brexit

National debate: The Lords

Like the House of Commons, the Lords occasionally holds debates on important national issues. The Lords tends to specialise in issues that have a moral or ethical dimension. In recent years, therefore, the Lords has held debates on such issues as assisted suicide, control of pornography, treatment of asylum seekers and refugees, stem cell research and the use of genetically modified (GM) crops. Such debates do not normally result in decisions but they help to inform decision-makers, especially as the Lords contains so many experts in these fields.

Social representation: The Commons

In a democracy, there is an argument that the legislative body that represents the people should reflect the people socially, meaning it should look like the people, to ensure that all sections of society are effectively represented in proportion to their size in society. There are many social elements that can be considered, but here we will confine ourselves to age, gender, race and ethnic origin, LGBT+ representation, and whether or not they were privately educated.

Figure 6.5 shows the composition of the Commons following the 2019 election. Although this is the most socially diverse Parliament elected in history, its domination by male, white MPs aged over 50, as well as by a disproportionately large number of privately educated MPs, suggests the House of Commons is far from socially representative, despite advances in recent years.

Synoptic link

You should consider the importance and nature of social representation in the House of Commons with the issues relating to social voting behaviour covered in Chapter 4.

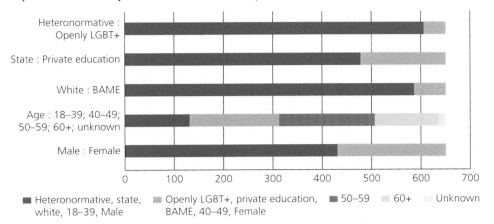

Figure 6.5 The social make-up of the House of Commons after the 2019 general election

Social representation: The Lords

The House of Lords suffers even more dramatic issues with representation, as shown in Figure 6.6. More than half the peers are over 70, perhaps not a surprise given the nature of life peerages, but still far from socially representative, while 62 per cent of peers were privately educated, compared with 6.5 per cent of people in the UK as a whole. The population of the UK is 80.5 per cent white and 19.5 per cent BAME, so as with the Commons, minority representation and diversity are lacking in the Lords, which undermines the argument that the Lords, not being subject to electoral pressures, can be made to be a more representative body than the Commons; in theory this may be true, but in practice this is far from a reality.

Knowledge check

How diverse are the Commons and Lords in terms of gender, race and sexuality?

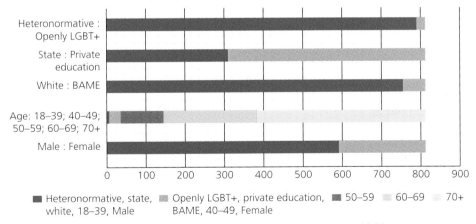

Figure 6.6 The social make-up of the House of Lords as of November 2019

Discussion point

Should Parliament be representative of the nation as a whole?

You may wish to consider the following points to help you in your discussion:
1 Why would being representative be beneficial?
2 What is the main purpose of representation in Parliament?
3 How could Parliament be made to be more representative of the nation and what would be the consequences of this?

Synoptic link

Consideration of how representative Parliament is should be undertaken in conjunction with the wider issue of representative democracy, discussed in Chapter 1.

Comparing the powers of the House of Commons and House of Lords

Comparing formal powers and functions

Here we examine the differing powers and functions of the two houses of Parliament. Table 6.3 shows which powers are shared by both houses, which are exclusively held by the House of Commons and which roles are largely reserved to the House of Lords. The House of Lords has no exclusive powers of its own.

Table 6.3 The respective powers and functions of the Houses of Commons and Lords

Powers and functions of the Commons only	Shared powers of the two chambers	Powers and functions largely of the Lords only
Examination and approval of the financial affairs of the government	Debating legislation and voting on legislative proposals	Examining secondary legislation and making recommendations for further consideration
Complete veto of legislation in certain circumstances	Proposing amendments to legislation	Delaying primary legislation for up to one year
Dismissal of a government by a vote of no confidence	Calling government and individual ministers to account	
Select committee examination of the work of government departments	Debating key issues of the day	
Final approval for amendments to legislation	Private members may introduce legislation of their own	

We can see from Table 6.3 that a large proportion of the powers and functions of Parliament are *shared* powers. Both houses appear to be doing the same thing. Indeed, the legislative process in each house is almost identical. It can be argued that the House of Lords examines legislation more thoroughly because all peers can be involved and because the party whips are less involved in forcing members to vote one way or the other. Otherwise, though, the same procedures apply. However, key differences exist between the two, with the Commons being the more dominant chamber.

The Parliament Acts

The powers of the House of Lords were significantly reduced by the Parliament Acts of 1911 and 1949. Before 1911, the Lords could block or effectively veto legislation indefinitely, but as a result of the 1911 Act, this power was removed and the Lords could only delay primary legislation for up to 2 years (reduced to 1 year in 1949). Since then, the Lords has lost its power to veto legislation and can only revise it, proposing amendments to the Commons that may then be rejected. If the Lords does insist on blocking legislation, the Parliament Act can be invoked and the Commons can bypass the Lords by delaying the vote for one year; this has been used to pass four pieces of legislation, as noted above.

Financial privilege

As a result of the Parliament Acts, the Lords has also lost its power to delay or amend money bills, or 'supply bills', that deal with taxation and the raising of money by the government. This is linked to the idea of democracy as the Lords does not answer to the taxpayers, therefore should not be able to demand the raising of taxation. Any bill that is designated as a money bill by the Speaker of the Commons must be passed without amendment by the Lords within one month, or it can receive royal assent without the support of the Lords. In this sense, the role of the Lords in passing money bills and the Budget is purely ceremonial and serves no practical purpose.

Confidence and supply

In order to function, the government needs the support of the House of Commons, by being granted '**confidence and supply**'. The supply side refers to the willingness of the House of Commons to authorise the necessary taxation and borrowing to allow the government to operate, while the confidence side refers to the Commons' faith in the government's ability to operate and get things done. Usually this results from the government having the support of a majority of MPs through its party, however a minority government may need to rely on another party granting it a confidence-and-supply agreement by pledging to support the government in passing key measures, like the Budget, but not joining in a formal coalition, as occurred with the DUP support for Theresa May's government between 2017 and 2019. In this situation, a three page agreement was signed between the Conservative Party and the Democratic Unionist Party of Northern Ireland, where in exchange for the 10 DUP MPs supporting key pieces of government legislation and supporting the government in passing a budget and in any votes of confidence, the government would provide an additional £1 billion in spending for the Northern Ireland Assembly, as well as supporting some DUP priorities, such as guaranteeing pensions would continue to rise at a minimum of 2.5 per cent each year. This agreement gave Theresa May's government a working majority of 328, but only on those issues agreed to, while the DUP MPs would continue to sit in opposition.

If a government fails to pass a 'supply measure' (meaning a Budget or taxation measure), or is defeated in a vote of no confidence, all members must resign from government. Before 2011, this would automatically lead to a general election, but the rules of the Fixed Term Parliament Act mean the existing House of Commons has 14 days to form a new government before an election is triggered. Such acts are rare, with the last government to lose a vote of no confidence being that of Jim Callaghan in 1979, though Theresa May did survive one in 2019 thanks to the confidence-and-supply agreement with the DUP.

The House of Lords has no such power and no role in the supply and confidence, meaning it has no role or power to remove a government from power, making it far less effective at holding the government to account.

Reasonable time convention

Although the Parliament Acts allow the Lords to delay legislation for up to a year, there has been a concern that the Lords could abuse this privilege and routinely take a year on all matters, which would hamper the effectiveness of government. As such, a reasonable time convention has emerged, which requires that the Lords consider government legislation in reasonable time and that it should aim to vote on it by the end of a parliamentary session. Although it is vague and open to interpretation, it is another way in which the ability of the Lords to scrutinise the government is restricted.

Despite the higher status of the House of Commons, it has to be said that the House of Lords did recover some of its former authority after the 1999 House of Lords Act. Since then it has been more active and also more willing to challenge the Commons, mainly because it has acquired greater legitimacy with the removal of most of the hereditary peers and because, since 2010, issues around winning party mandates have been less clear-cut. In cases where the government has a large majority in the

> **Key term**
>
> **Confidence and supply**
> The right to remove the government and to grant or withhold funding. Also used to describe a type of informal coalition agreement where the minority partner agrees to provide these things in exchange for policy concessions.

> **Knowledge check**
>
> Identify three key differences between the House of Commons and the House of Lords.

Commons, the only effective parliamentary scrutiny can sometimes come from the Lords, to avoid elective dictatorship, as the Commons will find itself simply doing as the government demands as the party in control has a large majority who will be loyal to the government. The Lords is free from such party control and can therefore provide effective scrutiny in a way the Commons cannot.

Legislation

The nature of legislation

When a law is introduced into Parliament it is described as a **legislative bill**, until it is finally passed and signed into law by the monarch, when it transforms into an official Act of Parliament. Legislative bills may be broken into three different types, as follows:

1 **Private Bills** If an organisation, for example a local authority or a church, wishes to take some action that the law currently forbids it from doing, it can apply for a Private Bill to be passed by Parliament to allow it to go ahead. Often this concerns the building of roads or bridges or various new uses of land. The same process may also allow organisations to make compulsory purchases of land or buildings for a building project. Private Bills are not normally considered by either house as a whole but are considered by committees of one house or the other. Members of the public and other interested parties may give evidence to these committees or may present petitions. It is rare for such legislation to attract any publicity and the bills usually concern only private interests.

2 **Private Members' Bills** As the name suggests, these are presented by individual or groups of MPs or peers. At the start of the year, members who wish to present such a bill enter their names in a ballot. Twenty MPs will be drawn from the ballot, but given how few days backbench MPs have control over the Commons and the amount of time available in a parliamentary session for them to discuss such bills, it is usally only the first seven to be drawn that are actually introduced. They are guaranteed at least one reading. Such bills have virtually no chance of being turned into law. This is either because it is difficult to persuade enough MPs or peers to turn up for a debate and **division** (a 'quorum' or minimum number is needed if the bill is to progress) or because it is opposed by the government. However, if a bill attracts the attention of ministers and seems to be desirable, it may receive government support. If it does, it will pass through the same procedures as a Public (or government) Bill, as shown below. MPs and peers know that their bills are unlikely to progress, but they use them to bring an issue to the attention of government in the hope that ministers might take it up later. Even if a Private Members' Bill does gain ministerial support, this can cause it problems. During the second reading of a Private Members' Bill, any MP can block it by simply shouting 'oppose'. This is something MP Sir Chistopher Chope did in 2018 when he shouted oppose in relation to two such bills with government support, one which would have banned 'upskirting' and another to give additional legal protections to police animals. In such a case, the bill has to be re-introduced on a future Friday Private Members Bills session, where it could be blocked again by 1 MP shouting 'oppose'. The power of one MP to do this is unique to Private Members' Bills and is not an issue that Public Bills have to deal with, highlighting once again the difficulties for Private Members' Bills in becoming law.

3 **Public Bills** Most bills fall into this category. They are presented by government and are expected to be passed without too much obstruction. Up to a year before they are drafted and announced, they are normally preceded by a **White Paper**

that summarises the proposal. At White Paper stage, a debate is held and a vote taken. Any potential problems are identified at this stage and, very occasionally, bills may be dropped if Parliament has serious concerns. Assuming all goes well, they follow the process shown below.

The legislative process

It is not necessary to grasp all the features of how laws are passed in the UK, but a basic knowledge is useful as it illustrates some of Parliament's roles. All Public Bills and Private Members' Bills must follow the procedure shown in Figure 6.7 to become Acts of Parliament.

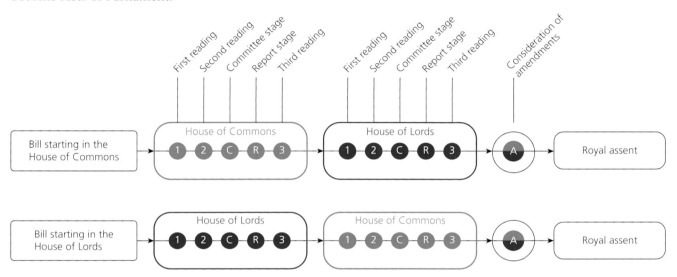

Figure 6.7 How a bill passes into law in Parliament

Source: www.parliament.uk/about/how/laws/passage-bill

- **First reading** A prepared bill is formally introduced to the chamber, either by a government minister if it is a Public Bill, or a private member if it is a Private Members' Bill. There is no debate or vote and this stage simply enters it into the legislative process.
- **Second reading** At this stage, the proposer of the bill must present more details and be subject to questions and debate about the nature and content of the bill. A vote is taken and Private Members' Bills will often go no further, but Public Bills are rarely defeated; only twice since 1945 has this happened at the second reading stage.
- **Committee stage** After the second reading, the bill passes to a **Public Bill Committee**, made up of between 16 and 50 MPs. Working in this smaller group, the MPs debate and consider each clause of the bill and suggest amendments to make the meaning clearer. They consider various aspects, including expert witnesses and reports. This is where the fine detail and meaning of the law are written. The make-up of Public Bill Committees is in line with party strength in the Commons, with members chosen and instructed on what to do by the whips, so the government has a tight control on this process. Finance bills and major constitutional bills are considered in a Committee of the Whole House as these are seen as too important to delegate to a smaller group.
- **Report stage** The Public Bill Committee reports back to the House of Commons and the whole chamber votes on the proposed amendments, which

Key term

Public Bill Committee
Committee responsible for looking at bills in detail.

may be accepted, rejected or altered. Other MPs may also introduce proposed amendments here. Once each amendment has been considered and voted upon, the final text of the bill is ready for final approval. If a bill has been determined to be an English only, or English and Welsh only, bill by the Speaker as part of EVEL, it proceeds to a vote by only those MPs eligible to vote on it, after which, like any other bill, it proceeds (if passed) to the third reading.

- **Third reading** The final version of the bill is presented and debated in the House of Commons, followed by a vote by the whole chamber. If successful, it passes to the House of Lords; if unsuccessful, it may return to the committee stage or be dropped entirely.
- **The House of Lords stage** The bill is introduced to the House of Lords as a first reading and then continues through the whole process in the same way as it did in the Commons, though the Lords tends to consider amendments as a Committee of the Whole House. If the Lords suggests any amendments, these must then be passed back to the Commons for approval or rejection and can be sent back and forth between the two chambers in a process often described as 'parliamentary ping-pong'. Once all variations have been resolved, the Lords votes and, usually, approves the final text.
- **Royal assent** Once both chambers of Parliament have passed the bill, it is sent to the monarch, who gives it royal assent by signing it into law. Although a monarch could refuse to sign it, and thus effectively veto it, this has not happened since the reign of Queen Anne in 1708. Once the monarch has signed the bill, it becomes an official Act of Parliament and part of the law of the land.

Knowledge check

What happens at each of the three reading stages of legislation?

The interaction between the House of Commons and House of Lords

Most Public Bills start life in the House of Commons, although occasionally they may be introduced in the Lords, especially if the Commons is very busy. Whichever way they begin, they must go through both houses if they are to become law. Normally the Lords follows the lead of the Commons and passes a bill without obstruction, thanks largely to the Salisbury Convention (discussed above). The only occasion when the House of Lords may be obstructive is if the government introduces a piece of legislation that was not stated in its manifesto or part of its electoral mandate.

Nevertheless, in recent years the Salisbury Convention has become less important. In 2010–15 the government did not enjoy an electoral mandate as it was a coalition of two parties, while between 2017 and 2019 the Conservative Party ruled as a minority government, meaning the Salisbury Convention once again did not apply. Between 2015 and 2017 Liberal Democrat peers suspended the Salisbury Convention, largely because the government was elected on such a small popular vote. As a result, the Lords passed many amendments to legislation in defiance of the government.

If the House of Lords proposes an amendment to legislation, it must go back to the Commons, where it can be overturned. However, if the Lords digs its heels in and insists on an amendment, it may force the government and the Commons majority into submission. This occurred in 2015 when a government proposal to cut the payments of tax credits was rejected in the Lords and the government had to concede defeat.

As we have seen, the House of Lords can delay legislation for a year by forcing the House of Commons to pass the same legislation in two consecutive sessions. This rarely results in a government defeat, but it does force government and the House of Commons to think carefully before introducing controversial legislation.

How Parliament interacts with the executive

The importance of parliamentary privilege

The term '**parliamentary privilege**' refers to the special protection that MPs and peers have when they are engaged in parliamentary business. It means that MPs and peers cannot be prosecuted or sued for libel or slander for any actions taken within the Palace of Westminster. In effect this means complete freedom of speech. Representatives may say anything they wish without fear of being arrested or of being sued for defamation. It dates back to the time, before the seventeenth century, when Members of Parliament feared being arrested if their words were seen to be threatening to the monarch or to the state. Since the seventeenth century members have had no such fear. This is a vital principle as it means that members can more effectively call government and ministers to account; they do not feel at all constrained. This is because:

- The fact that the government and the external legal system cannot interfere in Parliament means that members can feel secure and free to criticise government and other agencies of the state.
- The conduct of members is now regulated by the parliamentary commissioner for standards. Conduct is also regulated by committees on conduct and the Speakers of both houses.

This does not mean that members may act exactly as they please. Those who are considered to have abused parliamentary privilege may be disciplined for using provocative language, though this is a fairly weak check. For example, in 2011 an issue arose over the use of super-injunctions, which were legal bans on the press (or anyone else for that matter) reporting on a private issue or even being able to name the person who had requested such a protection from the courts. This made it a criminal offence to make any reference to the issue or in any way to identify the people who had taken out the super-injunction (normally wealthy celebrities, as super-injunctions typically incur high legal fees). However, in May 2011, shortly after the High Court had upheld the legal authority of a super-injunction, MP John Hemming rose to ask the following question:

> With about 75,000 people having named Ryan Giggs on Twitter it is obviously impracticable to imprison them all and with reports that Giles Coren also faces imprisonment... The question is what the Government's view is on the enforceability of a law which clearly does not have public consent?

The BBC and Sky News could then report on what Hemming had said in Parliament, effectively voiding the super-injunction. Had Hemming asked this question anywhere in the UK apart from Parliament, he would have broken the law, been in contempt of court and faced legal consequences. As it was, the then Speaker, John Bercow, did call him to order and remind him of the purpose of the debate being held, but he allowed Hemming to finish the question and took no further disciplinary action.

The naming of celebrities who were trying to prevent the press from reporting on alleged affairs may seem trivial, but it reveals a much wider and much more serious problem, as the use of parliamentary privilege can undermine the very concept of

John Hemming using parliamentary privilege to name recipients of super-injunctions in 2011, including Ryan Giggs

the rule of law and the ability of people to have a fair trial. In 2018, Peter Hain, sitting in the House of Lords, broke an injunction taken by Sir Philip Green, the owner of Topshop and other retail outlets, over allegations of sexual assault and misconduct. When the House of Lords took no action against Lord Hain, despite a formal complaint, the Lord Chief Justice of England and Wales (a senior judge, just below the Supreme Court level) claimed Hain's actions were 'a straightforward attack on the rule of law with no coherent justification'. His concern was that MPs and peers were increasingly using parliamentary privilege as a means of making themselves the sole arbiters of truth, without regard to the due process of law and the opportunity to hold a fair trial. This is a serious allegation and reflects a conflict between Parliament and the wider judiciary. It suggests that not only are MPs and peers exempt from the law, undermining the principle that everyone is equal under the law, but that some peers and MPs are using parliamentary privilege as a means of interfering in the work of the courts and undermining the working of the law in the UK.

Therefore, we may conclude that the use of parliamentary privilege is important in allowing MPs and peers the opportunity to discuss all issues freely and fully, without fear of reprisals, to ensure all issues and matters are considered in fulfilling its various functions. However, its misuse, no matter what the intentions, poses a serious threat to the rule of law and the strength of British democracy.

The role and significance of backbenchers in the Commons and Lords

MPs in the House of Commons

Recently, MPs at Westminster have had a generally poor reputation. Some of this has been self-inflicted, but a great deal is also misplaced. The truth is that much of their work goes unnoticed. The performance of backbench MPs, who sit on legislative committees and select committees and ask questions of ministers, is judged very differently from the performance of frontbench MPs who are the government ministers and leading opposition MPs and have much higher media roles and focus more of their time on executive functions, as outlined in Chapter 7.

For backbench MPs, who carry out the key work of Parliament in legislating, providing scrutiny of the frontbenches, and representation, their main roles include the following:

- Taking part in debates on legislation and voting in divisions.
- Speaking in general debates on government business.
- Speaking in backbench debates when national or constituency interests can be aired.
- Scrutinising proposed legislation at committee stage.
- Possibly being a member of a House of Commons select committee.
- Active membership of a campaign committee of MPs on a particular issue.
- Taking part in fact-finding missions, usually with groups of MPs and often abroad.
- Being a member of a committee formed by their own party to develop policy on a particular issue.
- Campaigning, lobbying and speaking on behalf of an outside interest or cause.

- Listening to grievances of constituents against a public body and sometimes acting to try to redress those grievances, including lobbying ministers and government officials.
- Holding regular surgeries so that constituents can meet and raise issues with them in person.
- Attending important events in the constituency, including listening to and perhaps joining local campaign groups.

This is a formidable list but, even so, some MPs also take on outside work, often as journalists and writers or as members of outside associations, though this has become much less common than it was before the 1990s, due to the growing workload of MPs and the increasing professionalisation of the work of the House of Commons.

Of course, an assessment of the significance of MPs is difficult to make because their work varies so much, and every individual MP is different. Some attend debates with great regularity; others may rarely be seen except when ordered to attend by the party whips. Committee membership, and attendance, also vary, as does the extent of constituency work undertaken.

Debate

How significant are MPs in the UK Parliament?

Limited significance

- MPs can be described as 'lobby fodder' or 'party hacks', who simply do as the party whips tell them uncritically and hope their loyalty will one day be rewarded by promotion to ministerial office.
- Backbench MPs are actually powerless in the face of the domination of the party front benches. They have little or no influence over legislation and fail to bring government effectively to account.
- Parliamentary debates are often sparsely attended, suggesting MPs lack interest in public policy.
- Parliament has very long recesses, giving MPs excessively long so called 'holidays', during which the executive can effectively work unchecked.
- MPs are often unknown in their constituencies.

Great significance

- There are numerous independent-minded MPs who are willing to put their beliefs and principles above narrow party interest. In the 2017–19 Parliament in particular, backbench MPs showed how significant they could be in blocking many of the government's measures.
- Since 2010, Parliament has been more willing to defy government and the select committees are becoming increasingly effective in calling government to account.
- Much of the work of MPs is carried out behind the scenes, often in committees. It is of great significance in the effective working of Parliament, but is much less visible and high profile than debates or the work done by the frontbenchers.
- MPs often use the long recesses to catch up on constituency work, ensuring a strong connection between the people and Parliament.
- Many, though not all, MPs undertake heavy workloads representing constituency interests, even if they are not well known.

In considering significance you should evaluate the impact that MPs have and consider the level of impact most MPs have; some will be excellent and some will be terrible, which should be noted, but try to base your judgement on the impact of what you might consider a typical MP.

The backbench MPs listed below are examples of individuals who have played a significant role in politics in the UK Parliament in recent years.

- **Mhairi Black (Scottish National Party, Paisley and Renfrewshire South)**
  The youngest member of the Commons, aged only 20 when she was elected in 2015, Black made an instant impact with a widely praised maiden (first) speech

in the House. Despite her youth, she was quickly made a member of the Work and Pensions Select Committee and specialised in issues concerning welfare and inequality. She has also been a prominent spokesperson for LGBT+ rights, and herself is an openly LGBT+ member of Parliament. Thanks to her work as a backbench MP, in 2019 she was moved to the SNP frontbench as their spokesperson on Scottish affairs.

- **Mike Freer (Conservative, Finchley and Golders Green)** Freer is known as a hard-working constituency MP. He has been a long-time champion of breast cancer screening and speaks often in the interests of the state of Israel. He is also a supporter of improved inter-community relations, encouraging links between the Jewish, Hindu, Sikh and Muslim communities in his constituency. He has served on the Work and Pensions Committee, the Scottish Affairs Committee and the Housing, Communities and Local Government Committee, and since January 2020 has sat on the Committee of Selection, which selects which backbench MPs will be allocated to the various other committees in the House of Commons.

- **Yvette Cooper (Labour, Normanton, Pontefract and Castleford)** Having been a Cabinet minister before 2010 and then in numerous frontbench positions as part of the official opposition, in 2015 she resigned as Shadow Home Secretary to return to the backbenches. In 2016 she was elected chair of the Home Affairs Select Committee, and has become a visible and important backbench MP, particularly in her role in the ongoing scrutiny of the Windrush scandal, which led the resignation of the then Home Secretary Amber Rudd in April 2018. In 2019 she introduced a Private Members' Bill, which narrowly passed despite government opposition, to prevent a 'no deal' Brexit, while also campaigning and working for rights and protections of refugees in the ongoing European refugee crisis.

Discussion point

Think about these situations and decide whether you think a backbench MP should follow the party line, follow their conscience, or do what they think their constituents would want them to do:

1　A proposal to send UK combat troops to a Middle East war zone.
2　A proposal to decriminalise cannabis.
3　A proposal to abolish zero-hours contracts.
4　A proposal to build a nuclear power plant in the MP's constituency.

Peers in the House of Lords

Apart from the frontbench peers who are either ministers or opposition spokespersons, the backbench peers have several roles:

- From time to time the House of Lords is asked to debate a great national issue. This is an opportunity for peers, particularly those who are former government ministers or civil servants, or heads of major organisations such as pressure groups, businesses or trade unions, to influence such debates. For example, Baroness Tanni Grey-Thompson, a former Paralympian and wheelchair racer, has used her experience to make speeches and contribute to debates on issues relating to sport governance, women in rowing, wheelchair services, and hotel facilities for people with disabilities.

- The Lords is an important part of the legislative process. Before 2010 the Lords rarely had much impact on legislation, largely because governments normally enjoyed large Commons majorities and could claim a strong mandate for their proposals. Between 2010 and 2019, however, governments had either no secure single-party majority or only a small advantage in the Commons. This meant that the government had to take potential opposition in the House of Lords seriously. Any controversial policy needed assured support in both houses of Parliament. This inevitably gave peers increased influence over both the principles and detail of legislation. The government does not have a majority in the Lords, so it must win over the support of crossbench or opposition peers if it is to avoid obstruction.
- As we have seen, peers take part in the scrutiny of legislation, whether from the government or from backbenchers of either house. It is on these occasions that their experience and knowledge can be most useful in improving legislation and in ensuring that minorities are protected and that all sections of society are fairly treated.
- Each government department has a representative, in the form of a junior or senior minister, who sits in the House of Lords. This enables peers to call government to account over its policies by asking questions of the minister. However, it is acknowledged that this is a relatively minor role. In addition, there are no adversarial select committees in the Lords as there are in the Commons, so the government does not need to fear excessively adversarial questioning.

Many individual peers have carved out an influential role for themselves. Here are three examples:

- **Lord Adonis (Labour)** Andrew Adonis is a former academic and minister who is an expert on economics, education and transport issues. He continues to play a leading role in advising both main parties on these issues.
- **Lord Dannatt (crossbencher)** Richard Dannatt was formerly chief of the general staff and thus the UK's most senior soldier. He now lends his huge knowledge of military matters to the work of the House of Lords. In 2015–16 he campaigned for UK ground troops to be redeployed in Iraq to fight against ISIS (Daesh).
- **Baroness Floella Benjamin (Liberal Democrat)** Floella Benjamin is a former children's presenter, a businessperson and activist. She is one of the most active and hard-working peers, raising awareness of, and working on, issues relating to education and children as well as healthcare issues and issues relating to immigrant children. In 2013, she marked International Women's Day by making a speech about the dangers and impact of violent pornography on children and the role it played in the objectification of women.

The three peers described above, along with many others like them, are all full- or part-time parliamentarians. They are as influential as any backbench member of the House of Commons. As well as their knowledge and experience, they have a high media profile, more so than most MPs, which alone gives them importance in the political system. Yet, for every active and influential peer, there are also many who are inactive and almost invisible. They are not accountable, so there is no check over the quality and quantity of their work.

The work of select committees

Much of the work of Parliament in calling government to account is conducted by **select committees**. The whole house is a rather clumsy way of achieving this,

Activity

Look up the websites of the following peers and identify what occupational experience they bring to the Lords and what committee roles they may have:
- Baroness Lane-Fox
- Baron O'Donnell
- Baroness Corston
- Lord Winston

Key term

Select committee Committee responsible for scrutinising the work of government, particularly individual government departments.

so smaller committees are used that can focus on aspects of government work. Committees can also escape the rather formal, ritualised procedures of Parliament and operate in a less formal but more effective manner. Departmental select committees were first set up in the Commons in 1979, though the Public Accounts Committee (PAC) dates from the nineteenth century. There are many select committees in both houses, but in this section we concentrate on those committees whose specific role is calling government to account.

House of Commons select committees

While all MPs (other than government ministers) must serve on legislative (Public Bill) committees from time to time, not all sit on important select committees. Whereas Public Bill committees are *ad hoc*, meaning they come into existence only to deal with a specific bill, select committees are permanent and seats on them are highly prized; indeed, since 2010, the chairperson of each select committee has been elected by the whole House and it raises the status and prestige of the backbench MP who is chosen as chair, as well as seeing them awarded an additional salary for the role.

There are many kinds of select committee in both houses of Parliament, but here we consider the most important types, all in the House of Commons. These are the Public Accounts Committee and 19 departmental select committees, the Liaison Committee, and the Backbench Business Committee. The first three of these committees exist to call government to account, while the Backbench Business Committee gives backbenchers some control over the parliamentary agenda.

The Public Accounts Committee

The Public Accounts Committee is the oldest committee in the Commons, dating back to 1861. Arguably it is also the most influential of all parliamentary committees. Its role is to examine the public finances. In the words of Parliament's own website, the committee's job is as follows:

> This committee scrutinises value for money — the economy, efficiency and effectiveness — of public spending and generally holds the government and its civil servants to account for the delivery of public services.

Source: **www.parliament.uk**

This is a very wide remit. In effect, it also includes the collection (though not the rates) of taxation and how well that is done. Taxation and spending are fundamental activities of government, so their examination is crucial. The committee conducts investigations into various aspects of the government's finances, particularly how it allocates and spends money on public services. It can call witnesses, who are obliged to attend. Witnesses may be ministers, civil servants, government officials, outside witnesses and representatives of interested bodies, as well as experts in a particular field.

It is powerful for a number of reasons:

- Its chair is always a member of the main opposition party.
- The chair has great prestige, not to mention a higher salary than other MPs.
- The chair and members are elected by all MPs and so are not controlled by party leaders.
- Its members, despite being party supporters, tend to act independently, ignoring on the whole their party allegiance. This means the government has no advantage on the committee even though it has a majority of members on the committee.

- Its reports are often unanimous in their conclusions, so the committee stands above party politics.
- It has a high profile in the media. Many of its important hearings are broadcast as news items.

The committee rose to even greater prominence between 2010 and 2015 under its chair, Margaret Hodge. She introduced a new device, which was to call senior civil servants to give evidence, as well as ministers and outside witnesses. Above all, however, she was determined to publicise major issues and to question public policy where the committee felt that taxpayers were not getting value for money.

For the 2019 Parliament, the PAC's party membership is shown in Table 6.4. Recent reports from the PAC demonstrate its importance. They are shown in Table 6.5.

Table 6.4 Party membership of the Public Accounts Committee, 2019 Parliament

Chair	Meg Hillier (Labour)
Conservative members	9
Labour members	5
SNP members	1
Liberal Democrat members	1

Table 6.5 Key PAC investigations

Year	Investigation	Conclusion and recommendations	Impact
2015	Into the effectiveness of cancer care by the NHS	Highly critical of variations in cancer treatment in different regions and for different age groups Criticised low cure rates and increased waiting times for treatment	The publicity caused the government to launch a review into cancer treatment and set up an independent cancer task force to improve the delivery of cancer treatment across the UK by 2020
2016	Into the tax affairs of Google	Google's payment of back tax of £130 million for 10 years was considered far too low Concluded that HMRC should investigate ways of better regulating the tax affairs of multinational companies and making them more transparent	The investigation led to a government consultation into strengthening tax avoidance sanctions and deterrents, following which it introduced a finance bill in 2017 to give HMRC more powers to deal with tax avoidance
2019	Into serious and organised crime	Critical of the Home Office's attempts to move the focus of policing from street-level criminals to investigating the networks and organised crime that lie behind them Recommended revised initiatives, developing more effective databases and that police services prioritise identifying and investigating organised criminal activity	In January 2020, Home Secretary Priti Patel announced an additional £750 million to recruit 6000 extra police officers to combat street crime, as part of the government's wider pledge to recruit an additional 20,000 officers by 2023
2020	Into gambling regulation and the issues of problem gambling and the vulnerable	Highly critical of the Gambling Commission and the Department for Digital, Culture, Media and Sport for failing to understand the impact of problem gambling and its weakness in standing up to the gambling industry at a time of rapid change Recommended that the Department and Commission should investigate the patterns and impact of problem gambling and produce plans for how to tackle this growing problem	In 2020 the government announced proposals to ban the use of credit cards on online gambling sites to prevent people from incurring large gambling debts

We can see that these and other investigations by the PAC deal with important aspects of public finance. The publicity achieved by the reports forces government to respond. Accountability is therefore effectively achieved.

> ### Study tip
>
> It is worth considering what it is that forces a government response. In the cases listed in Table 6.5 it seems that the PAC investigations directly resulted in government action, but in each case, the PAC report was publicised by the media and met with public outrage. The PAC's power rests on its ability to gain public support for its investigations, which pressures the government into acting.

Departmental select committees

Departmental select committees (DSCs) have existed since 1979. Along with the PAC, they are a vital way in which Parliament can call government to account. Each of the 19 committees investigates the work of a government department. Their features are as follows:

- Cross-party whips determine which committee chairs will be awarded to each party (the total to be in line with party representation in the Commons).
- Backbench MPs who wish to chair one of the committees allocated to their party put themselves forward for a vote of the whole House of Commons (the chair also receives an additional salary).
- The remaining seats on the committee are allocated in proportion to party size in the Commons; members are elected by a secret ballot of their party members.
- Membership varies between 11 and 14.
- The governing party usually has a majority on each committee.
- The chairs may be from any political party.
- The small parties have a scattering of members.
- Like the PAC, they act largely independently of party allegiance and often produce unanimous reports.
- Also like the PAC, they can call witnesses who may be ministers, civil servants and outside witnesses such as pressure group representatives or experts.
- Their reports and recommendations are presented to the whole House of Commons and receive considerable publicity.

The impact and effectiveness of the departmental select committees are certainly growing alongside the PAC. The chairs of the committees have become important, influential parliamentarians, and governments now feel they need to respond to their criticisms and recommendations.

Some examples of important reports from DSCs are shown in Table 6.6.

Table 6.6 Key departmental select committee reports

Date	Committee	Investigation	Conclusion and action
2015	Treasury	Into proposals for stricter regulation of the banking sector	Insisted that government should implement the recommendations of the Parliamentary Commission on Banking Standards This pushed policy forward on banking regulation
2016	Business, Innovation and Skills	Into alleged bad working practices at Sports Direct	The company was forced to pay compensation to its workers for paying below minimum wage
2016	Work and Pensions	Into the collapse of British Home Stores and the loss of much of the employees' pension fund	The company was reported to the Pensions Regulator
2018	Home Affairs	Into the causes and issues surrounding the Windrush scandal that denied many of the Windrush generation and their descendants fundamental rights	The committee was highly critical of the cultural, organisational and administrative changes that led to rights being denied to British citizens, and was highly critical of the Home Office's lack of knowledge or openness about the issue. It recommended an apology and immediate reforms of procedure and focus be introduced, but as of summer 2020 these have not yet been implemented
2020	Transport	Into the impact of the Covid-19 pandemic on the aviation industry and its employees	Condemned the way British Airways had treated its employees during the pandemic Urged the end of the 14-day quarantine by June 2020 and that airlines not be hasty about large-scale redundancies

In recent years, the DSCs have started to extend their work to consideration of matters of general public interest, not just the performance of government departments. Such investigations often take place in the hope and expectation that government will consider new legislation to deal with the problems revealed. As seen in Table 6.6, in 2016 the Business, Innovation and Skills Committee investigated alleged cases of unacceptable management practices at the Sports Direct company, while in the same year the events surrounding the takeover and subsequent failure of British Home Stores were investigated by the Work and Pensions Committee.

Activity

Choose any three departmental select committees. Check their websites and identify and summarise their latest report.

The Liaison Committee

This committee was created in its present form in 2002. It consists of the chairs of all the departmental select committees as well as several other committees. Apart from overseeing the work of House of Commons select committees, its main function is to call the Prime Minister to account. Twice a year the Prime Minister must appear before the Liaison Committee. Overall, this committee has not been very effective, apart from the brief stint under Andrew Tyrie. The lack of power of the Liaison Committee is exemplified by the fact that Boris Johnson cancelled several planned appearances, reflecting the lack of teeth that the committee really has.

Boris Johnson being questioned by the Liaison Committee during the unusual circumstances of the Covid-19 pandemic in 2020

The Backbench Business Committee

The committee was set up as part of the **Wright Reforms** of 2010. It is made up of elected backbench MPs. Its main role is to determine what issues should be debated on the one day a week allocated to backbench business. Before 2010 most of the parliamentary agenda was controlled by the government and the main opposition party leadership. Giving up one day a week to backbench business was, therefore, a major departure.

The subject matter of such debates comes from several sources. These include:

- When an e-petition on the Downing Street website achieves 100,000 signatures
- On the initiative of one of the select committees
- From a request by an MP or group of MPs
- Requests emerging from national and local campaigns

Recent debates include the following examples:

- The effect of Covid-19 on Black, Asian and minority ethnic communities
- Many debates over the issue of Brexit and leaving the European Union
- Improving cancer care
- The future of the BBC (from the Digital, Culture, Media and Sport select committee)
- Harvey's Law (the result of the police's poor handling of a dog killed in a road accident; an e-petition demanded a law requiring police to treat such incidents more seriously and to keep pet owners better informed following accidents)

But the most celebrated example of the committee's work occurred in 2011. An e-petition had been held to order the publication of all the documents relating to the 1989 Hillsborough football disaster. With over 100,000 signatures, the petition forced a parliamentary debate. As a result of this the government was forced to release previously secret papers about the disaster. The affair resulted in a new inquest and several inquiries into Hillsborough.

The impact of House of Commons select committees

All select committees in the Commons suffer the same basic weakness: they do not have the power to enforce decisions and recommendations. All they can hope to do is to bring publicity to issues, to call government to account and to recommend various courses of action. Nevertheless, they are becoming an increasingly significant feature of parliamentary politics.

The great strength of the select committees is that the MPs who make up their membership tend to be independent-minded. They operate outside the normal constraints of party loyalty so that ministers and government officials increasingly have to take notice of what they say. Hearings are televised so that any controversy receives media coverage. While the questioning of ministers on the floor of the House of Commons tends to be weak and easily countered, the various select committees described above are playing a growing part in Parliament's role of demanding accountability. Questioning can be intense and prolonged, and MPs do not accept weak answers. Sometimes the hearings can resemble cross-examination in a court of law. The investigations into the affairs of Philip Green (former owner of British Home Stores) and Mike Ashley (boss of Sports Direct) in 2016 became compelling TV viewing.

The role and significance of opposition parties

It is often said that the democratic working of government in the UK depends on the existence of a strong, effective opposition in Parliament. Indeed, the 'official' opposition is recognised in the UK Constitution. The Leader of the Opposition receives a minister's salary and takes part in all official ceremonies. The Leader of the Opposition also has the privilege of asking most of the questions at the weekly Prime Minister's Question Time (up to six, with priority given to them above all other MPs' questions). The opposition also has control over part of the parliamentary agenda. There are 20 days, known as 'opposition days', which are devoted to debates on issues chosen by the opposition. The largest opposition party is the official opposition, but other, smaller parties are also part of the opposition, though with fewer privileges.

Before 2017 the interplay between government and opposition was relatively simple. The government presented one set of policies and the opposition challenged those proposals with which it disagreed and presented alternatives. When the next general election came round, the electorate was presented with a clear choice between the two. Even under coalition government between 2010 and 2015 the opposition played its traditional role.

In 2017, the arrival of a minority government presented opportunities for the opposition parties to be an effective check on government. Unable to rely on the support of a parliamentary majority, the government was faced with the need to build a coalition of support for each policy initiative. Opposition members could now expect to have genuine impact on policy. This, in addition to the seriousness of the issue of Brexit, led to a period of chaos in Parliament, especially as the Conservative Party fractured, with the government experiencing three of the biggest defeats in history, surviving a vote of no confidence and nothing being passed other than extensions to the Brexit deadline. The 2019 general election result, with a more unified Conservative Party winning an 80-seat majority, seems to have redressed this balance, returning the country to single-party rule and weakening the role of the opposition.

The role of the official opposition and opposition parties can be summarised by the following main functions:

- The main opposition party is a 'government in waiting'. It must always be ready to take over if a government falls or resigns, and it must be ready to fight for power at the next election.
- All opposition parties have the key function of calling the government to account by critically examining policies and decisions and by questioning ministers.
- Opposition parties often seek to defend the interests of sections of society that they feel are being ignored or discriminated against in government policy. Thus, for example, Labour often defends the rights of workers, nationalist parties defend regional interests and the Greens defend those suffering environmental problems.
- The main opposition parties also have a ceremonial function at occasions such as visits by foreign heads of state, Remembrance Sunday and the like. The leaders of opposition parties are seen on these occasions in order to ensure all variations of political opinion are represented at national rituals.
- The main opposition party has a share in organising the business of the UK parliament. A significant minority of parliamentary time is given over to debates and business that they wish to hold.

Study tip

Although the official role of the opposition remains constant, its importance ebbs and flows over time, normally in direct correlation to the power of the government; a stronger government sees a weaker opposition, a weaker government sees a stronger opposition.

Calling ministers to account

There are three main ways in which government ministers are called to account:

1. On 'questions to ministers days', every minister has to take their turn to appear on the floor of the House of Commons (or Lords if they are a peer) to face questions from members. They also must respond to written questions. It is not just government policy that is under scrutiny; ministers may also have to answer questions about the way in which individuals and groups have been treated by an organ of the state. These are key occasions, when members pursue the grievances of their constituents or promote the interests of a group or association they may represent.

2. In departmental select committees or the Public Accounts Committee, ministers are subjected to close questioning, sometimes hostile, about their policies and about the performance of their department. This can be a challenging occasion for them and their civil servants and advisers. The select committees have, indeed, largely taken over from 'questions to ministers days' as the main way in which government is called to account.

3. The Prime Minister must submit themselves to Prime Minister's Question Time every week. This is a largely ritualised occasion, conducted more for publicity than effective democracy. It says more about the personal qualities of the Prime Minister than about the government's policies. Twice a year the Prime Minister also appears before the Liaison Committee containing the chairs of the main select committees.

Activity

Find an account of the most recent PMQT. What were the main issues raised by the Leader of the Opposition?

It remains true that ministers have a distinct advantage over MPs and peers when answering questions. Usually, they have notice of questions and so can use their army of civil servants and advisers to help them with answers. Many ministers also become highly skilled in avoiding becoming trapped by difficult questioning. In contrast, MPs and peers lack a great deal of research back-up and may be seen as 'amateurs' competing with professionals. The select committee system, however, has gone some way to redressing this imbalance.

How effective is the UK Parliament today?

We have seen that the relationship between government and Parliament fluctuates. However, it is possible to attempt an assessment of how effective Parliament is. To do this, it is necessary to determine what 'effective' actually means. We can say that effectiveness implies the following:

- It holds government properly to account.
- It provides democratic legitimacy for government initiatives.
- It scrutinises legislation thoroughly and seeks to improve it.
- It prevents government from exercising power beyond its electoral mandate.
- It is an effective vehicle for the representation of constituencies.
- It represents the interests of various sections of society.
- It represents the national interest.

Considering these requirements, we can determine how well Parliament is performing its functions today.

Is Parliament effective?

Yes

- The select committees are increasingly significant and effective in scrutinising the work of government.
- Ministers must still face questioning in both houses.
- Parliament provides strong democratic legitimacy.
- The House of Lords does an increasingly effective job at scrutinising legislation, often improving legislation and blocking unfair or discriminatory aspects of proposals.
- Experts in various fields in the Lords use their knowledge to good effect.
- Both houses increasingly check the power of government, especially when the governing party does not have a commanding majority in the Commons.
- FPTP allows MPs to effectively represent constituents.
- Many outside views and groups are represented in the Lords and in the Commons.
- When free votes and debate are allowed, Parliament shows itself to be highly effective at debating and acting on national issues.

No

- MPs still lack expertise, knowledge, research back-up and time to investigate government thoroughly.
- PMQT remains a media 'event' rather than a serious session.
- The House of Lords cannot provide democratic legitimacy as it is neither elected nor accountable.
- Legislative committees in the Commons are whipped, making them largely ineffective.
- The power of prime ministerial patronage and control by party whips still means that many MPs are unwilling to challenge government.
- MPs' care of their constituencies varies from MP to MP.
- There is still no effective mechanism for removing poorly performing MPs.
- When there is a clash between party policy and the interests of groups and causes, party loyalty often wins out.

 The key here is to consider the word 'effective' and decide, overall, if Parliament today meets the key requirements. You will need to make a clear judgement, so try to consider what you feel to be the most important function and use that to judge the overall extent of effectiveness.

Summary

Having read this chapter, you should have knowledge and understanding of the following:
→ How the House of Commons and the House of Lords are structured and their composition
→ The legislative, representative and scrutiny functions of the House of Commons and House of Lords
→ The key differences in powers and roles of the Commons and Lords
→ How the legislative process works
→ The work and significance of individual MPs and peers
→ How Parliament relates to the executive
→ The factors and situations that may impact the role and importance of Parliament
→ The significance of parliamentary privilege and why it has been criticised in recent years

Key terms in this chapter

Backbenchers MPs or Lords who do not hold any government office.

Confidence and supply The right to remove the government and to grant or withhold funding. Also used to describe a type of informal coalition agreement where the minority partner agrees to provide these things in exchange for policy concessions.

House of Commons The primary chamber of the UK legislature, directly elected by voters.

House of Lords The second chamber of the UK legislature, not directly elected by voters.

Legislative bill Proposed law passing through Parliament.

Opposition MPs and Lords who are not members of the governing party or parties.

Parliament The British legislature made up of the House of Commons, the House of Lords and the monarch.

Parliamentary privilege The right of MPs or Lords to make certain statements within Parliament without being subject to outside influence, including the law.

Public Bill Committee Committee responsible for looking at bills in detail.

Salisbury Convention The convention whereby the House of Lords does not delay or block legislation that was included in a government's manifesto.

Select committee Committee responsible for scrutinising the work of government, particularly individual government departments.

Further reading

Websites

The best website for information about all aspects of Parliament is: **www.parliament.uk** Debates about the reform of Parliament can be found on such sites as: **www. unlockdemocracy.org** and **www.ucl.ac.uk/constitution-unit** The Hansard Society is also very useful: **www.hansardsociety.org.uk** For up-to-date information and statistics on parliamentary revolts, consult: **www.revolts.co.uk**

Books

Norton, P. (2013) *Parliament in British Politics*, Palgrave
Rogers, R. and Walters, R. (2015) *How Parliament Works*, Taylor & Francis
Russell, M. (2013) *The Contemporary House of Lords: Westminster bicameralism revived*, OUP
Russell, M. (2017) *Legislation at Westminster: Parliamentary actors and influence in the making of British law*, OUP

Practice questions

1

Source 1

Reform of the House of Lords

After most of the hereditary peers were removed from the House of Lords in 1999, the behaviour and status of the upper house began to change. In that year the house became largely an appointed body. Many of its members are experts in their field and take their role as legislation revisers very seriously. At the committee stage of a bill's passage through the Lords, many peers contribute to improving the legislation. So, although it remains an undemocratic body, it could be said to be more effective than ever before. The scrutiny offered by the experts in the Lords offers an important check on the government, a thing that has only increased since peers must regularly attend the Lords or face removal.

Nevertheless, calls for its replacement by an elected body remain strong. For democrats it is not acceptable that half the legislature should be unelected and unaccountable. Supporters of this reform ignore the fear that an elected house would fall under the control of party leaderships. As things stand, the Lords is extremely independent of party control and can therefore provide more meaningful opposition and can more effectively call government to account. An elected chamber might also mean that many useful specialists and experts would be lost to politics.

Using the source, evaluate the view that the House of Lords is in need of major reform.

In your response you must:
- *compare and contrast different opinions in the source*
- *examine and debate these views in a balanced way*
- *analyse and evaluate only the information presented in the source.* (30)

2

> **Source 2**
>
> In recent years, we have seen the importance of select committees rise, with many high-profile politicians seeking a prized chair on one of the departmental select committees, making them a viable alternative to a government career. Former ministers, like Yvette Cooper and Hilary Benn, now stand for their committee and use it as a platform to scrutinise the government and hold it to account. Who can forget the Home Affairs select committee questioning Amber Rudd, leading to her resignation as a Cabinet minister? The Backbench Business Committee too has succeeded in giving MPs a voice and an opportunity to debate serious issues free from government control and manipulation. The Liaison Committee has, when given the opportunity, put the Prime Minister in awkward situations, ensuring he or she must answer for their actions and subjecting them to strict scrutiny, and the Public Accounts Committee continues to keep a close eye on government spending by issuing reports and investigating major issues.
>
> It is in select committees, not the rowdy Prime Ministerial Questions, nor the sparsely attended Ministerial Questions, that the real work of scrutinising the government is done. National debates are rare and do little to change the opinion of the government, the chamber is so lightly whipped that there is little it can do as a body to check the government, while the Lords lacks any real will or power to hold the government to account. The select committees are key and they need more power to prevent an elective dictatorship.

Using the source, evaluate the view that Parliament is effective in its role of scrutinising the government.

In your response you must:
- *compare and contrast different opinions in the source*
- *examine and debate these views in a balanced way*
- *analyse and evaluate only the information presented in the source.* (30)

3 Evaluate the extent to which backbench MPs are effective in fulfilling their various roles.

In your answer you should draw on relevant knowledge and understanding of the study of Component 1: UK politics. You must consider this view and the alternative to this view in a balanced way. (30)

4 Evaluate the extent to which the House of Lords plays a meaningful role in the passage of legislation.

In your answer you should draw on relevant knowledge and understanding of the study of Component 1: UK politics. You must consider this view and the alternative to this view in a balanced way. (30)

Study tip

Coverage of Core and Non-Core Political Ideas are available in other textbooks to complete your study for Components 1 and 2.

5 Evaluate the extent to which the House of Lords remains an important part of Parliament.

In your answer you should draw on relevant knowledge and understanding of the study of Component 1: UK politics. You must consider this view and the alternative to this view in a balanced way. (30)

6 Evaluate the extent to which Parliament may be considered an ineffective institution.

In your answer you should draw on relevant knowledge and understanding of the study of Component 1: UK politics. You must consider this view and the alternative to this view in a balanced way. (30)

7 The Prime Minister and the executive

On 24 May 2019, on the road outside 10 Downing Street, Theresa May announced that she would be stepping down as leader of the Conservative Party and would step down as Prime Minister once the party had chosen a successor. After less than 3 years as Prime Minister, it was perhaps not a surprise that she had had enough. Having been chosen to lead the Conservative Party to replace David Cameron following the Brexit referendum in 2016, she was appointed Prime Minister by the Queen and promised to deliver Brexit for the people. The triggering of Article 50 in March 2017 was something of a high point, but then the loss of her parliamentary majority, the growing divisions within the Conservative Party over Brexit and the threats of withdrawal of support from her allies the DUP saw her face one of the most challenging periods in modern political history. She experienced 33 defeats in the House of Commons, more than her five predecessors combined had experienced over the course of 38 years, including two of the largest defeats in parliamentary history. Attacked by the press, unpopular with the public, unable to achieve anything in Parliament, she stepped away to see if anyone else could restore order to British politics.

It was not Parliament, the general public or even the Queen that would choose the next prime minister, a person who would steer the UK through incredibly turbulent waters. That job was left to the Conservative Party. First the MPs narrowed the choice down to two candidates and then the wider party membership, comprising just 180,000, elected Boris Johnson over Jeremy Hunt, to be the new Conservative Party leader. Once the result of the party leadership election was announced, May travelled to Buckingham Palace to resign as Prime Minister. A little later, Johnson travelled to Parliament and was officially invited by the monarch to form a government, making him the new Prime Minister.

Boris Johnson at the Conservative Party manifesto launch, 2020

So, power passed quietly from one set of hands to another. It was all very civilised and understated. In one short ceremony, the Queen's ancient prerogative powers were transferred to the new Prime Minister. Afterwards, Johnson returned to Downing Street to begin the process of exercising the first of the Queen's powers, that of patronage. He set about appointing a new set of ministers. It all happened without a general election or a vote in Parliament. Little seemed to have changed: Johnson experienced more monumental defeats, was ordered by Parliament to extend the Brexit deadline that he so vehemently opposed, and expelled 21 of his own MPs from the Conservative Party. Like May, he soon called an early election, but unlike May, he secured a majority of 80 MPs in the newly elected House of Commons. The issues had not changed, but his electoral success gave him far greater authority to lead and fulfil his aim of 'getting Brexit done'.

Objectives

In this chapter you will learn about the following:

➜ The structure and roles of the central 'executive' of the UK
➜ The power and authority of the different components of the executive
➜ The sources of prime ministerial power and authority, and the nature of the Prime Minister's powers
➜ The nature, roles and limitations of the Cabinet, and how the 'Cabinet system' works
➜ The nature and importance of collective responsibility, and whether it is a negative or positive principle
➜ The nature and importance of individual ministerial responsibility
➜ The nature of the relationship between the Prime Minister and the Cabinet
➜ How the power of the Prime Minister can be assessed
➜ The extent to which Prime Ministers and the executive can dictate events and determine policy

The structure, role and powers of the national executive

The structure of the executive

The term '**executive**' refers to one of the three branches of government, standing alongside the legislature and the judiciary. While the role of the legislature is to pass, or confirm, laws, it is the job of the executive branch to execute those laws or make them happen. In this sense the executive is the branch of government responsible for making policy decisions, taking actions and running the country. While parliamentary sovereignty means the executive must answer to Parliament and seek parliamentary approval for its actions, the executive is, usually, responsible for determining policy and making it a reality, and so can appear to be the dominant branch of the UK government. This is especially true when a prime minister has a large majority and strong control over their party in the House of Commons.

Synoptic link

The executive in the UK is drawn from Parliament, so the relationship between the executive and Parliament is central to understanding UK politics. This is addressed further in Chapter 8.

The executive in the UK is comprised of a number of key elements.

- **The Prime Minister** Appointed by the monarch and head of the executive.
- **The Cabinet** The senior people in the executive, mostly made up of senior **ministers** who are responsible for running **government departments**, like the Home Office, and making policy decisions. Members are appointed by the Prime Minister.
- **Junior ministers** Members of the executive who assist the senior minister (or Secretary of State) in the running and policy decisions of a government department. For example, the Education Secretary will usually have three junior ministers, each one taking responsibility for an aspect of education policy: early years and primary education; secondary education; universities and higher education. Junior ministers are also appointed by the Prime Minister but are not part of the Cabinet.
- **The civil service** Permanent officials who carry out the day-to-day running of government departments. Their job is to advise the ministers and enact policy decisions made by ministers; they do not make policy decisions themselves as they do not answer to Parliament or the people. As such, members of the civil service are expected to act in a neutral fashion, standing outside the party battle, and are forbidden from serving the political interests of the government. This means that members of the civil service are not allowed to be members of a political party and can only help the government with government work, not party work, so members of the civil service can help organise a press conference for a government minister to make a public announcement, such as those over Covid issues in 2020–2021, but can not play any part in organising a party event, such as helping to organise a party conference.

The UK is unusual in that many of the organs of the executive are described as being under the control of the monarch, for example 'Her Majesty's ministers' or 'Her Majesty's Treasury'. However, this is an illusion. In practice, the executive branch is under the control of the Prime Minister (using the powers of the **royal prerogative**) and the Cabinet.

Synoptic link

The role of the Prime Minister and the use of the royal prerogative has developed by constitutional convention and is an example of the evolutionary and flexible nature of the UK Constitution, discussed in Chapter 5.

This is the key structure of what is sometimes described as 'the core executive' in the UK. By convention, the Prime Minister, all members of the Cabinet and all junior ministers must be members of Parliament, either in the House of Commons or the House of Lords, because they make the big political decisions and must answer to Parliament as the representatives of the people and be held to account through elections. Typically (though it does vary) this means the executive is comprised of about 120 MPs and peers (since 2019 the figure has been 116 in total, with 92 MPs and 24 peers), who are obliged to support it in Parliament.

The civil service, in contrast, is simply responsible for making the ideas and decisions made by the executive a reality and is not supposed to serve any political purpose. However, in recent years there has been a growth in special advisers (known as SpAds)

Key terms

Cabinet The Prime Minister and senior ministers, most of whom lead a particular government department.

Minister An MP or member of the House of Lords appointed to a position in the government, usually exercising specific responsibilities in a department.

Government department A part of the executive, usually with specific responsibility over an area such as education, health or defence.

Royal prerogative A set of powers and privileges belonging to the monarch but normally exercised by the Prime Minister or Cabinet, such as the granting of honours or of legal pardons.

Study tip

Although officially the term 'government' refers to all three branches of government, commentators' and popular opinion means when people discuss or write about 'the government' they are specifically referring to the executive. As such, the terms 'government' and 'executive' can be used interchangeably.

who are employed as **civil servants** but have the role of giving political advice to a minister. Although it is still the minister who makes the final decision, if a special adviser appears to be shaping or directing official policy, their position can be questioned and lead to controversy. This has been seen recently with the role of Nick Timothy and Fiona Hill, who had been employed as Joint Chiefs of Staff for Prime Minister Theresa May in 2016. They were employed as civil servants, but took charge of organising the 2017 general election campaign for the Conservative Party, even writing and making decisions over the manifesto, much to the anger of Conservative Party members. The pair were forced to resign as a result of the political failure of the 2017 election, something that civil servants would not be expected to take responsibility for. Similarly, Dominic Cummings was employed as Chief of Staff by Boris Johnson, led his election campaign in 2019 and appeared to be making decisions about government policy, rather than offering neutral advice to the Prime Minister, until he was forced to resign in late 2020.

Synoptic link

Special advisers often come from, or are linked with, think tanks and often go on to work as lobbyists once they have left their advisory role, a key element covered in Chapter 1.

Stating how large the executive as a whole is would be very difficult (see Table 7.1). In addition to the 120 MPs and peers who comprise the government ministers, there are thousands of special advisers and civil servants who make up the executive branch. The total number of people who have *some* direct influence over government policy-making is probably in the region of about 4000. However, the number of key players who form the core executive is somewhere in the region of several hundred people in total (Table 7.1).

Table 7.1 The structure of the executive

Key individual or body	Role	Supporting elements
The Prime Minister	Chief policy-maker and chief executive	Cabinet Cabinet secretary Private office of civil servants Policy unit
Cabinet	Approving policy and settling disputes within government	Cabinet committees Cabinet Office Cabinet secretary
Treasury	Managing the government's finances	Senior civil servants Special advisers Think tanks
Government departments	Developing and implementing specialised policies	Civil servants Special advisers Think tanks

The role of the executive

The main role of the executive in the UK is to govern and ensure the country operates effectively. As such, its main role is making decisions and coming up with ideas, either as part of its wider policy objectives outlined in its manifesto, or in reaction to events as they unfold, such as an act of war, a financial crisis or a global pandemic. The roles of the executive can therefore be summarised as follows.

Proposing legislation

The executive will develop legislative proposals for a first reading in Parliament, based on the policy decisions of the ministers and the expert advice of the civil service. As much of the legislative role of Parliament is spent dealing with government-proposed legislation, this is a crucial role that sets the parliamentary agenda.

> ### Synoptic link
>
> The role of the executive in proposing legislation is a fundamental part of the process of passing legislation outlined in Chapter 6.

Proposing a Budget

The executive has to calculate how much money it will need to run the country and carry out its proposed policies. As such, the executive must try to work out how much money it might be able to raise in the next financial year, from taxation, duties, investments and loans, by making an educated guess about how the economy will perform and how prosperous the UK will be. Based on these estimates, spending amounts will be allocated to the various government departments so that they can carry out their work. The Budget must be approved by Parliament as much of it is raised by public taxation, so it is a delicate balancing act between providing the services people need and want and taxing incomes and products, which will have an impact on economic performance. If the proposed Budget fails to pass Parliament, the executive has no funds and must, by convention, resign. This rarely happens though as the executive usually has a clear majority in the House of Commons and the Lords may not reject a Budget passed by the House of Commons. As such, the role of the Budget is more about speaking to voters and pledging money for what they want while avoiding unpopular increases in taxation, which can be politically damaging.

Making policy decisions within laws and the Budget

In addition to making policy and budgetary proposals to Parliament, the executive will need to make decisions about how to enact or enforce legislation and make changes to government spending. When a piece of legislation is passed, it often contains many clauses and terms; the executive must decide how to enact these in reality, and how to communicate this information or, in some cases, whether or not to use or enforce part of the law passed. Sometimes this is referred to as **secondary legislation**, where parts of an Act are amended or adapted to make them more workable in the real world, or delegated legislation, where Parliament grants the executive the power to make decisions without getting prior approval from Parliament. Similarly, issues may arise in which finances need to be amended and money moved around to respond to an unforeseen issue or crisis, such as the response to the global pandemic of 2020 and furlough scheme. The executive will still have to answer to Parliament for its actions, but these decisions are not the same as proposing primary legislation or an initial Budget and they allow the executive to respond quickly to an issue.

> ### Key term
>
> **Secondary legislation**
> Powers given to the executive by Parliament to make changes to the law within certain specific rules.

In addition to the main roles of the executive outlined above, it has a number of other important functions:

- Conducting foreign policy, including relations with other states and international bodies.
- Organising the defence of the country from external and internal threats.

- Responding to major problems or crises such as armed conflict, security threats, economic difficulties or social disorder.
- Controlling and managing the forces of law and order, including the police, courts, armed forces and intelligence services.
- Organising and managing the services provided by the state.

The power of the executive

The sources of prime ministerial authority and power

The office of Prime Minister originated in the early eighteenth century, but it is only held by convention, not statute law. So, on what basis has it become the most powerful office in the UK?

This is quite a complex set of circumstances. They are as follows.

Traditional authority

The monarch is no longer a political figure but *in theory* they have considerable powers, known as prerogative powers (also known as the 'royal prerogative'). As the monarch cannot exercise these powers, they delegate them to a Prime Minister. These powers are not constrained and so can be freely exercised by the Prime Minister personally. When exercising these powers, the PM is representing the whole nation, which means the Prime Minister is effectively the temporary head of state.

Boris Johnson meets the Queen in 2019 to formally be appointed as Prime Minister

There is a ritual followed that *appears* to show the monarch summoning and appointing her chosen Prime Minister following each general election (the PM goes to the palace to 'kiss hands', the formal act of being offered and accepting the post), but this is an illusion. It is simply a representation of the reality that both the monarch and the new Prime Minister understand. Nevertheless, the new Prime

Minister does inherit the **traditional authority** of the monarch. The monarch's approval, though merely formal, does grant the Prime Minister authority.

Party

The Prime Minister is always the leader of the largest party represented in the House of Commons following a general election. In this case, the PM's authority comes from the people through the leading party. If a party changes its leader, the new leader will automatically become Prime Minister. The monarch will summon that leader to the palace to confirm this. No election is necessary. This does occur from time to time, usually when the existing leader loses the confidence of their party, as occurred in the following cases:

- 1990: The Conservative Party replaced Margaret Thatcher with John Major.
- 2007: The Labour Party replaced Tony Blair with Gordon Brown.
- 2016: Theresa May replaced David Cameron after he resigned.
- 2019: Boris Johnson replaced Theresa May after she resigned.

Being party leader enables the Prime Minister to have the power to take the lead in policy-making. For as long as the PM can carry their party with them, they therefore become chief policy-maker.

Discussion point

Evaluate the view that it is democratically acceptable that the position of Prime Minister can change hands without any election taking place, as occurred in 2007, 2016 and 2019.

Three key areas to discuss are:
1. What aspects made these three changes democratic and/or undemocratic?
2. How else might a change in prime minister be brought about?
3. What was the attitude of the general public to such changes?

Parliament

Each new parliament, including the losing parties, recognises the authority of the Prime Minister to lead the government. There is no formal procedure to confirm this as there is in many other political systems; it is simply 'what happens'. Parliament has no formal procedure for replacing one prime minister with another. Although the ability of the Prime Minister to pass a Budget could be seen as a sort of formal approval from Parliament, all Parliament can officially do is to dismiss the whole government through a vote of no confidence.

The Prime Minister also gains significant power over Parliament as they are the leader of the largest party in Parliament. Clearly the larger the government's parliamentary majority, the more power the PM derives from this fact, but *all* prime ministers gain some power from it. If a government is unable to secure the passage of its legislation and financial plans through the House of Commons, it will lose power. MPs are always aware of this and so those who represent the governing party tend to support their prime minister most of the time to ensure the survival of their government. An illustration of this occurred in 1995. The Conservative Prime Minister John Major became concerned and angered by the disloyalty of a number of his own backbench MPs. He therefore resigned as party leader (but not as Prime Minister). In the subsequent leadership election, he was re-elected. This was a great boost to both his authority and his power. He had re-asserted some degree of control over Parliament.

Patronage

Patronage refers to the power an individual may enjoy to make important appointments to public offices. Having this ability grants power because it means that those who aspire to high office will tend to be loyal to the person who has the power to appoint them. Once appointed, that loyalty remains, not least because disloyalty may end in dismissal. The Prime Minister enjoys patronage over hundreds of appointments, including government ministers, peers and the heads of various state bodies. It means that the majority of MPs and peers in the Prime Minister's party will tend to be loyal to them. This gives the PM great power — they exert considerable influence over their party, their MPs and the whole of Parliament.

The people

The Prime Minister is not directly elected by the people. Nevertheless, during a general election campaign, the people are being asked to choose between alternative candidates for the position as well as for a party. We can therefore say that a prime minister does enjoy a degree of authority directly from the people. This causes a problem for prime ministers who rise to their position without a general election taking place.

> **Synoptic link**
>
> During a general election, the role of the party leader, outlined in Chapter 4, is crucial as they are appealing to the public to choose them to be the Prime Minister.

The powers of the UK Prime Minister

The key to understanding the position of Prime Minister is the nature of the royal prerogative. Before the powers of the monarch were curbed in the seventeenth century, the monarch enjoyed what were known as 'prerogative powers'. These were powers that could not be controlled by Parliament or any other body. These powers were mainly to wage war, to make foreign treaties, and to appoint ministers and other people to public office. The monarch could not levy taxes without the permission of Parliament and had to have their spending plans approved. As to other laws, including criminal law, the monarch could propose such laws, but they needed parliamentary approval.

During the seventeenth to nineteenth centuries, as the political authority of the monarchy began to wane, the question arose: What is to be done with the prerogative powers? If we no longer accept that the monarch can exercise them in a democratic system, who should exercise them? Parliament was much too big and too fragmented to do such things as conduct foreign policy or negotiate treaties or appoint ministers. The obvious candidate was government itself, the Cabinet in particular, but even the Cabinet was not capable of giving *singularity of purpose*. There needed to be a single figurehead. As a result, the position of prime minister gradually evolved during the eighteenth century.

> **Knowledge check**
>
> What do Alec Douglas-Home, James Callaghan, John Major, Gordon Brown, Theresa May and Boris Johnson all have in common?

> **Study tip**
>
> Try not to use 'power' and 'authority' interchangeably. They have different meanings. Use 'authority' when you are referring to the *right* to exercise power, and use 'power' when you mean the *ability* to achieve political ends.

The main role and powers of the Prime Minister today can be summarised thus:

- The Prime Minister has complete power to appoint or dismiss all government ministers, whether in the Cabinet or outside the Cabinet. The Prime Minister also has a say in other public appointments, including the most senior civil servants.
- The Prime Minister has power to negotiate foreign treaties, including trade arrangements with other states or international organisations.
- The Prime Minister is **commander-in-chief** of the armed forces and can commit them to action. However, it should be noted that this power has come under challenge in recent times. It is now accepted that the Prime Minister should only make major military commitments 'on the advice and with the sanction of Parliament'. Nevertheless, once armed forces have been committed to action, the Prime Minister has general control of their actions.
- The Prime Minister conducts foreign policy and determines relationships with foreign powers. In this sense the Prime Minister represents the country internationally.
- The Prime Minister heads the Cabinet system (see below), chooses its members, sets its agenda and determines what Cabinet committees should exist and who should sit on them.
- It is generally true that the Prime Minister sets the general tone of economic policy. Usually this is done alongside the Chancellor of the Exchequer, who is normally a very close colleague.

The Prime Minister has a number of formal and informal powers, as outlined in Table 7.2. Most of the formal powers derive from the royal prerogative, while the informal ones tend to derive from the other sources of prime ministerial authority. It is worth noting that while all prime ministers are granted the same powers, circumstances can affect how able they are to exercise these powers, something that will be considered later in this chapter.

> **Useful term**
>
> **Commander-in-chief** This term describes the person who has ultimate control over the deployment of the armed forces, including the security and intelligence services. In the UK the Prime Minister holds this position, delegated by the monarch.

Table 7.2 The powers of the Prime Minister

Formal powers (the powers that all prime ministers enjoy)	Informal powers (the powers that vary from one individual to another)
Patronage: • Appointing ministers • Appointing judges and peers • Granting other honours Chairing of Cabinet, including setting the agenda Foreign policy leader Commander-in-chief Signing foreign treaties and international agreements Ability to call an early general election if Parliament approves with a two-thirds majority or passes a vote of no confidence Power to recall Parliament	Controlling and setting government policy Controlling and setting the legislative agenda Economic leadership Media focus and platform National leadership in times of crisis

> **Knowledge check**
>
> Identify, with examples, three powers held by the Prime Minister.

How has the prerogative power of commander-in-chief changed?

Before we consider these powers, we need to discuss the role of commander-in-chief. Until the twenty-first century, it was generally accepted that the Prime Minister had the sole power to commit UK armed forces to action. While the Prime Minister might consult with their Cabinet and invite a parliamentary debate, it was acknowledged that the final decision belonged to the PM. There have been several modern examples of this prerogative power being exercised, as shown in Table 7.3.

All this appeared to change abruptly in 2013. It was revealed that the Syrian government was using chemical weapons against civilian populations in the civil war there. In response Prime Minister David Cameron stated his desire to intervene, using UK air power. On this occasion, however, he sought the approval of Parliament. He did not need this approval constitutionally, but he felt it was politically important to seek it. To Cameron's surprise, the House of Commons voted against such action. He respected the decision and cancelled any proposed intervention. It appeared that centuries of the prerogative power to command the armed forces had been set aside. Parliament seemed to be taking over military policy.

> Miliband: Can the Prime Minister confirm to the House that he will not use the royal prerogative to order the UK to be part of military action, given the will of the House that is expressed tonight?
>
> Cameron: I can give that assurance… I also believe in respecting the will of this House of Commons… It is clear to me that the British Parliament, reflecting the views of the British people, does not wish to see military intervention. I get that, and I will act accordingly.

Source: The exchange between opposition leader Ed Miliband and Prime Minister David Cameron at the end of the debate on intervention in Syria, Hansard, 30 August 2013

Two years later, in December 2015, Cameron again asked Parliament for approval for air strikes in Syria, this time against ISIS/Daesh. Parliament gave its approval and the strikes began. However, the fact that Cameron felt the need to consult MPs demonstrated the vulnerability of his position.

Table 7.3 Recent examples of the exercise of prerogative power

Date	Example
1982	Margaret Thatcher sent a task force to 'liberate' the Falkland Islands in the South Atlantic from Argentine occupation
1999	Tony Blair committed British ground troops to intervene in the Kosovo war in the Balkans
2000	Tony Blair sent troops to Sierra Leone in West Africa to save the elected government from an armed insurgence
2003	Tony Blair committed UK forces to assist the USA in the invasion of Iraq to depose Saddam Hussein
2011	David Cameron committed the Royal Air Force to air strikes in the Libyan civil war to save the 'democratic' rebels
2017	Theresa May called an early general election and successfully secured the decision with a two-thirds majority vote in the House of Commons
2019	Boris Johnson appointed Nicky Morgan as a peer so she could continue to be the Culture Secretary, until she stepped down in February 2020
2020	Boris Johnson signed the EU 'divorce treaty' to formally break the ties between the EU and the UK and begin the transition period

The Cabinet

The Cabinet sits at the centre of power in the UK political system. Indeed, the UK system of government used to be commonly described as '**Cabinet government**'. This is not to say that it is where all important decisions are made. It is not. Rather, it means that all official government decisions and policies must be cleared by the Cabinet if they are to be considered legitimate. In that sense the Cabinet holds a similar position to the UK Parliament. In order to be implemented and enforced, all laws must be approved by Parliament. In the case of policies and government decisions (which often lead to law-making), they must be approved by the Cabinet if they are to be considered *official policy*. In the case of both, Parliament and Cabinet approval may well be brief and may require little meaningful debate, but such formal approval is essential. Occasionally, of course, conflict and real disagreement may occur in both Parliament and Cabinet, but often such approval is merely ritualised. Cabinet is therefore sometimes described as a mere 'rubber stamp', simply confirming decisions that have already been made by the Prime Minister, working with their their special advisors in Downing Street, or perhaps a handful of close associates, or even influential think tanks.

> **Useful term**
>
> **Cabinet government**
> A situation where the main decision-making of government takes place in Cabinet. In modern history this is not normally the case. Its main alternative is the expression 'prime ministerial government'.

The nature of the Cabinet

The Cabinet consists of between twenty and twenty-five senior government ministers. The precise number of members is in the hands of the Prime Minister. Indeed, the Prime Minister controls much of the work and nature of the Cabinet. It is one of their key roles. The Prime Minister also personally appoints all Cabinet members and may dismiss them. The PM is not required to consult anyone else when making appointments or dismissals. Most of the members are senior ministers in charge of large government departments. A few may not have specific ministerial responsibilities but are considered important enough members of the party to sit at the centre of power. All Cabinet members must be members of either the House of Commons (and therefore also MPs) or the House of Lords (as peers). In practice, most are MPs.

Cabinet positions are listed in Table 7.4 in approximate order of seniority, with the top four posts traditionally being considered the 'four great offices of state'. Table 7.5 identifies other senior ministers who attend Cabinet meetings but are not full members of the Cabinet.

Table 7.4 Cabinet offices in approximate order of seniority

Position
Prime Minister
Chancellor of the Exchequer
First Secretary of State and Secretary of State for Foreign and Commonwealth Affairs
Home Secretary
Chancellor of the Duchy of Lancaster
Lord Chancellor/Secretary of State for Justice
Secretary of State for Defence
Secretary of State for Health and Social Care
Secretary of State for Business, Energy and Industrial Strategy
Secretary of State for International Trade
Secretary of State for Work and Pensions

Position
Secretary of State for Education
Secretary of State for Environment, Food and Rural Affairs
Secretary of State for Housing, Communities and Local Government
Secretary of State for Transport
Secretary of State for Northern Ireland
Secretary of State for Scotland
Secretary of State for Wales
Leader of the House of Lords/Lord Keeper of the Privy Seal
Secretary of State for Digital, Culture, Media and Sport
Secretary of State for International Development
Minister Without Portfolio

Table 7.5 People who also attend Cabinet meetings but are not full members

Position
Chief Secretary to the Treasury
Leader of the House of Commons and Lord President of the Council
Chief Whip
Attorney General for England and Wales

Several other ministers are also invited to attend Cabinet meetings and take part in discussions but are not Cabinet ministers. When final decisions are being made, their view is not invited. One of them will always be the chief whip of the governing party.

Individuals may also be invited to address the Cabinet if they have special knowledge or important views, but they will not take part in full discussions. One civil servant always attends to record the minutes (what is agreed). This is the Cabinet secretary, the UK's most senior civil servant. They are a key adviser to the Cabinet and to the Prime Minister personally. In June 2020 it was announced that Sir Mark Sedwill would step down from this role to be replaced by Simon Case in September 2020, as Prime Minister Boris Johnson sought to overhaul the working and structure of the civil service.

A number of other features of the Cabinet are also noteworthy:

- Only members of the governing party are Cabinet members. The only exception is with coalition government, as occurred in 2010–15. In that case there were both Conservative and Liberal Democrat members.
- Cabinet normally meets once a week, usually on a Thursday, and a meeting rarely lasts more than 2 hours.
- Additional emergency Cabinet meetings may also be called.
- The Prime Minister chairs the meetings unless abroad or indisposed, in which case their deputy may take over, though when this occurs Cabinet may not meet at all.
- The proceedings of the Cabinet are secret and will not be revealed for at least 30 years.
- Cabinet does not usually vote on issues. The Prime Minister always seeks a general consensus and then requires all members to agree to that consensus decision. Any member who wishes to dissent publicly will normally be required to resign and leave the Cabinet.
- The Prime Minister sets the final agenda.
- The Prime Minister approves the minutes made by the Cabinet secretary. These are a record of formal decisions made and key points raised for consideration.

- Cabinet decisions are released to a strictly limited number of civil servants and ministers. Media releases are also sent out, but with no details of the discussions.
- Cabinet members receive an enhanced salary, well above that of junior (non-Cabinet) ministers and MPs.
- Members of the Cabinet are bound by the convention of *collective responsibility*. This is described later in this chapter.

The role of the Cabinet

Perhaps surprisingly, the role of the Cabinet is both changeable and unclear. Indeed, like the role of the Prime Minister, its existence is merely an unwritten constitutional convention. To some extent, what it does may vary from one prime minister to another. It may also depend on political circumstances, and the factors affecting this relationship will be addressed later in this chapter. Yet, despite the variability of the Cabinet's position, it does have several functions that are common to all administrations in the UK. These are as follows:

- In some emergency or crisis situations the Prime Minister may revert to the collective wisdom of the Cabinet to make decisions. They may take a leading role in the discussion but will also invite comments from their close colleagues. Military situations are the most common example, such as UK intervention in the Syrian civil war and in the war against the Taliban in Afghanistan. Even a determined prime minister will normally inform the Cabinet of their intentions, as Tony Blair did before joining the US-led invasion of Iraq in 2003 and Margaret Thatcher did before sending a task force to liberate the Falkland Islands in 1982. The fact that Cabinet meetings are held in secret helps when military and security matters are at stake.
- Cabinet will discuss and set the way in which policy is to be *presented* to Parliament, to the government's own MPs and peers, and to the media. It helps to present a united front when all ministers describe and justify decisions in the same manner.
- Occasionally disputes can arise between ministers, very often over how government expenditure is to be shared out. Normally the Prime Minister and Cabinet secretary will try to solve such disputes, but, when this is not possible, the Cabinet acts as the final 'court of appeal'.
- Most government business must pass through Parliament, often in the form of legislation. The Cabinet will settle the government's agenda to deal with this. It is decided what business will be brought before Parliament in the immediate future, which ministers will contribute to debates and what tactics to adopt if votes in either house are likely to be close. The chief whip's presence is vital on these occasions.

Despite the need to carry out these functions from time to time, most of Cabinet's time is taken up with ratifying decisions reached elsewhere. Ministers are informed in advance of such proposals. Their civil servants prepare brief summaries of what is being proposed and any likely problems that might arise. If ministers decide they have some misgivings about proposals, they normally raise them with the Prime Minister or Cabinet secretary before the meeting, not during it. Despite what the popular press often claims, Cabinet 'rows' are rare. Any negotiations that need to be done are normally settled outside the Cabinet room.

So, the Cabinet is a kind of 'clearing house' for decisions. Usually, little discussion is needed. The Prime Minister will check that everyone can support a decision and it

Study tip

Although it is often said that the Cabinet is at the 'centre' of government, this does not mean it is where most decisions are made. Most decisions today are made elsewhere, so be careful not to confuse these two realities.

The structure and workings of the Cabinet

As we have seen, most decisions are made outside Cabinet and they only need to be formally approved in a full Cabinet meeting. Therefore it is better to think of a 'Cabinet system', rather than simply 'the Cabinet'. Where do these decisions originate? The answer is from a variety of sources.

The Prime Minister

The Prime Minister, together with their advisers, policy units, close ministerial allies and senior civil servants (known collectively as 'the **Downing Street Machine**'), will develop proposals of their own. It is extremely rare for the Cabinet to seriously question a prime ministerial initiative. When ministers intend to oppose the Prime Minister, they usually resign, an event that is invariably extremely dramatic. Perhaps the most remarkable example was when Sir Geoffrey Howe resigned from Margaret Thatcher's Cabinet in 1989, largely over her European policies. Howe's resignation and farewell speech in the Commons helped to bring Thatcher down the following year. Tony Blair lost two Cabinet colleagues (Robin Cook and Clare Short) in 2003 over his Iraq policy. But such events are rare. While a single Cabinet resignation can be embarrassing, it is not often a disaster for the Prime Minister. However, when it becomes endemic it can be highly damaging to a prime minister, as was seen when a total of 11 Cabinet ministers resigned from Theresa May's Cabinet over an 18-month period from November 2017 to May 2019, mostly over the issue of Brexit, damaging her authority and control over the Cabinet.

Cabinet committees

Most detailed policy is worked out in small committees consisting of Cabinet members and junior ministers. Most of these **Cabinet committees** are chaired by the Prime Minister or a very senior minister, such as the Chancellor. The committees present their proposals to full Cabinet and they are usually accepted, though they may sometimes be referred back to committee for amendments and improvement.

The Chancellor of the Exchequer

Almost always supported by the Prime Minister, economic and financial policy is presented to the Cabinet by the Chancellor, often as a *fait accompli*. Indeed, the annual Autumn Statement (in November) and the Budget (in March) are usually only revealed to the Cabinet on the eve of their presentation in Parliament.

The Budget must be passed by Parliament in the months following its presentation. This is largely a formal process but occasionally there has been dissension. In March 2017, for example, Philip Hammond's proposal to increase national insurance for the self-employed was resisted by all opposition parties and by a number of Conservative rebels, so the measure was quickly dropped.

Individual ministers

Policies that involve a government department specifically, but which require wider approval, are presented to Cabinet by the relevant minister aided by their civil servants. It is here that dissent is most likely, though a minister who is backed by the Prime Minister is in a good position to secure approval.

Groups of ministers

Policies are often developed by various professional advisers, policy units and think tanks. These may be adopted by various ministers who then bring the ideas to Cabinet, usually after securing the approval of the Prime Minister and Chancellor. If other ministers have problems with such proposals, they are usually voiced well in advance.

We shall see below that the variety of sources of policy coming into Cabinet helps the Prime Minister to control government in general. Prime ministers see all proposals in advance and have the opportunity to block policies of which they do not approve. They also control the Cabinet agenda so they can simply avoid discussion of ideas they do not like. Most prime ministers, most of the time, can manage the Cabinet system to promote their own policies and block those they wish to oppose. It was notable that Theresa May struggled to control her Cabinet, as we have seen, which perhaps explains why the Johnson Cabinet has been made up wholly of pro-Brexit ministers, meaning that to some extent it lacks the broad range of diverse opinions which are often seen as a strength in the deliberations of Cabinet.

Boris Johnson chairs a cabinet meeting, November 3 2020

The powers of the UK Cabinet

The Cabinet has a number of important roles but, surprisingly perhaps, it has relatively few powers of its own. This is largely because the Prime Minister has their own rival powers. However, we can identify a number of powers that the Cabinet has, whatever the Prime Minister may try to do. These are as follows:

- **Legitimising government policy** The Cabinet legitimises government policy and interprets what government policy actually is. The Prime Minister will have a say in this, but ultimately it is a Cabinet power to organise the presentation of official policy. When it comes to a vote in the Cabinet, the PM has one vote and so can be outvoted (though if a PM suspects they will lose such a vote, they will usually choose not to hold it).
- **Setting the legislative agenda** Though the Prime Minister has influence, it is a specific power of the Cabinet to determine the government's legislative agenda.
- **Supporting the Prime Minister** Cabinet does not have absolute power to remove a prime minister. There is no such thing as a 'vote of no confidence' in the Cabinet. Nevertheless, Cabinet can *effectively* drive a prime minister out of power by refusing to support them in public. The removal of a prime minister

has two main procedures: either forcing the prime minister to resign through public criticism (as happened to Tony Blair in 2007 and Theresa May in 2019) or provoking a leadership contest in the governing party that the prime minister may lose (as happened to Margaret Thatcher in 1990).

- **Deciding on government policy** The Cabinet has the power to overrule a prime minister if it can summon up enough political will and sufficient support for an alternative policy. In 2015, for example, Prime Minister David Cameron was forced by his Cabinet to suspend collective responsibility in the EU referendum campaign to allow ministers to express their personal views.

Apart from those described above, the Cabinet does not really have any powers of its own. Government power is effectively shared between the Prime Minister and Cabinet.

Ministerial responsibility

Ministerial responsibility comes in two forms: **collective responsibility** and **individual responsibility**. They are both conventions that ensure the executive works as a single entity and that it is clear who takes responsibility for the actions and conduct of each aspect of the executive.

> ### Synoptic link
>
> The concepts of both collective and individual ministerial responsibility exist as conventions in the UK Constitution, and can be interpreted flexibly, as shown in Chapter 5.

Collective responsibility

What is collective responsibility?

Governing in the UK is a collective exercise on the whole. While the Prime Minister does have their own prerogative powers, for example over foreign and security policy, decisions are taken *collectively* by the executive. This means that *all* ministers (whether in the Cabinet or not) are *collectively responsible* for all executive policies and decisions. Even though most policy is created by the most senior members of government, there is a convention that all ministers will defend and publicly support all official policy. It is part of the 'deal' when they take office. This is known as the doctrine of collective ministerial responsibility. It has five principles:

1 Ministers are collectively responsible for all government policies.
2 All ministers must publicly support all government policies, even if privately they disagree with them.
3 If a minister wishes to dissent publicly from a government policy, they are expected to resign as a minister first (and return to the backbenches).
4 If a minister dissents without resigning, he or she can expect to be dismissed by the Prime Minister.
5 As Cabinet meetings are secret, any dissent within government is concealed.

Why is collective responsibility important?

The principle of collective responsibility within government is a great support to prime ministerial power and this is perhaps its main significance. A prime minister's authority is greatly enhanced by the fact that they will not experience open dissent from within the government. It is also important that the government presents a united front to the outside world, including Parliament and the media. Specifically, the government knows it can rely upon the votes of all ministers in any close division in the Commons. This is known as the *payroll vote* and, in theory, means the Prime Minister can rely on about 120 votes in the House of Commons before any vote, or division, is taken.

It can also be said that collective responsibility reduces the possibility of open dissent. Critics say that it 'gags' ministers and prevents them from expressing their opinions. Supporters of the principle, on the other hand, say that the secrecy of the system means that ministers *can* express their views honestly *within* Cabinet, knowing that their disagreement is unlikely to be publicised.

Three key exceptions to collective responsibility
In recent years collective responsibility has been suspended for three different reasons:

- The first occurred when UK government was a **coalition** between the Conservatives and Liberal Democrats in 2010–15. Clearly it would have been impossible for ministers from two quite different parties to agree on every policy. Nobody would have believed them had they made such a claim. A special arrangement was therefore made. The coalition arrived at a Coalition Agreement, which included all the policies the two-party leaderships decided should be common to both sets of ministers. Collective responsibility applied to the Coalition Agreement, but some areas of policy were not included. For example, the renewal of the Trident nuclear submarine missile system was excluded. Coalition ministers were allowed to disagree publicly on this issue. The same exception was applied to the question of intervention in the Syrian civil war.
- The second suspension of collective responsibility became necessary when it was decided to hold a referendum on UK membership of the EU in June 2016. During the campaign, Conservative ministers were free to express views counter to the official government position — that the UK should remain in the EU. Several Cabinet ministers, including former Justice Secretary Michael Gove, and former leader of the House of Commons Chris Grayling, openly campaigned against the official government line. A similar arrangement had been made the last time there was a referendum on UK membership of the European Economic Community (EEC) in 1975.

Useful term

Coalition A type of government, rare in the UK but common in the rest of Europe, where two or more parties share government posts and come to agreement on common policies. Coalitions occur when no single party can command a majority in the legislature.

Study tip

Make sure you can confidently distinguish between collective responsibility and individual ministerial responsibility in relation to UK government.

Michael Gove and Boris Johnson openly campagined for Vote Leave

- The third reason, though not an official suspension, was seen between 2017 and 2019, when Theresa May failed to exert collective responsibility as various Cabinet ministers became openly critical of her Brexit deal. This was more out of necessity than any real policy, as May was trying to find a consensus between two deeply divided sides and forcing any of the key Cabinet ministers to resign, despite their open hostility to her deal, may have led to a loss of support and an even earlier end to her premiership. Although more than 50 ministers in total would resign in this period, it is noticeable how publicly key Cabinet ministers criticised official policy without being forced to resign. This suggests that the doctrine depends more on the strength and authority of the Prime Minister than any hard and fast rule.

In summary we can say that collective responsibility can be viewed both negatively and positively.

Debate

Is collective responsibility a negative aspect of the UK political system?

It is negative

- It puts too much power into the hands of the Prime Minister.
- It means that ministers cannot be openly honest about their views on policies. This may stifle debate within government.
- Resignations under the doctrine are dramatic events that may seriously undermine government.

It is positive

- It creates a government that is united, strong and decisive.
- The public, Parliament and the media are presented with a clear, single version of government policy.
- Though ministers cannot dissent publicly, the confidentiality of the Cabinet means that ministers can engage in frank discussions in private.

 The key to evaluation for this debate is the importance it holds for the political system as a whole, meaning you must decide whether having a publicly unified government is a greater positive for the UK political system than the opportunity for open and public debate from the senior members of government.

Individual ministerial responsibility

The nature of individual responsibility

As we have seen, ministers are collectively responsible for government policies. However, each minister is also *individually* responsible for matters that affect their department separately. Ministers are also individually responsible for their own performance as a minister and their conduct as an individual. The doctrine of individual ministerial responsibility used to be a significant feature of governing in the UK, but in some ways in recent years it has declined in importance. The features of the principle are these:

1 Ministers must be prepared to be accountable to Parliament for the policies and decisions made by their department. This means answering questions in the House, facing interrogation by select committees and justifying their actions in debate.
2 If a minister makes a serious error of judgement, he or she should be required to resign.
3 If a serious error is made by the minister's department, whether or not the minister was involved in the cause of the error, the minister is honour-bound to resign.
4 If the conduct of a minister falls below the standards required of someone in public office, particularly if they break the **ministerial code** of conduct, they should leave office and may face dismissal by the Prime Minister.

Useful term

Ministerial code The rules of conduct as determined by the current Prime Minister. They set out how the PM expects all members of their government to behave, covering issues of collective responsibility, engagement with Parliament, avoiding conflict of interests, ethical considerations and the way in which all ministers are to engage and treat members of the civil service. Any MP can request an investigation by the Cabinet Secretary of any minister for breaching the ministerial code, though whether such an investigation takes place is entirely at the discretion of the PM.

Synoptic link

The accountability of ministers is a key aspect of how democracy operates in the setting of UK parliamentary politics. You should therefore consider individual ministerial responsibility in the light of Parliament's role of calling government to account. See Chapter 6.

Cabinet resignations due to individual ministerial responsibility

Priti Patel

In November 2017 reports emerged that the then International Development Secretary, Priti Patel, had arranged and attended meetings in August 2017 with Israeli politicians, including the Prime Minister Benjamin Netanyahu, without approval or authorisation from the Foreign and Commonwealth Office (FCO). This was a breach of the ministerial code of conduct as any meeting between an official representative of the UK executive and a foreign government has to be authorised by the FCO in order to ensure consistency and full knowledge of the diplomatic situation. After an initial apology to Prime Minister May, Priti Patel was allowed to stay in post, but when details of two other similar meetings between her and Israeli officials in September 2017 emerged, she was recalled from a meeting in Uganda and, after a 45-minute meeting with the PM, offered her resignation, claiming that 'while my actions were meant with the best of intentions, my actions also fell below the standards of transparency and openness that I have promoted and advocated'.

Amber Rudd

Between 1948 and 1973, people born as 'subjects of the British Empire' were given British citizenship and the right to live and work in the UK, leading to half a million people travelling to and settling in the UK as part of the Windrush generation. While restrictions were imposed after 1973, all those who had settled in the UK before then were granted full British citizenship, as were their children. However, in 2017 a scandal emerged that showed that in a bid to impose tougher immigration targets and tackle illegal immigration, the Home Office had been responsible for many of the Windrush generation being treated as illegal immigrants, with benefits and services being cancelled, some being detained, others losing their jobs, being denied re-entry into Britain, and at least 83 being wrongly deported, partly as a result of a wider target to deport 10 per cent of illegal immigrants. When challenged about this by the Home Affairs select committee in March 2018, the then Home Secretary, Amber Rudd, stated that she was not aware of any such targets existing. However, a letter soon emerged proving that she was fully aware of the 10 per cent target. In her resignation letter, Ms Rudd stated that she had unintentionally misled Parliament and was resigning for that reason.

Sajid Javid

The relationship between the Prime Minister (working from 10 Downing Street) and the Chancellor (working from 11 Downing Street) is perhaps the most important political relationship in the UK. Although the Chancellor is appointed by the PM, their position in charge of the Treasury means they have a rival power base and can, at times, be seen to exercise almost as much power and control over government policy because they are the one who determines the spending and taxation that impact on all other departments.

In many instances, the Chancellor has been almost autonomous in their power. This was most clearly shown during the Brown and Blair years of government, and led to problems and rivalries that would eventually undermine Blair's leadership. A similar dynamic was beginning to emerge in 2020 between Prime Minister Boris Johnson and his Chancellor, Sajid Javid, with Javid regularly clashing with Johnson's special adviser, Dominic Cummings. In February 2020 Johnson, acting on the recommendation of Cummings, told Javid to replace all his advisers with people chosen by Johnson and Cummings or to leave his post. Sajid Javid chose to resign, stating in his resignation letter that he 'was unable to accept those conditions and I do not believe any self-respecting minister would accept those conditions'.

The erosion of ministerial responsibility

The principles of individual ministerial responsibility were given earlier in this chapter. The first principle, that ministers must offer themselves to be accountable to Parliament, certainly operates successfully and is a key principle of UK government. It is fully described in Chapter 6.

The second and third principles, however, have largely fallen into disuse. There is no specific way in which Parliament can remove an individual minister. Parliament and its select committees can criticise a minister and call for their resignation, but whether or not they go is entirely in the hands of the Prime Minister. There was a time when ministers resigned as a matter of principle when a serious mistake was made, but those days have largely passed. The last time a Cabinet minister resigned as a result of errors made was when the then Home Secretary Amber Rudd resigned as a result of Home Office failings relating to the Windrush scandal and inaccurate information she provided to the Home Affairs select committee as a result (see the case study above). In her resignation letter, Ms Rudd said she took 'full responsibility' for the fact she was not aware of 'information provided to [her] office which makes mention of targets'. This was a rare event indeed. Before and since, many ministers have experienced widespread criticism and have apologised for errors made, but have neither resigned nor been dismissed.

This erosion of the principle does not, however, extend to the fourth type of responsibility, that which concerns personal conduct. Here, when ministers have fallen short of public standards, they have been quick to resign or been *required* to resign by the Prime Minister. Some recent examples are shown in Table 7.6.

Table 7.6 Ministerial resignations/removals following personal misconduct

Year	Minister	Position	Reason for resignation
2010	David Laws	Treasury Secretary	Irregularities over his expenses claims
2011	Liam Fox	Defence Secretary	Employing a personal friend as adviser at public expense
2012	Chris Huhne	Energy Secretary	Conviction of a serious criminal offence
2012	Andrew Mitchell	Chief whip	Allegedly insulting a policer officer in Downing Street, using abusive language
2017	Michael Fallon	Defence Secretary	Allegations over personal behaviour
2017	Priti Patel	International Development Secretary	Conflict of interest over meetings with Israel
2017	Damien Green	First Secretary of State	Breaching the ministerial code after making 'misleading' statements about pornography allegations
2019	Gavin Williamson	Defence Secretary	Sacked in relation to a leak to the press of a security council meeting

We can see from Table 7.6 that ministers are vulnerable when it comes to their personal conduct. Perhaps ironically, it is more likely that a minister will lose their job because of personal conduct rather than the quality of their performance in office.

The erosion of ministerial responsibility

Gavin Williamson

In the wake of the Covid-19 pandemic and the closure of schools in March 2019, A Level and GCSE examinations were cancelled. The Secretary of State for Education, Gavin Williamson, tasked his civil servants working in the Department of Education and the government's exam regulator Ofqual with devising a system to award grades to candidates based on a concept of 'Centre-Assessed Grades' that would be a robust and fair system to make awarded grades comparable to other years. As the Education Secretary, final approval of any system and the process rested with Gavin Williamson as part of the code of individual ministerial responsibility, not with the civil servants who developed the system. However, the grades awarded met with public outcry over the unfairness of how they were awarded and many pupils were seen to be treated unfairly. The minister blamed the algorithm that had been devised by Ofqual for being overly restrictive and replaced the system created by awarding grades based on original teacher-based estimates. Under the ministerial code, it would have been expected that the minister responsible, Gavin Williamson, should have taken responsibility and resigned from his position. However, he refused to resign and was supported by the Prime Minister and it was in fact the Head of Ofqual that resigned while the Education Secretary continued in post.

Chris Grayling

In 2018 the railway companies introduced a series of new timetables to make the railway network more efficient and assured the Transport Secretary at the time, Chris Grayling, that there would be no major disruption. Chris Grayling allowed the reforms, but the changes led to chaos and many passengers being stranded or unable to travel to work while services were heavily disrupted for months. Although blame for the failures could be placed on the railway companies, an interim report by the railway regulator, the Office of Road and Rail (ORR) and a report issued by the Transport Select Committee in December 2018 both recorded that no one took charge when it became clear that the timetable was causing problems and there was a clear lack of leadership by the Department of Transport. The failures to respond to the crisis were largely placed on the Secretary of State and under ministerial responsibility, it would have been expected for him to resign. Although he apologised for the chaos and acknowledged there had been problems, Chris Grayling said that his mistake was in taking the initial report by the railways at face value and refused to resign. He continued as Transport Secretary until July 2019 and only left for the backbenches when Theresa May stepped down as Prime Minister.

Priti Patel

In March 2020 Sir Philip Rutnam, the most senior civil servant working in the Home Office, resigned in protest of what he claimed was bullying by the then Home Secretary Priti Patel, claiming that she had created a culture of fear. In a statement issued to the BBC, Sir Philip claimed that Priti Patel had been advised early on not to shout and swear at staff and that she must treat members of the Civil Service with respect. This appeared to be a clear breach of the ministerial code of conduct with set out expected standards of behaviour for all ministers whilst in office, including 'consideration and respect' for civil servants and other colleagues. Prime Minister Boris Johnson asked Sir Alex Allan, the Prime Minister's independent advisor on the ministerial code (a civil service position) to investigate the allegations made against Priti Patel. In his report issued in November 2020, Sir Alex found that Priti Patel had 'not consistently met the high standards required by the ministerial code of treating her civil servants with consideration and respect' and cited examples of swearing and shouting at staff and that 'her approach on occasions has amounted to behaviour that can be described as bullying in terms of the impact felt by individuals'. Although this appeared to be a clear case of a breach of individual ministerial responsibility that would traditionally have seen a minister resign or be removed, Priti Patel issued an apology but claimed she was not aware of the impact of her behaviour and that she had not been supported by the staff at the time. The Prime Minister supported her saying that he did not believe she was a bully and that there were mitigating circumstances and said he did not believe she needed to resign. In protest at the decision, Sir Alex Allan resigned claiming he could no longer work as the prime minister's independent advisor on the code.

The Prime Minister and the Cabinet

The power of the Prime Minister and the Cabinet

Earlier in this chapter, we reviewed the main roles and powers of the Prime Minister and the Cabinet. While the formal or legal powers held by each remain constant from one prime minister to another, their ability to exercise those powers varies based on a range of circumstances. Different circumstances can arise to make one prime minister appear more powerful than another and as the authority of a prime minister increases, so the authority of the Cabinet to influence them decreases, and vice versa. Here we will analyse and evaluate the changing nature of the relationship between the office of Prime Minister and the Cabinet. Some prime ministers may use the Cabinet as an important sounding board for ideas and policy initiatives. John Major and David Cameron, for example, used it in this way. Other prime ministers, notably Tony Blair and Margaret Thatcher, had little time for Cabinet discussion and tended to use it simply to legitimise decisions made elsewhere. Margaret Thatcher (1979–90), indeed, was notorious for downgrading Cabinet to a rubber stamp for her own ideas. One of her ministers, Nicholas Ridley, expressed her style thus:

> Margaret Thatcher was going to be the leader in her Cabinet. She wasn't going to be an impartial chairman. She knew what she wanted to do and was not going to have faint hearts in the Cabinet stopping her.

Source: Quoted in Hennessy, P., *The Prime Ministers*, Allen Lane, p. 400

Table 7.7 shows some of the powers that influence the balance of power between the Prime Minister and the Cabinet.

Table 7.7 Who has the upper hand — Prime Minister or Cabinet?

The powers of the Prime Minister	The powers of the Cabinet
The Prime Minister is perceived by the public to be government leader and representative of the nation. This gives them great authority	If the Cabinet is determined, a majority of members can overrule the Prime Minister
Prime ministerial patronage means the Prime Minister has power over ministers and can demand loyalty	Ultimately the Cabinet can remove the Prime Minister from office, as happened to Margaret Thatcher (1990) and Tony Blair (2007)
The Prime Minister chairs the Cabinet and controls its agenda, which means they can control the governing process	The Cabinet may contain powerful ministers with a large following who can thwart the will of the Prime Minister. Tony Blair was rivalled by Gordon Brown in 2005–07, and David Cameron was challenged by several influential Eurosceptics in 2010–15
The Prime Minister enjoys prerogative powers and so can bypass the Cabinet on some issues	If the Prime Minister has a small or non-existent majority in the Commons, the Cabinet becomes more important in implementing their agenda

Factors governing the selection of Cabinet ministers

Selecting the members of a cabinet is one of the key roles played by a prime minister and is often a reflection of their authority and ability to exercise their personal power. If they get this wrong, they will suffer difficulties ranging from poor policy-making to constant threats to their own position. It may seem simple, to choose the people for the job, but there is more to it than that. Essentially, prime ministers have three ways of constructing a cabinet:

1 To pack the Cabinet with the Prime Minister's own allies. This ensures unity and bolsters the Prime Minister's power, but it may lack critical voices who can improve decision-making. After 1982 this was the tactic adopted by Margaret Thatcher (1979–90), an especially dominant Prime Minister with a great singularity of purpose. Tony Blair (1997–2007) adopted a similar approach.

2 To pick a balanced Cabinet that reflects the different policy tendencies in the ruling party. When Theresa May became Prime Minister in 2016 she chose such a Cabinet, which included some of her former adversaries such as Boris Johnson, David Davis, Andrea Leadsom and Liam Fox. It was especially important for her to include members who were both in favour of and against leaving the EU. She did, though, keep some key allies close to her, including Chancellor Philip Hammond and Home Secretary Amber Rudd. John Major (1990–97) was forced into choosing a similarly varied Cabinet.

3 To pick a Cabinet of the best possible people. Such a Cabinet has not been seen since the 1960s and 1970s when Harold Wilson (1964–70, 1974–76) and James Callaghan (1976–79) assembled a group 'of all the talents'.

Prime ministers have complete patronage powers so they can **reshuffle** their cabinets at will. Some prime ministers have changed the personnel in this way annually. Dismissing and appointing new ministers is a device prime ministers can use for asserting and re-asserting their authority, as well as ensuring the quality of government, but they are not entirely free in how they make their choices. The factors that can influence a prime minister's selection of Cabinet members are listed in Table 7.8.

> **Useful term**
>
> **Reshuffle** Occurs when a prime minister changes the make-up of their government. A major reshuffle is when a number of Cabinet members are dismissed, appointed or have their jobs changed.

Table 7.8 Factors affecting Cabinet appointments

Factor	How it influences prime ministerial selection	Examples
Party unity	A prime minister may wish to ensure a balanced Cabinet that reflects different political wings of their party. The more divided the party, the more difficult this becomes. However, an authoritative PM may back one group over another for ideological reasons	Theresa May appointed a mixture of hard and soft Brexit supporters in her Cabinet Boris Johnson removed many 'soft Brexiteers' from his Cabinet
Experience	A prime minister must decide whether or not to have senior, experienced, heavyweight figures in their Cabinet. These can be more difficult to control, but can add weight and standing to the Cabinet A new PM may find having experienced ministers useful for giving advice on how government operates. Experienced ministers with prominent reputations (known as 'big beasts') on the backbenches can become powerful opponents and the focus point for opposition	David Cameron appointed former Conservative leaders William Hague and Iain Duncan Smith to his coalition Cabinet in deference to their standing in the party Theresa May removed George Osborne and Boris Johnson removed Philip Hammond from their first cabinets as they were too 'big' and could be a potential rival to the PM
Ability	The PM must identify talented MPs and give them an opportunity to apply their talents. A PM may well promote an able junior minister to a Cabinet post or move them to a senior Cabinet post. Ministers who are less able may be demoted or removed	Rishi Sunak was made Chancellor in 2020, having proven his ability as Chief Secretary to the Treasury and in other junior posts before then In 2012, Andrew Lansley was removed as Health Secretary as he was felt to be out of his depth and unable to implement the desired reforms

Factor	How it influences prime ministerial selection	Examples
Allies and advisers	A prime minister may want to have close allies in senior positions in the Cabinet, to help them formulate policy within the Cabinet and to add advice and support in Cabinet meetings and decisions	Gordon Brown had Peter Mandelson made a peer so he could become a member of his Cabinet to offer advice and assistance David Cameron appointed his main political ally, George Osborne, as Chancellor
External pressure	Although a PM is technically free to make whatever selections they wish to the Cabinet (as long as they are currently in Parliament), they also have to consider public attitudes, expressed through the media, or demands made through Parliament. However, a PM can ride out such public criticism in support of an ally	Amber Rudd was forced to resign following public and parliamentary outrage at the Windrush scandal Jeremy Hunt survived public and parliamentary calls for his resignation as Culture Secretary over the phone-hacking scandal and then as Health Secretary
Coalition agreement	As part of the coalition agreement, the then PM, David Cameron, had to have the Liberal Democrat leader, Nick Clegg, as his Deputy Prime Minister and to allow Clegg to appoint four other Cabinet positions from the Liberal Democrat MPs	When Liberal Democrat MP Chris Huhne was forced to resign as Secretary of State for Energy and Climate Change, he was replaced by another Liberal Democrat, Ed Davey Despite controversy surrounding Vince Cable's impartiality relating to Rupert Murdoch and the merger of BSkyB, David Cameron could not remove him from his Cabinet post
Diversity	A PM may wish to make their Cabinet more reflective of the nation in order to get a wider range of opinions. This could involve appointing people from a diverse range of backgrounds, geographic regions, racial backgrounds, gender and sexuality backgrounds, and so on	It is notable that since 1997 there has been something of an increase in the number of female and BAME Cabinet ministers; this may, however, simply be a reflection of greater diversity within Parliament

Prime Minister and Cabinet: A changing relationship

We can divide the recent history of the relationship between prime ministers and their Cabinets into four periods, as shown below.

Up to the 1960s

The Prime Minister was seen as 'first among equals', in other words the dominant member of the Cabinet but not able to command government completely. Prime ministers were aware that they had to carry their Cabinets with them and so had to allow genuine debate among ministers. This was often characterised as 'Cabinet government'.

1960s–2010

This period is often described as one of 'prime ministerial government'. Prime ministers were expected to dominate government completely. There had to be a Cabinet and decisions had to be legitimised by the Cabinet, but it was not expected that the Cabinet would act *as a collective body*, but rather that it should collectively support the Prime Minister. Successive prime ministers found ways of dominating the Cabinet or simply sidelining it so that it was relatively insignificant. This is not to say that *individual* ministers could not be powerful, but that the Cabinet as a body was not powerful. Former Labour Cabinet member Mo Mowlam summed this up in 2001 when describing how Prime Minister Blair managed the Cabinet. 'Mr Blair makes decisions,' she said, 'with a small coterie of people, advisers, just like the President of the United States. He doesn't go back to Cabinet, he isn't inclusive in terms of other Cabinet ministers' (*Guardian*, 17 November 2001).

Three styles of prime ministerial domination stood out in this period:

- Harold Wilson (1964–70 and 1974–76) would allow Cabinet ministers to play a prominent public role and develop their own policies, but when it came to deciding the overall direction of government policy, he manipulated Cabinet by controlling the agenda and discussions, and by reaching agreements with ministers outside the meetings.
- Margaret Thatcher (1979–90) dominated the Cabinet through the force of her will and by ruthlessly removing or marginalising her opponents.
- Tony Blair (1997–2007) marginalised Cabinet. He adopted a style known as '**sofa politics**' whereby he would develop ideas with a few advisers and senior ministers outside the Cabinet in informal discussions and then present the Cabinet with a *fait accompli*. This went further than Wilson, as ministers would have policy imposed on their departments and the Prime Minister himself would take the dominant public role on key issues.

The Cabinet during weakened single-party rule, 2010–19

Before continuing, it is important to understand what happened in the unique period when there was a coalition government in the UK between 2010 and 2015.

After the 2010 general election, no party enjoyed an overall majority in the House of Commons. It was therefore necessary to form a coalition that could command such a majority. The alternative would have been a minority government. Minority government is a daunting prospect. Such a government has to build a majority of support among MPs for each individual legislative proposal. This is extremely difficult and the government constantly faces the imminent prospect of defeat. Minority governments have survived in Scotland and Wales, and there was a brief period of minority (Labour) government in the UK from February to October 1974, but they are rare exceptions. So, in 2010, when there was a **hung parliament**, a coalition was quickly agreed between the Conservative and Liberal Democrat leaderships. The arrangements for coalition were as follows:

- As leader of the larger of the coalition partners, David Cameron was to be Prime Minister. Nick Clegg, leader of the Liberal Democrats, was to be Deputy Prime Minister.
- A period of negotiation followed during which an agreed set of policies was developed — the Coalition Agreement.
- Cabinet places were apportioned to the two parties in the ratio 22:5 Conservatives to Liberal Democrats. The Liberal Democrats were given five specific ministerial positions. Non-Cabinet posts were apportioned on a similar basis.
- David Cameron would control appointments or dismissals to the twenty-two Conservative posts and Nick Clegg controlled the five Liberal Democrat posts.
- Collective responsibility applied to all policies included in the Coalition Agreement. On other policies, ministers from the two parties were permitted to disagree publicly.

David Cameron still marginalised most of the Cabinet, working with an '**inner cabinet**'. This consisted of himself, Chancellor George Osborne, Liberal Democrat leader Nick Clegg, and Danny Alexander, Osborne's Liberal Democrat deputy. They were collectively known as the Quad. Cabinet is too big to serve the Prime Minister constantly, so such inner groups of senior ministers are common.

Activity

Research the coalition Cabinet appointed in 2010. Identify which positions were filled by Liberal Democrat ministers. Why do you think they were granted those positions and not others?

Ironically, the coalition proved to be something of a brief 'golden age' for the Cabinet. Suddenly, after years of becoming less and less significant, being increasingly marginalised within government and ignored by prime ministers, the Cabinet was important again. This was largely because the Cabinet now had roles it had never had before:

- Disputes within the coalition were inevitable. The Cabinet was one of the key places where these could be resolved.
- Presentation of policy became difficult, so the Cabinet had to develop ways in which agreements between the parties could be explained.
- If there was a dispute as to whether a policy had in fact been agreed between the coalition partners (and would therefore be subject to collective responsibility), Cabinet would be called on to clarify the issue.

Eventually, as the 2015 general election approached, the coalition Cabinet weakened and began to fragment. However, the government did, against many predictions, last for 5 years and it was the temporary restoration of Cabinet government that helped to maintain stability.

Following Cameron's resignation in 2016, Theresa May attempted to dominate the government machinery despite her small parliamentary majority but, having failed to retain that majority in the June 2017 election, it was clear that she would have to govern with the full co-operation of her Cabinet. It was her failure to do so, in part, that resulted in her decision to step down. Boris Johnson also struggled with this issue when he took over in 2019, in part precipitating his decision to call an early general election.

2019–the present

Since winning a clear majority in 2019, it would appear that Boris Johnson has returned to something more like the prime ministerial government of the 1960–2010 era. Not only is the Cabinet filled with pro-Brexit allies and supporters, but in 2020, when then Chancellor Sajid Javid refused to accept the Prime Minister replacing his special advisers with ones chosen by the PM's inner circle, he was forced to resign and replaced by the more compliant Rishi Sunak. The situation may change as pressure mounts on the government following the Covid-19 pandemic, the resulting recession and the ongoing consequences of Brexit, but for now (late 2020) the Prime Minister seems to have returned to being the dominant force in British politics; Cabinet, as a result, has been increasingly marginalised.

Factors affecting the balance of power between the Prime Minister and the Cabinet

Now that we have examined the position of the UK Prime Minister, including the circumstances that determine how much authority and power the holder of the office may enjoy and their relationship with the Cabinet, we can attempt an overall assessment of prime ministerial power.

Earlier in the chapter we considered the powers that a prime minister holds: patronage; chairing of the Cabinet; foreign policy leader; commander-in-chief; signing foreign treaties; and the ability, with Parliament's consent, to call an early election or simply to recall Parliament; as well as the informal powers of being a national spokesperson, setting the legislative agenda and setting the tone for government policy. Despite some recent constitutional reforms, such as the creation of the judicial appointments commission and the Fixed Term Parliament Act of 2011, these powers are similarly held from one prime minister to the next. These powers may have limitations on the extent to which a prime minister may exercise them, as outlined in Table 7.9.

Table 7.9 A summary of UK prime ministerial power

Permanent powers	Permanent limitations
Exercise of the royal prerogative	The ability to exercise these powers depends on the PM's relationship with Parliament and other factors
Patronage	Forced to promote senior party members who may be rivals
Foreign policy leader	Must consult Parliament
Party leader	Can be removed if loses party confidence
Parliamentary leader	May not be able to rely on the parliamentary majority
Chair of Cabinet	Can be removed by a majority in Cabinet

Yet, if all modern prime ministers hold the same formal powers and the same basic limitations, why are some prime ministers seen as being strong and powerful leaders, while others are seen as weak and ineffective? The answer comes down to a range of political factors that determine how able a prime minister is to exercise their powers. A prime minister who holds most, if not all, of these political advantages will appear strong and powerful, while one who holds few, if any, of these advantages will often be seen as weak and ineffective. This can lead to some prime ministers who hold these advantages being seen as acting like a '**presidential government**' able to rise above parliamentary politics. Those without these advantages are far less presidential in style and bound far more by Parliament. It is therefore important to consider the political factors listed below and how they have affected the prime ministers you study, as well as considering and being able to explain which of these factors is likely to be the most important in determining the strength of a prime minister, to help develop your evaluation skills.

Having a large majority in the House of Commons

A prime minister with a large majority in the House of Commons can afford to dismiss rebels and MPs who disagree with their view, partly because they can afford to lose a number of their own MPs in a vote and still win, and partly because the unlikeliness of a rebellion working (as in, defeating the Prime Minister's proposed legislation) is so remote that most MPs do not bother rebelling or dissenting. Prime ministers with little or no majority in the House of Commons are more likely to come under pressure from rebellious MPs and will have to work much harder at securing the passage of legislation. Margaret Thatcher and Tony Blair suffered only four defeats in House of Commons votes in their respective 11 and 10 years as Prime Minister, in contrast to the 34 defeats of Jim Callaghan's minority government from 1976–79 and Theresa May's 33 defeats (and withdrawal of other key votes to avoid defeat) in her 3 years. With a large majority, prime ministers can find themselves relatively free from parliamentary constraints and can appear to act presidentially.

Attitudinal cohesion of their party

Having a unified party standing behind them, whether inside or outside of Parliament, can give a prime minister the necessary support to take tough and bold decisions

Knowledge check

What powers does the Prime Minister have that can be used to control the Cabinet?

Key term

Presidential government An executive dominated by one individual; this may be a president but is also used to describe a strong, dominant prime minister.

Study tip

It is easy to focus on comparing strong and weak prime ministers, but the strength of any prime minister can rise and fall within their time in office, so try to consider the factors that make a PM become stronger or weaker as well as comparing different prime ministers.

and, even with a slim majority in Parliament, they could achieve great success. On the other hand, if the party is divided over key issues, it will become much more difficult to manage, making the Prime Minister appear weak and ineffective while also allowing potential rivals to emerge as the head of rival factions. From 1983 to 1990 the Conservative Party was largely unified behind Margaret Thatcher, as was the Labour Party behind Tony Blair for his first two terms, giving both greater authority. Their successors, however, suffered from greater divisions within their parties, the Conservatives under John Major being deeply divided about the EU and the 'Blairite' and 'Brownite' wings of the Labour Party becoming more divisive under Gordon Brown. Similarly, the deep divisions over Brexit made it almost impossible for Theresa May to unify her party behind her Brexit deal, while Boris Johnson has a much more unified party since the 2019 election.

Securing an electoral mandate for manifesto commitments

A prime minister who can claim a clear mandate from the British public through a general election will be in a much stronger position when it comes to developing policy and passing legislation. MPs in the Commons are less likely to rebel against manifesto commitments that have secured a mandate, while the Salisbury Convention means the Lords will not oppose anything in a 'winning' manifesto. A prime minister who has failed to secure an electoral mandate is more likely to face opposition from the Lords and some of their own MPs, while the media is perhaps more likely to portray a PM who has not secured a mandate as weak. As such, despite having a large majority during the coalition, David Cameron faced problems from the Lords and both sets of MPs as neither the Conservative nor Liberal Democrat manifestos in 2010 had actually secured a mandate. It is also worth contrasting the lack of support May had (having failed to win a mandate in 2017) compared with Johnson (who did secure a clear mandate in 2019).

Being a first-term government

Being a first-term government can give a new prime minister something of a 'honeymoon' period. Any problems they face can be attributed to the previous party in government and they have no 'history' for which to apologise. If a prime minister or their party continues in office for multiple terms, then decisions and mistakes made will start to undermine their authority. As such, Tony Blair was at the height of his authority during his first term from 1997–2001, but by his third term, unpopular policies, particularly the decision to invade Iraq in 2003, had reduced his authority as PM. Gordon Brown then had to take responsibility for being Chancellor during the Blair years when the financial crash of 2008 exposed some of the failure of New Labour economic policy. Similarly, John Major and Theresa May were both tarnished with the reputations of their party when they took over and also by the role they had played in previous administrations.

Having 'prime ministerial coattails'

A prime minister who is largely popular may well convince voters to elect MPs from a party they might not ordinarily support; this is often referred to as having '**prime ministerial coattails**'. This means a lot of MPs owe their seats to the PM and so are more likely to be loyal and vocal supporters of the PM. MPs will also have one eye on future general elections and will support a PM who is likely to help secure their seats, but turn against one whom they see as an electoral liability. In the 1983 election and the 1997 election, Thatcher and Blair won a lot of seats in areas not traditionally seen as being supporters of their parties. While they remained popular for a time, when their reputations

Useful term

Prime ministerial coattails A term that refers to the idea of a PM being so personally popular that a lot of MPs win their seats on the back of the PM's popularity, so they are dragged into Parliament by metaphorically holding on to the coattails of the PM.

became more of a hindrance, likely to cost them seats, splits started to emerge and there was a decline in support for them, leading to their eventual removal from power.

Having lots of new MPs

Having a lot of new MPs will help a prime minister maintain authority as new MPs will be more dependent on guidance and support for how to work as an effective MP. This means they will depend on the support and influence of the whips' office and are far more likely to be compliant, as was the case for Tony Blair in 1997 and Boris Johnson in 2019. As MPs gain more experience, they are likely to become more independent and less willing to 'do as they are told', making it much harder for a PM to control them, as both John Major and Theresa May found to their cost. Although this is not always the case, as a number of the new intake of MPs from 2019 proved difficult for Boris Johnson to manage, particularly those from the 'red wall' who voiced objections in 2020 about restrictive measures in their constieuncies.

Low salience of issues

If a prime minister is lucky, they will be in a position where the issues they face are not controversial and do not evoke strong passions. This means there will be far less opposition to their policies and decisions, both inside Parliament and across the media, which will allow them to achieve a great deal. With the economy going well and most of New Labour's 1997 manifesto being fairly low salience, Tony Blair was able to achieve a lot and appear to be very powerful. However, highly salient issues, which do evoke strong passions, can undermine a PM as they will face far more passionate and vocal opposition. Such issues make it far harder to persuade people to their point of view. The poll tax in 1990, the decision to invade Iraq in 2003, and the most salient of issues – Brexit – have all reduced the power and authority of their respective prime ministers.

Fear of the alternatives

Finally, a prime minister may appear to be more secure and powerful if there is a fear of the alternatives. A PM can assert their authority over their MPs by threatening to resign, or actually resigning as John Major did in 1995, or by threatening to call a general election. If MPs fear losing their seats or fear who might become leader, they will be more likely to give the PM the support they need. If the alternatives become more appealing, then a prime minister's power is greatly reduced, as Blair found when by 2007 most Labour MPs and party members appeared to prefer Brown to him. The strength of the opposition is also important here – a strong opposition leader who might have a chance of winning makes the threat of an early election a genuine one to bring MPs into line; if the opposition is seen as being quite weak, it makes the threat less effective and so reduces the power of the Prime Minister. John Major used this to his advantage when trying to control his party in the 1990s, while the threat of Jeremy Corbyn never worked to bestow power on May.

Discussion point

What is the most important factor in determining the strength and power of a prime minister?

Points you should consider may include the following:
1 Which prime minister do you consider to be the strongest and why?
2 Which of the factors listed above are most common to the strongest prime ministers?
3 Which of the factors might have an impact on other factors (good or bad)?

Knowledge check

1 What circumstances might lead to the removal of a prime minister?
2 What are two limitations on the Prime Minister's position as foreign policy-maker?
3 When is the Prime Minister likely to be most dominant in Cabinet?

External factors

In addition to the internal political considerations described above, there are a number of *external* factors that determine how much power the Prime Minister can exercise. These include the following:

- Devolution, as it develops further, gradually erodes the power of both the Prime Minister and the UK government as a whole. As Scotland, Wales and Northern Ireland develop legislative powers, there will be large parts of the country outside central control.
- While the UK was a member of the EU, the powers of UK government were limited as large areas of policy were in the hands of the EU's Council of Ministers. As the UK leaves the EU, however, these powers will be repatriated, offering a considerable boost to the Prime Minister's and Cabinet's ability to shape policy and determine the course of events.
- Similarly, the UK's membership of the NATO, and especially the country's close relationship with the USA, limits the UK's foreign policy options. The Prime Minister and Cabinet must take into account the country's main allies when conducting foreign policy. The UK's involvement in Middle East affairs is an especially important example.
- Finally, as we have seen, events significantly affect the power of the Cabinet and the Prime Minister. This is mainly true of economic policy. In the early 1980s, early 1990s and after 2008, for example, economic policy-making was dominated by the problems of an economic recession.

Prime ministerial case studies

Case study

Margaret Thatcher, Conservative, 1979–90

By 1979, the people of the UK were looking for something different from the painful upheavals of the 1970s. As leader of the Conservative opposition, Margaret Thatcher had become the figurehead of the 'New Right' political movement within the Conservative Party. She was never as ideological as some of her New Right colleagues, and was perhaps more pragmatic than people recognise, but in her determination to introduce New Right economic measures and transform Britain, her force of personality and the legacy she left, she proved herself to be a highly consequential Prime Minister. Table 7.10 highlights some of the areas in which she showed great control and also areas where she showed lack of control.

Parliamentary majorities

1979: 43

1983: 144

1987: 102

State of the party

Up to 1982–83 the Conservative Party was fundamentally split. One section was described as 'traditional' or 'one-nation' Conservatives, who believed in centrist policies, taking into account the interests of business, workers and welfare recipients equally and trying to mediate between them. They were known by their opponents as 'wets' because they were seen as weak in dealing with the UK's economic problems.

→

The other section was known as the 'dries'. They were neo-liberals who believed in free markets, low direct taxation, privatising large industries, reducing welfare benefits to a minimum and reducing the power of trade unions. Thatcher led the 'dries'. From 1982 she purged the leadership of the party of the 'wets'. She dismissed her opponents from the government and replaced them with her allies. This left her at the head of a united party, agreed on her political vision.

Margaret Thatcher outside Downing Street after winning the 1979 election

Examples of key policy goals

- Privatisation of major formerly nationalised industries
- Tight control over the government finances, avoiding excessive debt
- Curbing the power of trade unions
- Reducing direct corporate and personal taxes
- Reducing government regulation of business and finance
- Strengthening the rules governing who could claim welfare benefits
- Emphasis on national defence and an active foreign policy
- Strongly confronting the Soviet Union

Style of leadership

Thatcher was an extremely dominant personality who refused to compromise with her opponents. She believed that those who were not 'with her' were against her. Her supporters called her principled and visionary, while her opponents called her stubborn and uncompromising.

Prominent events

Thatcher was very unpopular until 1982. She introduced measures that did not seem to be having a beneficial effect on the ailing economy. Then her fortunes turned. First, the Argentine government invaded the British Falkland Islands. Thatcher ordered a taskforce, which soon liberated the islands. This created her reputation as the 'Iron Lady'. At the same time, the UK economy began to improve. These events increased her authority and power, and she began to transform her party in her own image. Towards the end of the 1980s, the Cold War began to ebb and the USSR to weaken. She and American President Ronald Reagan were given much of the credit for 'defeating' communism. She also resisted moves to create a stronger political union in Europe, staunchly defending British interests.

Circumstances of loss of power

In 1988 Thatcher and her close advisers introduced the idea of a 'poll tax' to replace local property taxes (the rates). The idea of a flat-rate poll tax was hugely unpopular as it did not take account of people's incomes and so broke a fundamental principle of taxation — that it should be based on ability to pay. Despite opposition from all sides, including from inside her own party, Thatcher declared that she was determined to introduce it, firstly in Scotland. This led to riots and protests and an outpouring of public hostility towards the policy. Her opponents in the Conservative Party feared that this would cost many seats in the looming general election. A challenge was mounted against her leadership by Michael Heseltine. Some of her close allies abandoned her and she withdrew from leadership election in 1990. John Major replaced her.

Table 7.10 Examples of Thatcher's control and lack of control

Control	Lack of control
Following the 1983 election, she was able to remove the 'wets' or moderate Conservatives from her Cabinet and marginalise them	From 1979 to 1983 she had to maintain party balance and was obliged to have a number of senior 'wets' in her Cabinet
When Argentina invaded the Falkland Islands, she was able to take decisive military action and show leadership	When the USA announced a boycott of the 1980 Moscow Olympics, Thatcher wanted to impose the same on UK athletics, but public opinion meant she could not
In defying the miners' strike, Thatcher showed that she, not the unions, was in control of economic policy	Despite wanting to reduce taxation, the cycles of boom and bust meant she had to increase some taxes, notably VAT
She was able to privatise key industries such as electricity, gas and telecommunications	She was unable to privatise other areas, such as Royal Mail and the railways, due to public and party opposition
In imposing the poll tax on Scotland she showed her ability to impose her will on her party, Parliament and the UK	The resulting riots and media backlash forced her to make a policy U-turn and would force her from power

Tony Blair, Labour, 1997–2007

The Conservative Party had been in power for 18 years by 1997 and, to some commentators, it had looked like the Labour Party was dead and would never win office again. However, following an economic recession in 1992, a succession of scandals and growing divisions in the Conservative Party, the electorate had lost faith and confidence in it. Tony Blair was able to take advantage of this, creating a 'New Labour Party' based on a moderate, Third Way ideology, keeping his party unified, and simultaneously building a strong, modern media image to reassure voters. Table 7.11 shows examples of Blair's control and lack of control over his decade as Prime Minister.

Tony Blair meeting supporters after winning the 1997 general election

Parliamentary majorities

1997: 179

2001: 167

2005: 66

State of the party

From the early 1990s onwards, a tight-knit group of leading members of the Labour Party adopted a new set of policies designed to challenge the Conservatives and to modernise the UK. They became known as 'New Labour' and their beliefs were collectively known as the 'Third Way'. The Third Way was a path somewhere between the radical right-wing, neo-liberal policies of Thatcher and the more socialist ideas of the left wing of the Labour Party, combining the best elements of each.

The group was initially led by John Smith and contained such people as Tony Blair, Gordon Brown, Robin Cook and Peter Mandelson. In 1994, however, Smith died suddenly and Blair was elected leader of the Labour Party in his place. By then New Labour had taken over most of the party. The left-wingers (of whom Jeremy Corbyn was a leading member) were only a small minority. Blair, therefore, led a united party with a clear vision and strong determination to oust the Conservatives. It was to prove as cohesive and dynamic as the Conservative group that underpinned Margaret Thatcher's authority in the mid-1980s. New Labour remained united until events began to divide the party in 2003–04. By 2007, the split in the party was so severe that Blair had to go.

Examples of key policy goals

- An extensive programme of constitutional reform including devolution and the Human Rights Act
- Sharp, sustained increases in expenditure on health and education
- Increased welfare benefits for those genuinely unable to support themselves
- Introducing a national minimum wage
- Introducing tax credits, mainly to reduce child poverty
- Granting independence to the Bank of England to establish more rational financial policies
- Using government financial surpluses to reduce government debt
- An active foreign policy with major interventions in the Balkans war, the Sierra Leone civil war and Iraq
- Pursuing closer links with Europe but resisting joining the eurozone
- Reducing business taxes to promote economic growth

Style of leadership

Blair was as charismatic as Margaret Thatcher. However, unlike Thatcher, he was part of a collective leadership. The key policies adopted by the Labour government after 1997 were delegated to his leading cohort. Economic policy in particular was handled by Gordon Brown and domestic social policy by other senior ministers such as Jack Straw, David Blunkett, Harriet Harman and Frank Dobson. Blair himself concentrated largely on foreign policy. After 6 or 7 years, however, Blair's leadership became more singular and his popularity in the party waned. It was widely felt that he had over-reached his authority.

→

Prominent events

Rarely have the fortunes of a prime minister turned so dramatically on a single event as happened to Tony Blair. Up to 2003 the UK had enjoyed a sustained period of economic growth, public services such as health and education were improving, and Blair himself had initiated two successful overseas military campaigns in Sierra Leone and Kosovo. At the same time the peace process in Northern Ireland had come to a successful conclusion with the establishment of a power-sharing government in the province. Then Blair ordered the UK armed forces to join the US-led invasion of Iraq. The aftermath of the war was a disaster. In particular it was revealed that the evidence that Saddam Hussein's regime had accumulated weapons of mass destruction was false. Saddam was deposed but Iraq fell into widespread sectarian strife. As violence in the Middle East grew, Blair was seen in an increasingly negative light. At the same time, it appeared that inequality was growing in the UK. Those who believed that the Labour Party's role was to reduce inequality were dismayed. Internal opposition gathered around Gordon Brown and the party fell apart.

Circumstances of loss of power

By 2007 the momentum in the party for a change of leadership became irresistible. Tony Blair resigned before a divisive leadership contest completely destroyed party unity. He recommended that Brown should succeed him.

Table 7.11 Examples of Blair's control and lack of control

Control	Lack of control
In making decisions on military interventions, Blair showed he was in control of foreign policy	Although early conflicts were regarded as successful, as public opinion and the media turned against him for the invasion of Iraq, his reputation suffered and his standing and power were weakened
Blair operated a 'sofa Cabinet' and was largely able to impose his policy initiatives on the Cabinet	Gordon Brown was given huge independence and authority at the Treasury; Blair had to have Brown's support for any initiative
Blair was able to impose his policy agenda on Parliament during his first two terms thanks to his large majority; he suffered no defeats in the House of Commons in his first 8 years	Following his reduced majority in 2005, Blair suffered four defeats in the Commons over 8 months from November 2005 to July 2006
In invoking the Parliament Act to bypass the Lords over the issue of fox hunting, Blair showed his control over the Lords	In his first 8 years as PM, Blair suffered 353 defeats in the Lords
During his first two terms, Blair was able to keep his party largely unified behind him and impose his agenda on the party	After 2005, divisions between New Labour and Old Labour supporters began to emerge, as well as divisions between Blairites and Brownites, which would ultimately force Blair out of office

Case study

David Cameron, Conservative, 2010–16

By 2010 the Conservatives had been out of power for 13 years and maintained deep divisions over membership of the EU. Cameron had become party leader in 2005, despite very little experience of government, but his confidence and media presence, as well as the damage the financial crash of 2008 did to the Labour Party, helped him make the Conservatives the largest party, though he had to settle for a coalition and then a very slim minority, which impacted his ability to control events. Table 7.12 shows examples of Cameron's control and lack of control during his time as Prime Minister.

Parliamentary majorities

2010: no majority

2015: 12

State of the party

The party that Cameron inherited in 2005 when he became leader was both demoralised by its three consecutive election defeats and divided. It has remained divided since then and this was a major barrier to Cameron becoming a dominant leader. He is, by nature, a controller, but control often eluded him. The party did unite around the need for a programme of austerity (cuts in government spending) after the 2008 financial crisis so he was able to govern effectively. The internal divisions over the UK's relationship with the EU, however, constantly made his party difficult to lead and ultimately led to his downfall.

David Cameron giving his resignation speech, June 2016

Examples of key policy goals

- A programme of austerity – higher taxes and reduced public spending – to reduce the government's financial deficit
- Progressive social policies including the introduction of same-sex marriage
- Promoting more devolution, mainly to Scotland
- Reducing direct taxes on those with very low or very high incomes

- Targeted reductions in welfare benefits in order to encourage more people to find work
- Subsidies for pre-school childcare to help families with young children and encourage work
- Significant rise in the minimum ('living') wage
- Introducing sharp increases in university tuition fees
- Decision to hold a referendum on the UK's membership of the EU

Style of leadership

Cameron had problems exerting the personal power he would have liked to wield. To combat the barriers to his leadership he formed a strong bond with his Chancellor, George Osborne, and his Home Secretary and eventual successor, Theresa May. He kept his rivals close by avoiding the temptation to remove them from government. Thus such opponents as Michael Gove, Iain Duncan Smith and Boris Johnson remained near the centre of power.

Prominent events

Cameron's main achievement may well be seen as his government's success in bringing the UK out of recession and stabilising the financial system. He will also be notable for having kept together a coalition for a full 5 years and following this with an election victory. However, he was bedevilled by foreign policy setbacks, especially when Parliament restricted his freedom to intervene in the Syrian civil war. Despite this mixed picture, Cameron's term in office will probably be best remembered for one single event — the referendum on the UK's membership of the European Union. The fact that he lost that referendum will define his premiership in the same way the Iraq war defined that of Tony Blair.

Circumstances of loss of power

Having led the calls for a referendum on UK membership of the EU and campaigned strongly for the UK to remain a member, it was inevitable Cameron would have to resign following defeat in the referendum.

Table 7.12 Examples of Cameron's control and lack of control

Control	Lack of control
Despite being in a coalition, Cameron was largely able to dominate the Cabinet and follow his own agenda	He was unable to remove the Lib Dem Cabinet members who criticised him, like Vince Cable
Cameron was able to implement sweeping 'austerity' measures to deal with the growing national debt	The global financial crisis that started in 2008 and would lead to a crisis in the eurozone dominated the financial decisions he could make and limited his options
Having agreed to hold a Brexit referendum, he was able to force that through Parliament	The fact that he had to call the referendum showed his inability to control the Eurosceptic wing of his party and the growing threat of UKIP
He was able to campaign and 'win' the Scottish independence referendum in 2014 through personal persuasion and media appearances	Cameron had opposed a Scottish independence referendum, but political pressure from the devolved Scottish Parliament forced him to grant it
He was able to push through same-sex marriage in Parliament, despite some opposition within his own party	He failed to get parliamentary support for his planned intervention in Syria

What can we learn from the prime ministerial case studies?

The three prime ministers described in the case studies above experienced very different sets of circumstances, some of them random. However, some generalisations can be attempted. The common features that can be identified from recent history include these:

- The size of a prime minister's parliamentary majority seems to be critical. Blair and Thatcher both benefited from large majorities for much of their terms of office. A large majority helps a prime minister in two ways. One is that it gives them strong democratic legitimacy. The other is that it makes it easier for them to secure the passage of legislation.
- Events are crucial. However favourable or unfavourable the *political* circumstances may be, prime ministers can be made or broken by events outside their control.
- Prime ministers need to head a united party if they are to be truly dominant. David Cameron is a good example of a leader who was limited by splits in their party.
- The lesson of the premierships of both Thatcher and Blair is that prime ministers who seek to stretch their power too far can expect to be reined in. Prime ministers enjoy considerable authority and have great political and constitutional powers, but if they try to overstep their authority, powerful forces will act against them to prevent them becoming too dominant. This is sometimes described as the 'elastic theory'. The further prime ministers stretch their powers, the stronger the forces will be that restrain them.

Summary

Having read this chapter, you should have knowledge and understanding of the following:

→ What the central 'executive' of the UK actually is and how it operates
→ The functions and purpose of the executive branch
→ The various powers and responsibilities of the different elements that comprise the executive branch
→ The basis of prime ministerial power and the factors that affect the exercise of those powers
→ What the Cabinet is and how it can control the executive, as well as its limitations
→ The importance of collective responsibility and its positives and negatives
→ The role of individual ministerial responsibility and how it has impacted UK politics in recent years
→ The factors that affect the balance of power and relationship between the Prime Minister and the Cabinet
→ How to assess and evaluate prime ministerial power
→ The extent to which prime ministers and the executive can dictate events and determine policy

Key terms in this chapter

Cabinet The Prime Minister and senior ministers, most of whom lead a particular government department.

Collective responsibility The principle by which ministers must support Cabinet decisions or leave the executive.

Executive The collective group of Prime Minister, Cabinet and junior ministers, sometimes known as 'the government'.

Government department A part of the executive, usually with specific responsibility over an area such as education, health or defence.

Individual responsibility The principle by which ministers are responsible for their personal conduct and for their departments.

Minister An MP or member of the House of Lords appointed to a position in the government, usually exercising specific responsibilities in a department.

Presidential government An executive dominated by one individual; this may be a president but is also used to describe a strong, dominant prime minister.

Royal prerogative A set of powers and privileges belonging to the monarch but normally exercised by the Prime Minister or Cabinet, such as the granting of honours or of legal pardons.

Secondary legislation Powers given to the executive by Parliament to make changes to the law within certain specific rules.

Further reading

Websites

Factual and up-to-date information about the Prime Minister and the Cabinet can be found on these official websites:

- **www.gov.uk/government/ministers**
- **www.cabinetoffice.gov.uk**

Books

Buckley, S. (2006) *The Prime Minister and Cabinet*, Edinburgh University Press
D'Ancona, M. (2014) *In It Together: The inside story of the Conservative–Liberal Democrat coalition*, Penguin
Hennessy, P. (2001) *The Prime Minister: The office and its holders since 1945*, Palgrave — probably the best review of the office of Prime Minister

Practice questions

1

Source 1

The role of the Cabinet

The Cabinet was certainly the main body and centre of power before the 1960s, but since then we have seen it marginalised under most prime ministers. Thatcher's domination of her Cabinet and Blair's famous 'sofa Cabinet' highlighted that the Cabinet has become merely a rubber stamp for policy decisions that are made elsewhere and is no longer a relevant force in policy formulation. The idea that the Prime Minister remains simply 'first among equals' has also gone, with modern prime ministers instructing the Cabinet and dismissing those who dare to defy them. Even in running their own departments, they are more likely to have policies imposed on them from Downing Street than to develop their own. Indeed, the only meaningful role a Cabinet minister seems to have is taking the blame when things go wrong within their department, which is also about the only time the media pays them any attention, unlike the Prime Minister, who is always the centre of the media circus. The power of collective responsibility means the Cabinet is no more than a sounding board for the PM, as few are willing to sacrifice their ministerial salaries by openly defying and challenging the Prime Minister. As such, it is hard to see the Cabinet as anything other than an administrative tool for implementing the ideas of the leader, rather than a viable political entity that actually governs in its own right.

Using the source, evaluate the view that the Cabinet plays a meaningful role in the UK's political system.

In your response you must:
- *compare and contrast different opinions in the source*
- *examine and debate these views in a balanced way*
- *analyse and evaluate only the information presented in the source.* (30)

2

> **Source 2**
>
> **The power of the Prime Minister**
> How much power does the UK Prime Minister really have? On the face of it, they dominate the political system, but the reality may be very different.
>
> It is especially true that the power of the Prime Minister will vary according to circumstances. For example, it depends upon how large a majority the governing party enjoys. The media image of the Prime Minister can also change over time. Tony Blair, for example, began with a positive media and public image, but was widely mistrusted when he left office in 2007.
>
> But possibly the most serious limitation on prime ministerial power lies with events beyond his or her control. Gordon Brown, for example, faced a major financial and economic crisis within a few months of taking office.
>
> So the ability of the Prime Minister to determine their own destiny is severely limited. Despite all the powers they have — over patronage, foreign policy and policy-making, for example — they are at the mercy of forces outside their control.
>
> One thing is for certain: virtually all prime ministers *appear* to be more dominant when they are conducting foreign policy abroad than when they face hostile forces at home. This was especially true of Margaret Thatcher (known internationally as the 'Iron Lady') and Tony Blair, who specialised in foreign policy initiatives, notably in Kosovo, Sierra Leone and Iraq. Of course, though, it was a military intervention that ultimately led to his downfall. As the situation in Iraq began to deteriorate following the 2003 invasion, and the uncertain legal basis for the war was revealed, Blair's position in the country and inside his own party became untenable.

Using the source, evaluate the view that prime ministers rarely face serious limitations on their powers.

In your response you must:
- *compare and contrast different opinions in the source*
- *examine and debate these views in a balanced way*
- *analyse and evaluate only the information presented in the source.* (30)

3 Evaluate the extent to which prime ministers are able to dominate their cabinets.

In your answer you should draw on relevant knowledge and understanding of the study of Component 1: UK politics. You must consider this view and the alternative to this view in a balanced way. (30)

Study tip

Coverage of Core and Non-Core Political Ideas are available in other textbooks to complete your study for Components 1 and 2.

4 Evaluate the extent to which the concepts of ministerial responsibility continue to play a meaningful role in British politics.

In your answer you should draw on relevant knowledge and understanding of the study of Component 1: UK politics. You must consider this view and the alternative to this view in a balanced way. (30)

5 Evaluate the extent to which there are any effective limitations on the powers of a prime minister.

In your answer you should draw on relevant knowledge and understanding of the study of Component 1: UK politics. You must consider this view and the alternative to this view in a balanced way. (30)

6 Evaluate how accurate it is to describe modern prime ministers as presidents in all but name.

In your answer you should draw on relevant knowledge and understanding of the study of Component 1: UK politics. You must consider this view and the alternative to this view in a balanced way. (30)

Relations between the branches

On 24 January 2017, the Supreme Court handed down its ruling in one of the most high-profile cases it had heard in its brief history. In *R (Miller)* v *Secretary of State for Exiting the European Union*, the Supreme Court ruled that Parliament, and only Parliament, had the legal authority to trigger Article 50 and begin the process of the UK leaving the European Union.

The Miller case, referred to above, shows how different branches of government relate to each other and exemplifies some of the changes in power that exist between these branches. Here we had the Supreme Court being asked to review and check the constitutional legality of the proposed actions of the executive. In its ruling, the Court declared that Parliament, not the Prime Minister, had to decide to act on the referendum result and trigger Article 50, a decision that made it clear that Parliament was the supreme power in the UK political system. Why was this? Because sovereignty, the concept of supreme legal power, is held by Parliament, not the executive, so the Supreme Court was also clarifying the location of sovereignty in the UK, especially as part of the ruling rejected the applications from the devolved bodies of Scotland, Wales and Northern Ireland that they should have a role in triggering Article 50. Of course, at the heart of this ruling was the UK's relationship with the European Union, and how the process of leaving the EU would begin following the 2016 referendum.

This one case, then, neatly shows the intersection of the various elements of government and politics in the UK that will be covered in this chapter and why it is so important to understand how the relationship between these branches works and how it has evolved over the years. Politics, as the Miller case shows, is not about separate bodies making independent decisions, but about various forces working in relation to each other. The relations between the branches are what makes politics what it is.

Gina Miller speaking to the press after the High Court ruling on the process of Brexit, 2016

The Supreme Court

Background: The creation of the Supreme Court in the UK

Study tip

Northern Ireland and Scotland have separate legal systems, so the information in this section refers mostly to England and Wales only. However, the Supreme Court is still the highest court of appeal for the whole of the United Kingdom.

The Constitutional Reform Act 2005

In 2005 Parliament passed the **Constitutional Reform Act**. Among other measures, the Act was designed to improve and guarantee the independence of the UK judiciary. Before the Act it had been *claimed* that the senior judges in the UK were independent of political influences, but there was mixed evidence as to whether such independence was being upheld. After the Act, however, there were fresh guarantees in place that have largely removed those doubts.

The central feature of the Act was the establishment of the **Supreme Court**. The Court became active in 2009. Before the Supreme Court was established, the highest court in the UK was situated in the House of Lords and was, therefore, a part of the legislature. This was a strange anomaly, almost unheard of in the democratic world.

Before the Constitutional Reform Act 2005, the old system worked like this:

- Twelve 'Lords of Appeal in Ordinary', commonly known as the Law Lords, were members of the House of Lords. They were expected to be neutral crossbenchers but were free to take part in the business of the Lords.
- At the head of the Law Lords was the Lord Chancellor. He (all Lord Chancellors were men before 2016) had three roles. First, he was Speaker of the House of Lords, i.e. the chairman of its meetings. This made him an important member of the legislature. Second, he was also a Cabinet minister, responsible for the

direction and management of the UK legal system. This made him a political figure and a member of the executive. Third, he was the head of the judiciary, the most senior judge in the UK. Far from being an independent figure, therefore, he was a member of all three branches of government!

- The Lord Chancellor played a full part in advising the government on legal policy, in the appointment of senior judges and in deciding which of the Law Lords would hear each appeal case.
- When a case was brought to the highest court in the land, it would usually be heard by a group of Law Lords, normally five. So, when it was said that 'a case was heard by the House of Lords', it actually meant that it was heard by a small group of these senior judges, not the whole House.
- While the Lord Chancellor did play a leading role in appointing senior judges, the Prime Minister would have the final say if they felt so inclined.

The Constitutional Reform Act changed all this in a number of ways, described below.

Study tip

Do not confuse the position of Lord Chancellor with that of the Chancellor of the Exchequer, a different post charged with handling the nation's finances.

The establishment of the Supreme Court

As we have said, the Constitutional Reform Act was mainly designed to reaffirm and guarantee the independence of the judiciary in the UK. Its main provisions were as follows:

1 The Lord Chancellor was no longer head of the UK judiciary as had been the case for centuries. This was now the Lord Chief Justice, a non-political figure and a senior judge. The Lord Chief Justice is also known as the President of the Courts of England and Wales (Scotland and Northern Ireland have their own chief judges).
2 The position of Lord Chancellor still exists, and the holder combines the position with that of Justice Secretary in the Cabinet. However, the holder is no longer an active member of the judiciary.
3 The Lord Chancellor was no longer to be the Speaker of the House of Lords and ceased to sit in the House of Lords.
4 The Supreme Court was established. It would be composed of 12 senior judges known as Justices of the Supreme Court.
5 The head of the Supreme Court was to be known as the President of the Supreme Court.
6 When a vacancy occurred on the court, a special Selection Commission would be established, consisting of a number of senior law officers from the whole of the UK. The commission recommends a candidate to the Lord Chancellor.
7 The Act reaffirmed the principle that a Supreme Court Justice can only be removed by a vote in both houses of Parliament and only for misconduct, *not* as a result of their decisions. The salary of the judges is also guaranteed. This means they have security of both tenure and salary.

Thus, the independence of the judiciary and the Supreme Court in particular was finally codified in law. Furthermore, the Lord Chancellor was charged with the task of guaranteeing and maintaining the independence of the Supreme Court and the rest of the judiciary from political or public pressure.

Knowledge check

Who are the three most recent appointments to the Supreme Court?

The role of the Supreme Court

The Supreme Court is not directly involved in ruling on cases of guilt or innocence, but instead in making sure the law is being correctly applied and being followed equally by everyone, including by the government and its representatives. This means that the cases it hears have already been heard in a lower court, making it an appellate court (one that hears appeals). The Court only hears cases it believes are important. The reasons why the Supreme Court may allow a case to be brought to it include the following:

- It may be an important **judicial review** concerning the government or some other important body such as a school, a newspaper or the NHS. The Court may need to establish what legal powers such bodies have.
- The case may have implications for other citizens and bodies; in other words, it may create an important precedent to be followed elsewhere.
- It involves an important interpretation of law. It may be that lower courts have been unable to make a judgement about the meaning of law, or the law has been interpreted differently by separate lower courts. The Supreme Court will examine what Parliament's *intention* was when it originally passed the law.
- It may be a case that has attracted a great deal of public interest.
- A key issue of human rights may be at stake.

Activity

Research two recent cases heard by the Supreme Court and determine which of the above issues each relates to.

While the Supreme Court is the highest court in the UK, during the UK's membership of the European Union, cases could be appealed to the European Court of Justice if they related to areas of EU law, such as workers' rights, but this will no longer apply once the UK leaves the EU. Cases concerning human rights can be taken to the European Court of Human Rights in Strasbourg, France, though there is no guarantee that the UK government or Parliament will obey its judgments, as happened following repeated rulings about the rights of prisoners to vote.

Study tip

The European Court of Human Rights is not part of the European Union and so the UK leaving the EU will have no impact on the role of the ECHR in the UK. Do not confuse this with the European Courts of Justice, which oversee EU law and which only members of the EU are subject to.

Case study

The Supreme Court at work: *R (Miller)* v *Secretary of State for Exiting the European Union* (2016)

This case was heard originally in the High Court, Queen's Bench Division, in October 2016 and was brought to appeal in the Supreme Court later in the same year. It is possibly the most important constitutional case to appear in the courts in recent history and illustrates well the relationship between the judiciary and government, as well as the relationship between Parliament and government.

Gina Miller, a private citizen, requested a judicial review of whether the Secretary of State for Exiting the European Union (David Davis) had the prerogative power to trigger Article 50 of the EU, which would start the process of the UK's departure. Her legal argument was that Parliament is sovereign and this stands above the claimed prerogative power to bring the UK out of the EU. The government's position was that it *did* have such a prerogative power.

The High Court ruled that the government did *not* have such a prerogative power and that the sovereignty of Parliament had to be exercised in this case. This was because departure from the EU would affect the rights of UK citizens in some cases. This ruling was later upheld in the Supreme Court. It forced the government to seek parliamentary approval to trigger Article 50 and begin the process of leaving the EU. Two extracts from the High Court's judgment are especially significant:

> The courts have a constitutional duty fundamental to the rule of law in the same way as the courts enforce other laws.

and:

> We hold the Secretary of State does not have the power under the Crown's prerogative [i.e. the royal prerogative] to give notice pursuant to Article 50 of the Treaty of the European Union for the UK to withdraw from the European Union.

The High Court decision was emphatically confirmed by the Supreme Court in January 2017.

The Miller case illustrates a number of constitutional realities:

- That it is for the judiciary to determine the limits of the government's prerogative powers.
- That the rule of law is superior to political considerations.
- That Parliament is sovereign over such matters in that the decision to leave the EU affects the rights of UK citizens and so the government must obtain parliamentary approval.
- That referendums are not legally binding and their outcome must be confirmed by Parliament, not government.

Following the case, several newspapers proclaimed that the judiciary was attempting to defy the will of the people, but politicians of all parties rushed to defend the judiciary, arguing that its independence from such pressures must be protected.

Ensuring the rule of law is applied

The rule of law means all citizens should be treated equally under the law. Trials and hearings should all be conducted in such a way as to ensure that all parties gain a fair hearing and that the law is applied in the spirit intended. All courts, at all levels, have the task of ensuring that the rule of law is maintained, but decisions as to whether the rule of law has been correctly applied or abused will ultimately be determined by the Supreme Court. In this sense, it oversees the work of the lower courts in the UK.

Interpretation of the law

The Supreme Court ultimately confirms how all laws in the UK should be interpreted. This role arises because the precise meaning of a statute law or common law is not always clear. There will always be circumstances where those in court (disputants in a civil case, defenders and prosecutors in a criminal case, or plaintiffs in a common law case) come into conflict over what the law is supposed to mean. In such cases it is for Justices of the Supreme Court to interpret the meaning of law.

In cases involving the powers of government or its agencies, or the rights of citizens, such interpretations may be of great public significance. Here, **judicial precedents** become important. Once the Supreme Court has interpreted the law in a certain way, and if this is a new interpretation, other judges must follow the same interpretation.

Study tip

You should learn the details of the Miller case and use it to illustrate the relationships between the three branches of government.

Useful term

Judicial precedent A principle that when a judge in a court declares an important point of law, which includes the meaning of law or how it should be applied, that declaration must be followed by all other courts and judges in similar cases.

Synoptic link

Statute law and common law are key elements of the UK Constitution and so the Supreme Court has an important role in interpreting the meaning of the UK Constitution, as outlined in Chapter 5.

Conducting judicial reviews

Citizens may feel they have been mistreated by a public body, usually part of the state at central or local level. When this happens there is an opportunity to seek a judicial review by the courts. The review will examine whether the citizens' claims are justified. The purpose of the review for the citizens concerned is to establish the wrongdoing and may involve either compensation or simply the reversal of a decision. The quantity of cases involving judicial review has grown dramatically since the 1960s, especially after the Human Rights Act came into force in 2000.

Key term

Ultra vires Literally 'beyond the powers'. An action that is taken without legal authority when it requires it.

Typical examples of judicial review are cases where a member of the executive has not dealt equally with different citizens, or where there has been a clear injustice, or where government or a public body has exceeded its statutory powers by acting '*ultra vires*'.

Judicial review is a critical role because it helps to achieve two democratic objectives. One is to ensure that government does not overstep its powers. The second is to assert the rights of citizens. The courts were given an enormous boost in this area when the **Human Rights Act 1998** came into force in 2000. This meant that courts could review actions by government and public bodies that might contravene the European Convention on Human Rights.

Furthermore, since the introduction of devolution to the UK in 1998 and the various extensions of powers and privileges granted to devolved bodies, disputes have arisen over the relationship between devolved bodies and Westminster and what powers each has. As such, an important function of the Supreme Court (which comes under the issue of judicial review) is to rule on such disputes and clarify the legal rights and limits of the devolved administrations.

The number of applications for judicial review, across all courts, rose to a peak in 2013 when over 15,000 applications for review were made. However, most of these were refused. In 2013 the government restricted the cases that could apply for legal aid when seeking judicial review and raised the court costs, with the then Justice Secretary Chris Grayling saying he wanted to drive out 'meritless applications' that were used as a 'cheap delaying tactic' to government actions. As a result of these changes, the number of judicial reviews has now settled down to a reasonable level. In 2014, 4062 cases were heard, of which 36 per cent were successful and led to a change in a decision by a public body. Between 2015 and 2019 the number of applications for judicial review fell by 44 per cent.

How has judicial review been applied in the UK?

Acting *ultra vires*: *R (on the application of Miller)* (Appellant) v *The Prime Minister* (Respondent) (2019)

In September 2019 the new Prime Minister, Boris Johnson, announced his intention to prorogue, or suspend, Parliament for five weeks before issuing a new Queen's Speech to set out his government's agenda. However, opponents believed this was an attempt to ignore Parliament by suspending it in a crucial period up to the then deadline for leaving the EU of 31 October, arguing that Johnson was only suspending Parliament to prevent it from ordering him to extend the Brexit deadline. Gina Miller brought a case, with the support of 70 prominent figures, including MPs from across the parties and former PM John Major, arguing that in suspending Parliament the PM had exceeded his powers and acted illegally. The Supreme Court agreed, ruling that 'the decision to advise Her Majesty to prorogue Parliament was unlawful because it had the effect of frustrating or preventing the ability of Parliament to carry out its constitutional functions without reasonable justification' (source: **www.supremecourt.uk**). The ruling meant that Parliament was not prorogued, effectively allowing MPs to return to Parliament immediately. The Court did not stop the PM from proroguing Parliament again to hold a new Queen's Speech, as long as it did not prevent Parliament from carrying out its functions 'without reasonable justification'.

Reviewing the legality of parliamentary legislation: *Steinfeld and Keiden* v *Secretary of State for International Development* (2018)

Under the Civil Partnership Act (CPA) 2004, only two people of the same sex were permitted to enter into a civil partnership. The Marriage (Same Sex Couples) Act 2013 (MSSCA) made marriage of same-sex couples lawful. However, the CPA was not repealed when the MSSCA was enacted, meaning that while same-sex couples have a choice as to whether to enter into a civil partnership or to marry, different-sex couples do not. In June 2018, the Supreme Court ruled that the Civil Partnership Act 2004 was 'incompatible' with the European Convention on Human Rights as it applies only to same-sex couples and therefore amounts to discrimination, thereby putting pressure on the government to change the law and allow heterosexual couples to become civil partners. On 2 October 2019, Theresa May announced that the government would support legislation to guarantee the same rights for mixed-sex couples in civil partnerships. However, the focus on Brexit and other matters meant that, despite judicial and executive support, the law was not changed until December 2020, reflecting the limitations of judicial review as rulings have to wait for government and parliamentary action.

Establishing a legal precedent: *An NHS Trust* v *Y* (2018)

In July 2018 the Supreme Court ruled that legal permission was no longer needed to withdraw treatment from patients in a permanent vegetative state in some cases. For more than 20 years, doctors had been required to seek approval of a court, usually the Court of Protection, in a process that costs health authorities about £50,000 in legal fees even when relatives agree that withdrawal of treatment would be in the interests of the patient. The Supreme Court ruled that as long as doctors and the family agreed, there was no need for the courts to be involved and that treatment could be withdrawn without an application, effectively creating a new legal precedent for all lower courts to follow.

Evaluate the importance of judicial review in UK politics.

Three key areas to discuss are:
1. What types of cases and issues is the Supreme Court being asked to review?
2. Who is applying for cases of judicial review and what are their aims?
3. Why might the government want to restrict the number of applications for judicial review?

Hearing cases

Not all the Supreme Court judges hear the cases; there is normally a selection of five (though as many as 11 may sit on a key case). In such cases the judgment will need a majority (i.e. three out of five) of the judges to agree. Once the case has been decided, the law is firmly established. Only the European Court of Human Rights might seek to reverse the judgment if human rights are at stake. The judgments are published, including the reasons for them.

Table 8.1 illustrates the variety of cases heard by the Supreme Court.

Table 8.1 Recent Supreme Court cases of special interest

Case and date	Details	Principle at stake	Outcome
PJS v *News Group Newspapers* (2015)	An unnamed celebrity sought to prevent the media publishing details of his private life as it would infringe his privacy	Extent of the right to privacy and freedom of expression	The Supreme Court ruled the celebrity's privacy should be upheld and took precedence over freedom of the press
Vince v *Wyatt* (2015)	Ms Wyatt and Mr Vince divorced, after which Mr Vince became very wealthy. Ms Wyatt made a claim for considerable maintenance based on Mr Vince's new wealth, though the couple had been poor when married	Interpretation of family law	It was ruled that Ms Wyatt did have a right to make such a claim after many years, opening the door to many similar cases
Trump International Golf Club v *Scottish Ministers* (2015)	Donald Trump argued that a decision by the Scottish government to allow a wind farm to be built near his new golf course was beyond its powers	*Ultra vires*. Trump argued the government was acting beyond its legal powers	Trump lost the case. The Supreme Court agreed the Scottish government had acted within its powers
Schindler v *Duchy of Lancaster* (2016)	Whether British citizens who have lived abroad for over 15 years should be able to vote in the 2016 EU referendum	Rule of law (equality under the law). Were such citizens suffering discrimination?	The Court ruled the government had the right to deny them the vote for administrative reasons
Pimlico Plumbers Ltd v *Smith* (2018)	Whether workers in the 'gig economy' were self-employed or classed as employees entitled to workers' protections	Clarifying the meaning of the law	The Court ruled that workers in the gig economy were employees, clarifying and updating laws that were written before the creation of the 'gig economy'
Sutherland (AP) (Appellant) v *Her Majesty's Advocate* (Respondent) (Scotland) (2020)	Whether evidence gathered by so called 'paedophile hunters' breached a person's right to a private life	The right to privacy outlined in the HRA	The Court ruled that the rights of children take priority over those of any paedophile to engage in criminal conduct

The composition of the Supreme Court

Appointment process

The appointment process for Supreme Court Justices is set out in law and has been adapted since it was introduced to ensure it is a fair and rigorous process that recognises the views and opinions of all areas of the UK.

- First, a vacancy must occur, either because a Justice has chosen to step down, or because they have reached the age of 70 and must retire. It is possible that a Justice may be removed for misconduct, but that has not yet happened.

- Second, the Lord Chancellor (the Justice Secretary) must convene a Special Commission, made up of the President of the Supreme Court, another UK judge who is not a current Justice on the Supreme Court and a representative of each of the three legal appointing bodies in the UK (from England and Wales, Scotland and Northern Ireland). At least one of these representatives must be a 'non-lawyer'.
- Although not legally required, the commission will then advertise the position and encourage formal written applications from those eligible. To qualify a candidate must be:
  - a judge who has held 'high judicial office' for at least 2 years, or
  - a qualified practitioner (a lawyer or barrister at a high level) for at least 15 years (Lord Sumption became the first such appointment in 2012, with Lord Burrows also qualifying in this way through his academic work).
- The commission will consult with senior politicians and other senior judges and draw up a shortlist of eligible candidates.
- The commission will carry out interviews of the shortlisted candidates.
- The commission will produce a report with a recommendation and send this to the Lord Chancellor for consideration.
- There will be another round of consultations between senior judges and senior politicians, including the first ministers of Scotland and Wales, based on the recommendation of the commission.

The Justices of the Supreme Court in 2019

- The Lord Chancellor may accept or reject the commission's recommendation or ask the commission to reconsider its recommendation. The Lord Chancellor may only reject a candidate if they believe they are not suitably qualified to sit on the Supreme Court. If they reject the recommendation, the commission must repeat the process and may not submit the same name. If they are asked to re consider their recommendation, they may submit the same name. Even if the Lord Chancellor rejects or asks the commission to re consider again, then the name of whoever is recommended must be forwarded to the Prime Minister.

- Once accepted, the Lord Chancellor must pass the name to the Prime Minister.
- The Prime Minister must then submit the name to the monarch for formal approval.
- Once the monarch has given formal approval, the Prime Minister then makes a public announcement confirming the new appointment.

As we can see from this procedure, this is a modern appointments process, designed to find the best qualified people to sit on the Supreme Court, but a number of politicians also play a role, with the heads of the devolved bodies playing important roles in the selection process. The membership of the Supreme Court as of July 2020 is outlined in Table 8.2.

Table 8.2 Membership of the Supreme Court as of July 2020

Lord Reed	Lord Hodge
President of the Supreme Court, The Right Hon Lord Reed of Allermuir	Deputy President of the Supreme Court, The Right Hon Lord Hodge
Lord Kerr	Lady Black
Lord Lloyd-Jones	Lord Briggs
Lady Arden	Lord Kitchin
Lord Sales	Lord Hamblen
Lord Leggatt	Lord Leggatt

Key operating principles of the Supreme Court

The independence of the judiciary

An essential feature of any healthy democracy is that the judicial branch should be independent of the government and that there should be an absence of any bias. There are several reasons why this is so:

- If judges are not independent, there is a danger that the government will exceed its powers without legal justification. Without any effective check on government power, tyranny may result.
- Citizens need to feel certain that any legal cases with which they become involved will be dealt with on the basis of justice and the rule of law. It may suit government to discriminate against individuals or groups in society for its own benefit. An independent judiciary can prevent such discrimination. The citizens of a democratic state must feel that their rights will be effectively protected.
- It is important that judges are not influenced by *short-term* changes in public opinion, reflected by politicians. Independent judges can take a long-term view. For example, following a terrorist atrocity, there may be calls for curbs in individual liberties. However, such curbs may harm the cause of human rights in the long term.
- In some political systems the judiciary may appear to be independent, but in practice the judges are specially selected by the government to ensure decisions are friendly to that government. Independence, therefore, also implies that judges are selected on a neutral basis to prevent collusion between the judiciary and the government.

How is independence of the judiciary maintained?

There are four main ways in which **judicial independence** is guaranteed in the UK. These are: security of tenure; rules of sub judice (i.e. when something is the subject of an ongoing court case); the system of appointments; and the method of judicial pay.

Security of tenure

The first and key principle that attempts to ensure the political independence of the judiciary is security of tenure. This principle says that judges cannot be removed from office on the grounds of the kinds of decisions that they make. The only reason a judge can be removed is if they can be shown to be corrupt as a result of personal conduct incompatible with being a judge. It follows, therefore, that judges are free to make decisions without fear of dismissal, even if such decisions offend the government.

The rule of sub judice

Second, the rule of sub judice means that it is a contempt of court for any servant of the government to attempt to interfere with the result of a court case or even to comment on such a case in public or in Parliament. This rule is designed to prevent any political pressure being placed upon judges. Any such interference would be strongly criticised in Parliament and could result in legal action against the government member concerned.

Independent appointments

Since 2005, appointments to the Supreme Court are largely independent of the government, as detailed above.

Judicial pay

As employees of the state, there is a danger that a government might use the pay of the judiciary as a means of influencing them. For this reason, the salary of judges is decided by an independent body, the Senior Salaries Review Body, which makes recommendations to the Prime Minister on the pay of judges, as well as senior civil servants and senior military officers. These recommendations are publicised and nearly always followed by the PM. In addition, judicial salaries cannot be cut and there are strict and public rules about what expenses may be claimed, to ensure there can be no political manipulation of the pay of judges that might reduce their independence.

In case one is tempted to be sceptical about these safeguards, there is a great deal of evidence to suggest that the UK judiciary is indeed politically independent. Under both Conservative and Labour administrations, senior members of the judiciary have made a large number of judgments that have been clearly contrary to government interests. This has been especially true since the implementation of the Human Rights Act in 2000. Indeed, it is significant that politicians of *both* the main parties have criticised the judiciary on the grounds that it is politically biased. This implies that there is little or no political bias among judges.

The neutrality of the judiciary

In addition to being independent of political influence, the judiciary must be neutral in the way it hears and rules on cases. **Neutrality** relates more to the personal beliefs and attitudes of judges, rather than to external influences. There are a few reasons why judges must be neutral in their rulings.

- The rule of law is one of the key principles of the UK Constitution and UK democracy. As the rule of law demands that all people are equal in the eyes of the law, judges must consider all people equally, with no personal prejudice based on social class, race, gender, sexuality or age.
- Judges are only supposed to interpret the meaning of the law, not the 'fairness' of the law. This means they must remove any personal feelings or emotions from the case they are hearing. If the people believe a law is unfair, they must act through Parliament to change the law, not the judiciary.
- In a democracy, the people must have faith that their cases will be heard on a fair basis and without prejudice, so must have confidence in the neutrality of the judiciary.
- The neutrality of the judiciary is another way of ensuring its independence from political influence, by ensuring that the judges will not have any personal or political views that may influence their decision-making in relation to the government of the day.

How is neutrality of the judiciary maintained?

Rulings must be made on the basis of law

All judges, including the Supreme Court, must make their decisions based on the law and explain how they have reached their decisions. This is done when they write their formal opinion and announce it, demonstrating how they have reached the decision based on law rather than any personal opinions or sense of justice.

Peer review

Below the Supreme Court, any judicial opinion can be appealed and be reviewed by a higher court to ensure that it complies with judicial neutrality. When a case concerns aspects covered by the European Convention on Human Rights, the decision may be referred to the European Court of Human Rights for review. As the Supreme Court has no higher court in the UK to answer to on any other issue, cases are heard by groups of at least five judges. This means no single personal prejudice is likely to influence the final decision as any decision must be fully explained to other members of the Court meeting on a case.

Restrictions on group membership

Judges and Justices are restricted in which groups they may join. No serving Justice may join a political party or certain other groups that may cause a conflict of interest. Of course, Justices may still vote and have their personal views, but there may be no formal link to a political organisation. If a Justice had a conflict of interest that might impact their neutrality, they would have to declare it and possibly recuse themselves (excuse themselves from hearing a case) to ensure that neutrality was adhered to.

Training and experience

Finally, all senior judges must have enjoyed a lengthy career as a lawyer and be highly trained. This means that they are accustomed to the principle that cases must be judged on the strict basis of law and not according to their personal opinions. Indeed, junior judges who gain a reputation for lack of impartiality are most unlikely to be put up for promotion. The Justices of the Supreme Court are the most highly regarded and best qualified legal minds in the UK, giving them the highest level of experience at maintaining professional neutrality.

Are judges really independent and neutral?

The checks and procedures noted above are designed to ensure that the Justices in the UK are as independent and neutral as possible, but no system is perfect, particularly with regards to personal neutrality. Justices are, after all, human and have their own internalised views and prejudices. At a basic level, consider why people usually dress up in their smartest outfits for a session at court. If appearance does not matter, the clothes should not make a difference, yet the idea of appearing at 'your best' shows that people, at the least, believe judges will be swayed by personal appearance. Judges are also a product of their environment and may suffer from some degree of unconscious bias against people who do not share similar backgrounds. With most judges in the UK, including the Justices on the Supreme Court, coming from a narrow range, being mostly older white males, privately schooled and Oxbridge-educated, there is a risk that some prejudice may appear in their rulings. However, it is unclear whether these issues can ever be fully removed.

Perhaps the biggest threat to the independence and neutrality is the risk of politicians and the media attempting to politicise the judiciary. Following the Supreme Court's decision that Boris Johnson's suspension of Parliament in 2019 was unconstitutional and therefore had not happened, and the subsequent electoral victory for the Conservatives, Johnson announced the creation of the Constitution, Democracy and Rights Commission to look at reforming the Supreme Court and the whole judicial system. Although the commission is yet to officially report, one proposal floated is that each Justice should be confirmed by Parliament, much like the Senate confirmations of Supreme Court Justices in the USA. Such a move would make the Court a far more political body, so much so that Lord Reed, the current President of the Supreme Court, has said he would resign in protest at such a move.

Debate

Are Supreme Court Justices truly independent and neutral?

Arguments for neutrality

- The Constitutional Reform Act 2005 removed most threats to their independence.
- Justices cannot be removed by ministers as a result of their decisions. They have security of tenure.
- Justices cannot be threatened with loss of income if politicians are unhappy with their decisions.
- Ministers have little or no influence over which Justices are appointed. Appointments are largely independent of politics.
- There is no recent evidence of bias either in favour of or against governments. Both Labour and Conservative governments have equally been controlled by Supreme Court decisions.

Arguments against neutrality

- Although it is small, ministers can exert some influence over the final appointment of Justices on the Supreme Court.
- The neutrality of Justices is sometimes challenged on the basis that they come from a very narrow social background.
- Some Conservative politicians claim the Supreme Court contains too many lawyers of a liberal disposition who tend to favour rights over state security and law and order.
- Though currently disqualified from sitting in the House of Lords, Lords Reed and Kerr both hold life peerages and so will be able to sit in the Lords once they step down from the Supreme Court and may have some existing connections to fellow life peers.

The key to this question is the word 'truly', so your judgement should focus on that and whether judges can ever be truly independent and neutral. You could also consider independence and neutrality separately; it would show strong evaluation skills to say something like 'they are truly independent, but not truly neutral', if that is your opinion.

The Supreme Court and Parliament

As you saw in Chapter 6, Parliament in the UK is *sovereign*. This is a fundamental feature of the country's constitutional arrangements. This means that the judiciary is a *subordinate* body. The judges are simply not in a position to defy the will of the UK Parliament. Furthermore, the UK Parliament is omnicompetent. This means it is able to do whatever it wants, to pass any law and to expect to have that law implemented and enforced. No matter how abhorrent or undesirable the judges may feel a law is, they must enforce it. They may pass an opinion on the law and they may recommend change, but that is as far as it goes.

Synoptic link

The powers of Parliament and its sovereignty are explored fully in Chapter 6.

Justices must take into account the wishes of Parliament when interpreting law. When determining the real meaning of statute law, judges will look back at the original proceedings to establish what Parliament *intended*. It is not for the judges to decide what is desirable, but only what *Parliament thought* was desirable.

Of course, if the Justices make a ruling of which the government and/or Parliament does not approve, Parliament always has the option of amending a statute or passing a new statute in order to correct what the judges have done. Such a circumstance occurred in 2010. The Supreme Court ruled that the government did not have the power to freeze the bank assets of terrorist suspects. Prime Minister Brown was incensed but had to temporarily accept the judgment. In the event, though, a new statute was passed later the same year (the **Terrorist Asset–Freezing Act 2010**), granting such a power to the government. The will of Parliament ultimately prevailed. The Supreme Court could do nothing about it.

Currently, what powers the Supreme Court hold in relation to Parliament (and the executive) rest largely on the existence of the Human Rights Act, but as the Act is not entrenched (see Chapter 5) there is a possibility that it could be repealed and replaced by something less powerful that would swing power away from the judiciary and back to Parliament. The Conservative Party that won a majority in 2015 had a manifesto commitment to replace the Human Rights Act with a British Bill of Rights. While it was not in the 2017 or 2019 manifesto, the Johnson government has spoken about removing or replacing the Human Rights Act. Repeal of the Human Rights Act would remove the provisions of the ECHR from the jurisdiction of UK courts. The ECHR would still be part of the UK constitution, but any case could only be heard by the European Court of Human Rights, making it expensive and inaccessible for many. Furthermore, any changes to the ECHR (and therefore the terms of the current Human Rights Act) have to be agreed unanimously by all members of the Council of Europe, making it almost impossible for a government to alter the terms to suit its own goals.

If the Human Rights Act were to be replaced by a British Bill of Rights, then the terms and provisions could be amended by a simple vote in Parliament. This would greatly reduce the power of the judiciary and increase the power and authority of Parliament. Supporters of the Human Rights Act argue that this would be catastrophic for rights protection in the UK and leave the judiciary unable to protect the people from a over powerful Parliament. Supporters of the British Bill of Rights argue that the new Act would ensure rights protection rests with the democratically elected Parliament and not unaccountable judges as well as allowing a greater consideration for the margin of interest, by which rights should comply with the national interest. This would mean traditional elements of the UK constitution, such as the right to trial by jury and prisoners not having the right to vote, could be incorporated into a clearly stated British Act that would not be subject to external pressures. As such, the debates over the nature of the Human Rights Act (covered in Chapter 1) directly relate to the future balance of power between the Supreme Court and Parliament.

Case study

Sentencing issues

Judges have become increasingly involved in issues concerning sentencing in criminal cases. This is an intensely *political* issue and so has jeopardised the traditional neutrality of the senior judiciary. However, judges used to have a much freer hand in deciding what sentences to give out. The only major restrictions were homicide cases of various kinds, where a life sentence was mandatory, and the general use of maximum sentences determined by Parliament.

Since the mid-1990s, however, when growing crime rates became a major political issue, politicians have sought to take control away from judges and to force their hand in various ways. The judges have resisted on the grounds that they should be independent from government and that they are the best judges of each individual case. Politicians counter this by saying that judges are not accountable to the public. The public have shown a clear preference for more severe sentencing in general, so judges should be forced to respond to public opinion. The dispute goes on, but politicians appear to be winning by introducing *minimum sentences* for certain offences and for repeat offending. This takes away most of the flexibility that judges formerly enjoyed. This remains a current issue, especially with the treatment of offenders convicted of very minor drugs offences.

The Supreme Court and the executive

Until the 1970s, the relationship between the UK judiciary and the UK government was quite different from what it has become. The judiciary was perceived as a largely conservative body whose members came from the same social and political background as members of successive Conservative governments. The judiciary usually showed support for the power of the state in relation to its citizens. Judges were not expected to challenge the authority of government in any significant way. They saw themselves as servants of the state rather than an equal partner. This relationship has changed considerably for several reasons:

- The growth of judicial review since the 1960s (following the cases of *Ridge* v *Baldwin* (1964) and *M* v *Home Office* (1993))
- The rise of liberal ideology in the UK from the 1960s onwards, including the growth of what is sometimes known as the 'rights culture'
- The appointment of a series of liberal-minded senior judges since the 1990s
- The passage of the Human Rights Act in 1998, giving judges a codified statement of human rights, which could be used to protect citizens against state power
- The Constitutional Reform Act of 2005, which improved the independence of the judiciary in general

So, the UK judiciary, including the Supreme Court, no longer sees itself as subordinate to the executive. Judges are no longer reluctant to challenge state power and to assert the rights of citizens. In short, the judiciary has become something of a counterbalance to executive power.

On the other hand, government does have a claim to greater authority than the judiciary. Furthermore, as long as it can control its majority in Parliament, it can use the sovereignty of Parliament to reverse any decisions made by the judiciary. However much they may protest, the judges must, by law, enforce the will of Parliament. Table 8.3 summarises the rival claims made by the executive and the judiciary to enforcing justice and the rule of law.

Table 8.3 Who has the stronger claim to establishing justice and rights: the executive or the Supreme Court?

The claims of government	The claims of the judiciary
The executive is elected and accountable; judges are neither elected nor accountable	Justices do not allow political considerations to interfere with their protection of rights
The executive usually has a clear mandate to run the country and to protect its citizens	As qualified lawyers, Justices bring a totally rational bearing to questions of law and justice
The executive can respond to public opinion	Justices are expected to be immune from outside, populist influences
The executive has an overarching responsibility to protect citizens, even if it means setting aside individual rights in the interests of national security	Because they are not elected, Justices can take a long-term view, while politicians have to consider their short-term re-election prospects

How far can the Supreme Court influence the executive and Parliament?

Key term

Elective dictatorship
A government that dominates Parliament, usually due to a large majority, and therefore has few limits on its power.

Since 2005, when it was set up, there has been much attention paid to the Supreme Court. For many rights campaigners and commentators who fear the rise of executive power and the consequences of an **elective dictatorship**, the Court has largely been a success. As proof of this, many senior politicians have criticised the Court for being *too* independent and for challenging the authority of government, none more so than Johnson following the 2019 ruling over his suspension of Parliament.

Nevertheless, the Court still has its weaknesses, particularly as it has to recognise the sovereignty of Parliament. The UK has no entrenched constitution, so the Court has no power to enforce constitutional principles against the wishes of Parliament. Table 8.4 assesses the strengths and weaknesses of the UK Supreme Court.

Table 8.4 How can the Supreme Court influence the executive and Parliament?

Factors that allow the Supreme Court to influence the executive and Parliament	Factors that limit the ability of the Supreme Court to influence the executive and Parliament
The independence of the Court is guaranteed in law	It cannot activate its own cases but must wait for appeals to be lodged
It can set aside executive actions that contradict the ECHR or the rule of law	The sovereignty of Parliament means that its judgments can be overturned by parliamentary statute
It can interpret law and so affect the way it is implemented	It has no power to enforce its rulings, relying on the executive and legislative branch to implement its rulings
It cannot overrule the sovereignty of Parliament but it can declare proposed legislation incompatible with the ECHR, which is influential	Its power and status are granted by statute law, so could be overturned and altered by Parliament, meaning it is subject to the whims of a potentially hostile Parliament
With the UK leaving the EU, its judgments cannot be overturned by a higher court	With the UK leaving the EU, there is a possibility that the HRA could be repealed by Parliament, removing a key component of judicial power and influence in relation to the executive and Parliament

Synoptic link

This section should be read in conjunction with Chapters 6 and 7, which explore the nature of Parliament and the executive.

The relationship between the executive and Parliament

How Parliament can control the executive

As you will have seen in Chapters 6 and 7, despite the executive (comprising the PM and Cabinet) being made up of people who also serve in Parliament (either as MPs or peers), the two branches can find themselves in conflict with each other. At other times, the executive dominates Parliament, meaning Parliament basically does as the executive tells it to.

In the past, particularly under powerful executives like those of Thatcher and Blair, it was common to refer to the position of the executive in the UK as an elective dictatorship. This implied that having been elected with a mandate, the government became all-powerful and there was little Parliament was likely to do to thwart its will. The House of Lords was weak, and the majority of the House of Commons was obedient to their party's leadership and the whips. This reality has evolved somewhat following the constitutional reforms that bestowed greater power to Parliament (see Chapters 5 and 6) and the presence of coalitions, small majorities and minority governments between 2010 and 2019. In this sense there has been, in theory at least, a greater balance between the power and influence of Parliament and the executive.

There is a natural conflict in the relationship between the executive and Parliament. It consists of these two constitutional principles:

1 Parliament is sovereign.
2 The government usually has an electoral mandate to carry out its manifesto commitments.

So, when Parliament exercises its right of sovereignty, it is threatening the democratic legitimacy of the government. This conundrum is normally solved by the fact that the government usually enjoys a majority of the members of the House of Commons. This means that Parliament, the House of Commons at least, will not need to exercise its sovereignty as long as the government is operating within its mandate. In this sense, the sovereignty of Parliament becomes, effectively, the sovereignty of the elected government. This is partly the reason why the Johnson government that won a clear mandate for its pledge to secure Brexit without any further delay in 2019 was able to pass Brexit legislation without any major difficulties, compared to the May and Johnson governments of 2017–2019 that had no such clear mandate from the British public.

In addition to this reality, the powers of the House of Lords have been gradually brought under control. In particular, there are three limitations on the Lords:

1 The **Parliament Act 1911** prevented the House of Lords having any control over the government's financial arrangements (spending and tax). The Act also stated that if a law is passed in two consecutive years in the House of Commons, the Lords cannot block it.

2 The **Parliament Act 1949** stated that the delaying power of the Lords, as first specified in 1911, should be reduced to only one year.

3 The **Salisbury Convention** was developed in the 1940s. This states that the Lords must not block any piece of legislation that was contained in a winning government's last election manifesto. This means that the unelected House of Lords cannot thwart the will of the elected House of Commons and government.

So, there are great limitations on the powers of Parliament in relation to the executive. This begs the question: What controls are there on the executive? There are a number of answers to this question:

- If the government lacks an electoral mandate for a policy, the Commons may exercise a veto unless the government can persuade the majority of MPs to support it.
- Parliament, both houses, may amend legislation to change its character and to protect certain minorities.
- Parliament calls government and its ministers to account through regular Question Times and the work of select committees (see Chapter 6). This means they are constantly aware that errors or injustices will be exposed.
- In very extreme circumstances, Parliament could dismiss a government by passing a vote of no confidence in the government, thus forcing a general election.
- Backbench MPs may express their concern over proposed legislation through the whips' office. By voicing such opinions, MPs may be able to convince the government to change course without facing a public division. This makes the role of party unity crucial in determining the effectiveness of control.
- The restrictions are only conventions, so MPs may rebel against the elected government and vote against it and the Lords may defy the Salisbury Convention, something that was seen a great deal during 2019.
- If there is enough vocal opposition, the government may be pressured into withdrawing a contentious proposal rather than risking a parliamentary defeat.

How effective is Parliament in holding the executive to account?

It is important to remember that there is no hard-and-fast rule for the effectiveness of Parliament in holding the executive to account, so while it might be tempting to look for a 'it is' or 'it is not' answer, we have to remember that it all depends on circumstances. In Chapter 7 we considered the factors that lead to success for a prime minister, and such factors alter the balance of power. We saw this clearly over 2019 and into 2020. In the early days of the prime ministership of Boris Johnson, Parliament seemed to effectively control the executive, forcing him to ask the EU for extensions for the UK leaving the EU, despite the Prime Minister saying he would rather be 'dead in a ditch' than ask for such an extension. However, since the general election of 2019, Johnson has been able to power through with more policies and Parliament's effectiveness at holding the executive to account has become much more limited, with Johnson and other ministers, notably Priti Patel, cancelling appointments to see the Liaison Committee and select committees respectively, while the Commons has been much more willing to follow the lead of the executive and the Lords has felt unable to defy a manifesto that secured an 80-seat majority.

So, when it comes to evaluating the effectiveness of Parliament in holding the executive to account, we need to remember that it depends on circumstances and consider which of these circumstances are the most important. The main circumstances that allow Parliament to control the executive more effectively are as follows:

- If the executive has a small or no majority in the Commons.
- If the executive's party is divided.
- If the Cabinet is divided.
- If the issues being addressed are controversial.
- If the executive (mainly the PM) has failed to secure a clear mandate during an election.
- If the executive (mainly the PM) has a poor popular image.
- If Parliament is mostly filled with experienced and well-established MPs.
- If the executive (mainly the PM) is seen as a liability, rather than an asset, by their own MPs.

Between the 2017 and 2019 general elections, the Prime Ministers, first May and then Johnson, saw a Parliament with all of the above factors. In such circumstances, Parliament was able to resist and defy the executive, with the largest defeats for an executive for almost 100 years. This reflects the problems of trying to rule with a minority government. As Theresa May found, Parliament gains effective influence over the executive during a minority. This is because:

- most proposals have to be negotiated individually with members of parliament from all parties to try to secure a majority of support
- the government is constantly facing the possibility of defeat
- small numbers of MPs, in other parties or from their own, gain tremendous influence and can make demands from the executive in order to secure their votes.

In this situation, the balance of power shifts markedly towards Parliament and the authority of the executive is greatly reduced.

How the executive can control Parliament

As we have seen, Parliament enjoys great *potential* power. However, this is rarely exercised to the full. The normal reality in the UK is that the executive dominates Parliament. Usually, the key circumstances that enable the executive to dominate Parliament include the following:

- Leading a party with a large parliamentary majority.
- Having a united party.
- Securing a clear mandate for manifesto policies.
- Having a united Cabinet.
- Having lots of new MPs who will rely on the help and support of whips.
- The popularity of the leader in helping lots of MPs win seats they might not have done otherwise (sometimes called a 'coattails effect').
- Lots of uncontroversial issues.
- A popular media image for the PM and the rest of the executive.

The above circumstances make it exceedingly difficult for Parliament to challenge the executive, and so Parliament becomes much more docile, allowing the executive to exert much greater control. All these circumstances fell into place for Tony Blair, allowing him to dominate Parliament so effectively. Theresa May and Boris Johnson enjoyed none of these circumstances between 2017 and 2019 and were very weak as a result, but following the 2019 election Johnson could certainly claim the first six points and has therefore been able to control and dominate Parliament much more effectively. Table 8.5 outlines some examples of the political circumstances that have affected the balance of power in the past.

Study tip

For any effective answer to this question, you must evaluate consistently throughout your essay by always being clear what the most important factor is. If you were to decide that the size of the PM's majority is the crucial factor in determining relations, you should make this judgement clear and consistently relate all other points to this judgement.

Synoptic link

Parliamentary majorities and how they work are outlined in Chapter 6.

Table 8.5 The changing relationship between government and Parliament

Circumstances favouring executive power	Examples	Circumstances favouring parliamentary power	Examples
The government enjoys a very large single-party majority in the House of Commons	1983 1987 1997 2001 2019	The government has no majority or a very small majority	1992 (Conservatives elected with a 21-seat majority) 2010 (two-party coalition) 2015 (Conservative majority of 12 seats) 2017 (minority government)
The government is united around a dominant ideology	1983–89 (Thatcherism) 1997–2005 (Labour Third Way) 2019 (pro-Brexit)	The governing party is split on issues	1992–97 (John Major's Conservative Party split over Europe) 2010–15 (coalition period and Conservative split over Europe) 2017–19 (divisions over the nature of Brexit and a second referendum)
The opposition is fragmented or weak	1983–92 (Labour split on left–right ideas) 2015–20 (Corbyn's leadership split the Labour Party)	The government faces a strong, united parliamentary opposition	1994–97 (Conservative government faced a united 'New Labour') 2008–10 (Brown faced a more unified and united Conservative Party)
The government is led by a dominant leader	1979–89 (Thatcher) 1997–2003 (Blair)	The leader of the governing party has lost popular and parliamentary authority	1989–90 (Thatcher) 1994–97 (Major) 2003–07 (Blair) 2008–10 (Brown) 2017–19 (May)

Knowledge check

Table 8.5 indicates that dominant governments were elected in 1983, 1987, 1997, 2001 and 2019 with large parliamentary majorities. What were those majorities?

Beyond these circumstantial factors, the executive enjoys some procedural devices that can be used to persuade or pressure Parliament into doing as the executive demands. These devices include:

- **Patronage** The PM has control of all appointments to government, as well as dismissals from it. This gives them power over the MPs in their own party. MPs who regularly cause problems for the government are likely to lose their chance of being promoted to ministerial office. This tends to concentrate their minds on party loyalty.
- **Party whips** The party whips exercise control in the way they organise timetables and allocate offices and other administrative tasks to help their party's MPs. In extreme circumstances, an obstructive MP can be suspended from their party, which will damage their career. Whips also remind MPs about prime ministerial patronage and how important party loyalty is.
- **The national platform** The Prime Minister, as head of the executive, has a national media profile and can speak directly to the public. They may use this platform to put popular pressure on MPs by gaining support for their position and against Parliament's.

Of course, having these devices is one thing, but the ability to use them effectively to influence Parliament depends on the wider circumstances. During Theresa May's time as Prime Minister, many ministers resigned on principle and she could not afford to sack those who remained, while the party whips seemed able to do little to influence their own MPs. Indeed, when Boris Johnson removed the whip from 21 Conservative MPs in September 2019, it seemed to make no difference to the executive's ability to influence Parliament; if anything, it made it harder. When Theresa May made a national televised address and claimed that she was trying to deliver the will of the British public while Parliament was trying to block it, making it an issue of Parliament versus the people, nothing in Parliament seemed to change.

How has the balance of power between the executive and Parliament changed?

As we noted above, there are various political circumstances that can change the balance of power in the short term or over the life of a Parliament, which means the balance of power between the executive and Parliament is usually fluid. However, over the long term, there has been a general trend since the 1960s towards greater power for the executive at the expense of Parliament. Some of these changes have been the result of general circumstances, such as the increasing prominence of television and social media tending to focus on the figure of the Prime Minister or senior ministers, rather than the institution of Parliament, giving the executive greater status in the public's mind. Others have been more deliberate and involve specific constitutional reforms, usually as a reaction against the growing power of the executive and an attempt to rein it in.

As you go through the various changes below, consider how far the balance of power has changed. Parliament, of course, remains sovereign, but have any of the reforms really changed the balance of power, or does it simply depend more on the circumstances of the time?

Constitutional changes

Some recent constitutional changes have somewhat altered the balance of power between the executive and Parliament. These include:

- **The removal of most hereditary peers** Removing most of the hereditary peers and ensuring no party has overall control in the Lords has meant the Lords is more willing to defy the executive and assert itself more in checking the executive.
- **The creation of the Backbench Business Committee** This has allowed ordinary MPs to control more parliamentary time away from government control, holding debates and introducing Private Members' Bills that the government may not wish to support.
- **Election of select committee chairs** The introduction of elections for select committee chairs by secret ballot of the whole house has reduced the ability of whips to influence the chairs. The additional salary they now receive also attracts a higher standard of MP, who see it as an alternative career route to one in government.

- **Election of members of select committees** Although done by party, the fact that MPs now vote for their members on a select committee has again reduced the influence of the whips.
- **Growing power of the Liaison Committee** Although it was set up in 1980 to administer the select committee system, since 2002, by convention, the Prime Minister appears before the committee twice a year to be subject to scrutiny, giving Parliament greater opportunities to scrutinise the executive.
- **The Fixed Term Parliament Act** The Fixed Term Parliament Act of 2011 removed the prerogative power of the PM to call an early election when they want to. Now a PM must gain permission from two-thirds of MPs before they can call an early election, meaning Parliament has gained some power over the process from the executive.

Many of these reforms appear to show power pass from the executive to Parliament, but it is worth remembering that, in many ways, these are only marginal reforms. The Lords still cannot veto the executive, select committees still have a majority of government MPs sitting on them and have no power to enforce their decisions, and the PM, as Johnson has shown, can cancel meetings with the Liaison Committee (which can only question the PM anyway, not instruct them). The Backbench Business Committee typically only controls 23 days a year of Parliamentary business, and bills and debates will only have a meaningful role if the government supports the measure. As for the Fixed Term Parliament Act, on the two occasions a PM has requested an early election, it has been granted, demonstrating little real change in the balance of power. This would suggest that these constitutional reforms have only changed the balance of power at the margins and that the executive still controls most of the Parliamentary agenda and systems, while the checks Parliament has over the executive remain weak in practice.

Despite this, the Johnson majority executive has sought to undermine some of the Parliamentary controls that do exist. Before 2020 the Liaison Committee would always elect its own chair, although before 2010 the PM could nominate a preferred candidate. However, since March 2020 the position is elected by the whole House and so the executive can use the whips and its majority to influence the appointment. This led to Sir Bernard Jenkin, a Conservative ally of the Prime Minister, being appointed to chair the Liaison Committee. In its 2019 manifesto, the Conservatives also proposed to scrap the Fixed Term Parliament Act, suggesting the executive is trying to swing the balance of power further towards the executive.

Discussion point

Key political decisions can be made by the executive or by Parliament. In each of the following examples, which of the two is the more appropriate and democratic body to make the final decision?

1 Whether to intervene in a major international conflict.
2 Whether to agree trade deals with foreign states.
3 Whether to reintroduce capital punishment.
4 Whether to replace the House of Lords with an elected second chamber.
5 Whether to reform the electoral system for general elections.
6 Whether to call a general election if the government loses its parliamentary majority.

Has the balance of power between the executive and Parliament shifted more towards Parliament in recent years?

Yes

- Parliament is achieving considerable influence over foreign and military policy, even voting against some military interventions.
- Select committees are increasingly influential and have come more under backbench control.
- The Liaison Committee is increasingly willing to call the Prime Minister to account.
- Between 2010 and 2019 there was no large and decisive government majority.
- The House of Lords has become increasingly proactive and obstructive.

No

- In 2019 the Conservatives won a large and decisive majority.
- The government still relies on a large 'payroll vote' where all ministers, numbering over 100, are bound by collective responsibility.
- Government still controls the legislative programme and the Public Bill Committees that propose amendments.
- Prime ministerial patronage still creates loyalty among the government's own MPs.
- Government still has a huge advantage in resources (media profile, advice and research) over Members of Parliament.

 Here it is worth considering the changing nature over time, so you may wish to suggest that in the circumstances of 2010–19 the balance of power did appear to shift towards Parliament, but since the 2019 general election, it appears to be moving more towards the executive.

The UK and the European Union

The **European Union (EU)** began in 1952 with the European Coal and Steel Community being formed between France, West Germany, Italy, Belgium, the Netherlands and Luxembourg. This established a body that was **supranational**, meaning it would have decision-making authority separate from the main nations. So began the development of this independent body that would come to bind member states together in closer union and economic and political harmony.

Under the Treaty of Rome in 1957, the original six members would become the European Economic Community (EEC) in 1958, and develop a Common Agricultural Policy (CAP) in 1962 and then an internal customs union in 1968.

In 1973, the EEC allowed new members to join in the form of the UK, Ireland and Denmark, with other member states joining as the EEC continued to grow.

In 1985 the Single European Act was agreed, coming into force in 1987, with the creation of a single European market, governed by common rules and regulations. In 1992 the Maastricht Treaty was signed, which transformed the EEC into the EU, proposing greater economic and monetary union, in the form of the euro, while also developing greater political union between member states.

In 1997 the Amsterdam Treaty established an 'area of freedom, security and justice' across the EU, and in 1999 the euro became the official currency of 11 member states. Greater integration occurred as a result of the Nice Treaty of 2001, which established a European security and defence policy, though not an EU military.

In 2007 there was an attempt to introduce a formal constitution for the EU, but it was rejected by referendums in France and the Netherlands. This resulted in the creation of the Lisbon Treaty, which came into force in 2009. This effectively codified the proposals from the rejected constitution and brought them into law for the member states without

> **Key term**
>
> **European Union (EU)** A political and economic union of a group of European countries.

> **Useful term**
>
> **Supranational** A body or organisation that has authority over and above national governments.

the necessity of holding a referendum. Crucially for the UK, it also included the first mechanism by which a member state could leave the EU, in Article 50 of the Treaty.

Figure 8.1 The European Union member states in 2020

Throughout its history, the UK has been something of a difficult partner for the EU. Originally invited to join the European Coal and Steel Community, the UK rejected the opportunity, seeing itself as separate from Europe. As the consequences of the loss of empire and loss of economic competitiveness became clear, the Conservative Party began to see membership of the European Economic Community as the best way to ensure the economic security of the UK, though Labour traditionally saw it as a means of promoting the capitalists at the expense of workers. Eventually, in 1973, the Conservative Prime Minister Ted Heath successfully brought the UK into the EEC, but following the election of a Labour government in 1974, which was divided over the issue, the UK held a referendum to determine public support for continued membership in 1975. On that occasion, the remain side won.

Theresa May in 2017 signing the letter informing the EU that the UK would be triggering Article 50

While many Conservatives favoured the free-market benefits of the EEC, as the EEC transformed into the EU, with greater integration of policies, politics and monetary concerns, so many began to turn into 'Eurosceptics' who feared the potential loss of sovereignty and independence the UK would face with the growing power of the EU. Increasingly, the UK resisted attempts by other big members to integrate further. Finally, in 2016 the UK voted to leave the EU in a referendum and in 2017 Article 50 was finally triggered. Officially the UK left the EU for a transition period in January 2020 (Figure 8.1), but the tense and difficult relationship between the UK and the EU continues to dominate the political and economic future of the UK.

The aims of the EU

The EU was founded on the basis of the values of 'human dignity, freedom, democracy, equality, the rule of law and respect for human rights'. This may not seem controversial, but as we have seen, these terms are open to some interpretation and as such disputes over national sovereignty and the interpretation of these values have led to conflict between the UK and the EU (and other member states).

Beyond these values, the core aims of the EU were set out in Article 3 of the Maastricht Treaty of 1992, as follows:

- Promoting peace and the EU's values.
- Establishing a single European market.
- Developing cohesion between member states on economic, social and territorial issues.
- Creating a monetary union (the euro) as well as economic union.
- Establishing an area of freedom, security and justice without internal frontiers.
- Fighting discrimination and promoting equality.

So, the question is, how far have these aims been achieved as they relate to social policy, economic policy, political policy and the four freedoms of the single market?

How far has the EU achieved its aims?

The aims of the EU were clearly ambitious, but it is worth considering how far it has managed to achieve these aims in the nearly three decades that have passed since the Maastricht Treaty created it. Inevitably, the extent to which they have been achieved depends on the individual area.

Political aims

Politically, the EU's aims have been based on the spread of liberal democratic ideas. By requiring all member states to adopt the principles of liberal democracy and the European Convention on Human Rights, the EU has helped to establish its values across Europe, using the lure of the prosperity afforded by the single market to establish political values. This was perhaps most successful in the 1990s, when the fall of communism could easily have led to many former communist states turning towards a new form of dictatorship or conflict. That so many of these states wanted to join the EU, partly as a block against future Russian aggression, helped it to establish such values across Europe.

However, in recent years there have been issues both within and without the EU. States looking to join the EU, like Ukraine and Turkey, appear to have gone backwards in terms of human rights and democracy, with the EU becoming less of an incentive to these states to adopt its values. There have also been challenges within the EU, with member states like Poland and Hungary adopting measures in recent years that go against these values, with anti-gay rights legislation being passed, states attempting to control the judiciary, and other measures that undermine the core values of the EU.

Another political aim was to ensure peace across the EU, and in this aim it has been quite successful. Internally, the EU has enjoyed an unprecedented period of peace. There have been no wars or military conflicts between the member states since the creation of the EEC. Considering that throughout the nineteenth and early twentieth centuries there was a succession of wars between France, Prussia/ Germany, Italy and other states, including the horrors of the First World War and Second World War, this is a remarkable achievement.

Beyond its borders though, the EU has not achieved such success. The EU lacks a military so while member states may intervene in conflicts, the EU can do little more than impose sanctions, make threats and try to persuade member states to intervene. In the wars that afflicted the former Yugoslavia in the 1990s, the EU was a limited presence, while the Russian invasion of Crimea and eastern Ukraine since 2014 has seen nothing more than sanctions imposed, which appear to have little effect on Russia.

Economic policies

Generally, there has been an increase in economic cohesion, with rules and regulations being applied pretty equally across the EU and common standards being met. However, the financial crisis of 2008 put this under huge pressure as wealthier states in northern Europe had to 'bail out' economies in the south. Reactions to this have caused tensions between member states.

The euro was established in 1999 with 11 original members. It has become established as a major international currency and its use has increased over time, with most EU members now using the euro and all new members of the EU being obliged to accept the euro as their national currency. It has helped to standardise and speed up transactions across the eurozone and to facilitate trade as a result.

However, not all member states have joined the euro, most notably Sweden and Denmark along with the UK. Furthermore, the sovereign debt crisis of Greece and then Spain and Portugal following the financial crisis of 2008 showed the problems of the EU, with some states using a currency that was too strong for their economy, while other states, in particular Germany, had to spend money bailing out those states. This also led to resentment when states like Greece had to impose dramatic levels of austerity in order to remain within the eurozone.

Although resulting in wider economic unity, such as the abolition of exchange rates between member states, monetary union has come at the cost of a loss of national sovereignty. For this reason, the UK, Sweden and Denmark resisted wider economic union, while other states are becoming increasingly hostile to their loss of control and sovereign authority.

Social policies

The European Courts of Justice have played a key role in this area, building on the EU Charter of Fundamental Rights from 2000. This is not the same as the European Convention on Human Rights or the Human Rights Act, but concerns equality and rights for workers and other EU citizens. This has standardised rights in areas like maternity leave and pay, and prohibition of discrimination in employment and the receiving of services. However, how effectively these rights are adhered to in member states varies and there is little the EU can do to impose or force rights' protections on member states.

There also seems to be tension emerging over social cohesion, with some states in Eastern Europe appearing more socially conservative and reacting against the liberalism of western states, particularly over LGBT+ rights, racism and gender rights. Although nominally all states adhere to the principles of equality, there are dramatic variations across Europe and the EU seems able to do little to resolve this conflict. In terms of territorial cohesion, there is a degree of consistency largely thanks to the common market, and the EU has a seat and is consulted in international organisations.

The four freedoms of the single market

This is probably the main success story for the EU. This single market was established in the Maastricht Treaty on the basis of **four freedoms**:

Key term

Four freedoms (EU) The principle of free movement of goods, services, capital and people within the EU's single market.

- **Free movement of goods** Goods produced in one part of the EU can be freely sold in another part of the EU, as long as they meet the agreed standards. There will be no taxes or trade barriers between member states.
- **Free movement of services** A service, such as financial services or advertising, can be offered by any company in one part of the EU to another company in the EU without additional regulations or rules.
- **Free movement of people** A citizen of one member state is entitled to equal treatment in another member state. This means a citizen can travel, work and reside in any part of the EU without the need for a visa or work permit and they are entitled to any rights enjoyed by the people of the state in which they live, such as health benefits, welfare payments and education.
- **Free movement of capital** This removes any restrictions on financial investment from one part of the EU to another and allows money to move freely across the EU, with no fees or limits on buying currency or investing in other countries.

The free market has created an estimated 3–4 million additional jobs across the EU and allowed GDP across the EU and its member states to increase, even after the financial crash of 2008. Membership of the single market appears to have helped the former communist states develop into modern economies and establish a strong economic base.

However, elements of this policy have been controversial. While the UK benefited from three of the fundamental freedoms (goods, services and capital) largely thanks to London's role as a financial services sector, the free movement of people proved controversial, with many UK citizens citing unregulated EU migration as a key reason for voting to leave the EU, especially with the UK being the only state in the EU with healthcare that is 'free at the point of use'. This led to a perception of 'health tourism', with EU nationals legally travelling to the UK simply to receive free healthcare.

While the legal immigration from the EU and EU health tourism were concerns for the British, the migrant crisis involving illegal migrants seeking to cross the Mediterranean and settle in the EU also put the free market in danger, as countries from Sweden to Hungary reintroduced checks on border crossings to limit the number of illegal immigrants travelling into their country.

Knowledge check

What are the four fundamental freedoms of the free market?

An extension of the four freedoms was the establishment of an area of freedom, security and justice without internal frontier. This meant member states agreed to cooperate in areas relating to immigration, asylum cases, policing and rights. All member states accept the power of the European Court of Justice to interpret and establish EU law, while the European Arrest Warrant and Europol have enabled police forces to work together to ensure that a person who commits a crime in one country cannot escape justice by fleeing to another EU state. By and large, this has been a great success, but member states do not always implement the rulings of the European Courts of Justice and in the UK there has been a sense of unease at UK citizens being extradited to another EU state to face trial (although the UK government does support it when an EU citizen has to be returned to the UK to face justice).

The institutions of the EU

Before we consider the role of the EU in policy-making, it is important to know what the main institutions of the EU are and the part each plays in developing EU policies. The European Union has five main institutions:

- **The European Commission** This is the civil service of the EU. In the same way as our civil service, it is staffed by non-elected officials. Their main role is to develop and propose policies that will further the aims of the EU, to draft European legislation and to organise the implementation of EU policies.
- **The Council of Ministers** In fact this is a number of councils, each one dealing with one aspect of the EU's activities, such as finance, economy, agriculture, transport or foreign policy. Ministers from the elected governments of member states attend meetings. Their role is to negotiate final legislation and to ratify new laws. In effect they are the legally sovereign bodies of the EU.
- **The European Council** The heads of government of member states meet, normally twice per annum, to form this council. They ratify important decisions and occasionally agree new treaties. Along with the Council of Ministers described above, this is the sovereign body of the EU.
- **The European Parliament** MEPs are elected from member states. Most represent political parties. The European Parliament has a veto on appointments to the European Commission and can amend or even block legislation in some circumstances. It calls the Commission to account.
- **The European Court of Justice** Staffed by judges drawn from member states, the court is the highest court of appeal. It deals with disputes between member states, interprets EU law when it is disputed and can punish states that disobey EU law. Its rulings are binding on all member states.

The role of the EU in policy-making

In policy-making, the role of the EU varies. In some areas it is the main body for policy-making, in other areas it shares policy-making power with member states, and in some areas it has no authority. Any power the EU does have over a given policy area has been granted to it by the various treaties, so in this sense member states have given the EU some authority, or sovereignty, to make policy decisions that they are then obliged to follow. When exercising its **competencies**, the EU must ensure that it does not go beyond what has been granted by the treaties and that, beyond its exclusive competencies, it should only intervene if it would be more effective than national policy-making, say in dealing with a major crisis like that of immigration.

The different levels of policy-making competencies are as follows:

- **Exclusive competencies** Areas of policy that only the EU can make policy on (national governments have no say or role). This covers customs, external trade, monetary policy (for the eurozone), competition policy and marine conservation.
- **Shared competencies** In these policy areas the EU has priority, but if it chooses not to make laws or regulations in these policy areas then a member state can if it so chooses. This covers a wide range of policy areas including the single market, social policy, employment policy, territorial cohesion, environment, transport, energy, security, justice, freedom, and agriculture and fisheries.
- **Supporting competencies** These are policy areas in which the EU may only coordinate or support the actions and decisions of member states; it cannot make policy decisions itself. This covers areas like culture, industry, education, healthcare, and foreign and security policy.
- **Exclusive member-state competencies** These are policy areas in which the EU can play no role. This basically refers to any policy area that has not been specifically granted to the EU through treaties, such as the level of income tax in member states or public spending on social services.

As we can see from the list above, the EU plays some role in most policy areas, but it is on a sliding scale of authority. The key point is that in return for sharing their own sovereignty with other countries, member states are gaining influence over what happens in those other other countries. This is the nature of shared or pooled sovereignty.

The impact of EU policies on the UK

The development of the EU has seen many EU policies implemented in the UK. Many may appear uncontroversial, such as the requirements surrounding food labelling and animal-welfare rules, but others have had much more profound effects that have alienated sections of society from the EU. EU funding was also used to help regenerate low-income parts of the UK, such as Cornwall, parts of Wales, South Yorkshire and Northern Ireland, in projects to regenerate town centres in Wales, such as Aberdare, Pontypridd and Ebbw Vale, investment in job creation schemes and support for cultural structures, like funding for leisure centres in South Yorkshire and John Lennon Airport in Liverpool. Some of these policies are identified in the case studies that follow.

Looking at the case studies, it is tempting to see the EU as an external body that imposed policies on the UK; indeed, much of the media portrayal relating to the EU focused on this idea. However, when it was a member of the EU, the UK had representatives in all the institutions of the EU and, with the third-largest population, had the third-largest number of seats in the EU Parliament. Indeed, in December 2010 the UK played a leading role in blocking EU plans to extend the right to paid maternity leave to 20 weeks across all member states. The UK therefore played as much, if not more, of a role in developing policies, and resisting them, as any other member state.

EU policies that have impacted the UK

The Common Fisheries Policy

This policy grants all member states equal access to the waters of all member states, as well as setting quotas and regulations for the number of certain types of fish that are caught. The UK has some of the largest areas of prime fishing under its control and it is believed that allowing fleets from states such as Spain and Portugal to have free access to UK waters triggered a terminal decline in the British fishing industry. The quotas and regulations have failed to reverse the long-term decline of various fish stocks, but have led to fishermen having to throw dead fish back into the sea because they would otherwise exceed their quota, again causing hardship for the British fishing industry.

The Common Agricultural Policy

The CAP is a policy area that dominates the EU, taking up about 38 per cent of the EU's annual budget. It is a series of subsidies, designed to ensure a consistent food supply by the EU buying surplus stocks if prices fall below an agreed level. This protects farmers from being undercut by cheaper products and ensures arable land stays in production. UK farmers did benefit from such subsidies, but as the UK farming industry is relatively small compared with those of states like France, Italy and Spain, it is perceived that the UK, which was the second-largest contributor to the EU's budget (after Germany), was subsidising inefficient farming practices by 'European farmers'. Beyond this, concerns were raised about wasted food, the so called 'butter mountains', as the EU bought up food stocks that no one wanted.

The 'social chapter'

The 'social chapter' relates mostly to workers' rights. The UK initially had an opt-out from this, but in 1997 New Labour brought it into UK law. The social chapter was designed to protect workers' rights and ensure a certain quality of life, enforced by the EU. This would cover rules about the maximum number of hours a person could work each week (48), legally required breaks, holiday and sick pay, as well as issues relating to maternity leave and pay. Although the UK did have an opt-out in some areas, such as workers being allowed to work beyond 48 hours a week if they asked to, many small and medium British industries have suggested the cost of complying with such regulations has made Britain less competitive in the international market and seen factories and jobs go to other parts of the world, like India and China, where employment is cheaper. It is also a policy area that highlights a distinct cultural difference between the UK and other member states. The UK tends to favour free-market capitalism with a minimum role for the state and it opposes European-style social policy with strong protections for workers and a greater role for the state in employment areas.

Immigration policy

Based on one of the four freedoms, the free movement of people, the EU has always maintained a policy of a citizen of one state being treated as an equal citizen in whichever EU country they reside. The UK has tried to resist this, attempting to block the free movement of people from Romania and Bulgaria when these countries joined in 2007. The belief is that citizens from poorer countries would travel to take advantage of the UK's relatively generous welfare provisions, including access to the NHS. It has also seen a rise in families from Eastern Europe settling in the UK, requiring UK schools to adapt to a more multilingual system of teaching in some areas. This caused a great deal of resentment across some sections of UK society, though it is worth bearing in mind that the tax revenue from EU citizens working in the UK far exceeded the costs incurred by welfare recipients from the EU. In 2016, UCL academics Proessor Christian Dustmann and Dr. Tommaso Frattini calculated that immigrants from the EU were less likely to claim benefits and on average contribute more to the UK economy than native British citizens. As Dustmann put it at a press conference, 'these guys (EU citizens living in the UK) pay more in than they take out'. It is also the free movement of people that has allowed many UK citizens to settle overseas, in particular those who have chosen to retire in Spain and France.

The impact of the UK's exit from the EU

The constitutional impact

The position is clear. Upon leaving the European Union, the UK Parliament regained all its sovereignty. EU laws are no longer part of UK law and the UK is no longer subject to EU treaties. The European Court of Justice no longer has jurisdiction in the UK.

The political impact

Here the picture is less clear. The political repercussions of the UK people's decision to leave the EU will be felt for many years to come. However, a few conclusions can already be reached. These include the following:

- The top level of the Conservative Party who had supported the 'remain' side of the campaign lost power, notably David Cameron and George Osborne, the Chancellor of the Exchequer. The remaining supporters of a 'soft Brexit' were eventually removed from the party during the 2019 election.
- The issue of Brexit has dominated the political debate and caused deep divisions within the two main parties.
- The referendum revealed deep divisions in UK society, between young and old, England and Scotland, the cities and the countryside, the wealthy and the poor. The vote to leave was, in some ways, something of a protest against the 'political class' in Westminster and a populist movement against powerful vested interests in general.
- As Scotland voted overwhelmingly to remain inside the EU, there are renewed demands for Scottish independence so that Scotland can remain in the EU.
- The UK will now have to undertake a long-term programme of developing new political and trade links with other countries.
- The issue of immigration control, which is now in the hands of a newly independent UK, will be a key political issue for years to come.
- Parliament and the UK courts will have to spend a lot of time unpicking the elements of the EU that are embedded in nearly 40 years' worth of legislation and rulings.
- The position of Northern Ireland has become much less clear, with it having a separate and different relationship with the EU as a result of its land border with the Republic of Ireland.

Synoptic link

The way in which the media reported on legal immigration from Romania and Bulgaria in 2007 helped shape public perceptions towards EU migration, reflecting the importance of the media in politics, outlined in Chapter 4.

Study tip

You need to know the main effects of at least two EU policies and their impact on the UK political system and policy-making. Make sure you learn at least two of the policies outlined in this section.

What influence will the EU have after the UK leaves?

The UK is now set to become a completely independent sovereign state after the transition period ended in January 2021. However, at some stage there may well be a new treaty with the EU, not least because the EU remains the UK's biggest trading partner. If there is such a treaty it might have some of the following conditions:

- The UK may be obliged to allow workers to enter the workforce without hindrance.
- There may be reciprocal arrangements to allow people to move freely in and out of the UK from EU countries.
- If there is a trade deal, it may be that the UK will have to allow goods and services to be imported without any import taxes (tariffs), in return for tariff-free exports to the EU.

- There may be a reciprocal arrangement to allow the free movement of financial capital in and out of the EU.
- There may be other agreements governing international policing, security, drug enforcement, internet control, etc.

Thus, even though the UK is no longer a member of the EU, it may still need to accept one or more of these freedoms if it is to retain access to the European single market or have special trading arrangements with the EU.

Despite any such agreements, the UK will remain sovereign and will be able to cancel such treaty agreements. In the absence of any agreements, the EU will cease to have direct influence over the UK.

Synoptic link

The issue of the EU seems to have been pivotal in reshaping voting behaviour and party support in the UK, so should be considered in conjunction with Chapters 2 and 4.

The location of sovereignty in the UK political system

What is sovereignty?

Synoptic link

This section should be read in conjunction with Chapter 5 on the Constitution, as parliamentary sovereignty is a key element of the UK Constitution, and with Chapter 6, as it relates to the power and authority of Parliament.

Sovereignty is a difficult concept, but one that must be mastered if we are to understand fully the relationships between different parts of the political system. A general description of the term is 'ultimate power', but this does not tell us enough. A better and fuller explanation should include the following principles:

- Sovereignty implies a power that cannot be overruled.
- It is the source of all other political powers. In other words, a sovereign body can delegate its powers to others, but reserves the right to recover those powers.
- Sovereignty can only be removed or transferred, at least in a democracy, by some special procedures that usually involve popular consent.

In the UK, the most important application of sovereignty is parliamentary sovereignty. The official parliamentary website describes this principle as follows:

> Parliamentary sovereignty is a principle of the UK constitution. It makes Parliament the supreme legal authority in the UK, which can create or end any law. Generally, the courts cannot overrule its legislation and no Parliament can pass laws that future Parliaments cannot change. Parliamentary sovereignty is the most important part of the UK constitution.

Source: www.parliament.uk

The sovereignty of Parliament is the principal example of **legal sovereignty**. This is a constitutional reality that will be enforced in law. It may not, though, reflect **political sovereignty**. This refers to where real power lies as opposed to theoretical power.

A constitutional reality: Legal sovereignty

This refers to the single body, Parliament in the UK, whose laws will be supreme in all circumstances. It also means that political power can only be exercised if Parliament explicitly grants that power to a body. Parliament is also the single authority that can declare what the Constitution is. The UK's exit from the European Union is the final confirmation of parliamentary sovereignty. Previously, the EU's laws took precedence over parliamentary statutes (because Parliament allowed them to) but this is now in the past.

A political reality: Political sovereignty

Political sovereignty relates more to what the sovereign body can actually do in the face of public opinion and circumstances beyond its control. For example, legally, the UK Parliament could exercise its sovereignty by taking away the power of the Scottish Parliament, but politically it is unthinkable that it would do so. Similarly, we know in reality that it is the UK government that develops most national laws and not Parliament. The government is effectively sovereign, especially if it enjoys the democratic mandate granted by the electorate.

> **Key terms**
>
> **Legal sovereignty** The legal right to exercise sovereignty, i.e. sovereignty in theory.
>
> **Political sovereignty** The political ability to exercise sovereignty, i.e. sovereignty in practice.

> **Synoptic link**
>
> The use of referendums, discussed in Chapter 3, neatly shows the distinction between legal and political authority; a referendum is only advisory and has no legal power, but politically Parliament is obliged to follow the wishes of the people expressed in a referendum, so they are politically sovereign.

Sovereignty revisited

Some political commentators now argue that sovereignty is an outdated concept and that the distinction between legal and political sovereignty is meaningless. All that matters, they argue, is where real power lies. There are a number of examples that illustrate this philosophy:

- The key strategic decisions concerning the Constitution are now being made by referendum. In this sense it is the people who are sovereign. The fact that Parliament must ratify the result of a referendum because it is legally sovereign does not matter. Realistically, Parliament will never overturn a referendum outcome.
- Devolution means that real power over key areas now lies with devolved bodies and sovereignty in Parliament has had to adapt to this reality.
- It can be argued that by winning a general election, the UK government has been granted sovereignty, certainly to implement its manifesto. The fact that Parliament can veto legislation is not relevant as it is such a rare event.
- The situation with the protection of rights is similarly clouded. In theory Parliament can set aside the rights of the citizens under its sovereign powers. However, the Supreme Court, the European Convention on Human Rights, and the European Court that enforces it, effectively control the protection of rights.

These examples suggest that the location of sovereignty, politically at least, is somewhat fluid and can change over time.

The debates of Brexit in 2016 centred on whether the EU or Parliament should hold political sovereignty in the UK

The changing location of sovereignty

There is no doubt that sovereignty is on the move in the UK. Now that sovereignty has returned from the European Union, attention has shifted back to parliamentary sovereignty. The old certainty, that Parliament was sovereign and nothing else mattered, has gone. The location of sovereignty therefore varies:

- The people are sometimes sovereign: At elections and when a referendum takes place.
- The executive is often sovereign: When it has such a decisive majority in the House of Commons, and therefore a very strong electoral mandate, that it dominates the political establishment.
- The courts may be sovereign: When they are enforcing the European Convention on Human Rights.
- The devolved assembly of Northern Ireland and the parliaments of Scotland and Wales may be sovereign: When they are passing laws under powers granted to them by the UK Parliament.

This is a situation that simply did not exist before 1997. Until that time, the only question was whether Parliament or the executive was truly sovereign.

Knowledge check

Using information from this chapter or Chapter 5 on the Constitution, identify examples of the following:
- The people being sovereign during a referendum
- The courts exercising sovereignty over a rights issue
- Parliament exercising its legal sovereignty over government
- A responsibility of government where the Scottish Parliament is sovereign

Sovereignty today

The first observation we can make about sovereignty today, in the third decade of the twenty-first century, is that Parliament does appear to be restoring some of its past constitutional powers. It is claiming control over UK foreign and military policies, it has control over when general elections can be held, and it is increasingly determined to have a say over key political issues. These include agreements the UK makes with foreign powers and international organisations, and measures for which the government does not have a mandate.

The second current development is that sovereignty is increasingly being divided into different *functions*. In other words, sovereignty resides with whichever body or individual has ultimate power over a particular political issue. So, we may call this 'functional sovereignty'. This works like this:

- The people have sovereignty over key constitutional changes, such as devolution, the electoral system, the independence of Scotland and membership of the European Union.
- The Supreme Court is politically sovereign when human rights and civil liberties need to be defined.

- The devolved administrations are sovereign when decisions and policies concern *only* Scotland, Wales or Northern Ireland.
- The Prime Minister is sovereign when determining who shall form the government.
- Parliament is sovereign when the government is proposing a major military initiative.

As we have seen, all this can be changed if Parliament wishes, as Parliament remains *legally* sovereign. But the political reality looks very different today.

Debate

To what extent does the UK Parliament remain sovereign?

Sovereignty retained

- There has been no challenge to the legal sovereignty of Parliament.
- The Miller case confirmed that Parliament remains sovereign even after a referendum decision and above the executive.
- The UK's departure from the EU restores full parliamentary sovereignty over those areas that had been delegated to the EU.
- The UK does not have to conform to rulings by the European Court of Human Rights.
- Devolution can be reversed by Parliament.

Sovereignty threatened

- The executive continues to claim political sovereignty as long as it has a mandate.
- Devolution has been called 'quasi-federalism' in that it is unlikely that devolved powers will ever return to the UK Parliament.
- Though Parliament must confirm a referendum result, it is virtually unthinkable that Parliament would defy the will of the people.
- The European Convention on Human Rights is increasingly entrenched, making it difficult for Parliament to defy its terms.

Make an overall judgement about 'extent', considering which side of the debate is more convincing. It may be useful to divide sovereignty into its two parts by saying something like 'Parliament retains legal sovereignty but it has lost political sovereignty in a number of key areas'.

Summary

Having read this chapter, you should have knowledge and understanding of the following:
→ The importance of the UK Supreme Court to the UK political system
→ How the UK Supreme Court Justices are selected and how the Court operates
→ The relationship between the Supreme Court and the executive and Parliament, and factors that impact that relationship
→ The factors that affect and alter the relationship between Parliament and the executive in the UK
→ The extent to which the executive is able to control Parliament
→ The effectiveness of Parliament in controlling the government
→ The main aims and workings of the European Union
→ The impact the European Union has made on British politics, including the reasons for leaving
→ What sovereignty is and its various forms and how it applies across the UK
→ How the location of sovereignty has changed in recent history, and its current location in the UK

Key terms in this chapter

Elective dictatorship A government that dominates Parliament, usually due to a large majority, and therefore has few limits on its power.

European Union (EU) A political and economic union of a group of European countries.

Four freedoms (EU) The principle of free movement of goods, services, capital and people within the EU's single market.

Judicial independence The principle that judges should not be influenced by other branches of government, particularly the executive.

Judicial neutrality The principle that judges should not be influenced by their personal political opinions and should remain outside of party politics.

Judicial review The power of the judiciary to review, and sometimes reverse, actions by other branches of government that breach the law or that are incompatible with the Human Rights Act.

Legal sovereignty The legal right to exercise sovereignty, i.e. sovereignty in theory.

Political sovereignty The political ability to exercise sovereignty, i.e. sovereignty in practice.

Supreme Court The highest court in the UK political system.

Ultra vires Literally 'beyond the powers'. An action that is taken without legal authority when it requires it.

Further reading

Websites

The best website to use for further information on the Supreme Court is its own website: **www.supremecourt.uk**

For general information about judicial matters, look at the website of the Department of Justice: **www.justice.gov.uk**

The issue of parliamentary sovereignty is explained on Parliament's website: **www.parliament.uk**

The main pressure group in the field of human rights is Liberty. Its website contains information about issues that involve the courts, including some key rights cases: **www. liberty-human-rights.org.uk**

Another rights pressure group is Rights Watch UK: **http://hrw.org**

Books

Paterson, R.N. (2014) *Final Judgment: The last Law Lords and the Supreme Court*, Hart Publishing — a reasonably up-to-date book

Young, A. (2008) *Parliamentary Sovereignty and the Human Rights Act*, Hart Publishing an expensive but useful reference book

1

Source 1

A debate about the authority of the 2016 referendum

A Brexit supporter:

The UK is a democracy and it has clearly voted to leave the EU. The will of the people, as expressed by the referendum, is clear and must be respected by Parliament, the EU and the devolved institutions. The idea that the Scottish Parliament should have any say in the matter, or even block the will of a majority of UK people, is ridiculous. Parliament agreed to hold the referendum and just because a majority of MPs disagree with the result, does not mean they should reject the will of the people. The Prime Minister seems to understand this, so why don't their MPs?

A Brexit opponent:

The Supreme Court has spoken and it is clear that Parliament, not the executive, must decide whether or not to trigger Article 50 and leave the EU. Parliament is the constitutionally appropriate body to make such major decisions, not a populist referendum, the result of which is subject to misinformation about the power and institutions of the EU and the impact it has on the UK. Sovereignty will not 'be returned to Parliament' as some claim, because Parliament never stopped being sovereign. Why do those who campaigned to 'return sovereignty to Parliament' now seek to ignore this principle in order to achieve their goals? Parliament has the power and authority to do what is right, not what is popular, and it should use that authority now to prevent the UK leaving the EU.

Using the source, evaluate the view that sovereignty is no longer held in one place in the UK.

In your response you must:

- *compare and contrast different opinions in the source*
- *examine and debate these views in a balanced way*
- *analyse and evaluate only the information presented in the source.* (30)

2

Source 2

Conflict between the judiciary and the government

In recent decades there has been an increasing level of conflict between the senior judiciary in the UK and the government. Three main areas can be identified:

1 The judges very closely guard their role in determining sentences in criminal cases. They are determined to impose the rule of law and assure every citizen a fair hearing and equal treatment. Government, on the other hand, constantly tries to control sentencing, mainly by imposing minimum prison terms for particular crimes such as murder and carrying offensive weapons.

2 While government sees the security of the state as its first priority, the judiciary sees the rule of law and the protection of human rights as its key role. The problem is that these two objectives often come into conflict. In particular, suspected terrorists, ministers often claim, should not be treated in the same way as other citizens because they represent a public danger. The judges, however, refuse to treat such suspects differently. They have the same rights as the rest of the population, they say.

---

3   With the increasing threat of religious extremism, the government wishes to curb freedom of speech in cases where it is believed that public words by religious figures may be encouraging terrorism. The judges, armed with the European Convention on Human Rights, seek to protect freedom of expression, even when the views expressed are distasteful.

This leads us to ask the question of whether unelected but independent judges are better placed to protect rights than the elected and accountable government and Parliament.

Using the source, evaluate the view that the Supreme Court is effective at upholding the rights of citizens.
In your response you must:
- *compare and contrast different opinions in the source*
- *examine and debate these views in a balanced way*
- *analyse and evaluate only the information presented in the source.* (30)

3   Evaluate the extent to which the Supreme Court can control government power.
In your answer you should draw on relevant knowledge and understanding of the study of Component 1: UK politics. You must consider this view and the alternative to this view in a balanced way. (30)

4   Evaluate the extent to which the balance of power between the executive and Parliament has markedly shifted towards Parliament in recent years.
In your answer you should draw on relevant knowledge and understanding of the study of Component 1: UK politics. You must consider this view and the alternative to this view in a balanced way. (30)

5   Evaluate the extent to which Parliament remains sovereign in the UK.
In your answer you should draw on relevant knowledge and understanding of the study of Component 1: UK politics. You must consider this view and the alternative to this view in a balanced way. (30)

6   Evaluate the extent to which the European Union impacted the political system of the UK prior to January 2020.
In your answer you should draw on relevant knowledge and understanding of the study of Component 1: UK politics. You must consider this view and the alternative to this view in a balanced way. (30)

Study tip

Coverage of Core and Non-Core Political Ideas are available in other textbooks to complete your study for Components 1 and 2.

Index

free media   12, 14
French Revolution   23, 65
Friedman, M.   66
funding, political parties   15, 57–8

G

Gallup   158
gender, and voter behaviour   133–5
general elections *see* elections
Gladstone, W.   77
Gove, M.   265
government
   accountability   114
   authority and power   254
   Budget   253
   Cabinet   250–2, 259–77
   calling to account   243–4
   civil service   250–2
   coalitions   10, 77–8, 114–15, 186,
      265, 273–4
   collective responsibility   264–6
   competency   146
   decentralisation   8
   departments   250–2
   executive   250–6, 301–9
   first-term   276
   forms of representation   8–10
   individual responsibility   266–9
   levels of constituency   9
   levels of representation   8
   limits to power   13, 14
   mandate doctrine   149–50
   ministerial code   266–9
   ministers   250–2, 262–3
   minority   115
   multi-party   114–16
   reshuffles   271
   scrutiny   223–4, 229
   select committees   307–8
   single-party rule   10, 52
   whole community representation   10
   see also democracy; political parties;
      Prime Minister
Government of Wales Acts   179, 189,
   198–9
Grayling, C.   269
Great Reform Act 1832   23
Green Party   17–18, 84
   funding   58
   policies   85
   regional voting   140
Greenpeace   31
Grimond, J.   77

H

Hague, W.   69
Hammond, P.   146
Harman, H.   39
Hayek, F.   66

head of state   15
Heath, T.   310
Hemming, J.   233
hereditary peers   184, 203, 307
House of Commons   183–4, 203,
   213–21, 223–30, 232, 234–6, 275
House of Lords   14–15, 25, 61, 183–
   4, 189, 203, 215–22, 224–30, 232,
   236–7, 303–4
House of Lords Act 1999   184, 215,
   229
House of Lords Reform Acts   189
Houses of Parliament   212
   see also Parliament
Howe, G.   262
human rights   13, 14, 37, 79, 206
   binding rights   40–1
   causal representation   7
   collective   45
   individual   45
   legislation   38–41
   pressure groups   10, 41–4
   reforms   185
   responsibilities   44–5
   suspension of   41
Human Rights Act 1998   13, 14,
   38–42, 68, 179, 183, 185, 202,
   292, 300
   Article 8   46
hung parliament   273

I

ideological principles   9
immigration   316
Independent Group for Change   59, 70
Independent Labour Party (ILP)   71
independent representation   6
individualism   75
Industrial Revolution   64
injunctions   233–4
international aid   69
internet
   false information   14
   freedom of information   12, 14
   see also social media
internment   41
issue-based voting   147

J

Javid, S.   267, 274
Jenkins Commission   112, 186
Johnson, B.   53, 55, 83, 96, 141,
   151–2, 154, 166, 241, 249–50,
   271, 274, 308
judicial precedent   38, 291
judicial reviews   292–3
judiciary
   independent   13, 296–9
   neutrality   297–9

reform   183, 186–7, 202
   see also Supreme Court

K

Kennedy, C.   89
Khan, S.   196
Kinnock, N.   72–3, 154

L

Labour Party   9, 43
   age-based voting   136–7
   class-based voting   130–3
   constitutional reform   182–7
   economic policies   75–6, 82
   education-based voting   135–6
   ethnicity and voting   138–9
   factions   76–7
   far left   76
   foreign policies   76, 82
   funding   31, 57, 58–62
   gender-based voting   133–5
   gender-impact assessment   134
   ideology   73
   key features   55
   law and order policies   76, 82
   leadership   55, 90, 146, 150–2,
      255
   manifestos   53, 112, 148–9
   membership   17–18, 55
   New Labour   17, 72, 74–5, 112,
      156, 182
   Old Labour   73–4
   one member one vote   73
   opinion polls   158–9
   origins   71
   regional voting   140
   representation   54
   and trade unions   27, 31, 57–8,
      59–60, 71–5
   unity   90
   welfare policies   76, 82
   see also elections
law and order
   common law   38, 174, 180
   internment   41
   legislation   230–2
   policies   68–9, 76, 80, 82
   rule of law   13, 14, 174, 179–80,
      291
   statute laws   179
   terrorism   41, 68, 147
   see also constitutions; judiciary;
      Supreme Court
leadership *see* political parties:
   leadership
leadership debates   154
left–right divide   63–4
left-wing   62–3, 71
legislation   230–2